This book is dedicated to my wife Nina and my four sons
Peter, Steven, Bill, and Stelios, whose patience and love gave me
courage when the work seemed insurmountable

Acknowledgments

I am indebted to my assistant Nikos J. Farsaris who worked endless hours helping me put this manuscript together. This book would not have been possible without the collaboration of the individual contributors to whom I owe a lifelong obligation.

Preface

Chaos, up to a few years ago, was just an interesting field to researchers who studied the behavior of nonlinear dynamical systems of certain mathematical structure.

Chaotic systems are frequently viewed in phase space, which is a region of space where the current state of the system perpetually exists. Regions of space without the perpetual existence of chaotic dynamics are uninteresting because the points in these areas move off to infinity and do not contribute to the continuation of the chaotic process. The chaotic equations thus derived operate on coordinates in space and each iteration of the equations signifies the passage of the next time increment. This type of modeling (phase space coordinates) facilitates not only the presentation of trajectories or orbits that the chaotic system follows in the temporal evolution but also easily clarifies the concept of synchronization, which plays a central role in the communication properties of chaos. By synchronization of two chaotic systems, we define the situation by which a chaotic system is driven (coupled) by a phase signal of another chaotic system in such a way that these two systems eventually synchronize. In other words, the other phase variables of the two systems converge to the same value one to one. The concept of transmission of information from one system to the other has thus been developed due to this convergence. For example having constructed two chaotic systems that can be synchronized, we somehow modulate on one phase signal of one system the information signal to be transmitted, which drives the other system, and after synchronization, we subtract (demodulate) the information from the corresponding phase signal of the other coupled chaotic system. This technique is demonstrated in this book in various applications of communication systems. A great deal of introductory details are given in Chapter One. Chapter Two shows how chaotic signals are generated and transmitted. The design of chaotic transmitters and receivers is shown in Chapter Three whereas chaos-based modulation and demodulation techniques are presented in Chapter Four. In Chapter Five a chaos-based spreading sequence is shown to outperform classical pseudorandom sequences in two important cases such as selective and nonselective channels. In Chapter Six, several channel equalization techniques specifically designed for chaotic communications systems are developed by using knowledge of the systems dynamics, linear time-invariant representations of chaotic systems, and symbolic dynamics representations of chaotic systems. Finally a specific application is presented in Chapter Seven for optical communications. The fundamental concepts of chaos and its modeling are presented in Appendix A and Appendix B.

It is believed and hoped that this book will provide the essential reading material for those who want to have an integrated view of how an old concept such as chaos can open new roads in the communications field.

Contributors

Tassos Bountis
University of Patras
Patras, Greece

Sergio Callegari
Dipartimento di Elettronica Informatica
e Sistemistica
Bologna, Italy

Antonio Cândido Faleiros
Instituto Tecnológico de Aeronáutica
São José dos Campos, Brasil

Mahmut Ciftci
Texas Instruments
Stafford, Texas

Arthur Fleming-Dahl
Aerospace Corporation
Chantilly, Virginia

Waldecir João Perrella
Instituto Tecnológico de Aeronáutica
São José dos Campos, Brasil

Francis C.M. Lau
Hong Kong Polytechnic University
Hong Kong, China

Gianluca Mazzini
Universitá di Ferrara
Ferrara, Italy
Universitá di Bologna
Bologna, Italy

Tânia Nunes Rabello
Instituto Tecnológico de Aeronáutica
São José dos Campos, Brasil

Riccardo Rovatti
Universitá di Bologna
Bologna, Italy

Adalberto Sampaio Santos
Instituto Tecnológico de Aeronáutica
São José dos Campos, Brasil

Giahluca Setti
Universitá di Ferrara
Ferrara, Italy
Universitá di Bologna
Bologna, Italy

Christos Skiadas
Technical University of Crete
Chania, Crete, Greece

Peter Stavroulakis
Technical University of Crete
Chania, Crete, Greece

Chi K. Tse
Hong Kong Polytechnic University
Hong Kong, China

Gregory D. VanWiggeren
Measurement Research Lab
Agilent Laboratories
Palo Alto, California

Douglas B. Williams
Georgia Institute of Technology
Atlanta, Georgia

Nei Yoshihiro Soma
Instituto Tecnológico de Aeronáutica
São José dos Campos, Brasil

Contents

1

Introduction

Peter Stavroulakis

CONTENTS

1.1 Chaos Modeling

In the beginning, according to ancient Greek theology and philosophy, there was chaos.[1] The Greek philosophers believed that our ordered universe, the cosmos, was formed out of this chaos. Ancient Greeks did not give a specific definition to chaos even though it was related to infinity, disorder, and unpredictability. The long-term behavior of chaos, however, was well understood because, as Heraclitus said, "**Τα παντα ρει**," which means all things flow or all is flux, and the only interpretation of that is that the cosmos was formed out of this chaos. In other words, ancient Greeks believed what modern science discovered centuries later, that disorder can result in order under certain conditions and thus the Greeks taught us of the existence of attractors or limit cycles, as explained mathematically in Appendix A. This type of evolution has been proved implicitly by science because all natural processes have one direction of evolution, which is the increase of entropy or reaching the state of minimum energy. If we extrapolate this further, it is not impossible to believe that the present-day universe is an attractor of the Big Bang. Because order cannot be generated from nothing, it is necessary to believe that chaos does not mean absence of order but that there exists order in chaos. The order in chaos is not obvious, nor are the initial conditions that are necessary to arrive at some kind of order. This subtle conclusion is what has triggered the interest of scientists from ancient times to the present. This is exactly what scientists are using, the evolution of chaos, to study many phenomena and processes, including telecommunications, as we shall see in subsequent chapters.

Chaos is an aperiodic long-term behavior in a deterministic system that exhibits sensitive dependence on initial conditions.[1]

The three components of the definition are clarified as follows:

1. "Aperiodic long-term behavior" means that the system's trajectory in phase space does not settle down to any fixed points (steady state), periodic orbits, or quasi-periodic solutions as time tends to infinity. This part of the definition differentiates aperiodicity due to chaotic dynamics from the transient aperiodicity of, for example, a periodically oscillating system that has been momentarily perturbed.
2. "Deterministic" systems can have no stochastic (meaning probabilistic) parameters. It is a common misconception that chaotic systems are noisy systems driven by random processes. The irregular behavior of chaotic systems arises from intrinsic nonlinearities rather than noise.
3. "Sensitive dependence on initial conditions" requires that trajectories originating from very nearly identical initial conditions will diverge exponentially quickly. The meaning of this will be made clear in the following discussion.

The mathematical model developed, now called the Lorenz system, has been used as a paradigm for chaotic systems that satisfy the above definition. The Lorenz system consists of just three coupled first-order ordinary differential equations:

$$\begin{aligned}\dot{x}_1 &= \sigma(x_2 - x_1)\\ \dot{x}_2 &= -x_1x_3 + rx_1 - x_2\\ \dot{x}_3 &= x_1x_2 - bx_3\end{aligned} \tag{1.1}$$

Lorenz chose parameter values $\sigma = 10$, $b = 8/3$, and $r = 28$. With these choices for the parameters, the Lorenz system is chaotic, exhibiting the traits described in the definition given for chaos.

The second component of the definition is clearly satisfied by the Lorenz system because none of the parameters is stochastic. To demonstrate the aperiodicity of the system, a numerical simulation of the Lorenz system can be performed.[1] A time-series plot of the x_1 variable is shown in Figure 1.1. The initial conditions can be chosen arbitrarily.

From direct observation of the time series in Figure 1.1a, it seems reasonable to say that the x_1 variable is aperiodic. To be certain that this system is

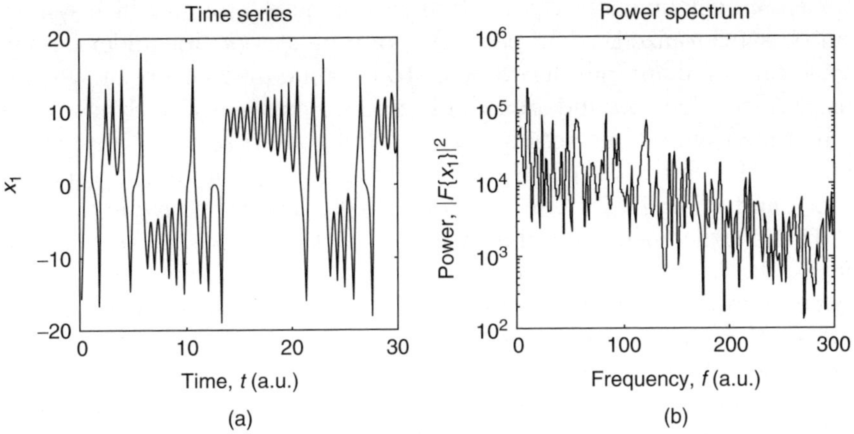

Figure 1.1 Both panels were generated using the Lorenz equations and the parameter values $\sigma = 10$, $b = 8/3$, and $r = 28$. (a) A representative time series for the x_1 Lorenz variable. (b) A typical power spectrum for the x_1 variable.

not just quasi-periodic, the power spectrum for the x_1 variable is shown in Figure 1.1b. The symbol F in the label of Figure 1.1b represents the Fourier transform operator. The power spectrum is very broad with no particularly conspicuous frequencies apparent. The irregular behavior shown for x_1 also occurs for the other variables as well, and does not diminish with increasing time.

The third component of the definition for chaos requires sensitive dependence on initial conditions. Again, the numerical simulation demonstrates that the Lorenz system satisfies this condition. The simulation can be run for two identical systems, x and y, starting from very nearly identical initial conditions. The only difference in their initial conditions was between the two variables, x_2 and y_2. Specifically, $y_2(t=0) = x_2(t=0) = 10^{-6}$. It is shown in Reference 1, Appendix A, and Appendix B that the magnitude of the difference between the two variables is a function of time, t. The linear t, even though started from nearly identical initial conditions, grows exponentially.

Another concept from chaos theory can be demonstrated using this numerical model by plotting x_2 against x_1. We obtain the well-known strange attractors as explained in Appendix A and Appendix B and in subsequent chapters as well.

1.2 Synchronization

It can be shown[1] that two identical chaotic systems can be synchronized if they are coupled together in an appropriate way.

Before beginning a discussion of the techniques through which systems can be synchronized, some powerful conceptual tools for understanding synchronization are introduced here. To illustrate the concept of synchronization, consider two independent Lorenz systems, x and y. Synchronization of the two systems occurs if $[x(t) - y(t)]$ goes to zero as t approaches infinity. The end result of this process is actual equality of the variables, x_1, x_2, and x_3, and y_1, y_2, and y_3, respectively, as they evolve in time.

Thinking geometrically, the dynamics of the composite system, comprising the two Lorenz systems, initially occurs in a six-dimensional phase space, with three dimensions associated with each individual system. As the two systems become synchronized, however, the trajectory of the composite system moves onto a three-dimensional subspace (or hyperplane) of the original six-dimensional phase space. This subspace is often called the synchronization manifold. It contains all of the points where $x_1(t) - y_1(t) = x_2(t) - y_2(t) = x_3(t) - y_3(t) = 0$.

Consider an n-dimensional chaotic system

$$\dot{u} = f(u) \tag{1.2}$$

Pecora and Caroll[3] proposed decomposing such a system into two subsystems,

$$\dot{v} = g(v, w) \quad (m \text{ dimensional}) \tag{1.3}$$

$$\dot{w} = h(v, w) \quad (k \text{ dimensional}) \tag{1.4}$$

where $n = m + k$, and together these subsystems are the drive system.

The subsystem v of Equation (1.3) is used to drive a response system with the same functional form as Equation (1.4). Thus, the response system is written as

$$\dot{w}' = h(v, w') \tag{1.5}$$

The coupling between the systems occurs through the variable v from the drive system, which is substituted for the analogous v'in the response system.

To determine whether the two systems will synchronize, the evolution of the drive response composite system along directions transverse to the synchronization manifold must be analyzed. To do so, the evolution of the difference between the two systems, $w_\delta = w - w'$, is analyzed as

$$\dot{w}_\delta = h(v, w) - h(v, w') \tag{1.6}$$

$$= D_w h(v, w) w_\delta \quad \text{for small } w_\delta \tag{1.7}$$

where $D_w h = \frac{\partial h(v, w)}{\partial w}$ is the Jacobian of the w subsystem.

To clarify the concept of synchronization and illustrate the Pecora–Carroll technique, the Lorenz system is analyzed again in light of the preceding discussion. The drive system is given by the Lorenz equations,

$$\dot{x}_1 = \sigma(x_1 - x_2) \tag{1.8}$$

$$\dot{x}_2 = rx_1 - x_2 - x_1x_3 \tag{1.9}$$

$$\dot{x}_3 = x_1x_2 - bx_3 \tag{1.10}$$

where again, $\sigma = 10$, $b = 8/3$, and $r = 28$.

This system is decomposed so that the x_2 variable is coupled to the response system; it plays the same role as the v subsystem in Equation (1.3). The response system, therefore, is given by

$$\dot{y}_1 = \sigma(x_2 - y_2) \tag{1.11}$$

$$\dot{y}_3 = y_1x_2 - by_3 \tag{1.12}$$

where the variable y_2 has been completely replaced by x_2. A simple graphical illustration of this sort of decomposition is shown in Figure 1.2.

Thus, the analog to Equation (1.7) is

$$\begin{pmatrix} \dot{e}_1 \\ \dot{e}_3 \end{pmatrix} = \begin{pmatrix} -\sigma & 0 \\ x_2 & -b \end{pmatrix} \begin{pmatrix} e_1 \\ e_3 \end{pmatrix} \tag{1.13}$$

where $e_1 = x_1 - y_1$ and $e_3 = x_3 - y_3$. In this case, the eigenvalues of the matrix fortunately do not depend on the drive variable, x_2. There is consequently no need to integrate numerically to determine the conditional Lyapunov exponents. The eigenvalues of the Jacobian, which are also the transverse Lyapunov exponents, are easily obtained; $\lambda_1 = -\sigma$ and $\lambda_2 = -b$. Both are negative at all times, and therefore, the error variables e_1 and e_3 converge to zero as $t \to \infty$. Equation (1.13) is not an approximation, and consequently, it proves that the systems will synchronize as $t \to \infty$, regardless of the initial conditions. The synchronization is quite fast relative to

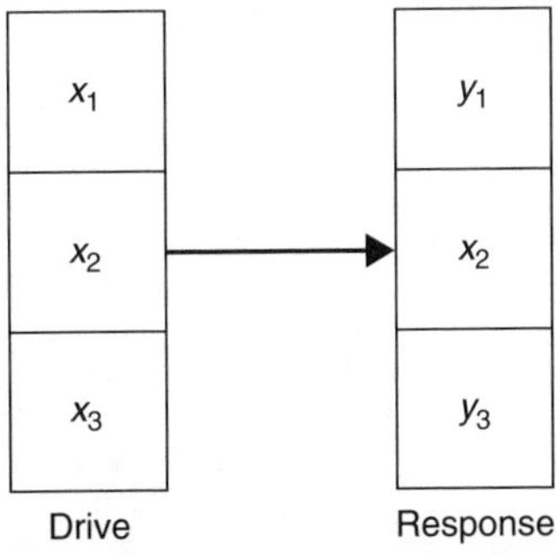

Figure 1.2 Synchronization through decomposition into drive and response systems. The variable y_2 has been completely replaced by x_2.

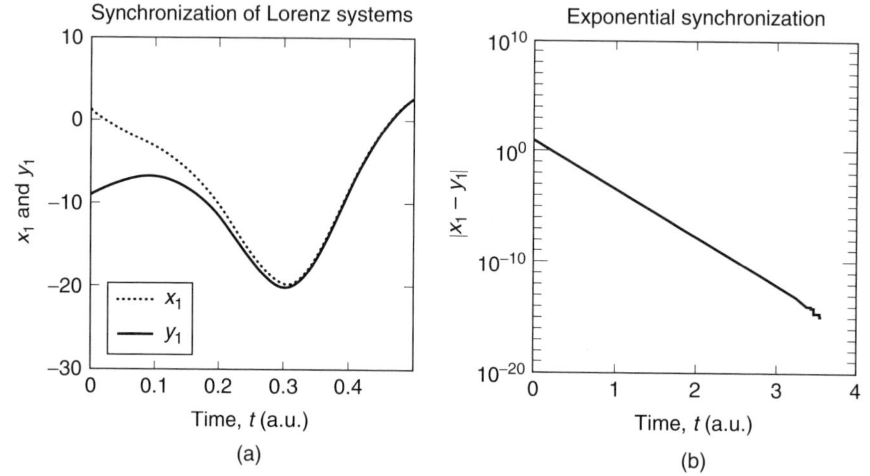

Figure 1.3 (a) Rapid synchronization of two coupled Lorenz systems even when they are started from very different initial conditions. (b) Synchronization occurs exponentially quickly. The slope can be calculated from Equation (1.13).

the speed of the oscillations of the system shown in Figure 1.1. Although Figure 1.3a indicated an exponential divergence of trajectories started at nearly identical initial conditions, Figure 1.3b shows that the trajectories between these two coupled systems converge exponentially quickly.

The slope of the line on this log-plot gives the exponent for this convergence. One could predict that this exponent ought to be the Lyapunov exponent associated with e_1, which was earlier determined to be $\lambda_1 = -\sigma = -10$. Converting σ to an exponent of 10, rather than of e, gives a value of $\sigma_{10} = -4{,}34$. This value matches the slope of the line shown in Figure 1.3b extremely well.

Other schemes for synchronization include the error feedback.[1] The simplicity of this method makes it amenable for experimental realizations of synchronization. Another of its advantages is that only small amounts of coupling between two systems are required for synchronization.

$$\begin{aligned} \dot{u} &= f(u) \\ y &= h(u) \\ y &= h(u) \end{aligned} \tag{1.14}$$

and the matching response system is written as

$$\begin{aligned} \dot{u}' &= f(u') + g(y - y') \\ y' &= h(u') \end{aligned} \tag{1.15}$$

Equation (1.15) clearly shows the manner in which the feedback is to be included. The same type of analysis leads to Equation (1.7).

Synchronization occurs for appropriate choices of feedback variable y and the coupling function g. As before, the real parts of the resulting conditional Lyapunov exponents must be negative for the systems to synchronize. The magnitude of the feedback term, $g(y - y')$, goes to zero as the systems become synchronized. Thus the error feedback technique allows chaotic systems to become predictable.[1]

The synchronization schemes that have been discussed thus far have one drive system and one response system. The error feedback method, however, permits a generalization to mutual coupling. In mutual coupling, both systems influence each other. Two systems, mutually coupled through an error-feedback signal, can be represented as follows:

$$\begin{aligned} \dot{u} &= f(u) + g(y' - y) \\ y &= h(u) \\ \dot{y}' &= f(u') + g(y - y') \\ y' &= h(u') \end{aligned} \tag{1.16}$$

For mutual coupling, each system is both a drive and response system.

1.2.1 Generalized Synchronization

The preceding discussion has focused on the concept of identical synchronization, a type of synchronization in which two identical coupled systems exhibit identical chaotic dynamics. A less restrictive manner of synchronization, called generalized synchronization (GS), has been investigated recently.[1] GS can occur between systems with mismatched parameters or even systems that are functionally dissimilar. The resulting dynamics of these coupled systems are not similar.[1]

1.3 Chaotic Communication Techniques

The relationship between synchronization and communication is not unique to chaotic communication. In AM radio communication, a message is used to modulate the amplitude of a specific frequency sine-wave carrier. A receiver tuned to that particular carrier frequency is able to recover the message. Recalling that synchronization (for nonchaotic systems) requires a common frequency for the two systems, the transmitter and receiver can be considered synchronized. In digital telecommunication systems, too, the transmitter and receiver must be synchronized. The receiver must sample bits with the proper timing to recover the message accurately. A chaotically fluctuating carrier of information represents a generalization of the more conventional sinusoidal or digital carriers. As with more traditional techniques, chaotic communication is also reliant on synchronization. The

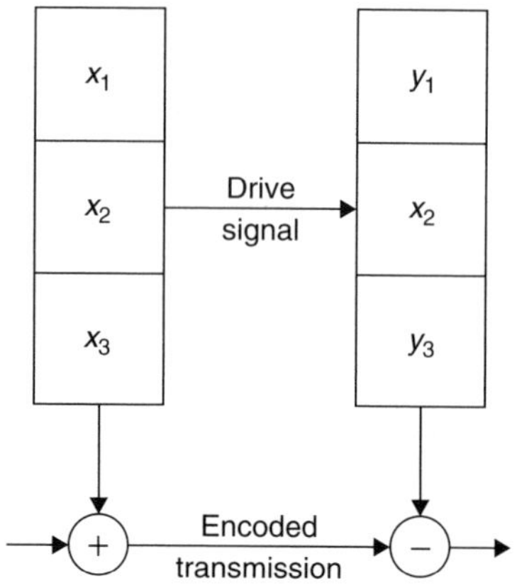

Figure 1.4 Representation of a simple chaotic communication technique.

message to be communicated is simply carried by a chaotic waveform. A receiver, synchronized to the transmitter, is able to recover the message from the chaos. The application of chaotic synchronization to communication was suggested in Reference 3. The first configuration suggested is also one of the easiest to explain. A representation of the idea, using the Lorenz system as its basis, is shown in Figure 1.4. As illustrated in the figure, a small-amplitude message is added to the much larger chaotic fluctuation of the x_3 variable and transmitted to the receiver. Because the message has a smaller amplitude than the chaotic signal, it is effectively hidden, or masked, by the chaotic signal during transmission. The other communication channel is used to transmit the x_2 variable to the receiver where it is substituted for the receiver's own y_2 variable, as is shown in Figure 1.4. This substitution of the x_2 for the y_2 variable causes the entire y system to synchronize to the x system. This synchronization enables the original message to be recovered from the masking signal by simply subtracting y_3 from the encoded transmission, $x_3 + m(t)$.

A small amplitude message, $m(t)$, is masked by a larger amplitude chaotic signal, $x_3(t)$ during transmission. Because the two systems are synchronized, subtraction of y_3 from $x_3 + m(t)$ recovers the message that was buried in the chaotic transmitted signal. In the technique described above, two channels between the systems are required to communicate a message. A pictorial representation of the method is shown in Figure 1.5.

A small-amplitude message, $m(t)$, is masked by a larger amplitude chaotic signal, $x_1(t)$. The composite signal, $x_1(t) + m(t)$, drives the receiver system, which synchronizes to the transmitter system. The message is recovered by subtracting $y_1(t)$ from the composite signal. As in Figure 1.4, a small-amplitude message is transmitted along with a much larger chaotic masking signal. In spite of the addition of the message signal, the receiver

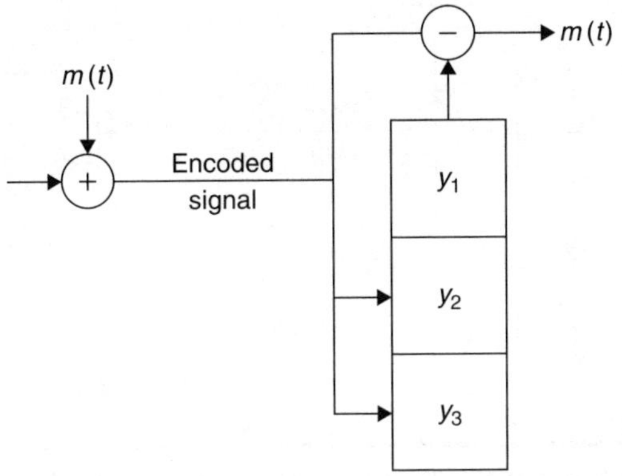

Figure 1.5 Representation of a chaotic communication technique that does not require separate signal transmission.

synchronizes to the dynamics of the transmitter. The small-amplitude message in the transmitted signal acts as a small noiselike perturbation to the larger chaotic driving signal. Because the synchronization manifold is stable, the noiselike message is not able to cause large deviations from the synchronization manifold. This principle also suggests that chaotic synchronization of two systems can be used as a way of filtering noisy chaotic signals.

The efficacy of the technique was demonstrated experimentally using a circuit implementation of two Lorenz systems.[1] As shown in the figure, the x_1 variable is substituted for y_1, but only in the equations governing the evolution of y_2 and y_3 and not y_1. Even with this partial substitution, however, synchronization of the two systems is globally asymptotically stable. A subtraction of the relevant signals from the transmitter and receiver permits message recovery. To demonstrate how a message can be carried by a chaotic signal and still be perfectly recovered, a specific realization of the "inverse system" can be provided here. In this approach, an invertible function of transmitter dynamical variables and information signal is used to drive the receiver system.

The receiver is able to invert the function and recover the message because it synchronizes exactly to the transmitter. Again, the Lorenz model is employed for the specific example provided here, though other chaotic systems can additionally be investigated. The transmitter, then, is just the modified Lorenz system:

$$\begin{aligned}\dot{x}_1 &= \sigma(s - x_1)\\ \dot{x}_2 &= rx_1 - x_2 - x_1x_3\\ \dot{x}_3 &= x_1x_2 - bx_3\end{aligned} \tag{1.17}$$

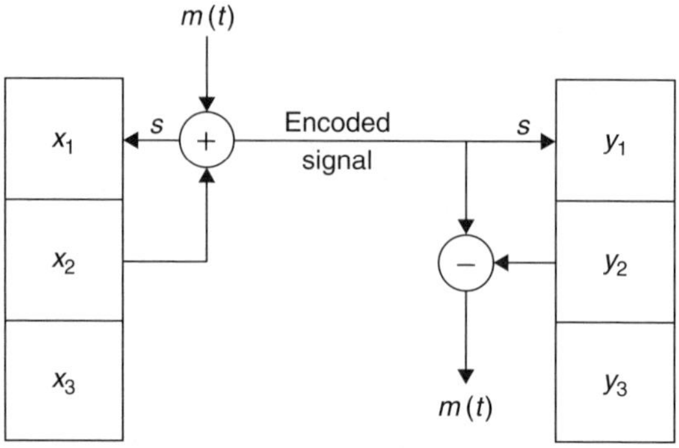

Figure 1.6 Realization of the "inverse system" approach involving two Lorenz systems.

where, as usual, $\sigma = 10$, $b = 8/3$, and $r = 28$. Also, the invertible function s has been chosen as:

$$s = x_2 + m(t) \tag{1.18}$$

where the message to be communicated is represented by $m(t)$. The second system has exactly the same form:

$$\begin{aligned} \dot{y}_1 &= \sigma(s - y_1) \\ \dot{y}_2 &= ry_1 - y_2 - y_1y_3 \\ \dot{y}_3 &= y_1y_2 - by_3 \end{aligned} \tag{1.19}$$

The configuration for this type of communication system is shown in Figure 1.6.

The two systems synchronize exactly, allowing potentially perfect reconstruction of the message. The equations for the error dynamics, $e = x - y$, are

$$\begin{aligned} \dot{e}_1 &= -\sigma e_1 \\ \dot{e}_2 &= re_1 - e_2 - e_1x_3 - e_3x_1 \end{aligned} \tag{1.20}$$

Clearly, $e_1 = 0$ as $t \to \infty$ and $e_1 = 0$ is a stable fixed point. With that fact, the remaining eigenvalues (Lyapunov exponents), $\lambda_{2,3}$, for e_2 and e_3 are

$$\lambda_{2,3} = \frac{1}{2}\left(-(b+1) \pm \sqrt{(b+1)^2 - 4(b + x_1^2)}\right) \tag{1.21}$$

Stable synchronization of the two systems results if these Lyapunov exponents have negative real parts. This condition is met for all times if $b > 0$ because x_1^2 is always positive. The systems, therefore, are guaranteed to synchronize because the parameter $b = 8/3$. This proof of synchronization is not affected by the amplitude or frequency of the message. When these systems have synchronized, the message can be easily recovered by subtracting y_2 from s.

Figure 1.6 shows the transmitted signal, $s = x_2 + m(t)$. The message signal has been chosen to be a simple sine wave with a peak-to-peak amplitude of 2 and a radian frequency of π. Thus, $m(t) = \sin(\pi t)$. This relatively small-amplitude signal is used to demonstrate the principle of chaotic masking, in which a larger chaotic signal hides a smaller message. More complicated messages with many frequencies and larger amplitudes could also be perfectly communicated in this way.

1.4 Communication Privacy

The principle of chaotic masking hints at the larger issue of communication secrecy, which arises naturally in a discussion of chaotic communication. The issue of communication secrecy has served as an important motivation for research involving communication with chaos. Much of this research has sought to use chaotic signals as a means of sending secret messages. Certainly, fundamental properties of chaotic systems seem to make them ideal for this purpose. Chaotic systems are inherently unpredictable. Their dynamics are aperiodic and irregular. A small message added to or modulated onto unpredictable, aperiodic, and irregular waveforms could be difficult to decipher without a second chaotic system, identical to the first, which can synchronize to the transmitter.

In his pioneering paper, "Communication Theory of Secrecy Systems," Claude Shannon[2] discussed three aspects of secret communication systems: concealment, privacy, and encryption. These aspects can be interpreted in the context of chaotic communication. By concealment, Shannon refers to such methods as invisible ink in which the existence of the message is concealed from an eavesdropper. Concealment of a message using chaotic carrier signals is possible because the carrier is irregular and aperiodic. The presence of a message in the chaotic fluctuations may not be obvious. According to Shannon, the second aspect, communication privacy, occurs for systems in which special equipment is required to recover the message. This situation is present with chaotic communication systems because an eavesdropper must have the proper receiver system, with matched parameter settings, to decode the message. Finally, encryption occurs naturally in chaotic communication techniques. In conventional encryption techniques, a "key" is often used to encrypt the message. If the transmitter and receiver share the same encoding key, the scrambled message can be recovered by the receiver. In chaotic systems, the transmitter itself acts as a "dynamical key." The receiver must be able to synchronize to the transmitter's dynamics, which requires matched systems and parameters, to recover the message. Using a chaotic carrier to encrypt a message dynamically does not preclude the use of more traditional encryption schemes as well. A chaotic signal, therefore, can act as an additional layer of encryption for information that has already been encrypted by conventional means. More details are included in Chapter 2.

1.4.1 Communication via Generalized Synchronization

The discussion in the section above focused on using identical synchronization as the basis for communication, but generalized synchronization can also be used. As part of this thesis research, a new communication scheme based on generalized synchronization is proposed. Its development was inspired by the generalized synchronization observed in the three-laser array. The new scheme illustrates some of the advantages that can be realized by using generalized, rather than identical, synchronization.

Generalized synchronization is more appropriate for communication schemes in the context of using synchronized cascaded chaotic systems as high-dimensional transmitters. The possibility of generalized synchronization arose if dissimilar systems were chosen for the cascade. However, generalized synchronization of the dissimilar systems is not necessary in order for the method to succeed, and consequently, it is not accurate to say that this method depends on generalized synchronization. The concepts presented in this chapter will be used one way or another to analyze the aspects of a variety of communication systems in subsequent chapters.

References

1. Vanwiggeren, G.D., Chaotic Communication with Erbium-Doped Fiber Ring Lasers, Ph.D. dissertation, Georgia Institute of Technology, Atlanta, Georgia, 2000.
2. Shannon, C.E., Communication theory of secrecy systems. *Bell Sys. Tech. J.*, 28, 657–715, 1949.
3. Pecora, L.M. and Caroll, T.L., Synchronization in chaotic systems, *Phys. Rev. Lett.*, 64, 821–824, 1990.

2

Chaotic Signal Generation and Transmission

Antonio Cândido Faleiros, Waldecir João Perrella, Tânia Nunes Rabello, Adalberto Sampaio Santos, and Nei Yoshihiro Soma

CONTENTS

2.1 Introduction

In the past decade, the demand for services experienced in the telecommunication sector has been astonishing. Communications, in a broad sense, became well established in the 1930s and 1940s by the telephone and in the following two decades by television. In any case, the degree of interaction and volume of information exchanged between the agents involved cannot be compared in the form in which it is carried out now, for instance with the integration of voice and image and digital data.

To satisfy that demand for real time communication the services to be provided must be as efficient as possible, for example, mobile communication has become an essential service especially in populated areas. To manage the large number of different users in a large metropolis, adequate allocation of bandwidth is vital and problems such as mutual interference need to be addressed.

Alternative approaches to the standard telecommunication system have been suggested in the literature and a feasible solution emerged in what is called chaotic telecommunication (C_bT^c). The potential use of C_bT^c is addressed in this chapter as well as the role it plays in communication and the theory on which it is based. No background is expected from the reader other than a first course in communications.

The C_bT^c is a relatively new field of interest in communications[1] and its major motivation is derived from the fact that, on one hand, chaotic systems present a complex behavior, and on the other, they are governed by some very few simple rules.[7] Moreover, small changes in input parameters imply large deviations in output. A direct application of chaos theory to telecommunication systems appears in a conventional digital spread spectrum, where the information is spread over a wider band by using a chaotic signal instead of the usual periodic sequence, called pseudo-noise (PN) sequence, the latter is generated, for instance, by linear shift registers. The problem with a linear shift register generator is that the price paid for making the period of the PN long increases sharply because a large amount of storage capacity and a large number of logic circuits are required. This imposes a practical limit on how large the period of the PN can actually be made. This can be overcome by the use of a digital chaotic sequence generator.[9]

In Section 2.2, the basic concepts of the chaos theory are introduced and in the remaining sections the theory of normal communication system and the major developments of chaotic telecommunication systems are described.

2.2 Chaos Theory

Chaos theory appears in a number of areas of science and it has applications varying from pure mathematics to biology.[13] A very simple example of a direct application of the chaos theory is described here.

In the area of secure communications, the method used to cipher a message is named stream cipher. Essentially, the analog signal produced by voice into a microphone is transformed into a sequence of zeros and ones. As a result of this transformation we have a digital signal. To protect a given message, the sequence of bits generated by the voice is operated together with a randomly generated sequence and the result is then transmitted. The operation just mentioned is the XOR bit a bit, defined on the set {0,1} in the following way:

$$0 \oplus 0 = 0, \quad 0 \oplus 1 = 1, \quad 1 \oplus 0 = 1, \quad 1 \oplus 1 = 0$$

The value 1 means TRUE and 0 means FALSE so the operation above, corresponds to the logical operator XOR (exclusive OR). To recover the original message, authorized receivers synchronously create the same random sequence that is XORed again with the received message.

It is worth mentioning that the methods used in practice do not produce a truly random sequence, but only what is called a pseudorandom sequence in the sense that anyone of these methods is deterministic. However, they are capable of producing sequences with good "randomness properties." An important field of research in this area is directed toward the design of more secure methods to produce bits with a good pseudorandom sequence.

A classic, efficient, and well-studied method of generating a sequence of pseudorandom bits is the linear feedback shift register. A shift register is a very simple electronic device, which produces a very fast pseudorandom sequence. Basically, this device is formed by a sequence of adjacent bits in a register and at each clock signal the sequence is shifted a position to the right. The rightmost bit is the output. To the left, one additional bit is introduced which is computed by a function, named feedback function, of the previous contents of the registers. The binary storage elements are called the stages of the shift register, and their contents are called the state of the shift register. After starting the shift register in any initial state, it progresses through some sequence of states, hence a periodic sucession ultimately results. This sucession is used in the XOR operation to produce the ciphered message.

When the feedback function is linear, the shift register is called a linear shift register and using the theory of polynomials over finite fields, it is possible to discover how to design a device that produces a sequence whose period is very long and has good randomness properties. But, it is important to stress that the linear feedback shift register is insecure for cryptographic purposes.

2.2.1 Discrete Dynamical Systems

Another way to produce a pseudorandom sequence is based on dynamic systems with chaotic behavior. To give a motivation to the subject, the

Verhult's model on population growth is given next, in which the number of individuals is measured at discrete times. This model is an example of a discrete dynamical system.

It is reasonable to assume that the total number (y_{n+1}) of members of the population at instant $n+1$ is a function of the population (y_n) at the instant n. This dependence can be written in the following way

$$y_{n+1} = f(y_n)$$

where f is a given real function. As a first approach, note that a linear law of population growth is not realistic because it leads to an exponential growth rate and cannot be sustained forever. For example, if

$$f(y) = ky$$

where k is a real constant, and $n=0$ at the initial instant of time, then

$$y_1 = f(y_0) = ky_0$$

$$y_2 = g(y_1) = ky_1 = k^2 y_0$$

$$y_3 = g(y_2) = ky_2 = k^3 y_0$$

and so on. After n steps of time one has:

$$y_n = k^n y_0$$

This is not a realistic model for almost all of the cases as the following simple analysis demonstrates. If $k > 1$ the population grow without limit according a geometric progression. If $k=1$ the population remains unchanged. If $k < 1$, the population size decreases steadily toward zero. Therefore, any realistic model can be modeled only by a nonlinear function.

What is expected in an environment with limited resources is that the population grows until a value Y of equilibrium and it decays if that limit is surpassed. In the formulation of a nonlinear model of population growth it is interesting to work with the variable

$$x_n = \frac{y_n}{Y}$$

where the population is a fraction related to the equilibrium point. The relation between the population fraction x_n at the instant n and the fraction x_{n+1} at the instant $n+1$ follows a law of the following type

$$x_{n+1} = f(x_{n+1})$$

where f is a real nonlinear function.

A well-known model that explains the population growth[11,12] is given here. Let x_0 the initial population fraction and x_n the same fraction after n instants of time (it can be seconds for bacteria, weeks for mice, years for humans, and so on).

In a first approach, let r be constant growth rate:

$$r = \frac{y_{n+1} - y_n}{y_n} = \frac{x_{n+1} - x_n}{x_n}$$

with a linear population dynamics given by:

$$x_{n+1} = (1 + r)x_n$$

and, as has already been shown before, it is not a reasonable model. To get more realistic results, Verhulst introduced a modification in the above law by assuming a growth rate proportional to $1 - x_n$. His dynamical law becomes:

$$c(1 - x_n) = \frac{x_{n+1} - x_n}{x_n}$$

where c is a positive real constant number. When x_n goes below 1 ($y_n < Y$) there is population growth. When x_n goes above 1 ($y_n > Y$), the population decreases. This is exactly what is expected in an environment with limited resources; an oscillation around stabilization in the equilibrium point. Solving the last equation for x_{n+1},

$$x_{n+1} = cx_n(1 - x_n) + x_n$$

To study the evolution of a population that follows this law it is important to understand the behavior of the function

$$f(x) = cx(1 - x) + x$$

at iteration $x_{n+1} = f(x_n)$, $n = 1, 2, 3, \ldots$. Given an initial state x_0, the evolution of the population, is governed by the sequence

$$x_1 = f(x_0), \quad x_2 = f(x_1) = f[f(x_0)], \quad x_3 = f(x_2) = f\{(f[f(x_0)]\}, \ldots$$

The sequence $\{x_0, x_1, x_2, x_3, \ldots\}$ gives the evolution of the population and it is called the **orbit** of x_0.

Let $n \geq 0$ an integer and f^n denotes the nth **iterate** of f, the function can be defined recursively by:

$$f^{n+1} = f^\circ f^n$$

where $^\circ$ is the composition between functions and $f^1 = f$. Thus, it is common to write

$$x_1 = f(x_0), \quad x_2 = f^2(x_0), \quad x_3 = f^3(x_0), \ldots$$

to denote the elements of the orbit of x_0.

With $c=1$ the orbit of $x_0=0.8$ and according to the Verhulst law this is

$$\{0.8, 0.96, 0.998, 0.999, 1.0, 1.0, 1.0, 1.0, 1.0, \ldots\}$$

which converges to 1. If $c=2.5$, the resulting orbit of $x_0=0.2$ is

$$\{0.2, 0.6, 1.2, 0.6, 1.2, 0.6, 1.2, 0.6, 1.2, \ldots\}$$

which diverges and after some instants of time it oscillates between 0.6 and 1.2. This orbit is *periodic* with a period of 2. For $c=3$ the orbit corresponding to $x_0=0.5$ is

$$\{0.5, 1.25, 0.31, 0.95, 1.08, 0.81, 1.26, 0.26, 0.85, 1.22, 0.38, \ldots\}$$

This orbit does not converge nor oscillate between two or more values. This behavior is referred to as chaos. These examples are known as *discrete dynamical systems* because they describe the evolution in time of a given population.

2.2.2 Chaos Modeling

In the last decade of the 19th century, Henri Poincaré initiated the research of the chaotic behavior of dynamical systems. At that time, Poincaré was studying the stability of the solar system. During his research he created a new branch of mathematics called topology. Today this subject has experienced great development and there has been intensive research on its potential use in secure communications.

The main goal of this section is the introduction of the main ideas of the dynamics of a sequence produced by a nonlinear function that presents a chaotic behavior.

The main question in population dynamics concerns the behavior of $f^n(x_0)$, such as its limit when $n \to \infty$. It will be seen that the answer to this question is quite difficult, even for simple maps.

For ease of comprehension instead of the Verhulst's model a slightly different function to the population growth must be considered:

$$f(x) = kx(1-x)$$

where k is a real constant. From now on, f will refer to the above function, which is topologically equivalent to the original one of Verhult's. Recall that two real functions f and g are topologically equivalent if there is a continuous and one-to-one function h defined on **R** onto **R**, such that, for each real x,

$$h[f(x)] = g[h(x)]$$

For the case, with $g(x)=cx(1-x)+x$ and $f(x)=kx(1-x)$ we have:

$$h(x)=\frac{1+c}{c}x$$

with $k=1+c$. It is possible to prove that topologically equivalent functions give rise to equivalent dynamics. Moreover, the chaotic properties of the orbits of f and g are preserved.

Consider the orbit of x_0,

$$x_0, x_1, x_2, \ldots, x_n, x_{n+1}, \ldots$$

where, as before, $x_{n+1}=f(x_n)$ for $n=0, 1, 2, \ldots$

If this sequence is convergent to a point, say, x^*, taking the limit on both sides of the equality $x_{n+1}=f(x_n)$ we get

$$x^* = \lim_{n\to\infty} x_{n+1} = \lim_{n\to\infty} f(x_n) = f(\lim_{n\to\infty} x_n) = f(x^*)$$

because f is a continuous function. Then, if an orbit is convergent, it converges to a point x^* such that

$$f(x^*) = x^*$$

Those points are the *fixed points* of f.

Looking at the graph of the function f in the plane xy (Figure 2.1), the fixed points are the x coordinates of the intersection of the graph of f with the straight line $y=x$.

The fixed points of $f(x)=kx(1-x)$ are the solution of the polynomial equation

$$kx(1-x)=x$$

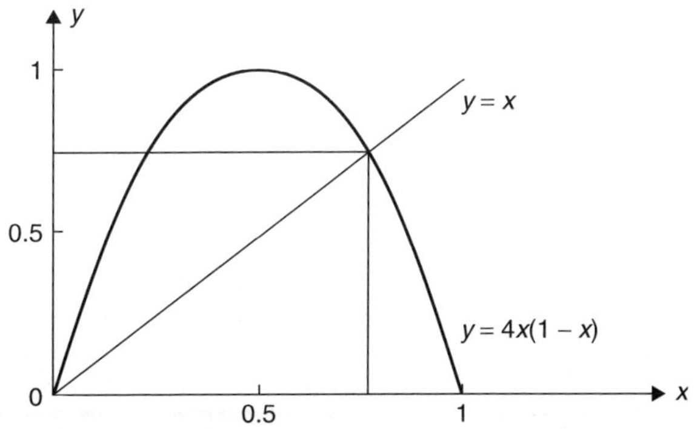

Figure 2.1 Graph of the function $y=f(x)=kx(1-x)$ with $k=4$ and the line $y=x$.

and are given by:

$$x = 0 \quad \text{and} \quad x = (k-1)/k$$

If x_0 is a point of the interval [0,1], it will be seen that the orbit,

$$x_0, x_1, x_2, \ldots, x_n, \ldots$$

determined by $f(x) = kx(1-x)$, can (i) converge to a fixed point, (ii) is periodic, or (iii) has chaotic behavior. This behavior will not depend on the initial value of the orbit but on the value of k.

The theoretical analysis of the behavior of the orbits may be very difficult and in such a case a graphical analysis can be an invaluable tool.

Graphical analysis is a technique that consists in drawing the graph of f and of the straight line $y = x$ in the plane xy. To draw the orbit of a point x_0 the following process is used: draw a vertical segment from (x_0, x_0) to $[x_0, f(x_0)]$ followed by a horizontal line from $[x_0, f(x_0)]$ to $[f(x_0), f(x_0)]$ going back to the line $y = x$. Remembering that $x_1 = f(x_0)$ is a polygonal line that stops at point (x_1, x_1). After that iteration the process is repeated again. The polygonal line with vertices over the points (x_1, x_1), $[x_1, f(x_1)]$ and $[f(x_1), f(x_1)]$ is drawn and this procedure continues with $x_2 = f(x_1)$, $x_3 = f(x_2)$, and so on.

Figure 2.2 show the orbit of $x_0 = 0.1$ under $f(x) = kx(1-x)$ with values $k = 2.4$ (left) and $k = 2.8$ (right). It is possible to see that the orbit goes off the fixed point zero and walks in the direction of the other fixed point $(k-1)/k$. The first point is **repelling** and the last is an **attracting fixed point**. A fixed point x^* is attracting if

$$|f'(x^*)| < 1$$

and repelling when

$$|f'(x^*)| > 1$$

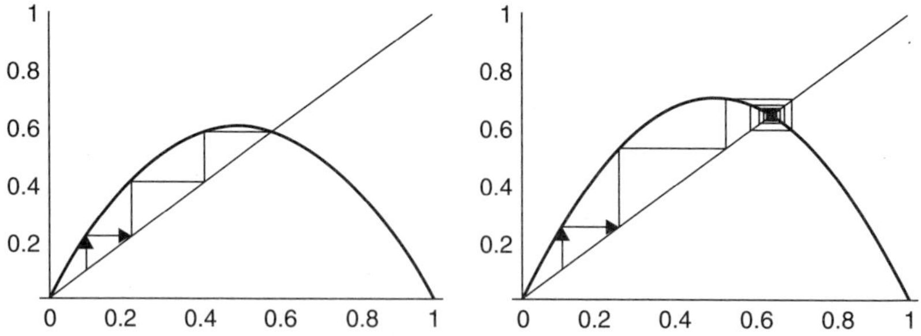

Figure 2.2 Here is the plot of $f(x)$, the line $y = x$ and the orbit of $x_0 = 0.1$. At left, $k = 2.4$ and, at right, $k = 2.8$. In both cases, it can be seen that the fixed point $(k-1)/k$ is attracting and zero is a repelling fixed point.

If x_0 is in the interval [0,1], its orbit under $f(x) = kx(1-x)$ with k in the interval [1,3] converges to a fixed point $(k-1)/k$ of f. Therefore, this point is an attracting fixed point and the zero is a repelling fixed point and hence the dynamics of f is completely determined in that case.

Figure 2.3 shows the orbit where $x_0 = 0.1$ under $f(x) = kx(1-x)$, with $k = 3.2$. In that case, the orbit is not convergent but oscillates between two different values: 0.513045 and 0.799455.

A point x_p is a **periodic point** of period n for f if and only if

$$f^n(x_p) = x_p, \quad \text{but } f^i(x_p) \neq x_p, \text{ for } 0 < i < n$$

A point x_e is an **eventually periodic point** of f if and only if there exists a positive integer n such that the point $f^n(x_e)$ is a periodic point and y_0 is not, itself, periodic.

In the example of Figure 2.3, it can be seen that the point 0.1 is periodic and the points 0.513045 and 0.799455 are periodic points of period two.

A periodic point (with period n) is an **attracting periodic point** if all orbits, sufficiently near them, converge to them. It can be proved that a periodic point x_p with period n is an attracting periodic point if

$$|(f^n)'(x_p)| < 1$$

A period point is a **repelling periodic point** if all orbits sufficiently near to them diverge from them. It can be proven that a periodic point x_p with period n is a repelling periodic point if

$$|(f^n)'(x_p)| > 1$$

A periodic point x_p of period n is a **neutral periodic point** if

$$|(f^n)'(x_p)| = 1$$

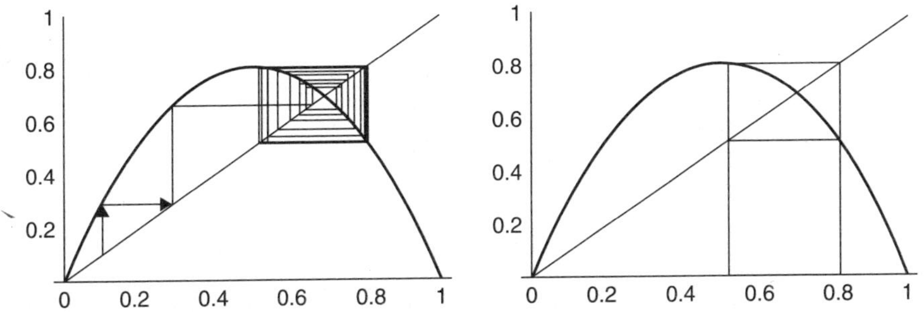

Figure 2.3 This figure show the orbit of $x_0 = 0.1$ under the function $f(x) = kx(1-x)$ with $k = 3.2$. On the right side, the first 100 points of the orbit where dropped. The orbit does not converge but it goes to a periodic orbit with period two.

The points 0.513045 and 0.799455 are attracting periodic points of period of two of the function $f(x) = kx(1-x)$ with $k = 3.2$. The point $x = 1$ is an eventually periodic point of f because $f(1) \neq 1$ but $f[f(1)] = 0 = f(1)$.

When k is such that $3 < k \leq 4$ there is a change in the behavior of the dynamics of the orbits. When $k = 3$, the attracting fixed point become a neutral fixed point. Increasing the value of k (above three) the orbits do not converge any longer but they become oscillating between two different values provided that the initial point x_0 lies within the interval $0 < x_0 < 1$, as was seen in the Figure 2.3.

Increasing the value of k, the dynamics becomes much more complex and the orbits converge to a periodic one whose period increase with k. There are some values in between that period that decays. When $k = 4$, the limit of the orbit is not periodic any more. The orbit is not convergent either to a fixed point or to a periodic orbit. When this happens, the orbit is named chaotic.

With $k = 3.5$, the orbits end in the periodic orbit {0.38, 0.82, 0.50, 0.87} of period four (see Figure 2.4 through Figure 2.7). With $k = 3.55$, the orbits go to the periodic orbit {0.37, 0.82, 0.50, 0.88, 0.35, 0.81, 0.54, 0.88} where it was kept to only two decimal digits. When $k = 3.566$ the final orbit has period of 16. But, it is interesting to note that when $k = 3.835$ the orbits goes to a period of one and with a period three the orbit is {0.49, 0.95, 0.15}.

To have a better visualization of the final values of an orbit, it is worthwhile to look at Figure 2.8, which is called a Feigenbaum diagram.[5] In this diagram, the value of k goes in the horizontal axis and the final values of the orbits that start at $x_0 = 0.1$ goes in the vertical axis. Looking at the diagram it becomes evident that, for k between zero and one, the orbit converges to zero. When k stays between one and three the orbits converge to the fixed point $(k-1)/k$. At the value $k = 3$, there is a **bifurcation**. From that point onwards the orbit does not converge anymore and it goes to an attracting

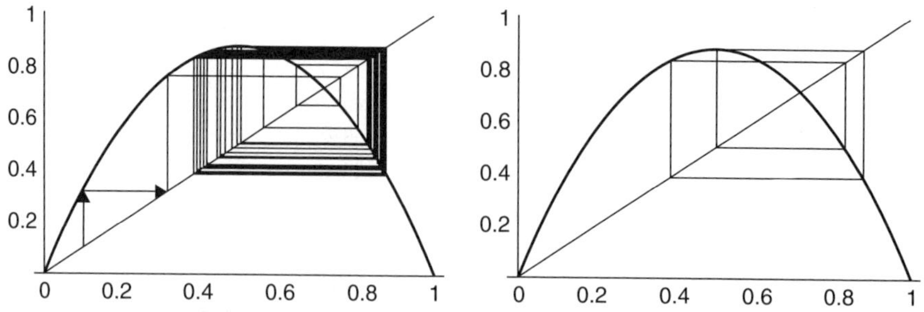

Figure 2.4 This figure shows the orbit of $x_0 = 0.1$ under the function $f(x) = kx(1-x)$ with $k = 3.5$. On the right, the first 100 points of the orbit where dropped. The orbit does not converge but it goes to a periodic orbit with period four.

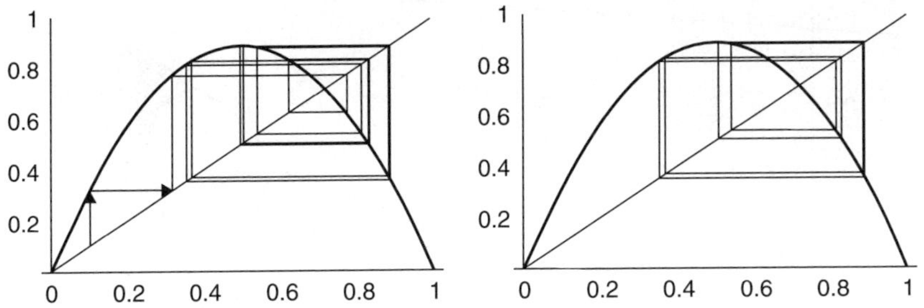

Figure 2.5 This figure shows the orbit of $x_0 = 0.1$ under the function $f(x) = kx(1 - x)$ with $k = 3.55$. On the right, the first 100 points of the orbit where dropped. The orbit does not converge but it goes to a periodic orbit with period eight.

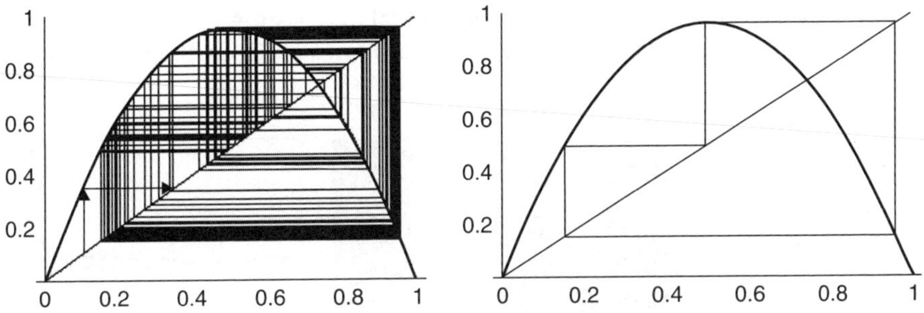

Figure 2.6 This figure shows the orbit of $x_0 = 0.1$ under the function $f(x) = kx(1 - x)$ with $k = 3.835$. On the right, the first 100 points of the orbit where dropped. The orbit does not converge but it goes to a periodic orbit with period three.

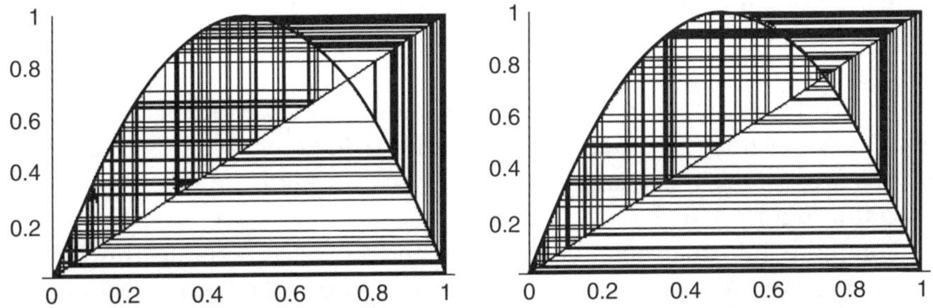

Figure 2.7 This figure shows the orbit of $x_0 = 0.1$ under the function $f(x) = kx(1 - x)$ with $k = 4$. On the left it is plotted the first 100 points of the orbit. On the right, the first 1000 points of the orbit were dropped and the next 100 points are shown. The orbit is not convergent either to a fixed point or to a periodic orbit.

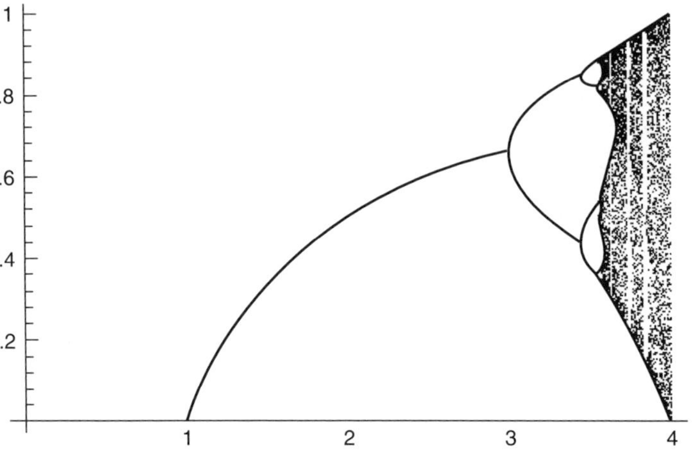

Figure 2.8 Feigenbaum diagram. The horizontal axis has the value of *k* and in the vertical axis are the final values of the orbit of $x_0 = 0.5$ without the first 500 points.

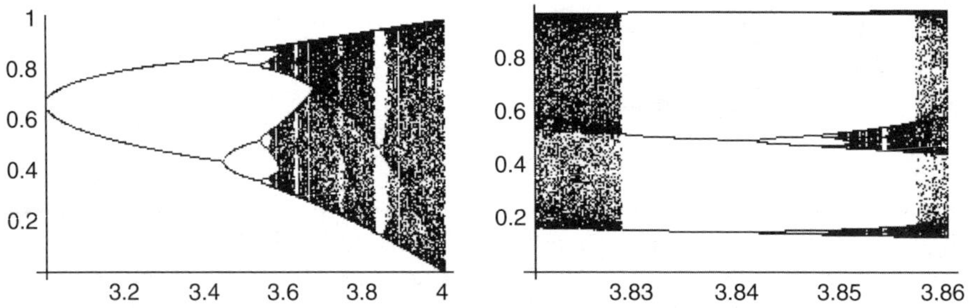

Figure 2.9 Amplifications of two interesting regions in the Feigenbaum diagram. On the left, the effect of period doubling in the orbits is shown. On the right, there is an amplification of the Feigenbaum diagram with *k* between 3.82 and 3.86. In that position there exist values of *k* whose period is three.

periodic orbit of period two. The orbits double when *k* increases. It is possible to see islands with periods less than one for lower values of *k*. These islands are visible in the amplifications of some regions in the Feigenbaum diagram (Figure 2.9).

For more informations about chaos and how to generate the figures, see, for instances, the excellent books from Gray and Glynn[14] and Peitgen et al.[15]

2.2.3 Chaotic Regime

In the last section it was hinted that the simple Verhult's model could produce unpredictable sequences in the following sense: small changes

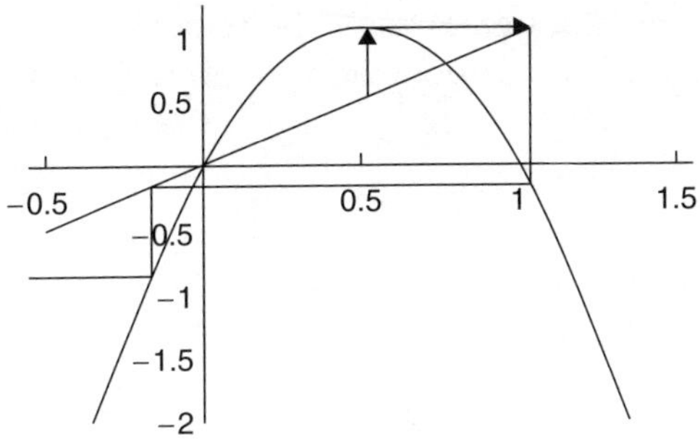

Figure 2.10 Orbit of $x_0 = 0.55$ under $f(x) = kx(1 - x)$ with $k = 4.2$. Note that $f(x_0) > 1$ and $f[f(x_0)] < 0$. After that, the orbit goes monotonically to $-\infty$.

upon the initial conditions, will result in great differences at the subsequent terms.

We examine now the chaotic regime and the model that corresponds to $k > 4$ (Figure 2.10). In this case, the maximum value of $f(x) = kx(1 - x)$ is larger than one. It is worth mentioning that if $x > 1$, then $f(x) < 0$ although for $x < 0$, $f(x) < x < 0$. Therefore, if x is a point in the interval $0 < x < 1$ such that $f(x) > 1$, then $f[f(x)] < 0$ and the orbit of x goes to $-\infty$. Hence, there exists an open subinterval A_0 of the interval [0,1], centered at 1/2 such that, for all $x \in A_0, f(x) > 1$ and hence $f^n(x) \to -\infty$, when $n \to \infty$.

Observe that A_0 is centered at $x_0 = 1/2$, which is the point where the function attains its maximum. The resulting set $A_1 = [0,1] - A_0$ is the union of two closed subintervals I_0 and I_1, whose image by f is the whole interval [0,1]. Hence, there are points $x \in A_1 = I_0 \cup I_1$ such that $f(x) \in A_0$. Then $f[f(x)] = f^2(x) > 1$ and therefore $f^n(x) \to -\infty$, when $n \to \infty$. Thus, there are two open subintervals I_2, I_3, of I_0 and two open subintervals I_4, I_5 of I_1, such that for all x, which belongs to one of these subintervals it becomes $f^2(x) > 1$ and its orbit goes to $-\infty$. Continuing this reasoning inductively, one can construct the subset

$$\begin{aligned} A_n &= \{x \in [0,1]; f^n(x) \in A_0\} \\ &= \{x \in [0,1]; f^{n+1}(x) > 1 \quad \text{and} \quad f^i(x) \in [0,1], \ \text{for } 0 \leq i \leq n\} \end{aligned}$$

$[0,1] - A_0 \cup A_1 \cup \ldots \cup A_n$ consists of the union of 2^{n+1} closed subintervals. Notice that the dynamics inside the subsets A_n is known. The behavior of

$$\Omega = [0,1] - A_0 \cup A_1 \cup \ldots \cup A_n \cup \ldots$$

remains to be analyzed where the orbits stay at [0,1].

The set Ω is a **Cantor set**, i.e., it is closed, does not contain any interval, and every point in Ω is an accumulation point of Ω.

A classic example of a Cantor set is the Cantor middlethird whose construction resembles what was done above and will be described next.

Example

(Cantor middle–third set) Consider the interval [0,1] and remove the open interval (1/3,2/3), obtaining $[0,1]-(1/3,2/3)=[0,1/3]\cup[2/3,1]$. Next, remove the open intervals (1/9,2/9), (7/9,8/9), to obtain $[0,1/9]\cup[2/9,3/9]\cup[6/9,7/9]\cup[8/9,1]$ and so forth, continue removing the middle thirds of each closed subinterval. The resulting set is the middle–third set.

What can be said about the dynamics inside Ω? For $k>2+\sqrt{5}$, the quadratic map is chaotic on Ω. But what does a **chaotic map mean**? There are many possible definitions of chaotic maps and the definition of Devaney is presented here.[2,4]

Definition

Let V be a set of real numbers. $F: V \rightarrow V$ is said to be chaotic on V if:

F has sensitive dependence on initial conditions, i.e., if there exists a $\delta>0$ such that, for any $x\in V$ and any neighborhood N of x, there exists $y\in N$ and $n\geq 0$ such that $|F^n(x)-F^n(y)|>\delta$.

F is topologically transitive, i.e., for any pair of open sets I, J in V there exists a $k>0$ such that $F^k(I)\cap J\neq\emptyset$.

Periodic points are dense in V.

It can be proven that function $f(x)=kx(1-x)$ is chaotic on Ω for $k>2+\sqrt{5}$ and it is chaotic on the *whole* interval [0,1] for $k=4$.

Chaotic maps have, therefore, three important features: (i) unpredictability because of the sensitive dependence on initial conditions, (ii) nondecomposability, i.e., it cannot be decomposed into two open invariant subsets under f due to the topological transitivity, and (iii) a regularity given by the density of the periodic points. To measure the sensitive dependence on initial conditions, the Lyapunov exponent is explained next.

Consider the discrete dynamic system

$$x_{n+1}=f(x_n)=f^n(x_0), \quad \text{where } f: I\rightarrow \mathbf{R}$$

Consider two initial points x_0 and y_0 and let $\delta = |y_0 - x_0|$ be the initial distance between these points. After an iteration the new distance is $\delta_1 = |y_1 - x_1|$, where $y_1 = f(y_0)$ and $x_1 = f(x_0)$. If L is defined by $\delta_1 = e^L\delta$, then L measures the exponential rate of expansion from the distance δ to the distance δ_1 after just an iteration. After n iterations the distance δ_n becomes

$$\delta_n = |f^n(x_0) - f^n(y_0)| = \delta e^{nL}$$

and it can be rewritten as

$$L = (1/n)\ln |f^n(x_0) - f^n(y_0)|\delta$$

Considering an infinitesimal initial distance, $\delta \to 0$, after an infinity number of iterations, the formula of L becomes:

$$k(x_0) = L(x_0) = \lim_{n\to\infty} \lim_{\delta\to 0}(1/n) \ln\{|[f^n(x_0 + \delta) - f^n(x_0)]/\delta|\}$$

which is the definition of Lyapunov exponent of chaotic discrete dynamical system. In the chaotic domain, the Lyapunov exponent is positive. The larger is this exponent, more sensitive to the initial conditions are the solutions of the discrete dynamical system. This exponent can be used as a tool to separate the chaotic behavior from the stable one.

An application to cryptography of this theory is the use of chaotic maps on an interval. Several attempts have been made to use the simulation of chaotic maps on a computer for encryption purposes, but the properties of chaotic maps that could serve to enhance trust in the security of these systems have been neglected. For example, if the proposed map is really chaotic on a whole interval I the occurrence of a periodic point is restricted to a set of measure null on I, and all are *repelling periodic points*. Moreover, when iterating the dynamical system on a computer, one has to bear in mind that a computer has only a finite precision. Therefore, the condition about topological transitivity cannot be fulfilled on a computer because the dynamical system on a computer is not situated on an interval of real numbers. It is necessary to adapt the definition of dynamical systems in a discrete and finite context.

In a computer, discrete iterations either end at fixed points or in cycles of a given length due to limitations on the computational space requirements. Thus, every chaotic system simulated on a computer will end up in a periodic cycle. The question to be answered is how to extend the definition of a chaotic dynamical system to these discrete iterations.

From this point on, the discrete iteration on the computer will be called chaotic whenever it is possible to extend it to a whole interval according to the definitions given previously. Yet, because the rounding errors made in each iteration are cumulative, and from the sensitivity of the chaotic system's initial conditions, the real (x_n) and calculated (y_n) trajectories will be very different. To solve this inherent numerical discrepancy there is a result called *shadowing lemma* derived by Coven et al.,[3] which claims that there exists a real trajectory (x_n) arbitrarily close to the trajectory (y_n), thus at least the statistical theoretical results remain true for "*chaotic*" discrete iterations.

The general use of a chaotic discrete dynamical system as a block cipher is to define a map $f_k: I \rightarrow I$, where the parameter k is the secret key and the initial value x_0 is a plaintext. After a chosen number of n iterations of f_k the value $f^n(x_0)$ is taken as the ciphertext.

One can also use a chaotic discrete dynamical system as a key stream generator, considering a chaotic map $f: I \rightarrow I$, and $x_0 \in I$ is kept secret. There exist two ways of deflecting a running key sequence from the $x_k = f^k\ (x_0)$. The first method is to use a threshold function such that

$$\theta(x) = 0 \quad \text{if } x < c \quad \text{and} \quad \theta(x) = 1 \quad \text{if } x \geq c$$

where c is a value chosen so that the binary sequence (x_k) exhibit good cryptographic features. The second way is to define a binary sequence by using the ith bit of $|x_k|$ in binary representation. In both cases the value x_0 serves as a secret key.

2.3 Chaotic Communication

Another very interesting application of the chaotic sequences appears in communications, because those sequences have the properties required for spread spectrum (SS), as will be seen next. The SS is a modulation technique that the information is spreaded in frequency by a sequence of bits, here called chips, totally independent of the information. The great advantage of this kind of modulation is that it permits that many different users have communications in the same band of frequency and at the same time. In this work, we will spread the information by using a periodic pseudo-sequence. This modulation is called direct sequence spread spectrum (DS-SS). The most important characteristics of the periodic sequence are: the autocorrelation and the cross-correlation. The autocorrelation is important in the synchronization between the periodic pseudo-sequence generate at the transmitter and at the receiver. The cross-correlation of the periodic pseudo-sequences must be zero to obtain communication between different users at the same band of frequency and at the same time. First, let us describe the required properties of chaotic sequences in comparison with the periodic pseudonumber sequences:

Noiselike waveform;
Wideband spectrum;

Advantage: sensitive dependence on the initial conditions, which is desirable for multiuser communications (different orthogonal sequences) and also for secure communications;

Advantage: infinitely long period without increasing the generator, which is desirable for multiuser communications and also secure communications;

Advantage: the generators can be built identically for both the transmitter and receiver by digital implementation;

Disadvantage: to synchronize the received chaos sequence with local generated at the receiver end is a complex study, see Parlitz.[10] The performance of SS system using NRZ chaos sequence is the same obtained with the SS system using PN sequence.

The following theory development is standard for the spread spectrum application, see for example Lam and Tantarana.[16]

Figure 2.11 shows a functional block diagram of the direct sequence spread spectrum with binary shift keying communication (DS-SS BPSK) analyzed. The binary data input is transformed in NRZ (nonreturn to zero) signaling pulse format. The spectrum of this signal is spread, in baseband, by PN-NRZ in the conventional SS communication system or by NRZ-chaos in the proposed communication system. A modulator BPSK, which shifts the spectrum to the assigned frequency range, follows the resulting spread

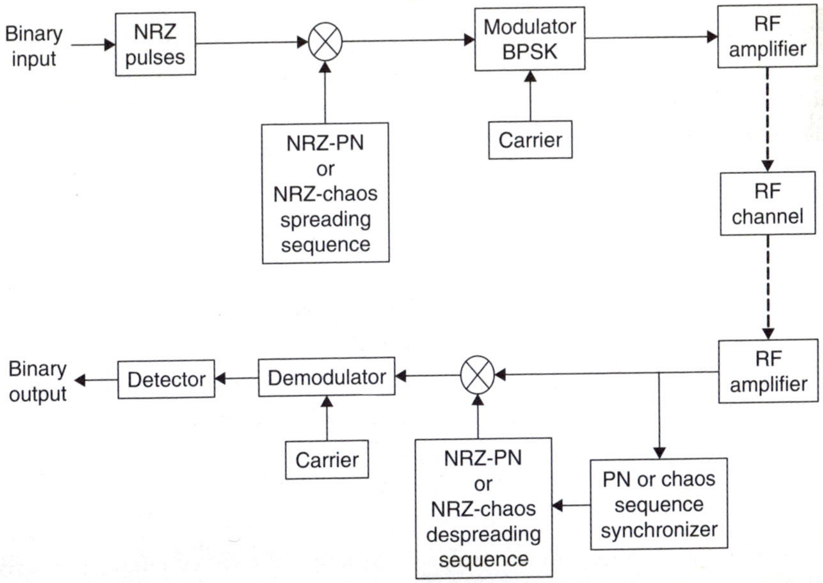

Figure 2.11 DS-SS — Conventional with NRZ-PN or NRZ-chaos sequences.

signal. The modulated signal is then amplified and sent through the radio frequency (RF) channel, i.e., a wireless communication channel, which has a limited bandwidth. It is assumed that the channel introduces only additive white Gaussian noise (AWGN). At the receiving end, the signal is amplified and filtered at the desired RF bandwidth and is fed to the demodulator, which shifts the spectrum to baseband. The resulting signal is despread by PN-NRZ in the conventional SS communication system or by NRZ-chaos in the proposed communication system. The PN-NRZ or NRZ-chaos generated at the receiver end must be in synchronism with the received sequence. The synchronizer does this. After this the binary output information is detected.

In the next section we will illustrate the chaotic telecommunication application in DS-SS using chaotic sequence, instead of a PN sequence, as proposed by Li and Haykin, 1995. The following conditions are assumed for the communication environment:

The channel is AWGN.
The binary input information and the spreading sequences are bipolar binary NRZ with values ±1 volt.
The chaotic sequence is generated by a tent map.[8]
A hard quantizer is used to generate the NRZ-chaos sequence. Haykin in his paper used other methods to generate the chaotic sequence.
Synchronism of the carriers is assumed and also in the spreading sequences.

To understand the principles of DS-SS systems we will need to study the properties of random binary sequences, the principle of modulation BPSK and its performance, and finally the DS-SS convention using PN sequences and the others using chaos sequences.

2.3.1 NRZ Random Signal

Defining a pulse $p_T(t)$ by

$$p_T(t) = \begin{cases} 1 & \text{for } 0 \leq t \leq T \\ 0 & \text{otherwise} \end{cases} \tag{2.1}$$

A NRZ random signal consists of a sequence of a binary data sequences $\{b_k = \pm 1\}$ into a waveform $p(t)$ resulting in the following expression:

$$x(t) = A \sum_{k=-\infty}^{\infty} b_k p(t - kT) \tag{2.2}$$

where T is the duration in seconds, A is the amplitude of each pulse NRZ and the bit rate is $R_b = 1/T$ bps (bits per second). Figure 2.12 illustrates an NRZ sequence with $T = 0.05$ s and $A = 1$ V.

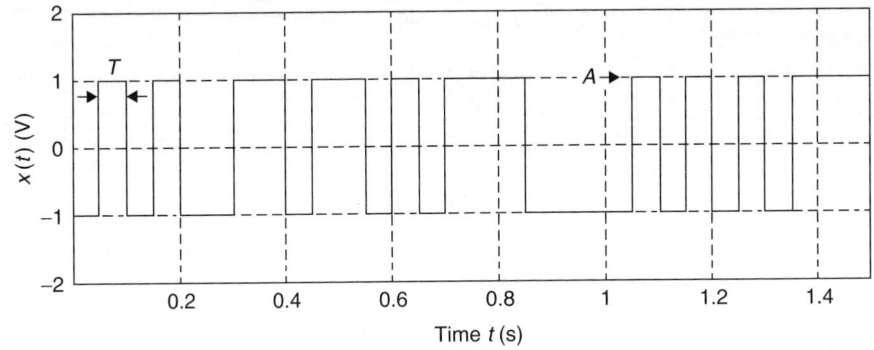

Figure 2.12 Example of a NRZ random sequence.

Statistical properties are easily characterized when the random variables b_k are independent and identically distributed with equal probabilities. It can be shown that the autocorrelation, $R_x(\tau)$, and power spectrum density (PSD), $S_x(f)$, are a pair of Fourier transforms, given by

$$R_x(\tau) = \begin{cases} A^2\left(1 - \frac{|\tau|}{T}\right) & |\tau| \leq T \\ 0 & \text{otherwise} \end{cases} \quad \Leftrightarrow \quad S_x(f) = A^2 T \,\text{sinc}(fT) \qquad (2.3)$$

Figure 2.13 illustrates these expressions for the NRZ random sequence given in the Figure 2.12. It is important to comment that we will model the data information, the PN, and the chaos signals as random signals. In this case, baseband communication, one important parameter is the required bandwidth. For these classes of signal the bandwidth is defined as the value of the first null of the sinc(·) function, defined as,

$$\text{sinc}(x) = \sin(\pi \cdot x)/\pi \cdot x \quad \pi = 3.14\ldots.$$

So the bandwidth is $B_x = 1/T$ Hz as is shown in Figure 2.13a and Figure 2.13b.

The principle of DS-SS communication systems is the spreading of the data information bandwidth, usually by many orders of magnitude, by the use of a random binary signal.

The conventional DS-SS system uses a pseudo-random signal is a deterministic signal and has almost the same statistical properties of truly random signal. It is usually called a PN sequence. A PN sequence is a periodic sequence of N bits, called chip. The time duration of each chip is T_c seconds. The period of the PN sequence is NT_c. A PN binary sequence has the same representation of the random sequence and is expressed by

$$c(t) = \sum_{k=-\infty}^{\infty} c_k p(t - kT_c) \qquad (2.4)$$

where $c_k = \pm 1$.

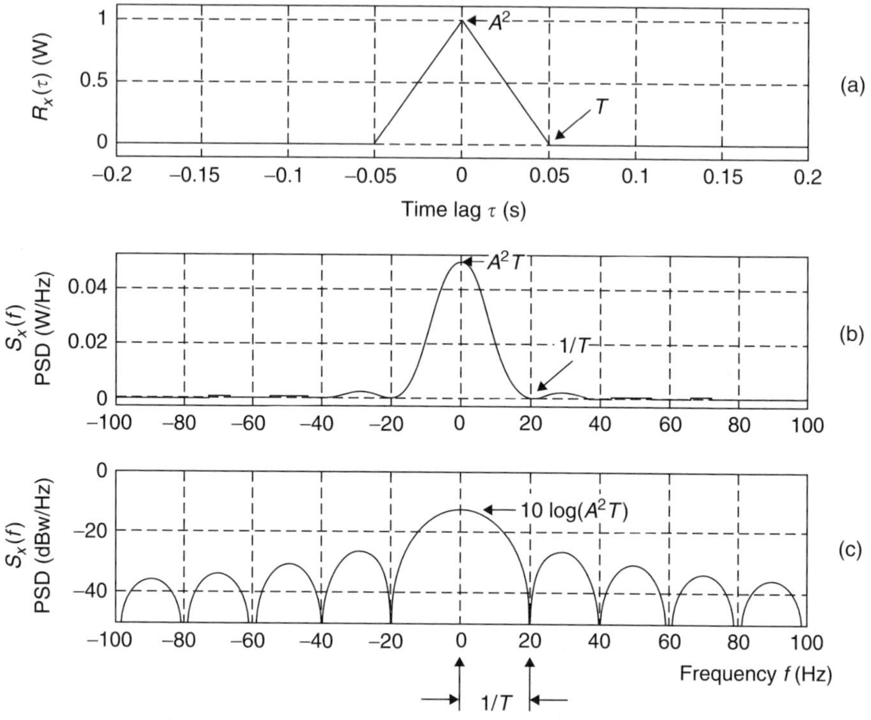

Figure 2.13 Autocorrelation and PSD corresponding to the random signal of the Figure 2.2. (a) Autocorrelation $R_x(\tau)$. (b) Power spectrum density $S_x(f)$ in W/Hz. (c) Power spectrum density $S_x(f)$ in dBw/Hz.

Figure 2.14a is an illustrative example of a PN sequence where the number of chips per period is $N = 5$, the duration of each chip is $T_c = 0.2$ s, and the duration of one period is $NT_c = 1$ s. Figure 2.14b shows the corresponding PSD of a random sequence (continuous curve) $C(f)$ where the bandwidth is $B_c = 5$ Hz and power spectrum at zero frequency is -7 dBw/Hz. As the NRZ-PN sequence is periodic its spectrum is formed by impulses at discrete frequencies, so, in Figure 2.14b the power spectrum at each discrete frequency is plotted in discrete curve. It is important to note that the envelope of the power spectrum of the PN sequence is the PSD of the random sequence.

In this work we use another kind of PN sequence that is an application of chaos theory to generate a spreading sequence.[8]

Later, each sample is hard quantized to generate the NRZ-chaos sequence. The tent map is the generator of the chaotic sequence, defined by

$$x_{n+1} = \begin{cases} mx_n + p & \text{for } x_n < X^t \\ m(1 - x_n) & \text{for } x_n \geq X^t \end{cases} \tag{2.5}$$

where $m = 1.9$, $p = 0.1$, and $X^t = 9/19$.

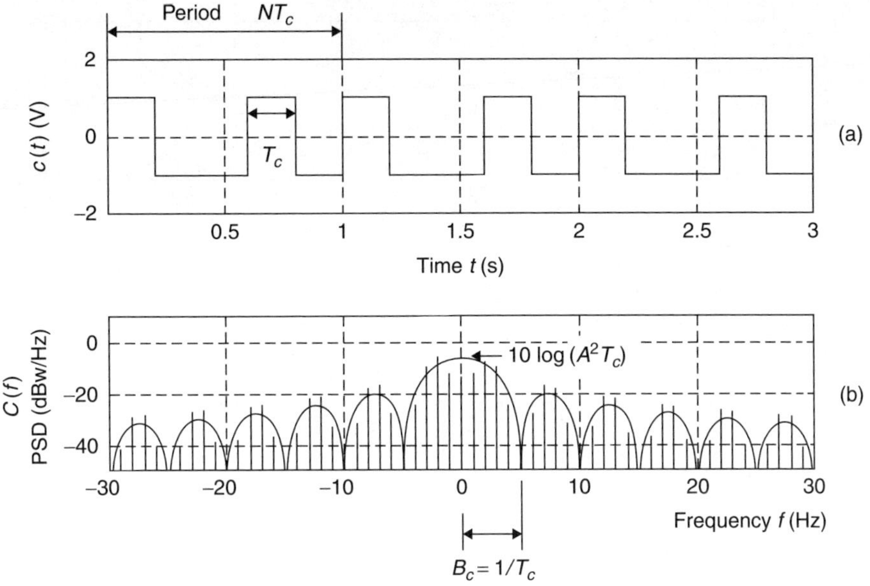

Figure 2.14 Conventional PN sequence. (a) Example of a NRZ-PN sequence. (b) PSD of the NRZ random sequence (continuous curve) and power spectrum of the NRZ-PN sequence (discrete curve).

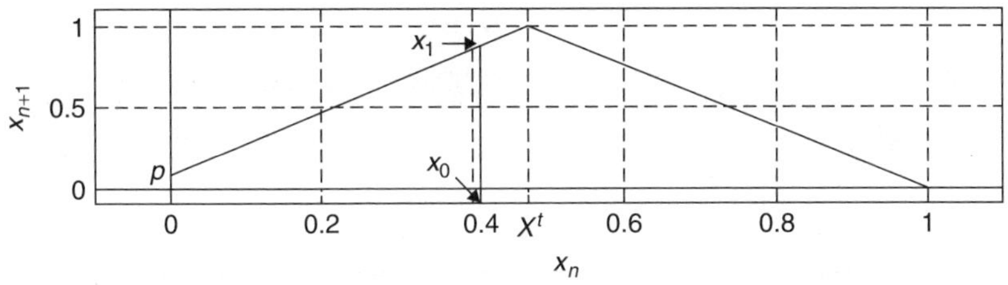

Figure 2.15 Tent map chaos generator.

The initial state of the iteration has to be started from a value in the following range: $0 \leq x_0 < 1$. In our simulation we started with $x_0 = \sqrt{2} - 1$. Figure 2.15 illustrates the tent map used.

Statistical properties of this chaotic generator were evaluated in Reference 6.

In this application, to have a sequence with zero mean value, its mean value is subtracted from the chaotic sequence. Figure 2.16a shows a chaotic sequence with 15 iterations. In Figure 2.16b each sample of the chaotic sequence is hard quantized to ±1 V. Figure 2.16c and Figure 2.16d illustrate the NRZ-chaos sequence and the respective power spectrum with the same

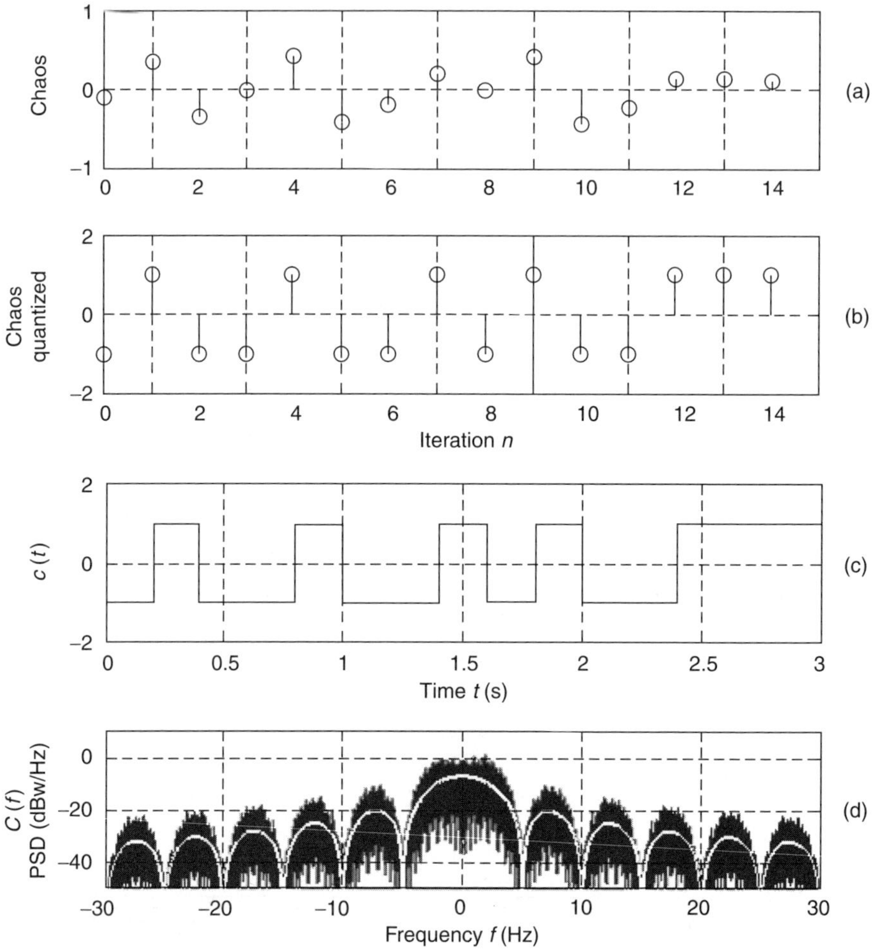

Figure 2.16 Example of chaos sequence. (a) Chaos sequence. (b) Chaos sequence hard quantized. (c) NRZ chaos sequence. (d) Spectrum.

condition as the NRZ-PN sequence of the Figure 2.14. Figure 2.16d plots the spectrum of the simulated NRZ-chaos sequence with 500 chips and, in white, the theoretical spectrum of a true random sequence. It is important to note that the NRZ-PN sequence is not periodic.

2.3.2 *Principle of DS-SS Baseband System*

The principle of a DS-SS communication system is the spreading of the data information bandwidth, usually by many orders of magnitude, by using a random binary signal. At the transmitter, the data information is modulated

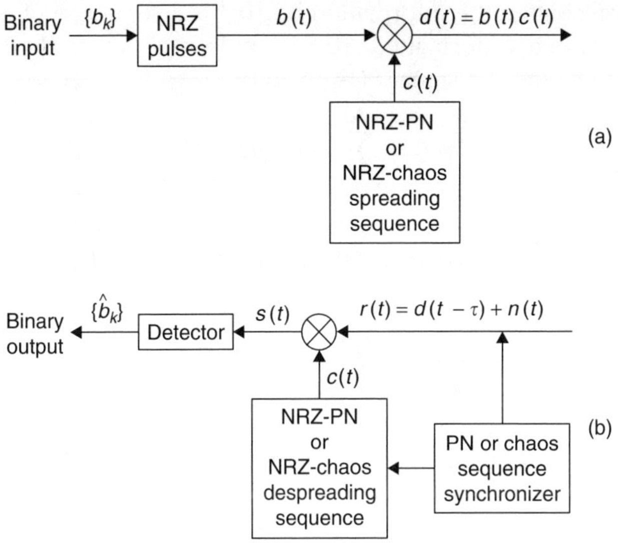

Figure 2.17 Principle of DS-SS system in baseband. (a) Transmitter: spreading. (b) Receiver: despreading.

by the random binary signal to widen the bandwidth of the transmitted signal. At the receiver, the widened signal is despread by using the same random binary sequence. The transmitter and receiver end must have a generator of these sequences and a method to synchronize them at the receiver end.

In this section we will analyze the DS-SS system in baseband, as shown in the Figure 2.17. We begin using the conventional NRZ-PN sequence and after using NRZ-chaos sequence as proposed by Haykin. We will analyze the waveforms at the transmitter and at the receiver end in time and frequency domains, assuming the following conditions:

Information, $b(t)$: bipolar binary NRZ, with values ± 1 V and bit rate $R_b = 1/T_b$ bps;

Spreading sequence PN or chaos, $c(t)$: bipolar binary NRZ, with values ± 1 V, chip rate $R_c = 1/T_c$ cps and a processing gain $\mathrm{PG} = T_b/T_c$;

Communication channel: AWGN with bilateral $\mathrm{PSD} = N_0/2$ W/Hz and autocorrelation $\delta(\tau)N_0/2$;

Synchronization: (a) spreading sequence and (b) bit.

Figure 2.17 illustrates the processes of the spreading at the transmitter and despreading at the receiver in baseband, using NRZ-PN or NRZ-chaos sequences.

At the transmitter, the information bit $\{b_k\}$ is a sequence of binary with values ± 1, which is formatted in NRZ pulses $p_{T_b}(t)$, with duration T_b seconds, generating the signal $b(t)$ given by

$$b(t) = \sum_{k=-\infty}^{\infty} b_k p_{T_b}(t - kT_b) \tag{2.6}$$

The spreading sequence, may be generated by a sequence of NRZ-PN or a NRZ-chaos pulses $p_{T_c}(t)$, with duration T_c seconds, resulting in the following signal:

$$c(t) = \sum_{k=-\infty}^{\infty} c_k p_{T_c}(t - kT_c) \tag{2.7}$$

where $\{c_k\}$ has also values ± 1 V.

The modulation process is done by multiplying $b(t)$ with $c(t)$ resulting in the spreading signal $d(t)$ given by

$$d(t) = b(t)c(t) \tag{2.8}$$

The spread signal $d(t)$ is sent through a communication channel and the received signal $r(t)$, shown in Equation (2.9), has two components: $d(t-\tau)$, which is the transmitted spread signal with a delay τ due to transmitter channel and $n(t)$, which is the channel noise.

$$r(t) = d(t-\tau) + n(t) \tag{2.9}$$

Using Equation (2.8) the received signal results in

$$r(t) = c(t-\tau)b(t-\tau) + n(t) \tag{2.10}$$

At the receiver we have two kinds of synchronism, between the received signal and the signal generated at the receiver to detect the information. The two kinds of synchronism are:

1. Between the transmitting and the receiving spreading sequences, that is, receiving received $c(t-\tau)$ and the generate at the receiver $c(t-\hat{\tau})$, where the delay $\hat{\tau}$ is obtained by a sequence synchronizer block cf. Figure 2.17b;
2. Between the bits of the receiving information, i.e., consists in the estimations of the instants of the transitions of the of information bits, $b(t-\tau)$, demodulated at the receiver. A block called bit synchronizer generates these instants.

From this point, all kinds of synchronization are assumed. In this case it is equivalent to consider $\tau = 0$ in the received signal as

$$r(t) = c(t)b(t) + n(t) \tag{2.11}$$

The process of despreading is obtained by multiplying $r(t)$ by $c(t)$, generated at the receiver, resulting in

$$s(t) = r(t)c(t) = c^2(t)b(t) + c(t)n(t) = b(t) + c(t)n(t) \tag{2.12}$$

The first component of $s(t)$ is the information data $b(t)$ and the second component is due to noise.

The resulting signal $s(t)$ is fed to the detector block that has the function of estimating the information data $\hat{b}_k$, the information is detected and the performance of the communication can be determined.

In the following sections we will illustrate the waveforms used in a DS-SS system in baseband, using NRZ-PN and NRZ-chaos sequences, and considering that the noise is not present. Finally, the performance of the communication will be determined.

2.4 Application of NRZ Chaos Sequence to DS-SS in Baseband

We will now illustrate the waveforms at the transmitter and at the receiver end in time and frequency domains assuming the following conditions:

Information, $b(t)$: bipolar binary NRZ, with values ± 1 V and a bit rate $R_b = 1/T_b = 1$ bps;

Spreading sequence PN or chaos, $c(t)$: bipolar binary NRZ, with values ± 1 V and a chip rate $R_c = 1/T_c = 5$ cps, resulting in a processing gain $\text{PG} = T_b/T_c = 5$;

Transmission channel: AWGN with bilateral $\text{PSD} = N_0/2$ W/Hz and autocorrelation $\delta(\tau)N_0/2$;

Synchronization: (a) spreading sequence and (b) bit.

This illustration will start by showing the conventional DS-SS using NRZ-PN sequence after showing the application of the NRZ-chaos sequence to the DS-SS. It is important to note that, in the frequency domain, the theoretical and simulated results are illustrated, using Matlab.

2.4.1 *Transmitter of DS-SS Baseband using NRZ-PN Sequence*

Figure 2.18 illustrates the process of spreading using a NRZ-PN sequence with a period of $P = 5$ chips and the duration of each bit of information. Figure 2.18a shows the data information $b(t)$ where the duration of each bit is $T_b = 1$ s and the rate is $R_b = 1/T_b = 1$ bps (bits per second). Figure 2.18b shows the NRZ spreading sequence $c(t)$ where the duration of each chip is

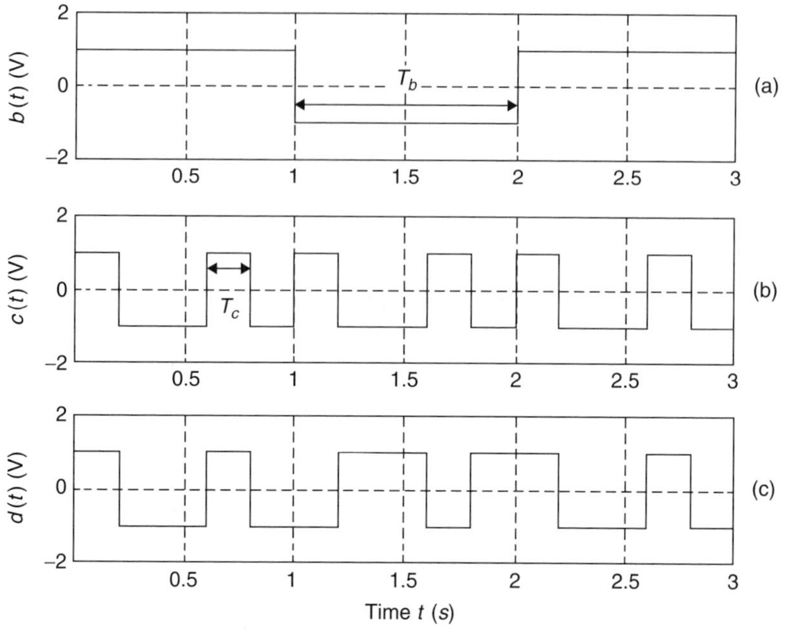

Figure 2.18 Principle of DS-SS system in baseband. (a) Data information *b*(*t*). (b) Spreading NRZ-PN *c*(*t*). (c) Spread signal *d*(*t*).

$T_c = 1/P = 1/5 = 0.2$ s and the rate is $R_c = 1/T_c = 5$ cps (chips per second). Finally, Figure 2.18c shows the spread signal $d(t)$, where each bit of information is multiplied by a period of the spreading sequence $c(t)$. Thus, when the bit of information is $+1$ the spread sequence is exactly one period of $c(t)$ and when the bit of information is -1 the spread sequence is the reverse of one period of $c(t)$, i.e., $-c(t)$.

The spreading process may be easily seen in the frequency domain. Figure 2.19a, Figure 2.19b, and Figure 2.19c shows the PSD of the data information, $B(f)$, and the NRZ-PN sequence, $C(f)$, and the spread signal, $D(f)$, respectively. These figures illustrate the PSD of the corresponding random sequence and the PSD obtained by simulation.

It is important to note that the PSD of data information of the Figure 2.19a has a bandwidth of $B_B = 1/T_b = 1$ Hz. The PSD is given by

$$B(f) = A^2 T_b \operatorname{sinc}(f T_b) \tag{2.13}$$

The PSD of the spreading NRZ-PN sequence modeled by a random sequence has a bandwidth of $B_C = 1/T_c = 5$ Hz and can be expressed as

$$C(f) = A^2 T_c \operatorname{sinc}(f T_c) \tag{2.14}$$

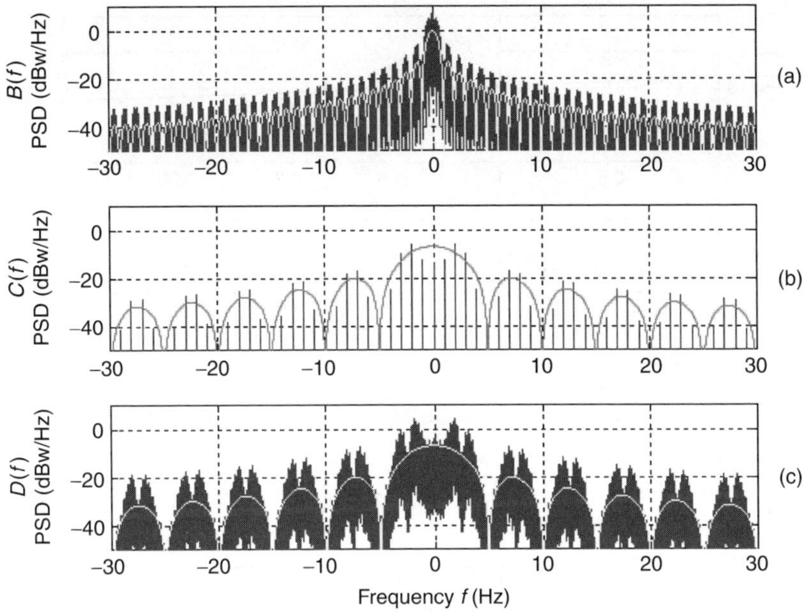

Figure 2.19 Power spectrum density. (a) PSD of *b*(*t*). (b) PSD of *c*(*t*). (c) PSD of *d*(*t*).

Finally, the PSD of the spread signal $D(f)$, also has a bandwidth of $B_D = 1/T_b = 1$ Hz and can be expressed as

$$D(f) = A^2 T_c \operatorname{sinc}(f T_c) \tag{2.15}$$

The process of the widening the bandwidth can be observed by noting that the information bandwidth is spread to the bandwidth of modulated signal $d(t)$, i.e., from $B_B = 1/T_b = 1$ Hz to $B_D = 1/T_c = 5$ Hz.

We can define a spectrum-spreading ratio in baseband, normally called processing gain, as

$$\text{PG} = \frac{\text{Bandwidth of SS signal}}{\text{Bandwidth of information}} = \frac{B_D}{B_B} = \frac{T_b}{T_c} \tag{2.16}$$

The PG is an important performance parameter of a SS system because a higher PG implies better anti-interference capabilities, for example in multiple user cellular systems and in antijamming military systems.

2.4.2 *Transmitter of DS-SS Baseband using NRZ-Chaos Sequence*

The process of DS-SS system with NRZ-chaos spreading sequence is very similar to the NRZ-PN sequence. The unique difference is the fact that the NRZ-chaos sequence is not periodic. All the other parameters are maintained. Figure 2.20 and Figure 2.21 show the spreading process in the

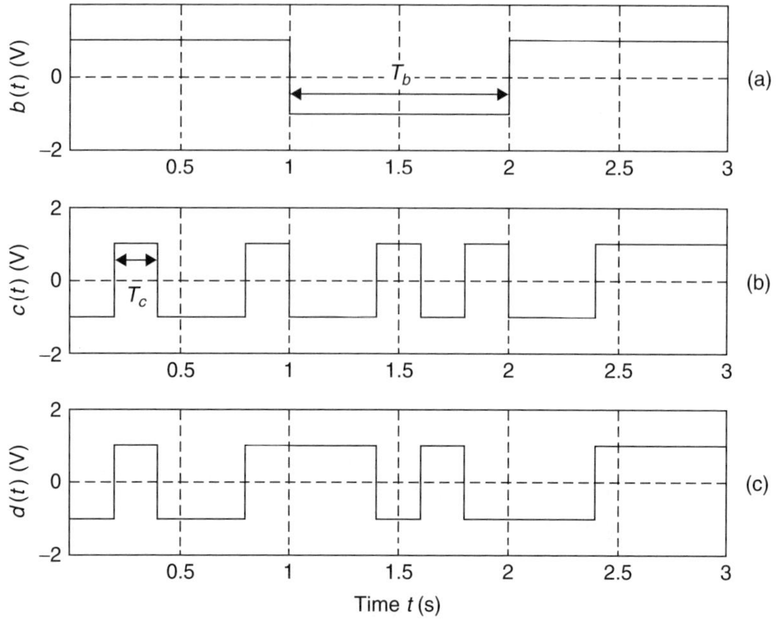

Figure 2.20 Principle of DS-SS system in baseband with chaos sequence. (a) Data information $b(t)$. (b) Spreading NRZ-chaos $c(t)$. (c) Spread signal $d(t)$.

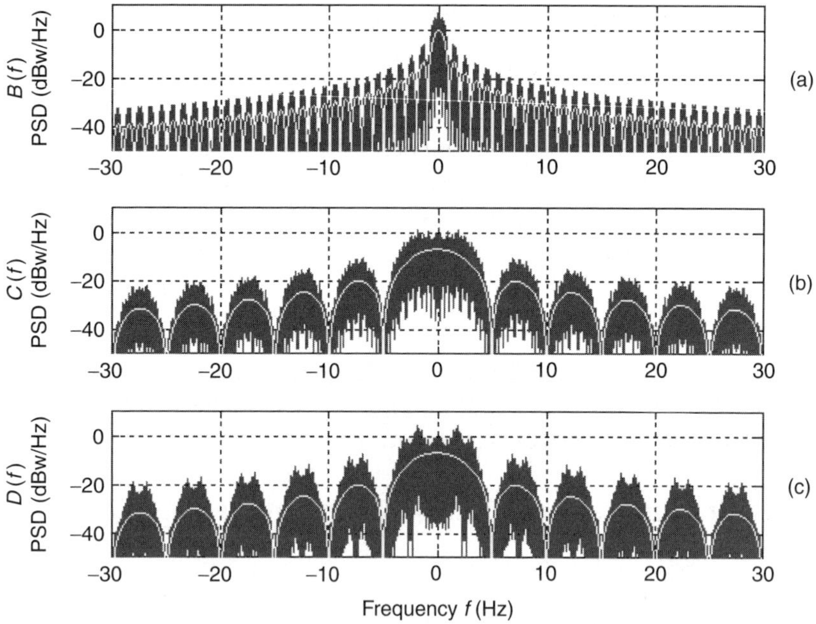

Figure 2.21 Power spectrum density. (a) PSD of $b(t)$. (b) PSD of $c(t)$. (c) PSD of $d(t)$.

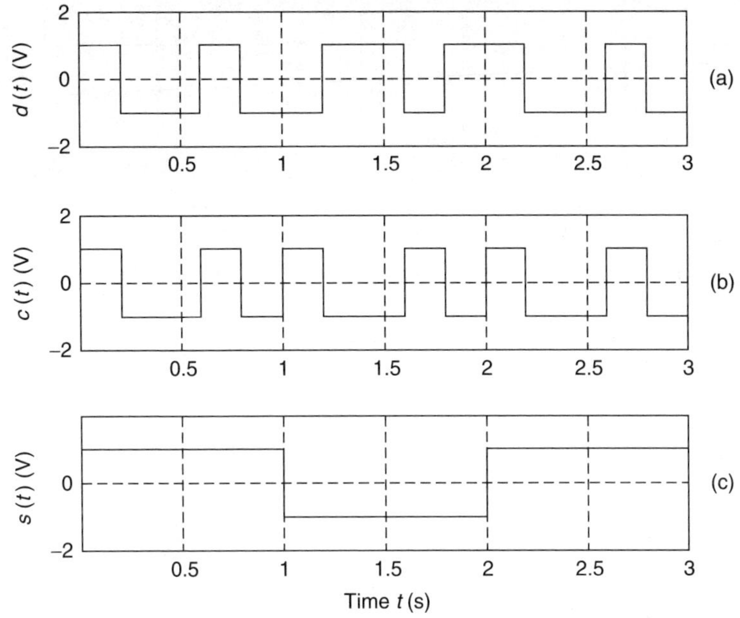

Figure 2.22 Principle of DS-SS system in baseband with periodic sequence. (a) Spread signal $d(t)$. (b) Spreading NRZ-PN $c(t)$. (c) Despread signal $s(t)$.

time and frequency domains, for the same conditions used for NRZ-PN sequence.

In Figure 2.20a, as the NRZ-chaos sequence is not periodic, each bit of information is folded by a different sequence of chips, in five different chips. As a consequence of the nonperiodic sequence the PSD of the NRZ chaos can be modeled by a random sequence, as shown in the Figure 2.21b.

2.4.3 Receiver of DS-SS Baseband using NRZ-PN Sequence

Figure 2.22 shows the waveforms of the despreading process using an NRZ-PN sequence. Figure 2.22a shows the received spread signal $d(t)$, Figure 2.22b shows the NRZ-PN sequence assuming synchronization and Figure 2.22c shows the despread sequence $s(t)$, which is exactly the same information sequence transmitted as $b(t)$ according to Equation (2.12). We can also note the despreading process observing that the bandwidth of $d(t)$ and $s(t)$ are $B_D = 1/T_c = 5\,\text{Hz}$ and $B_S = 1/T_b = 1\,\text{Hz}$.

2.4.4 Receiver of DS-SS Baseband using NRZ-Chaos Sequence

The despreading process with an NRZ-chaos sequence is similarly made as with an NRZ-PN. Figure 2.23 shows the unique difference between

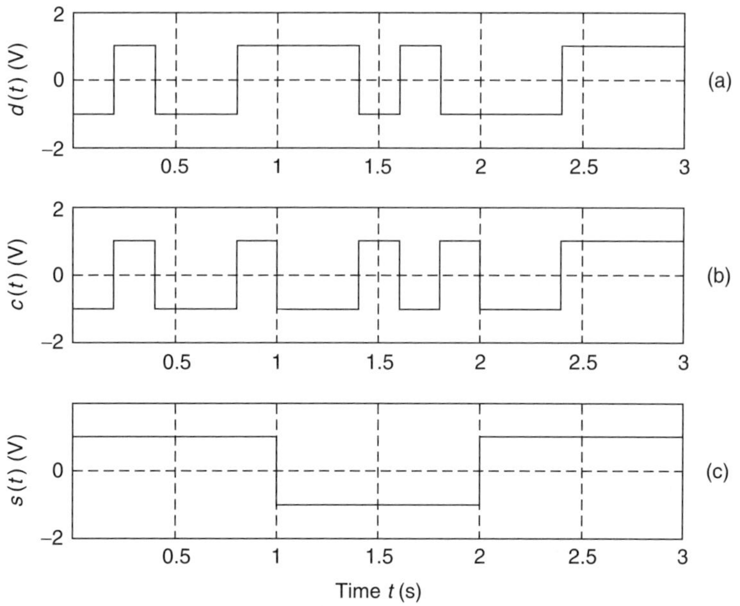

Figure 2.23 Principle of DS-SS system in baseband with chaos sequence. (a) Spread signal $d(t)$. (b) Spreading NRZ-chaos $c(t)$. (c) Despread signal $s(t)$.

the chaotic and periodic sequences is that in the former one, each bit of information is despread by a nonperiodic sequence (Figure 2.23b).

2.5 Performance DS-SS Baseband Communication

Although the communication using DS-SS system is bandpass modulated usually BPSK, the performance of a DS-SS in baseband is illustrative and is the same as the bandpass modulated in BPSK. The performance of a digital communication is measured by the bit error rate (BER) in a function of signal to noise ratio (SNR). Figure 2.24 shows the expanded block diagram of the receiver detector shown in Figure 2.17b.

Assuming that the communication channel is AWGN and has propagation delay of zero, the receiver signal is given by

$$r(t) = d(t) + n(t) \tag{2.17}$$

where the noise $n(t)$ is modeled as a gaussian random process with a mean equal to zero, and auto correlation and PSD given by

$$R_n(\tau) = \frac{N_0}{2}\delta(\tau) \quad \Leftrightarrow \quad S_n(f) = \frac{N_0}{2} \tag{2.18}$$

where $N_0/2$ is the bilateral PSD in W/Hz of the noise.

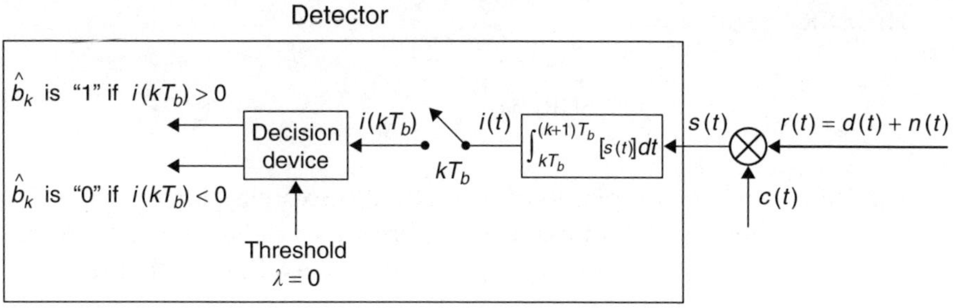

Figure 2.24 Block diagram of the detector of DS-SS system in baseband.

After the despread process, the signal $s(t)$ at the input of the detector is given by the Equation (2.12) and is repeated here for convenience:

$$s(t) = r(t)\,c(t) = b(t) + c(t)n(t) \tag{2.19}$$

The first component of $s(t)$ is the information data $b(t)$ and the second component is the noise channel.

The goal is to find the signal-power to noise-power ratio at the output of the matched filter or correlation filter sampled at instants of time kT_b. This ratio is denoted SNR_i. The resulting signal $i(kT_b)$ is fed to decision device that produces an estimate $\hat{b}_k$ of the information transmitted b_k using the following expression:

$$\begin{cases} \text{if } i(kT_b) > 0 & \text{then } \hat{b}_k = 1 \\ \text{if } i(kT_b) < 0 & \text{then } \hat{b}_k = 0 \end{cases} \tag{2.20}$$

The output of the matched filter or correlation filter consists of two components: d_k, due to the desired signal $d(t)$, and n_k, due to the channel noise $n(t)$.

It is well known that the matched filter or the correlation filter maximizes the SNR_k at sampled instants kT_b and that the value of the component d_k is given by

$$d_k = \pm E_b \,\mathrm{V} \tag{2.21}$$

where the signal of the component d_k is due to the received symbol "±1" and E_b is the signal energy of each bit. Applying the definition of calculation of a energy of signal to one bit of a polar NRZ signal $d(t)$ with amplitude $\pm A$ V and duration T_b s, the energy E_b results in

$$E_b = \int_{-\infty}^{\infty} [d(t)]^2 dt = \int_0^{T_b} (\pm A)^2 dt = A^2 T_b \;\; \mathrm{J\ (joules)} \tag{2.22}$$

The noise component n_k is a random variable with zero mean value and variance given by

$$\sigma_{n_k}^2 = \frac{E_b N_0}{2} \,\mathrm{W} \tag{2.23}$$

The SNR_k results in:

$$\text{SNR}_k = \frac{d_k^2}{\sigma_k^2} = \frac{2E_b}{N_0} \tag{2.24}$$

The BER, which is a function of the SNR and expressed in terms of E_b/N_0, measures the digital communication performance. For the application, the BER of a DS-SS system binary bipolar NRZ in AWGN channel noise is

$$\text{BER} = Q\left(\sqrt{\frac{E_b}{N_0}}\right) \quad \text{where } Q(z) = \int_z^{\infty} \frac{1}{\sqrt{2\pi}} e^{-\frac{x^2}{2}} dx \tag{2.25}$$

It is important to note that the spread spectrum does not present any advantage with respect to AWGN because the same performance would be obtained if the communication were done without spread spectrum. Another important result is the performance of the DS-SS system bandpass modulated BPSK.

Next, the waveforms at the receiver, implemented with a correlation filter assuming synchronism, NRZ-PN sequence and without noise, will be shown.

After the despreading process shown in Figure 2.25a, Figure 2.25b, and Figure 2.25c, the despread information sequence $s(t)$ is fed to the detector, which was implemented by a correlation filter. The correlation filter integrates each bit, according to the Equation (2.26). At the end of each bit, the signal $i(t)$ is sampled at $(k+1)$th instant, resulting in $i[(k+1)T_b]$ which corresponds to the kth bit. The resulting sample goes to the decision device that estimates the kth transmitted bit, i.e., $\hat{b}_k$. After the decision, the sample $i[(k+1)T_b]$ is dumped to zero to start the integration of a new bit. This process is called integrate and dump.

$$i(t) = \int_{kT_b}^{t} s(t')dt' \tag{2.26}$$

In Figure 2.25d, the waveform of $i(t)$ is shown and the end of each bit is illustrated with a dot that corresponds to the kth bit. According to Equation (2.21) and Equation (2.22), the values of $i(kT_b)$ are $\pm E_b = \pm A^2 T_b$ V. $A = 1$ V and $T_b = 1$ s, $i(kT_b) = \pm 1$ V represented with dots was used in Figure 2.25d. Figure 2.25e shows the detected bits of information, $\hat{b}_k$, which is exactly the same transmitted sequence b_k because we do not take the noise into consideration. We observe that the sequence detected has a delay corresponding to one bit, i.e., T_b seconds.

Figure 2.26 shows the performance of the DS-SS system in baseband using NRZ-PN and NRZ-chaos sequences. The theoretic curve, given by

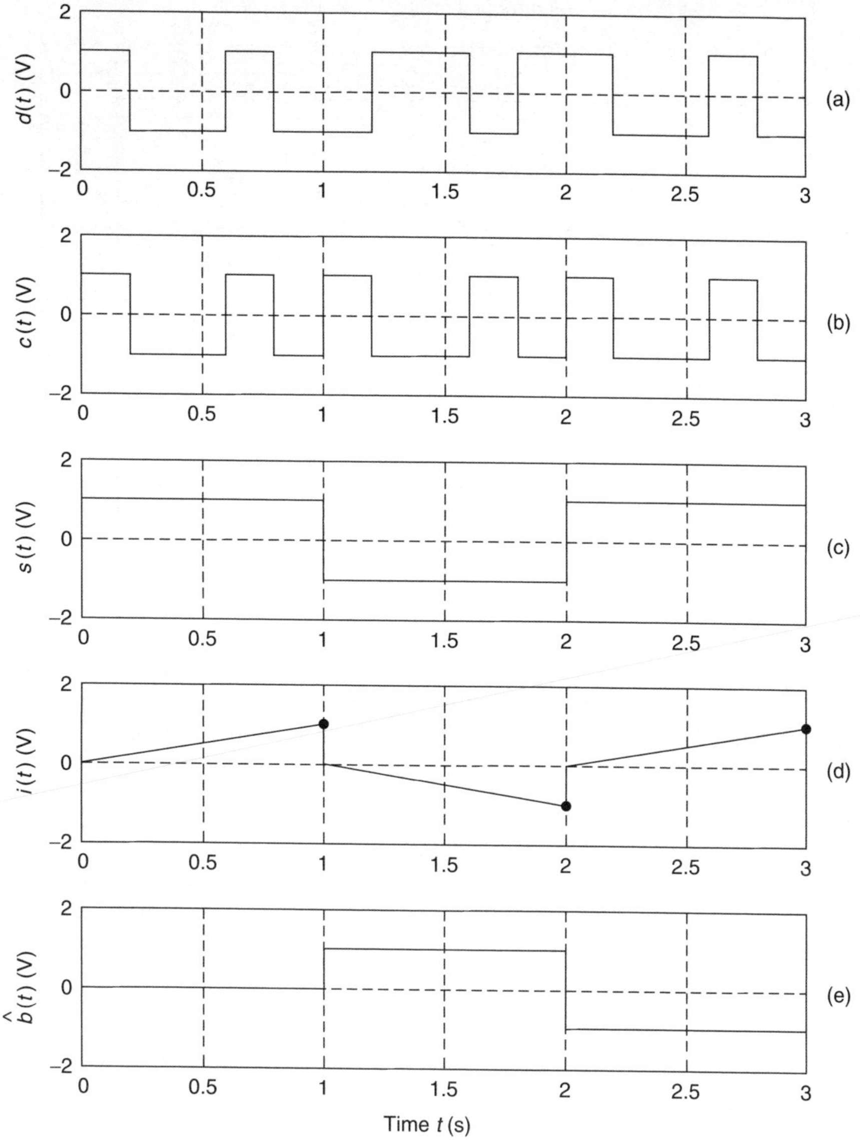

Figure 2.25 Waveforms of the receiver of DS-SS system of the Figure 2.24. (a) Received $r(t)$ without noise. (b) Locally generated NRZ-PN $c(t)$. (c) Despread signal $s(t)$. (d) Integrated and dump signal $i(t)$. (e) Detected signal $\hat{b}(t)$.

Equation (2.25), is plotted. The estimated BER, using NRZ-PN and NRZ-chaos sequences, was developed with a Matlab program that simulates 10,000 bits for each SNR. Comparing the transmitted with the detected bits, BER was estimated. The square and circle dots correspond to the BER when

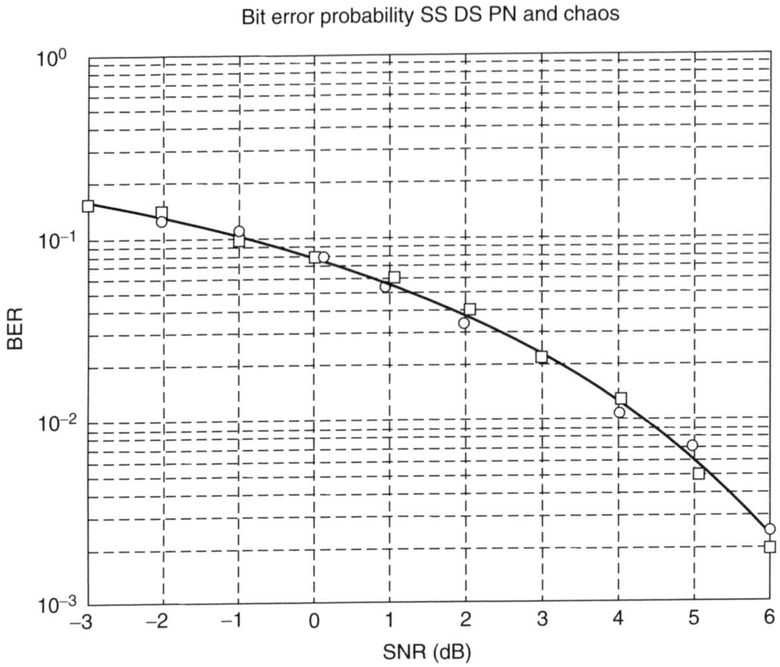

Figure 2.26 BER in function of E_b/N_0 in dB: curve, theoretical BER; square dots, simulated in DS-SS NRZ-PN; circle dots, simulated in DS-SS NRZ chaos.

NRZ-PN and NR-chaos sequences were used, respectively. The simulated results agree with the theoretical results.

It is important to note that the performance of the DS-SS system bandpass modulated BPSK, which will be described in the next section, is exactly the same.

2.6 DS-SS Bandpass System with BPSK Modulation

In this section, the DS-SS system with BPSK modulation, shown in the Figure 2.11, is analyzed. To start, the conventional NRZ-PN sequence is used and then a NRZ-chaos sequence, as proposed by Haykin. The waveforms at the transmitter and receiver end will be analyzed in time and frequency domains, assuming the following conditions:

Information, $b(t)$: bipolar binary NRZ, with values ± 1 V and bit rate $R_b = 1/T_b$ bps;

Spreading sequence PN or chaos, $c(t)$: bipolar binary NRZ, with values ± 1 V, chip rate $R_c = 1/T_c$ cps and a processing gain $\mathrm{PG} = T_b/T_c$;

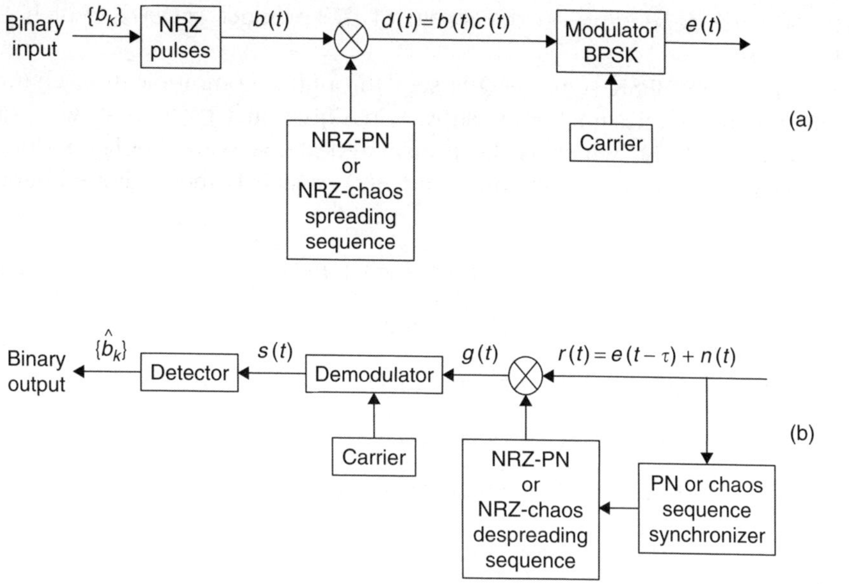

Figure 2.27 Block diagram of the DS-SS system with BPSK modulation. (a) Transmitter. (b) Receiver.

Sinoid carrier: with amplitude A V, frequency f_c Hz and phase θ_0 rd;
Communication channel: AWGN with bilateral PSD $= N_0/2$ W/Hz and autocorrelation $\delta(\tau)N_0/2$;
Synchronization: (a) carrier, (b) spreading sequence, and (c) bit.

Figure 2.27a and Figure 2.27b show a block diagram of the DS-SS transmitter and receiver, respectively, with binary phase shift keying BPSK modulation. In the next sections the waveforms with the conditions already described will be illustrated, using NRZ-PN and NRZ-chaos sequences.

At the transmitter end, shown in Figure 2.27a, the information data $b(t)$ and the spreading sequence, $c(t)$, are the bipolar NRZ sequences, given by Equation (2.6) and Equation (2.7), respectively. The spreading process, expressed by Equation (2.8), is the same realized in baseband, which is repeated here for clarity:

$$d(t) = b(t)c(t) \tag{2.27}$$

The spread signal, $d(t)$, modulates the carrier using BPSK, producing a DS-SS BPSK signal, given by

$$e(t) = Ab(t)c(t)\sin(2\pi f_c t + \theta_0) \tag{2.28}$$

where A is the amplitude, f_c is the frequency, and θ_0 is the initial phase of carrier. The BPSK modulation is obtained by changing the phase of carrier

(0 or π rd) depending on the signal of the product $b(t)c(t)$, which can be ± 1.

The DS-SS BPSK signal $e(t)$ is sent through a communication channel and the received signal $r(t)$, shown in Equation (2.29), has two components: $e(t-\tau)$, which is the transmitted signal with a delay τ due to propagation path in the channel and $n(t)$, which is the additive channel noise:

$$r(t) = e(t-\tau) + n(t) \tag{2.29}$$

Substituting the Equation (2.29) the received signal results in

$$\begin{aligned} r(t) &= Ab(t-\tau)\,c(t-\tau)\sin\left(2\pi f_c(t-\tau)+\theta_0\right) + n(t) \\ &= Ab(t-\tau)\,c(t-\tau)\sin\left(2\pi f_c t + (\theta_0 - 2\pi f_c\tau)\right) + n(t) \\ &= Ab(t-\tau)\,c(t-\tau)\sin\left(2\pi f_c t + \theta_r\right) + n(t) \end{aligned} \tag{2.30}$$

where θ_r is the carrier phase due to propagation delay and given by $\theta_0 - 2\pi f_c\tau$.

At the receiver there are three kinds of synchronism used to detect the information. The three kinds of synchronism are:

1. Between the spreading sequences: the received, $c(t-\tau)$ and those generated at the receiver, $c(t-\hat{\tau})$, where the delay $\hat{\tau}$ is obtained by the sequence synchronizer block shown in Figure 2.27b;
2. Between the carriers: the received, $\sin(2\pi f_c t + \theta_r)$, and those generated at the receiver, $\sin(2\pi f_c t + \hat{\theta}_r)$, where $\hat{\theta}_r$ is the phase obtained by a carrier synchronizer. In this case the demodulation process is called coherent demodulation;
3. Between the bits of the receiving information, i.e., consists in the estimations of the instants of the transitions of the of information bits, $b(t-\tau)$, demodulated at the receiver.

From this point on it is assumed that all signals are synchronized. This is equivalent to $\tau = 0$ in the received signal as represented in the following equation:

$$r(t) = Ab(t)c(t)\sin(2\pi f_c t) + n(t) \tag{2.31}$$

The process of despreading is obtained by multiplying $r(t)$ by $c(t)$, which is generated at the receiver, resulting in

$$\begin{aligned} g(t) = r(t)c(t) &= Ab(t)c^2(t)\sin(2\pi f_c t) + c(t)n(t) \\ &= Ab(t)\sin(2\pi f_c t) + c(t)n(t) \end{aligned} \tag{2.32}$$

The first component of $g(t)$ is the information data $b(t)$ modulated in BPSK by despread and the second component is due to noise.

The resulting signal $g(t)$ feeds the demodulation block where it is multiplied by the carrier, $A\sin(2\pi f_c t)$, resulting in

$$\begin{aligned} s(t) = g(t)A\sin(2\pi f_c t) &= A^2 b(t)\sin^2(2\pi f_c t) \\ &\quad + Ac(t)n(t)\sin(2\pi f_c t) \\ &= \frac{A^2}{2}b(t) - \frac{A^2}{2}b(t)\cos(4\pi f_c t) + Ac(t)n(t)\sin(2\pi f_c t) \end{aligned} \tag{2.33}$$

The demodulated signal has three components:

The first component, $\frac{A^2}{2}b(t)$, is the information sequence in baseband;

The second, $-\frac{A^2}{2}b(t)\cos(4\pi f_c t)$, is the information sequence modulated in BPSK but at a frequency two times the carrier frequency, f_c;

The third, $Ac(t)n(t)\sin(2\pi f_c t)$, is due to the noise channel.

The resulting signal $s(t)$ feeds the detector block, which is the same, as shown in Figure 2.24. The detector is implemented with the correlation filter, and its output, $i(t)$, is sampled at the instants $(k+1)T_b$, giving information of the kth bit. The output is given by

$$i((k+1)T_b) = \int_{kT_b}^{(k+1)T_b} s(t)dt \tag{2.34}$$

The output, $i((k+1)T_b)$, due to the three components of $s(t)$, is denoted by i_1, i_2, and i_3. The result of the first component, i_1 is

$$i_1 = \int_0^{T_b} \frac{A^2}{2}b(t)dt = \pm\frac{A^2T_b}{2} = \pm E_b \tag{2.35}$$

where the E_b, the energy of each bit, can be calculated by the Equation (2.22), as

$$E_b = \int_{-\infty}^{\infty} [d(t)]^2 dt = \int_0^{T_b} (\pm A\sin(2\pi f_c t))^2 dt = \frac{A^2 T_b}{2} \text{ J (joules)} \tag{2.36}$$

The second component, i_2, is

$$i_2 = \int_0^{T_b} -\frac{A^2}{2}b(t)\sin(4\pi f_c t)dt = 0 \tag{2.37}$$

where the result is zero because the period is a multiple of the period of the carrier.

The third component, i_3, is a random variable and is given by

$$i_3 = \int_0^{T_b} Ac(t)n(t)\sin(2\pi f_c t)dt \tag{2.38}$$

As i_3 is a random variable it is important to calculate its mean and variance. The mean is zero because the noise $n(t)$ has zero mean. The variance is calculated using the statistical expectation of the square of i_3, thus:

$$\begin{aligned}
\sigma_{i_3}^2 = E(i_3^2) = E\left\{\int_0^{T_b} Ac(t)n(t)\sin(2\pi f_c t)dt\right. \\
\left.\times \int_0^{T_b} Ac(u)n(u)\sin(2\pi f_c u)du\right\} \\
= \int_0^{T_b}\int_0^{T_b} A^2 E[n(t)n(u)]c(t)c(u) \\
\times \sin(2\pi f_c t)\sin(2\pi f_c u)dt\,du \\
= \frac{A^2 N_0}{2}\int_0^{T_b}\int_0^{T_b} \delta(t-u)c(t)c(u)\sin(2\pi f_c t)\sin(2\pi f_c u)dt\,du \\
= \frac{A^2 N_0}{2}\int_0^{T_b} c^2(t)\sin^2(2\pi f_c t)dt = \frac{A^2 T_b N_0}{4} = \frac{E_b N_0}{2}
\end{aligned} \tag{2.39}$$

The SNR at the output of the correlation filter, using the definition given by Equation (2.24), results in

$$\mathrm{SNR}_k = \frac{i_1^2}{\sigma_{i_3}^2} = \frac{2E_b}{N_0} \tag{2.40}$$

The resulting SNR_k is the same obtained in the baseband communication. Thus the performance is also the same, as mentioned.

In the following sections, the waveforms obtained in DS-SS BPSK, using NRZ-PN and NRZ-chaos sequences, will be illustrated, considering no noise present.

2.6.1 *Application of NRZ Chaos Sequence to DS-SS BPSK*

Figure 2.27 shows the waveforms in time and frequency domains assuming the following conditions:

Information, $b(t)$: bipolar binary NRZ, with values ± 1 V and a bit rate $R_b = 1/T_b = 1$ bps;

Spreading sequence PN or chaos, $c(t)$: bipolar binary NRZ, with values ± 1 V and a chip rate $R_c = 1/T_c = 5$ cps, resulting in a processing gain $\mathrm{PG} = T_b/T_c = 5$;

Sinusoidal carrier: with amplitude $A = 1$ V, frequency 10 Hz and phase $\theta_0 = 0$ rd;

Transmission channel: AWGN with bilateral PSD $= N_0/2$ W/Hz and autocorrelation $\delta(\tau)N_0/2$;
Synchronization: (a) carrier, (b) spreading sequence and (c) bit.

The illustration will start by demonstrating the conventional DS-SS BPSK using NRZ-PN sequence and then the application of NRZ-chaos sequence to the DS-SS BPSK. It is important to note that, in the frequency domain, the theoretical and simulated results are illustrated, using Matlab.

2.6.2 *Transmitter of DS-SS BPSK using NRZ-PN Sequence*

Figure 2.28 and Figure 2.29 show the waveforms and PSDs, respectively, of the transmitter shown in Figure 2.27, when the DS-SS BPSK uses a NRZ-PN sequence to spread the information.

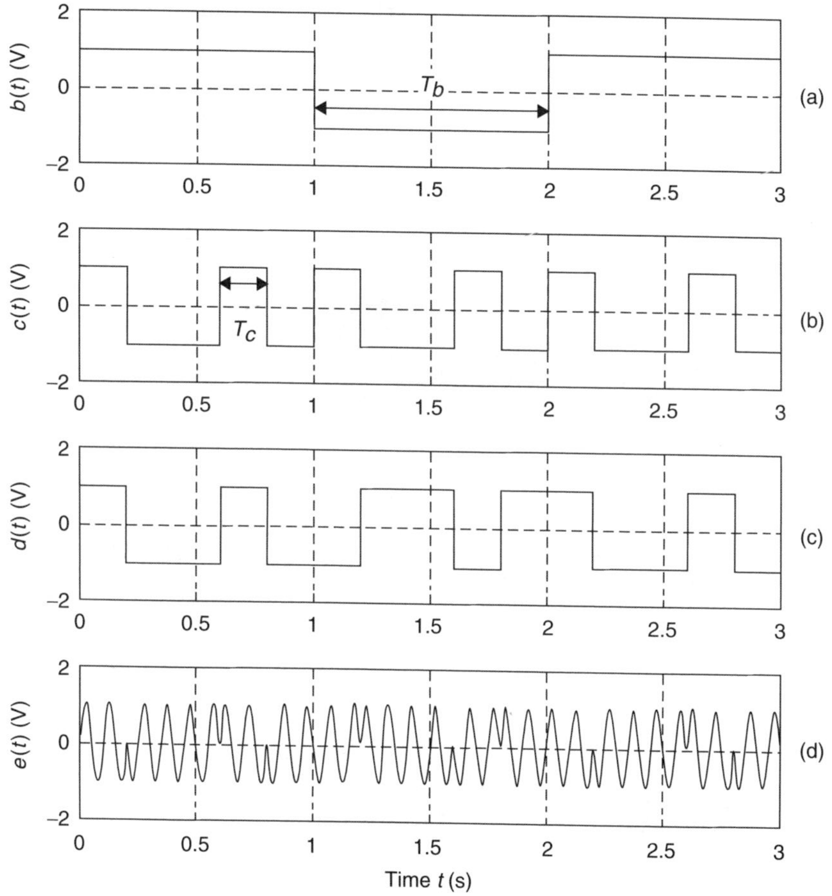

Figure 2.28 Time waveforms of DS-SS system BPSK. (a) Data information $b(t)$. (b) Spreading NRZ-PN $c(t)$. (c) Spread signal $d(t)$. (d) BPSK spread signal $e(t)$.

The Information data $b(t)$ and the NRZ-PN spreading sequence, $c(t)$, shown in Figure 2.28a and Figure 2.28b, are given by Equation (2.6) and Equation (2.7), respectively. Figure 2.28c shows the spreading process expressed by Equation (2.27). The result is the same of that in baseband.

The spread signal, $d(t)$, modulates the carrier using BPSK, generating a DS-SS BPSK signal, given by Equation (2.28).

Figure 2.28d shows the DS-SS BPSK signal, $e(t)$, where the phase of carrier varies (0 or π rd) depending on the sign of the spread signal, $d(t)$, which can be ± 1.

Figure 2.29 shows the simulated and modeled PSD of the waveforms represented in Figure 2.28. In Figure 2.29a and Figure 2.29b the PSD of $B(f)$ and $D(f)$, are in baseband and exhibit the same properties of that in Figure 2.19. The modeled PSDs in Figure 2.29a and Figure 2.29b are given by Equation (2.12) and Equation (2.15), respectively. The bandwidth of $b(t)$ is $B_b = 1\,\text{Hz}$ and of $d(t)$ is $B_d = 5\,\text{Hz}$. The modeled PSD of the DS-SS BPSK, shown in the Figure 2.29c, is given by

$$E(f) = \frac{A^2 T_c}{4}\left\{\text{sinc}^2[(f - f_c)T_c] + \text{sinc}^2[(f + f_c)T_c]\right\} \tag{2.41}$$

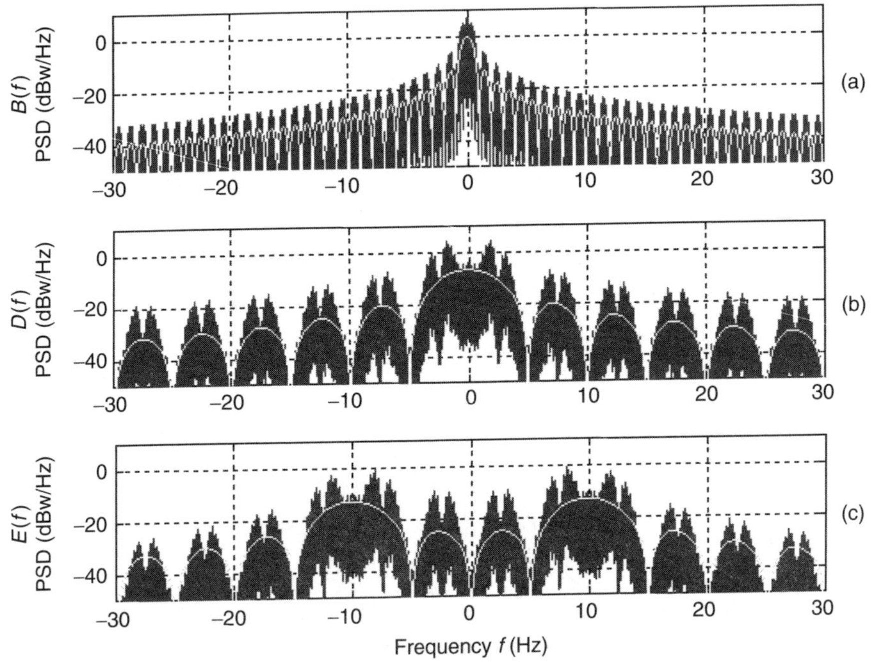

Figure 2.29 PSD of the DS-SS BPSK. (a) PSD of the data information *b*(*t*). (b) PSD of the Spread NRZ-PN *d*(*t*). (c) PSD of DS-SS BPSK signal e(*t*).

where f_c is the carrier frequency. The magnitudes of the PSB of $E(f)$ at $\pm f_c$ are $10\log(A^2T_c/4) = -13$ dbw/Hz. The bandwidth of $e(t)$ is $B_e = 2/T_c = 10$ Hz and the processing gain is the same, i.e., PG = 5, with the following definition:

$$\text{PG} = \frac{\text{Bandwidth of Modulated SS signal}}{2\,(\text{Bandwidth of baseband information})} \tag{2.42}$$

2.6.3 Transmitter of DS-SS BPSK using NRZ Chaos Sequence

Figure 2.30 and Figure 2.31 show the time waveforms and the PSD of the transmitter signal, when the DS-SS BPSK uses a NRZ-chaos sequence to spread the information.

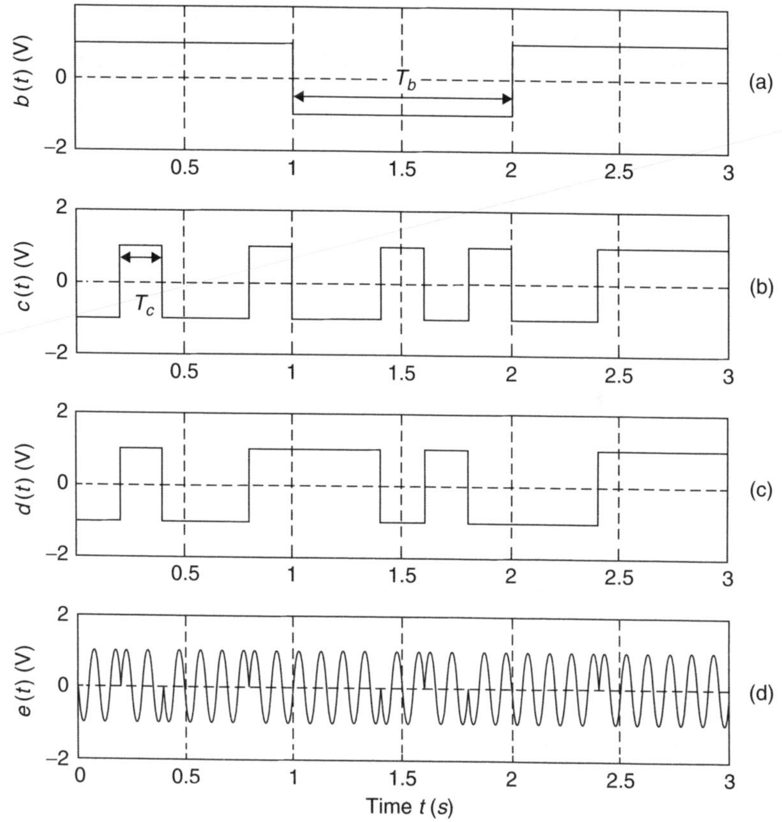

Figure 2.30 Time waveforms of DS-SS BPSK using NRZ-chaos. (a) Data information $b(t)$. (b) Spreading NRZ-chaos $c(t)$. (c) Spread signal $d(t)$. (d) BPSK spread signal $e(t)$.

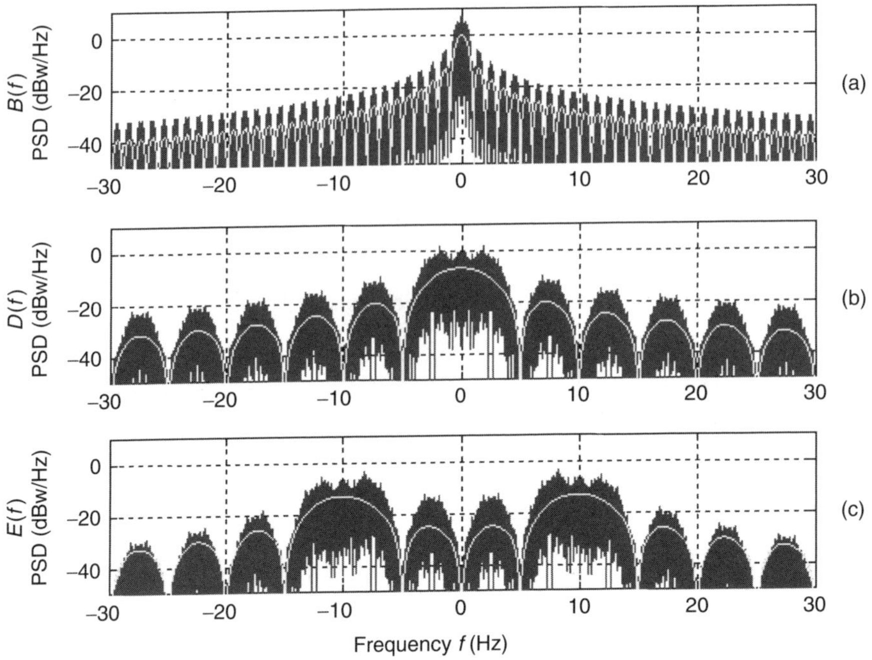

Figure 2.31 PSD of the DS-SS BPSK using NRZ-chaos. (a) PSD of the data information *b*(*t*). (b) PSD of the spread NRZ-chaos *d*(*t*). (c) PSD of DS-SS BPSK signal *e*(*t*).

The process of DS-SS BPSK system with NRZ-chaos spreading sequence is very similar to the NRZ-PN sequence. The unique difference is the fact that the NRZ-chaos sequence, $c(t)$, is not periodic. All other parameters are maintained.

2.6.4 Receiver of DS-SS BPSK using NRZ-PN Sequence

Figure 2.32 and Figure 2.33 show the time waveforms and the PSDs of the receiver signals, when the DS-SS BPSK uses a NRZ-PN sequence. Synchronism is assumed and noise is present.

Figure 2.32a shows the received signal DS-SS, modulated in BPSK at the chip rate $R_c = 1/T_c = 5$ cps. Figure 2.33a shows that the respective PSD is centered at the carrier frequency, $f_c = 10$ Hz, and the bandwidth is, using the Equation (2.42), $B_e = 2/T_c = 10$ Hz. Figure 2.33b shows the despreading NRZ-PN sequence $c(t)$. Figure 2.32c shows the despread signal $g(t)$, which is still modulated in BPSK at the bit rate $R_b = 1/T_b = 1$ bps. Figure 2.33c

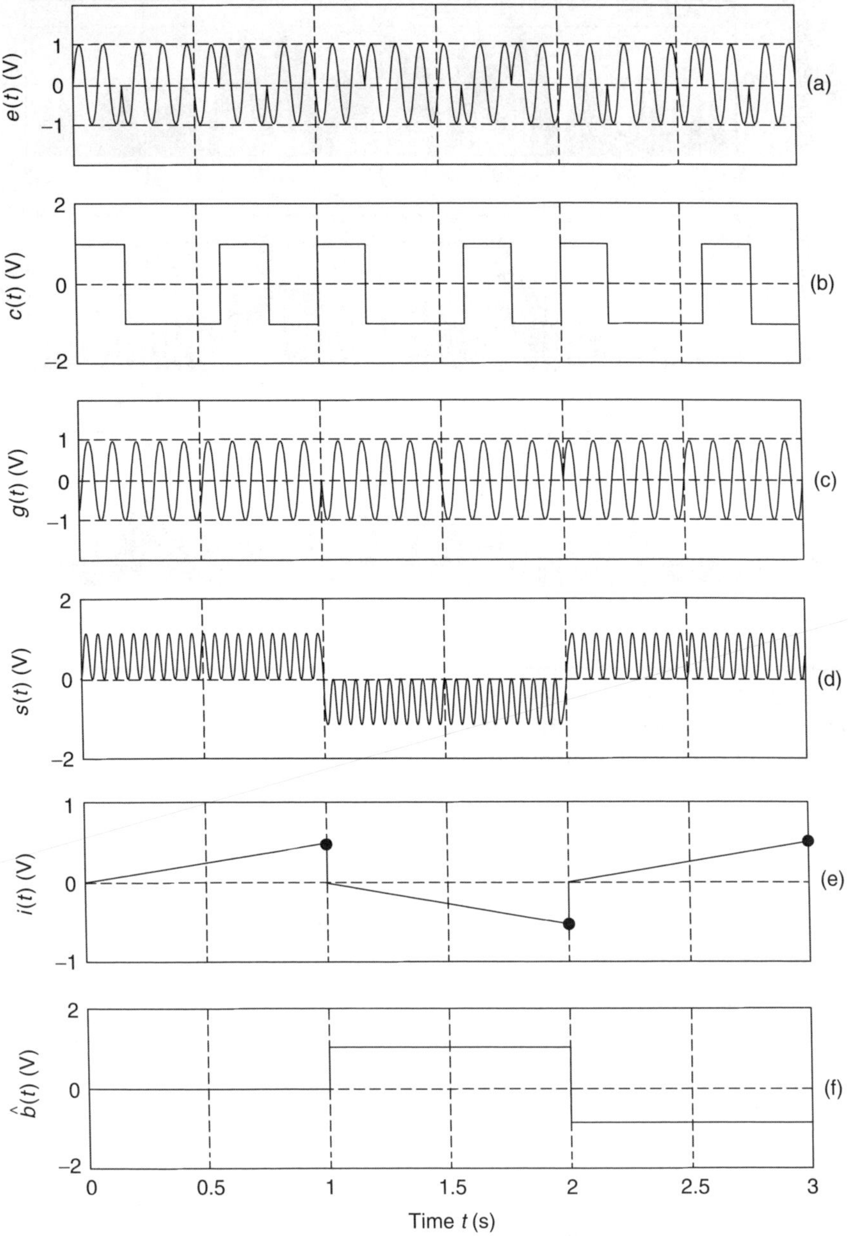

Figure 2.32 Time waveforms of receiver DS-SS BPSK with NRZ-PN sequence. (a) Received DS-SS BPSK signal e(t). (b) Spreading NRZ-PN signal $c(t)$. (c) Despread BPSK signal $g(t)$. (d) Demodulated BPSK signal $s(t)$. (e) Integrate and dump signal $i(t)$. (f) Detected signal $\hat{b}(t)$.

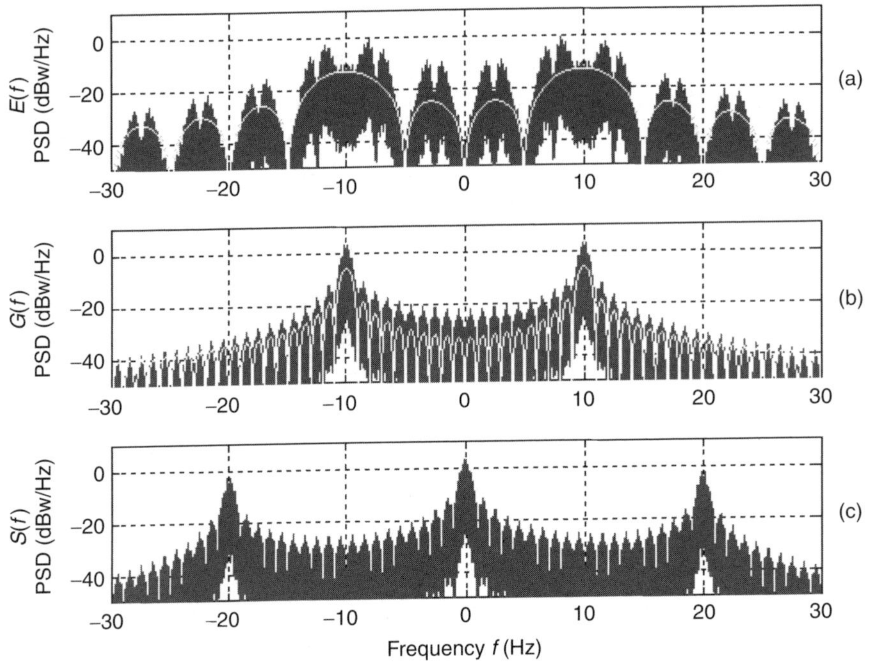

Figure 2.33 PSD of the receiver DS-SS BPSK with NRZ-PN sequence. (a) PSDS of received DS-SS BPSK signal e(*t*). (b) PSD of despread BPSK signal *g*(*t*). (c) PSD of the integrate and dump signal *i*(*t*).

shows that the PSD of the $g(t)$ is the same as the despread signal modulated in BPSK because its bandwidth is $B_g = 2/T_b = 2\,\text{Hz}$. Figure 2.32d shows the demodulated signal $s(t)$, which has two components, given by Equation (2.33). The first component, $\frac{A^2}{2}b(t)$, is the information sequence in baseband and the second, $-\frac{A^2}{2}b(t)\cos(4\pi f_c t)$, is the information sequence modulated in BPSK but at a frequency two times the carrier's, f_c. This fact can be seen in the PSD (Figure 2.33c), where the first component has a PSD centered at zero frequency and corresponds to the received information $b(t)$ at the bit rate $R_b = 1/T_b = 1$ bps and the second has a PSD centered at two times the carrier, $2f_c = 20\,\text{Hz}$, and modulated in BPSK at the bit rate of the information, i.e., $R_b = 1/T_b = 1$ bps.

The signal $i(t)$ (Figure 2.32e) corresponds to the process of integrate and dump at the output of the correlation filter of the detector shown in Figure 2.24. At the end of each bit the sample $i((k+1)T_b)$ corresponding to the kth bit is shown with a dot. According to Equation (2.35) and Equation (2.36), the values of $i(kT_b)$ are $\pm E_b = \pm A^2 T_b/2$. $A = 1\,\text{V}$ and $T_b = 1\,\text{s}$ was used, thus $i(kT_b) = \pm 0.5\,\text{V}$ as shown in Figure 2.32e. Figure 2.32f shows

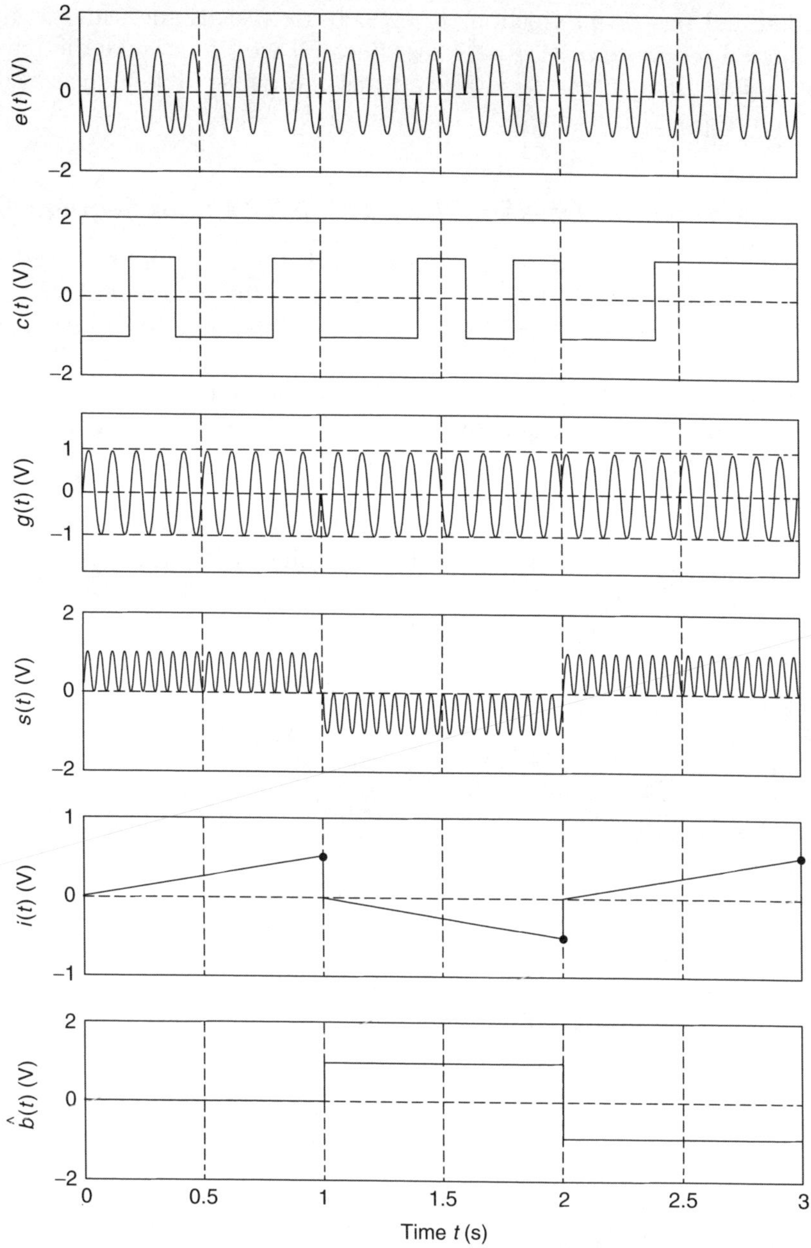

Figure 2.34 Time waveforms of receiver DS-SS BPSK with NRZ-PN sequence. (a) Received DS-SS BPSK signal $e(t)$. (b) Spreading NRZ-PN signal $c(t)$. (c) Despread BPSK signal $g(t)$. (d) Demodulated BPSK signal $s(t)$. (e) Integrate and dump signal $i(t)$. (f) Detected signal $\hat{b}(t)$.

the detected bits of information, $\hat{b}_k$, which are exactly the same as the transmitted sequence b_k (Figure 2.28a) because we do not take the noise into consideration. It can be observed that the detected sequence has a delay corresponding to one bit, i.e., T_b seconds.

2.6.5 Receiver of DS-SS System with NRZ Chaos Sequence

The receiver using an NRZ-chaos sequence is the same as that used in last section with NRZ-PN. Figure 2.34b shows that the unique difference is that the NRZ-chaos sequence is not periodic. All the other parameters are maintained.

Figure 2.34 and Figure 2.35 show the time waveforms and the PSDs of the signals of the receiver shown in Figure 2.27, when the DS-SS BPSK uses a NRZ chaos sequence. It also assumes synchronism and that the component of noise is present.

In this section we showed that it is possible to use of chaos sequences to DS-SS system as proposed by Li and Haykin. Itoh extended this application on to Frequency Hopping SS system.[17] Another application of NRZ chaos sequence could be is in stream cipher cryptography.

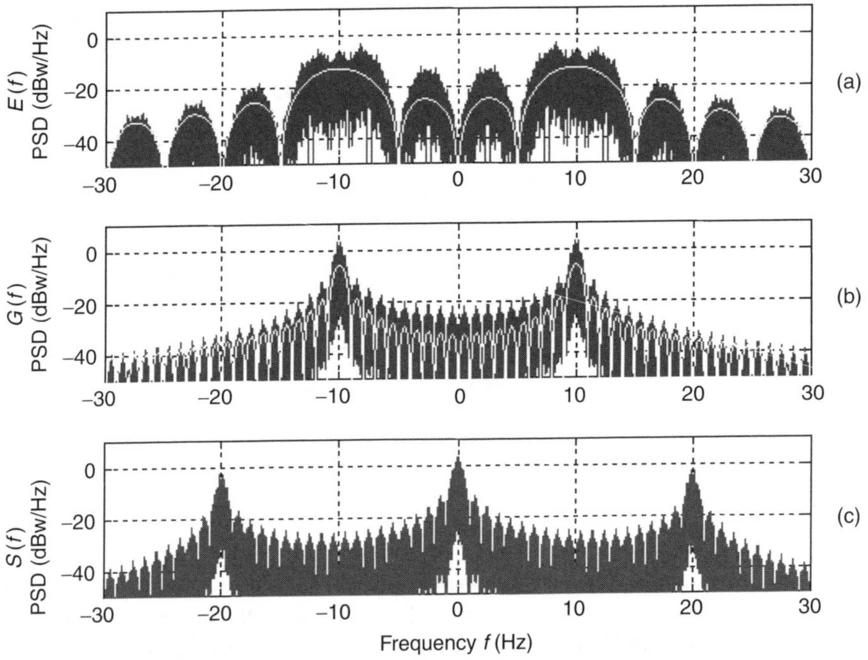

Figure 2.35 PSD of the receiver DS-SS BPSK with NRZ-PN sequence. (a) PSDS of received DS-SS BPSK signal e(*t*). (b) PSD of despread BPSK signal *g*(*t*). (c) PSD of the integrate and dump signal *i*(*t*).

References

1. Abel, A. and Schwarz, W., Chaos communications-principles, schemes, and system analysis, *Proc. IEEE*, 90(5), 691–710, 2002.
2. Barnsley, M.F., *Fractals Everywhere*, Academic Press, New York, 1988.
3. Coven, E., Kan, I., and York, J.A., Pseudo-orbit shadowing in the family of tent maps, *Trans. Am. Math. Soc.*, 308(1), 227–241, 1988.
4. Devaney, R.L., et al., *Chaos And Fractals: The Mathematics Behind The Computer Graphics, Proceedings of the Symposia in Applied Mathematics*, Vol. 39, American Mathematical Society, Providence, RI, 1988.
5. Feigenbaum, M.J., The metric universal properties of period doubling bifurcations and the spectrum for a route to turbulence, *Ann. NY Acad. Sci.*, 357, 330–333, 1980.
6. Kilias, T., Kutzer, K., Mögel, A., and Schwarz, W., in *Proc. IEEE-ISCAS Tutorial*, Seattle, April 1995.
7. Kolumbán, G. and Kennedy, M.P., Recent results for chaotic modulation schemes, ISCAS *IEEE Int. Symp. Circuits Syst.*, 3, 141–144, 6–9, May 2001.
8. Kolumbán, G., Vizvári, B., Mögel, A., and Schwarz, W., Chaotic Systems: A Challenge for Measurement and Analysis, in *Proc. IEEE-IMTC 1996*, Brussels, pp. 1396–1401, June 1996.
9. Li, B.X. and Haykin, S., A new pseudo-noise generator for spread spectrum communications, ICASSP-95, *Int. Conf. Acoustics, Speech, and Signal Process.*, 5(9–12), 3603–3606, May 1995.
10. Parlitz, U., *Analysis and Synchronization of Chaotic Systems, Digital Communication Devices based on Nonlinear Dynamics and Chaos*, Third Physical Institute, Georg-August-University Göttingen, WS 2001.
11. Verhulst, P.F., Notice sur la loi que la population suit dans son accroissement, Correspondence Mathématique et Physique 10, pp. 113–121, 1838.
12. Verhulst, P.F., Nouveaux Memoires de l'Academie Royale des Sciences, des Lettres et des Beaux–Arts de Belgique 18, pp. 1–38, 1845.
13. Wolfram, S., *A New Kind of Science*, Wolfram Media, London, 2002.
14. Gray, T.W. and Glynn, J., *Exploring Mathematics with Mathematica*, Addison-Wesley Publishing Company, The Advanced Book Program, Redwood City, CA, 1991.
15. Pietgen, H.-O., Jürgens, H., and Saupe, D., *Chaos and Fractals: New Frontiers of Science*, Springer-Verlag, NY, 1992.
16. Lam, A.W. and Tantarana, S., *Theory and Applications of Spread-Spectrum Systems*, IEEE/EAB Self-Study Course, NJ, 1994.
17. Itoh, M., Chaos-based spread spectrum communication systems, in *Industrial Electronics, Proc. ISIE 98 IEEE Int. Symp.*, 2, 430–435, 1998.

3

Chaotic Transceiver Design

Arthur Fleming-Dahl

CONTENTS

3.1 Introduction

This chapter analyzes the design of a transceiver using some of the principles discussed in Chapter 1 and in Appendix A and Appendix B. Data recovery and synchronization are both accomplished in the receiver estimation engine, whose design is based on models of both the transmit sequence probability density function (PDF) and the chaotic attractors used. The combination of probability calculations with chaotic dynamics enables accurate estimates of the transmitted data from noisy received values with as little as two chaotic iterates per data bit. In addition, the achievement of chaotic synchronization from this same set of calculations obviates the need for the restrictions inherent in the synchronization methods used in the prior art as follows.

Chaotic systems need not be split into stable and unstable subsystems, nor be invertible, for chaotic synchronization to occur. The need for trial-and-error subsystem determination, followed by Lyapunov exponent calculations to determine stability or instability is avoided. The technique set forth in this design uses two versions of the same strange attractor to transmit the binary logical data states of 0 and 1. The parameters controlling the chaotic system are chosen such that the continuous range of values associated with both versions of the attractor completely overlap. This overlap causes every received value to be valid for either version of the attractor, confounding the possibility of data recovery for the unintended listener.

Modeling the transmit PDF with Gaussian functions is beneficial in the presence of an additive white Gaussian noise (AWGN) channel because the PDF of the received sequence is the convolution of the transmit PDF with the channel PDF, and the convolution of Gaussian functions is Gaussian. The resulting received PDF, therefore, has a closed-form representation that facilitates the determination of the signal-to-noise ratio (SNR), which in turn is used in the estimation of the maximum a posteriori (MAP) transmitted value for the currently received iterate. The transmit sequence is a chaotic-shift-key (CSK) modulation method and is modeled as a random choice between the Henon and mirrored-Henon time sequences. The attractors of both systems are presented, and the models using four parabolas each are designed. A repelling fixed point on the Henon attractor is identified and its possibility for use in the implementation of this communications explored. The PDFs for the Henon, mirrored-Henon, and transmit sequence are found from multimillion iterate data runs and modeled using a DC level and weighted Gaussian functions. The Henon, mirrored-Henon, and transmit sequence power spectral densities (PSDs) are determined, and finally the chaotic reconstruction plots for the transmit sequence corrupted by AWGN are explored.

3.1.1 The Chaotic Communications System

The top-level communications system block diagram is depicted in Figure 3.1, showing the transmitter, lossless nonfading AWGN channel, and receiver. In the transmitter, the chaotic system is specified as the Henon map, and the alternate version of the same strange attractor labeled the mirrored-Henon map. Typical data sequences have a stream of logical 0 and 1 values, each occurring with a probability of ½. In these simulations, the data bits are therefore generated as a random bit stream, and the logical values of 0 and 1 toggle a switch between the Henon and mirrored-Henon chaotic systems. Although not explicitly shown in Figure 3.1, the output of the switch is passed through a digital-to-analog converter (D/A), followed by a power amplifier and possibly frequency translation to transmit the encoded chaotic sequence. There are practical issues of bit alignment, sampling synchronization, D/A and A/D (analog-to-digital converter) quantization effects, and channel phenomena such as losses, fading, Doppler, and non-Gaussian noise that must be addressed in a working hardware implementation of this chaotic communications technique.

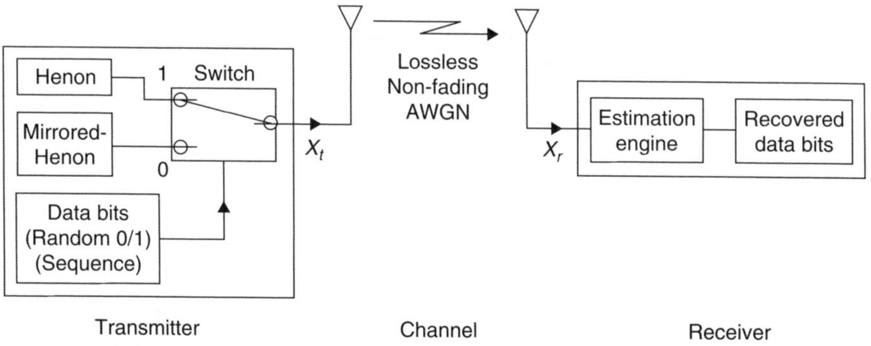

Figure 3.1 Top-level chaotic communication system using the Henon strange attractor.

These are beyond the scope of this thesis and represent subject matter for the further investigation of this method.

The receiver in Figure 3.1 shows the estimation engine, followed by the data bit calculation block. Transmitted value and transmitted chaotic system decisions are made on every received value. Data bits containing a certain number of chaotic iterates are recovered by combining the appropriate individual iterate decisions to reach a bit decision.

3.2 Transmitter Design

The transmitter is based on the Henon chaotic system, which was proposed by the French astronomer Michel Henon in 1976.[2] The controlling equations are

$$x_b[k+1] = -ax_b[k]^2 + y_b[k] + 1 \tag{3.1}$$

$$y_b[k+1] = bx_b[k], \quad k = 1, \ldots, K \tag{3.2}$$

where Michel Henon's chosen values are $a = 1.4$ and $b = 0.3$[2] and $k =$ iterate index with K total iterates in a time series.

The resulting system is chaotic, with an attractor that looks like a crescent and occupies all four quadrants of a Cartesian coordinate system. Investigations in the literature sometimes work with modified Henon systems, where the parameters a and b have values other than Michel Henon's and also may have a multiplier in front of the y term and/or a different additive constant in Equation (3.1).

3.2.1 *The Strange Attractor Models*

The standard attractor investigation techniques involve generating a chaotic time series by starting at a single initial point and iterating through the chaotic equations until sufficient information has been obtained to visualize an attractor. Because the transmitter uses the Henon chaotic system, which is two-dimensional, this investigation used MATLAB image windows instead of iterations on a single point to find the attractor.

3.2.1.1 Henon Attractor

A quick generation of an accurate attractor was achieved by using image calculations on the Henon map. More details are found in Appendix A and Appendix B. Instead of starting with a single point and generating a plot from the time series, an entire section of the basin of attraction was chosen as the initial condition. This region quickly changed its morphological appearance into a form resembling the Henon attractor. Figure 3.2

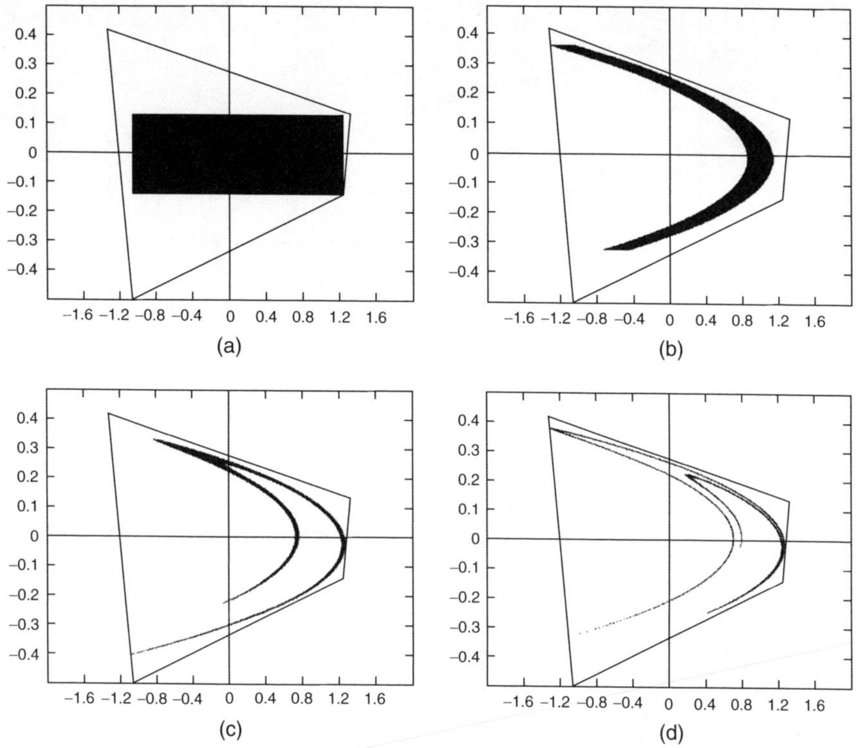

Figure 3.2 Image windows of Henon strange attractor initial array and first three iterates. (a) Initial array in Henon basin of attraction. (b) First iterate of initial array. (c) Second iterate of initial array. (d) Third iterate of initial array.

shows this process, and Figure 3.3 depicts the Henon attractor. This process works because there is a trapping region for chaotic systems, also called a basin of attraction, within which no orbit can escape.[2] In general, the stretching and folding of the chaotic dynamics will cause many orbits to escape to infinity, but there exists a region of phase space that stretches and folds back onto itself ad infinitum. This region of phase space is called the basin of attraction. This entire trapping region maps back onto itself with a morphologically altered appearance, and multiple iterations of the chaotic transform result in the strange attractor. Letting R represent the basin of attraction and H represent the Henon transform,

$$H(R) \subset R \tag{3.3}$$

so the Henon transformation of R lies entirely within R. Repeated applications of this region produce subsets of the region, and so no orbits can escape.

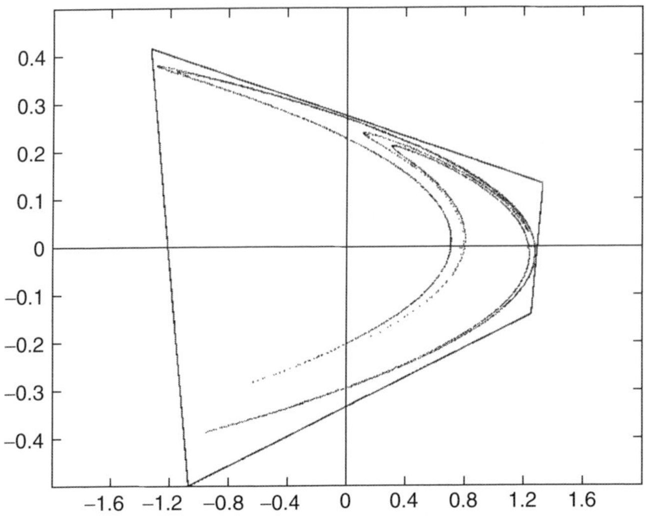

Figure 3.3 Henon strange attractor.

The choice of a rectangular subset of the basin of attraction for the process shown in Figure 3.2 was one of convenience. MATLAB works well with matrices, and the chosen one was easy to generate because it fit closely into two corners of the trapping region quadrilateral and was a fairly large symmetrical matrix. Any other area of arbitrary shape wholly contained within the basin of attraction would have generated the same result, albeit with more or fewer iterations depending on its size.

Figure 3.3 is the superposition of iterates 5 to 10,000 of this initial set. The first few iterates of the Henon transform were eliminated because they contained excess area, tending to obscure the attractor detail. Iterates in the vicinity of 10,000 had only a few points left to plot because the finite precision of the internal representation of numbers caused the transforms of distinct numbers to be indistinguishable to the computer.

Two things became readily apparent. First, the range of x-values for the Henon system was approximately symmetric, covering roughly $(-1.28, 1.28)$. The y-values have the same approximate symmetry over a smaller range because they are scaled versions of the x-values, as can be seen in Equation (3.2). This symmetry allows a second version of the same attractor with an identical range of x-values, yielding the desired transmit characteristic of completely overlapping transmit values for both logical data states. Second, the structure of the main-attractor features appeared roughly parabolic. Simple geometric equations can therefore be used to model the major-attractor sections, yielding a means by which to map a probability-based receiver decision onto the chaotic dynamics represented by the attractor.

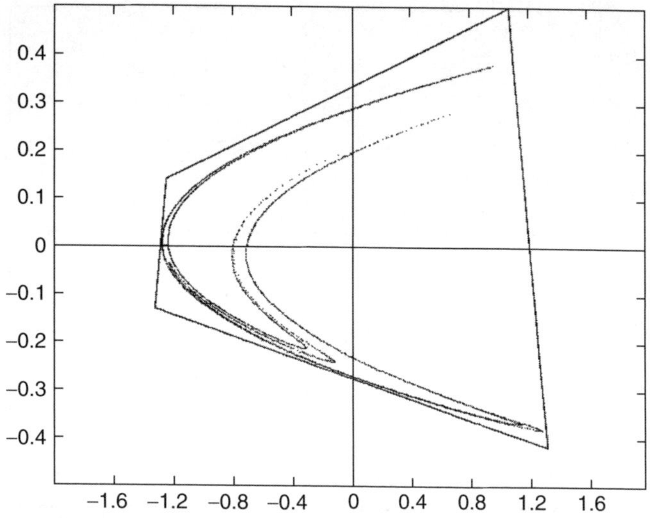

Figure 3.4 Mirrored-Henon strange attractor.

3.2.1.2 Mirrored-Henon Attractor

To create a second version of the same attractor for the second logical state, a mirror image through the x- and y-axes yields intersecting crescents with completely overlapping x-value ranges. Letting

$$x_m = -x_h \quad \text{and} \quad y_m = -y_h \tag{3.4}$$

$$x_m[k+1] = 1.4x_m[k]^2 + y_m[k] - 1 \tag{3.5}$$

$$y_m[k+1] = 0.3x_m[k] \tag{3.6}$$

The attractor formed by these equations is shown in Figure 3.4, along with the corresponding basin of attraction. Figure 3.5 shows the superposition of the two versions of the strange attractor and highlights the identical ranges of values for both the Henon and mirrored-Henon systems. The complete overlap of value ranges is indicative of the fact that there is not a one-to-one correspondence between a transmit value or a range of transmit values and digital message value, as exists for most other communications systems. This feature yields security for the communications scheme because the observation of a received value or range of values does not reveal in a direct manner which logical data state has been sent. The sequence of received values in a data bit must be processed and the attractor from which the sequence was generated found to determine the logical data state of the bit being received. There are areas where the two attractors intersect each other, and it will be difficult to distinguish between the Henon and mirrored-Henon chaotic dynamics in these regions. In addition,

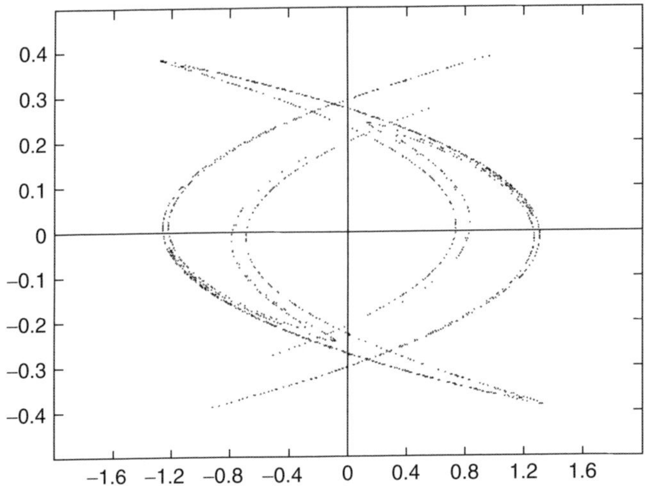

Figure 3.5 Superposition of the Henon and mirrored-Henon attractors.

there are precise points of intersection for which the correct decision cannot be guaranteed even in the absence of noise.

The poorer quality of the image in Figure 3.5 compared to Figure 3.4 and Figure 3.3 is a puzzle. The matrices for the two attractors contain values of zero and one for image window pixels that are off and on, respectively. Adding the two matrices together results in another matrix of zeros and ones, except for four pixels receiving the value of two. The resulting matrix has been verified to contain approximately twice the number of nonzero values of either individual image matrix. Because pixels can only be either on or off in these images, it is unclear why the attractor image quality should be different between the individual and superimposed attractor images. The Henon attractor was modeled with four parabolas to achieve a fairly accurate representation of the structure. Image tools were used again to plot the attractor and parabolas on the same image window to evaluate the match of a given parabola to the appropriate attractor section. The Henon attractor was broken into sections according to a large (major) crescent pair, with a smaller (minor) crescent pair contained inside the major section. These four primary features were deemed to model the attractor well, and so four parabolas were found. The parabola equations and valid y-values are listed below, and the plots of each parabola shown in Figure 3.6.

Henon Parabola	Y_{min}	Y_{max}	Equation	
1	−0.2846	0.3659	$\tilde{x}_{b1} = -15.15(y - 0.014)^2 + 0.710$	(3.7)
2	−0.1911	0.2215	$\tilde{x}_{b2} = -14.75(y - 0.018)^2 + 0.800$	(3.8)
3	−0.2154	0.2378	$\tilde{x}_{b3} = -16.60(y + 0.014)^2 + 1.245$	(3.9)
4	−0.3862	0.3780	$\tilde{x}_{b4} = -16.05(y + 0.013)^2 + 1.280$	(3.10)

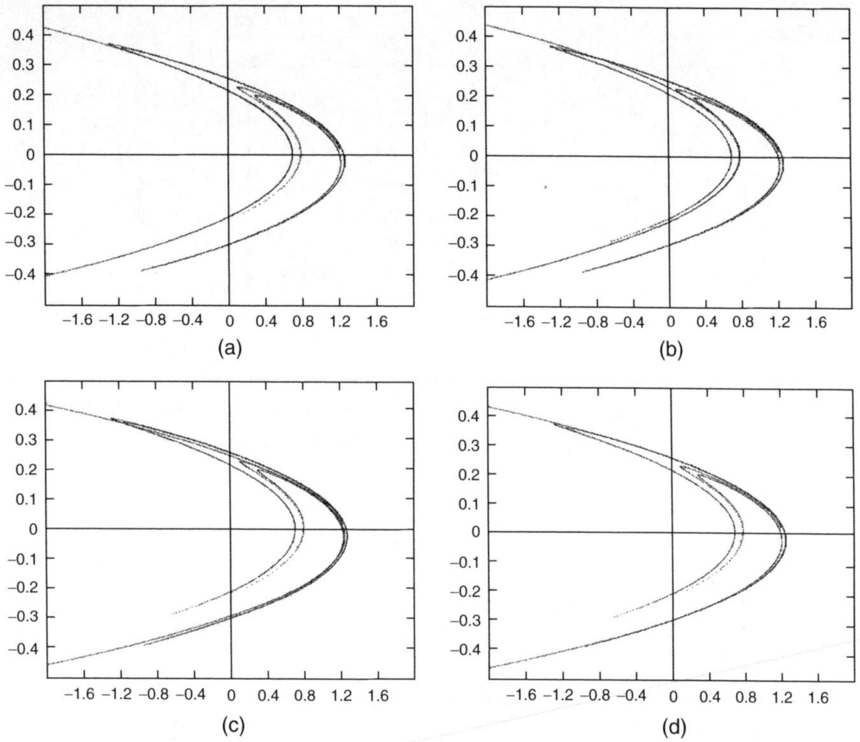

Figure 3.6 Parabolas modeling the Henon strange attractor. (a) Parabola 1 — Innermost major section. (b) Parabola 2 — Innermost minor section. (c) Parabola 3 — Outermost minor section. (d) Parabola 4 — Outermost major section.

Mirrored-Henon Parabola	**Y_{min}**	**Y_{max}**	**Equation**	
1	−0.3659	0.2846	$\tilde{x}_{m1} = 15.15(y + 0.014)^2 - 0.710$	(3.11)
2	−0.2215	0.1911	$\tilde{x}_{m2} = 14.75(y + 0.018)^2 - 0.800$	(3.12)
3	−0.2378	0.2154	$\tilde{x}_{m3} = 16.60(y - 0.014)^2 - 1.245$	(3.13)
4	−0.3780	0.3862	$\tilde{x}_{m4} = 16.05(y - 0.013)^2 - 1.280$	(3.14)

The use of image tools to develop the attractor parabola models restricted their appearance options such as the use of darkened lines because an image window pixel is either on or off, and complicated the determination of a quality of fit metric. The appearance could not be altered to highlight the parabola, but a measure of the fit was obtained by the following method. A time sequence of Henon (x, y) values was generated and the y-values used as input to the four parabola equations. This created four corresponding x-values for each Henon x-value, and the closest parabola was chosen for each Henon point. The resulting set of errors was obtained by subtraction of only the x-values for each parabola (because the Henon

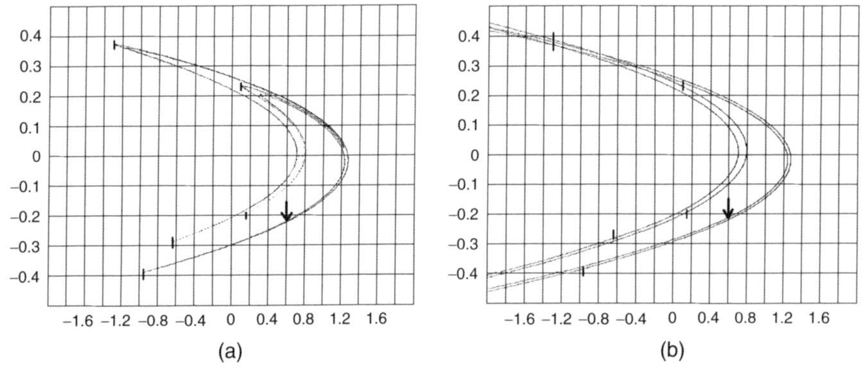

Figure 3.7 Parabolic regions of validity. (a) Major and minor section limits on the Henon attractor. (b) Major and minor section limits on the parabolas.

y-values were used in parabola equations), and the mean-square-error (MSE) was calculated for each parabola fit. Increasingly longer time sequences were used until the results stabilized, and a 500,000-point time sequence generated the following results.

	parabola 1	**parabola 2**	**parabola 3**	**parabola 4**
MSE	0.00018	0.00046	0.00047	0.00017

The regions of validity for the four parabolas were determined by the limits of the major (large pair of crescents) and minor (small pair of crescents) sections on the attractor. These regions are shown in Figure 3.7, where part (a) shows the Henon attractor and part (b) shows the four modeling parabolas. Referring to Figure 3.7a, the vertical dash marks delineate the limits of the major and minor sections on the attractor. These limit marks are transposed directly onto the parabolas in Figure 3.7b, identifying almost all termination points on the parabolic regions of validity. There is, however, an arrow in the vicinity of point (0.6, −0.22). This arrow marks the limit of the parabola-3 region of validity, because this area of the attractor is about equidistant between parabola-3 and parabola-4, and the model transitions to parabola-4 at this point.

3.2.1.3 *Chaotic Dynamics on the Various Parabolic Sections*

Each of the model parabolas was broken into sections according to quadrant of the Cartesian coordinate system and traced through an iteration of the Henon system equations to determine which parabolic sections they became after a chaotic iteration. Examples are shown in Figure 3.8. It was anticipated that this knowledge would aid in the determination of the transmit value.

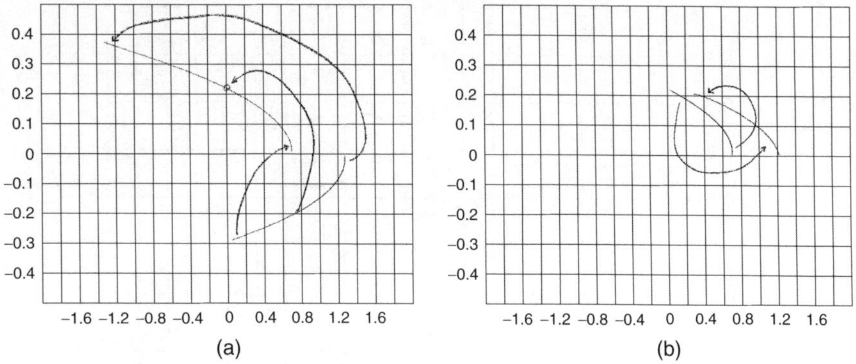

Figure 3.8 Examples of parabola sections transposed through the Henon equations. (a) Parabola-4, Quad. IV → Parabola-1, Quad. I & Parabola-1, Quad. II. (b) Parabola-1, Quad. I → Parabola-3, Quad. I.

Each final receiver decision of the current transmit value is mapped by a minimum Euclidean distance metric onto the closest attractor section via the parabolic model equations. Assuming an accurate decision, knowledge of the current parabolic section and its transposition through the system equations was at one time used to identify the range of valid received values on the next iterate. As the receiver design progressed, however, it turned out that this information is embedded in other areas of the receiver calculations, and so these tracings are no longer explicitly used. An interesting discovery that resulted from this investigation is the identification of a fixed point on the Henon attractor, which is discussed in the next section.

3.2.1.4 The Henon Fixed Point

Tracing the parabolic sections through the Henon map uncovered a repelling fixed point in the Henon attractor. Figure 3.9 shows the image plot that highlighted this characteristic. A fixed point on a chaotic attractor is one that repeats itself through chaotic iterations ad infinitum, i.e., it maps exactly onto itself. A *stable* or *attractive* fixed point is one toward which a neighborhood around the fixed point moves under iteration of the chaotic equations, so chaotic iterates on the neighborhood decrease the distance from the fixed point. An *unstable* or *repelling* fixed point is one for which the distance between itself and any point in a surrounding neighborhood increases through iteration of the chaotic system.

Within the resolution of MATLAB 4.2.c, the Henon fixed point is (**0.63135447708950, 0.18940634312685**). This point has been found to the 14th decimal place. The first iterate changes the x-value by one digit in the 14th decimal place, with the y-value unchanged. After 45

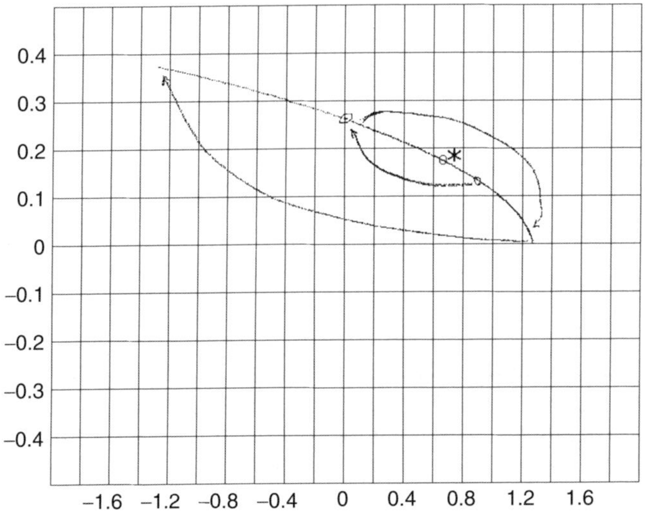

Figure 3.9 Parabolic transposition through the Henon map highlighting the fixed-point (star) Parabola-4, Quad. I → Parabola-4, Quad. I, and Parabola-4, Quad. II.

iterations of the Henon equations, the result is still within 2% of the starting values.

It was thought that this point might have allowed the development of a technique for synchronizing the transmitter to the receiver. At first startup, when the receiver has no knowledge of the transmitter's current chaotic state, initializing the receiver to the repulsive fixed point would cause iterations of the system map to linger in the vicinity of the initial point. The transmitter was anticipated to approach the neighborhood of the fixed point at some time during these initial iterations, and at that time the receiver could achieve a fast lock to the transmitter's chaotic state. This approach was found to be impractical for two reasons. First, there is a great variability in the number of iterates required to approach a small neighborhood of the fixed point from an arbitrary starting condition. Reaching this neighborhood at other than the first 50 or so iterates did nothing to enhance the synchronization performance of the receiver. Second, is the fact that noise could cause the received value to be in the neighborhood of the fixed point when the chaotic dynamics actually were not very close, causing the initialization of false tracking and the concomitant errors. Although not useful for this investigation, the Henon fixed point might yet be useful in the implementation of a chaotic communications system. It is possible that the receiver could align with the transmitter through a training period, as is done in adaptive equalization schemes for time-varying channels.[3] This training period could involve the 45 iterates of the Henon fixed point that remained within 2% of the starting value. Future studies of this possibility may lead to an adaptive implementation of a chaotic communication system.

3.2.2 The Transmit PDF Model

The transmit sequence is chosen between the Henon and mirrored-Henon free-running maps by the logical value of the data, with a logical 0 choosing the mirrored-Henon value and a logical 1 choosing the Henon value for this investigation. A typical digital data stream contains about equal occurrences of zeros and ones, so the transmit data sequence was modeled as a random sequence of zeros and ones with equal probability of occurrence. The PDFs for the Henon, mirrored-Henon, and transmit sequences were found empirically using multimillion iterate data runs. The weak law of large numbers states that the probability sample characteristics approach the theoretical characteristics as the number of samples increases, and it was found that 100,000 samples yielded a good general shape but that several million were required to find high quality PDF traces for model generation. All three PDFs were modeled, but the transmit PDF is the only one deemed to be of practical value in the receiver estimation engine for this design.

Starting from any given initial point within the appropriate basin of attraction, the first 100 or so iterates were ignored as transient although the chaotic trajectory moved onto the attractor. A time series was then recorded for 2,000,000 iterates for the Henon and mirrored-Henon systems, and 4,000,000 iterates for the transmit sequence.

It was desirable to define the range of values according to the anticipated values for the received data sequence with noise contamination. This range of values was based on a worst-case SNR of 0 dB, at which point the noise power and signal power are equal. It is generally difficult to work with a higher noise power than signal power, and most techniques approach a 0.5 BER in these cases. It was decided to use 4096 data bins to perform the various PDF calculations and the subsequent receiver processing on all of the PDFs so found.

It will be seen in Section 3.3.1.1 that the received data PDF is the convolution of the transmit PDF and the PDF of the additive noise in the channel. In the general case of a received PDF without closed form representation, the Fast Fourier Transform (FFT)-based fast convolution technique can be used to find the received PDF for any SNR value and any combination of transmit and noise PDF waveforms. It is therefore important to define the bin range with a sufficient region of insignificant values to make the aliasing contamination of the desired linear convolution result caused by the inherent circular convolutional characteristic of the FFT-based method negligible. In addition, the greatest computational efficiency of the FFT is realized when its input data length is a power of two. Zero-padding could be used to address the circular convolution concern. However, the increased lengths of the FFT calculations load down the computer system resources if 4096 bins are spread over a smaller range of received values such that zero padding is required, and the calculation fidelity decreases if fewer bins are used over the range of significant values.

The situation was resolved by using a range of $[-5.12, 5.12]$ with 4096 bins, yielding a bin width of 1/400, or 0.0025, and a bin density of 400 bins/UnitRange. To empirically find the many PDFs in this investigation, the MATLAB histogram function was used for each PDF calculation to get the number of occurrences in the time series that fell into the range of numbers for each data bin.

Letting b = bin index
B = number of data bins
K = number of iterates in the time series
${}_bn$ = number of hits in data bin b
${}_br$ = relative number of occurrences in data bin b
w = bin width, which is constant for all bins
p = continuous probability density function
${}_bp$ = PDF value of data bin b (quantized probability density function)
P = probability

$$ {}_br = \frac{{}_bn}{K} \tag{3.15} $$

where

$$ \sum_{b=1}^{B} {}_bn = K \tag{3.16} $$

and ${}_bn$ is found easily by using MATLAB's histogram function.
The fundamental definition of PDF requires that

$$ P_\infty = \int_{-\infty}^{\infty} (p(x))dx = 1 \tag{3.17} $$

For the present formulation, this translates into the requirement

$$ \sum_{b=1}^{B} {}_bpw = 1 \tag{3.18} $$

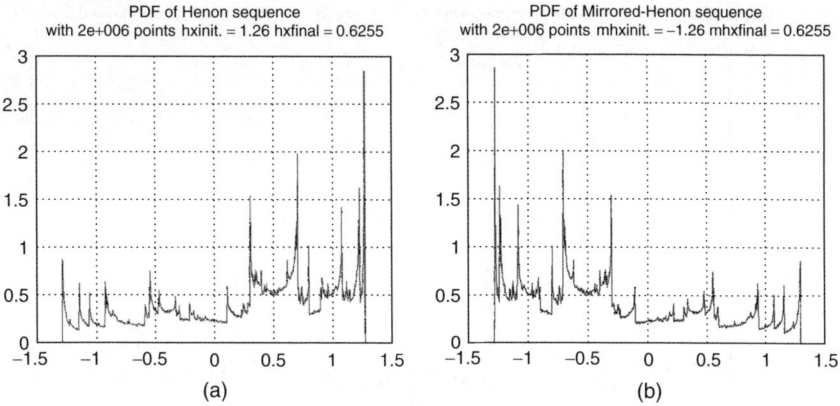

Figure 3.10 Henon and mirrored-Henon empirical PDFs. (a) Henon PDF — 2,000,000 iterates. (b) Mirrored-Henon PDF — 2,000,000 iterates.

The PDF can be found from the relative number of occurrences as follows:

$$\sum_{b=1}^{B} {}_b r = \sum_{b=1}^{B} \frac{{}_b n}{K} = \frac{K}{K} = 1 \tag{3.19}$$

$$\sum_{b=1}^{B} {}_b p w = w \sum_{b=1}^{B} {}_b p = \sum_{b=1}^{B} {}_b r \tag{3.20}$$

$${}_b p = \frac{{}_b r}{w} \tag{3.21}$$

3.2.2.1 The Henon PDF

A time series of 2,000,000 iterates gave reasonably smooth PDF traces for the Henon system. The resulting plot is shown in Figure 3.10.

3.2.2.2 The Mirrored-Henon PDF

Similarly, the PDF trace for the mirrored-Henon map is based on 2,000,000 iterates and is shown in Figure 3.10. It is a mirrored version of the Henon PDF. The Henon and mirrored-Henon PDFs were modeled with Gaussian functions, but these models have served no practical purpose in the receiver design. It is possible that a usefulness will become apparent in the future, but the model with a currently realized benefit is the transmit PDF of the next section.

3.2.2.3 The Transmit PDF

Because the logical 0 and 1 values of a typical data sequence each have a probability of occurrence of ½, the transmit sequence is modeled as

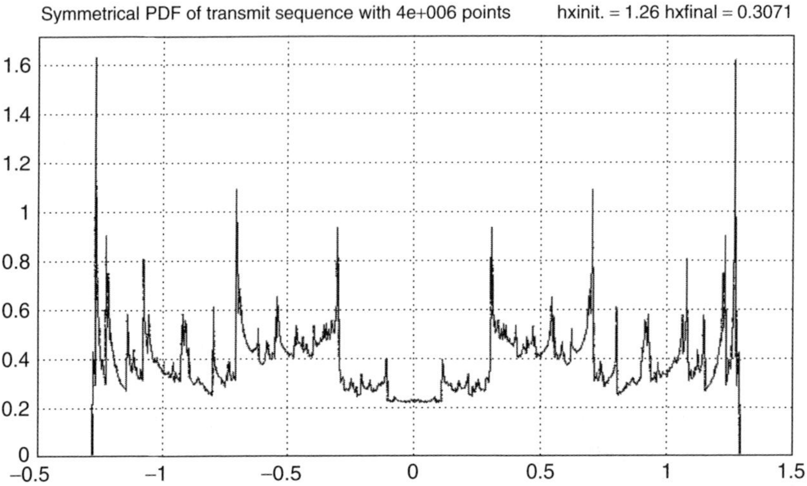

Figure 3.11 Transmit sequence empirical PDF — 4,000,000 iterates.

an arbitrary choice between the Henon and mirrored-Henon time series for each data bit duration. Data runs of 4,000,000 iterates were taken to empirically construct a smooth transmit PDF as shown in Figure 3.11. This transmit PDF can be found from the Henon and mirrored-Henon PDFs via multiplying each by its corresponding probability of occurrence (½ for both in this case) and adding them together. Actually, all PDFs can be generated from a single Henon or mirrored-Henon PDF.

3.2.2.4 The Transmit PDF Gaussian Model

The transmit PDF was modeled as a DC level plus a summation of Gaussian functions. There is a choice as to window only the DC level and let the Gaussian functions remain infinite, or to set up the DC level and Gaussian functions first and window the entire PDF model. In either case, the window must be defined from observations of the empirical PDF trace coupled with the bin range and bin width choices outlined in Section 3.2.2. The maximum positive bin center value for which there is a nonzero transmit PDF value is 1.2841, and the bin limits are $[1.2828, 1.2854]$. Because the transmit PDF is symmetrical, the corresponding negative bin limits are $[-1.2828, -1.2854]$. The window is, therefore, defined as

$$\begin{aligned} W_{dc}(x) &= U(x-(-H_{\max})) - U(x-H_{\max}) \\ &= U(x-(-1.2854)) - U(x-1.2854) \qquad (3.22) \end{aligned}$$

where $U(x_0) = \begin{Bmatrix} 0 \Rightarrow x < x_0 \\ 1 \Rightarrow x \geq x_0 \end{Bmatrix}$ = the unit step function and $H_{\max} = 1.2854$ is the maximum Henon magnitude for the specified bin structure.

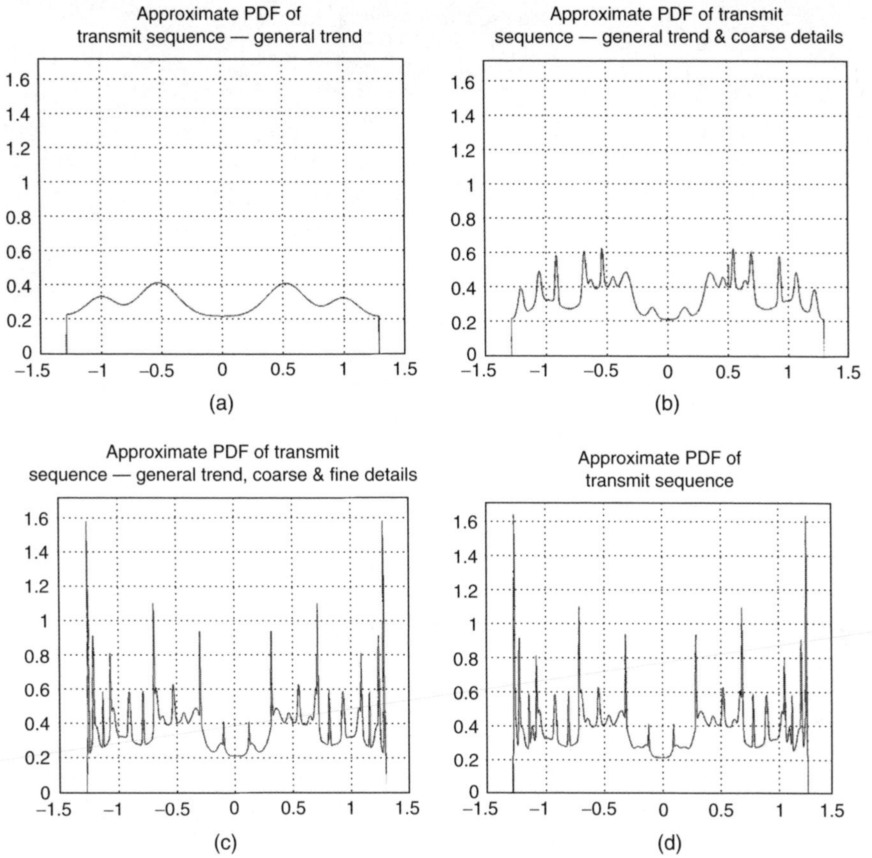

Figure 3.12 The evolution of the transmit PDF approximation. (a) DC level plus general trend. (b) Coarse details added. (c) Fine details added. (d) Attractor approximate — final details added.

The differences between the two are very small in terms of establishing the PDF trace model. It turns out that windowing both the DC level and the Gaussian functions require the DC level (0.2160) to be greater by 0.0002 than the method of windowing only the DC value (0.2158). The small degree of difference is because the Gaussian functions used to model the transmit PDF have very little area outside the valid DC region. The choice between windowing methods cannot be made on considerations stemming from the construction of a PDF model, but SNR estimation calculations involving the transmit PDF in Section 3.3.1.2 drove the decision to window only the DC component.

Figure 3.12 shows the evolution of the transmit PDF approximation. First the DC level and four Gaussian functions structure the general trend with MSE = 0.0057. In part (b), some coarse details are added in with 18

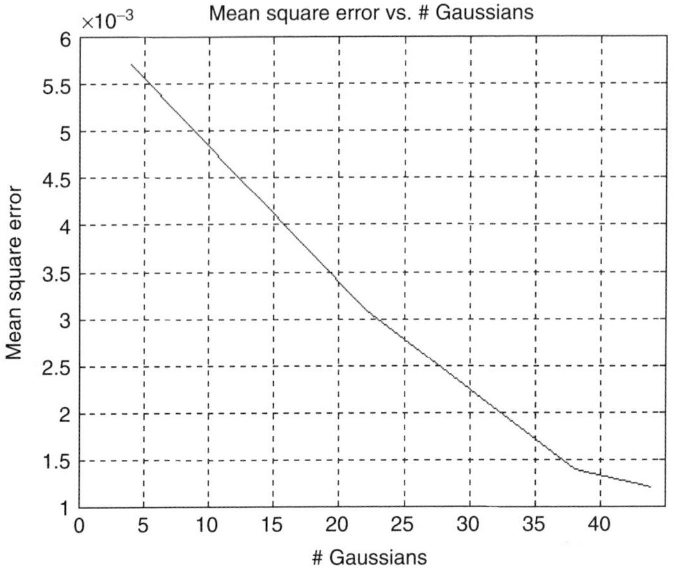

Figure 3.13 MSE vs. number of Gaussians in model.

more Gaussian functions (22 total) with MSE = 0.0031, and part (c) has some detailed fine structure consisting of 16 Gaussian functions (38 total) and MSE = 0.0014. Part (d) shows the final attractor approximation with the final details added in using six Gaussian functions (44 total) that fill in some gaps and have a final MSE = 0.0012.

A graph of the MSE vs. number of Gaussian functions is presented in Figure 3.13. The first line segment between 4 and 22 Gaussian functions continues almost unchanged between 22 and 38 Gaussian functions. The addition of the final six Gaussian functions has introduced an obvious knee in the curve, indicating that a point of diminishing returns has been reached. After this point it is no longer useful to trade off fidelity gained against the added complexity and computational burden incurred by additional refinements to the model.

The final tally is a windowed DC level with 44 weighted Gaussian functions to construct the transmit PDF model. Because of the symmetry of the structure, only 22 Gaussian functions (mean and variance) had to be defined, along with the corresponding 22 weighting factors. The empirical attractor is overlaid on the model in Figure 3.14. The close match with the distinguishing features of the true PDF can be readily seen. A metric against which to measure the match between the empirical-PDF and the approximation is the MSE, whose value is 0.0012. The approximation total probability, which is the area under the PDF approximation, is 0.9999 (see Figure 3.15).

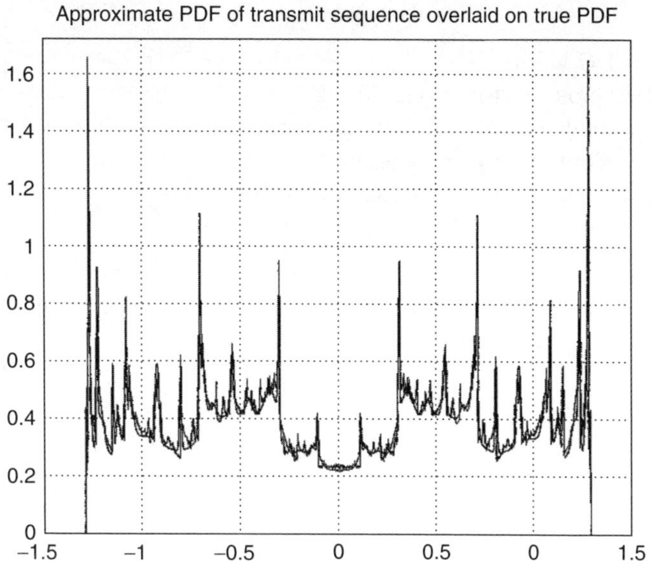

Figure 3.14 Transmit PDF — empirical overlaid on model.

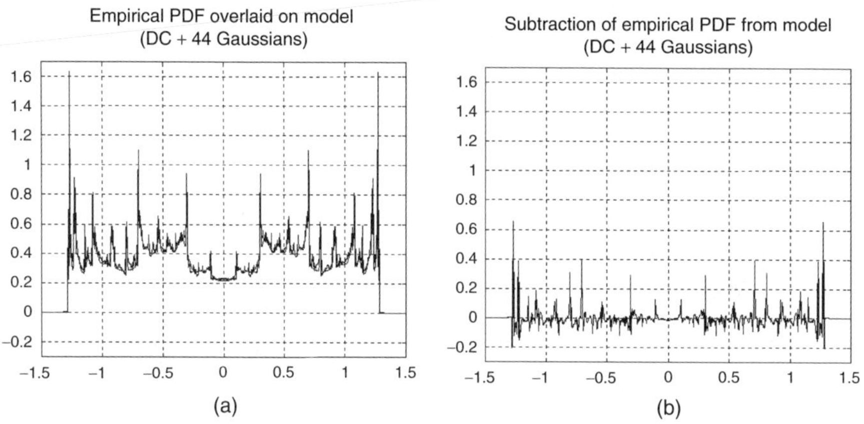

Figure 3.15 Transmit PDF — empirical and model (a) overlaid and (b) subtraction.

The equation for the transmit PDF approximation model with windowing only for the DC level is

$$\tilde{p}_t(x) = cW_{dc}(x) + \sum_{l=1}^{L} \frac{b_l}{\sqrt{2\pi}\sigma_l} e^{-\left(\frac{x-m_l}{\sqrt{2}\sigma_l}\right)^2} \tag{3.23}$$

where

c = DC value
l = Gaussian functions' index
L = total number of Gaussian functions
b_l = weights of the Gaussian functions
m_l = mean values of the Gaussian functions
σ_l^2 = variances of the Gaussian functions

and the values for the approximation depicted here are listed below, with a one-to-one correspondence between the values occupying identical positions in the vectors b, m, and σ

$c = 0.2158$
$L = 44$

$$b = [\begin{matrix} 0.0036 & 0.0030 & 0.0026 & 0.0098 & 0.0070 & 0.0024 & 0.0045 \\ 0.0040 & 0.0045 & 0.0038 & 0.0010 & 0.0085 & 0.0077 & 0.0069 \\ 0.0099 & 0.0037 & 0.0050 & 0.0030 & 0.0181 & 0.0048 & 0.0318 \\ 0.0770 & 0.0770 & 0.0318 & 0.0048 & 0.0181 & 0.0030 & 0.0050 \\ 0.0037 & 0.0099 & 0.0069 & 0.0077 & 0.0085 & 0.0010 & 0.0038 \\ 0.0045 & 0.0040 & 0.0045 & 0.0024 & 0.0070 & 0.0098 & 0.0026 \\ 0.0030 & 0.0036] & & & & & \end{matrix}$$

$$m = [\begin{matrix} -1.2600 & -1.1200 & -0.2000 & -1.2709 & -1.2280 & -1.1455 & -1.0802 \\ -0.8017 & -0.7063 & -0.3080 & -0.1091 & -1.2100 & -1.0600 & -0.9200 \\ -0.6900 & -0.6400 & -0.5400 & -0.4515 & -0.3400 & -0.1300 & -1.0000 \\ -0.5250 & 0.5250 & 1.0000 & 0.1300 & 0.3400 & 0.4515 & 0.5400 \\ 0.6400 & 0.6900 & 0.9200 & 1.0600 & 1.2100 & 0.1091 & 0.3080 \\ 0.7063 & 0.8017 & 1.0802 & 1.1455 & 1.2280 & 1.2709 & 0.2000 \\ 1.1200 & 1.2600] & & & & & \end{matrix}$$

$$\sigma = [\begin{matrix} 0.0071 & 0.0099 & 0.0255 & 0.0028 & 0.0047 & 0.0028 & 0.0042 \\ 0.0050 & 0.0028 & 0.0028 & 0.0028 & 0.0212 & 0.0170 & 0.0099 \\ 0.0141 & 0.0184 & 0.0090 & 0.0170 & 0.0410 & 0.0283 & 0.1131 \\ 0.1570 & 0.1570 & 0.1131 & 0.0283 & 0.0410 & 0.0170 & 0.0090 \\ 0.0184 & 0.0141 & 0.0099 & 0.0170 & 0.0212 & 0.0028 & 0.0028 \\ 0.0028 & 0.0050 & 0.0042 & 0.0028 & 0.0047 & 0.0028 & 0.0255 \\ 0.0099 & 0.0071] & & & & & \end{matrix}$$

3.2.2.5 *The Transmit Power*

It should be noted that the Henon and mirrored-Henon PDFs were skewed positive and negative, respectively, indicating correspondingly positive and negative mean values for the systems. The mean and variance for the systems have been estimated from million-iterate data runs to be

Henon	**Mirrored-Henon**
$\hat{\bar{x}}_h = 0.2558$	$\hat{\bar{x}}_m = -0.2558$
$\hat{\sigma}_h^2 = 0.5210$	$\hat{\sigma}_m^2 = 0.5210$

The transmit PDF is seen to be symmetrical about zero, however, because of the random selection between the Henon and mirrored-Henon time sequences according to the data bits. Because the Henon and mirrored-Henon values ranges have complete overlap, the range of transmit values is the same as the range in either individual system. In terms of the original systems,

$$\bar{x}_h = \frac{1}{K}\sum_{k=1}^{K} x_h[k] \tag{3.24}$$

$$\sigma_h^2 = \frac{1}{K}\sum_{k=1}^{K} (x_h[k] - \bar{x}_h)^2 = \frac{1}{K}\sum_{k=1}^{K} x_h^2[k] - \bar{x}_h^2 = \overline{x_h^2} - \bar{x}_h^2 \tag{3.25}$$

$$\bar{x}_m = \frac{1}{K}\sum_{k=1}^{K} x_m[k] \tag{3.26}$$

$$\sigma_m^2 = \frac{1}{K}\sum_{k=1}^{K} (x_m[k] - \bar{x}_m)^2 = \frac{1}{K}\sum_{k=1}^{K} x_m^2[k] - \bar{x}_m^2 = \overline{x_m^2} - \bar{x}_m^2 \tag{3.27}$$

For the transmit sequence,

$$P_h(x_h) = P_r(x_h \text{ is sent}) \quad P_m(x_m) = P_r(x_m \text{ is sent}) \tag{3.28}$$

$$P_h(x_h) = P_m(x_m) = 0.5 \tag{3.29}$$

$$x_t = (x_h)(P_h(x_h)) + (x_m)(P_m(x_m)) \tag{3.30}$$

$$\bar{x}_t = \frac{1}{K}\sum_{k=1}^{K} x_t[k] = P_h(x_h)\left(\frac{1}{K}\sum_{k=1}^{K} x_h[k]\right) + P_m(x_m)\left(\frac{1}{K}\sum_{k=1}^{K} x_m[k]\right) \tag{3.31}$$

$$\bar{x}_t = \frac{\bar{x}_h + \bar{x}_m}{2} = \frac{0.2558 - 0.2558}{2} = 0 \tag{3.32}$$

$$\sigma_t^2 = \frac{1}{K}\sum_{k=1}^{K} (x_t[k] - 0)^2 = P_h(x_h)\left(\frac{1}{K}\sum_{k=1}^{K} x_h^2[k]\right) + P_m(x_m)\left(\frac{1}{K}\sum_{k=1}^{K} x_m^2[k]\right) \tag{3.33}$$

Substituting from Equation (3.25) and Equation (3.27) into Equation (3.33) yields the transmit power equation

$$\sigma_t^2 = P_h(x_h)(\sigma_h^2 + \bar{x}_h^2) + P_m(x_m)(\sigma_m^2 + \bar{x}_m^2) = \frac{(\sigma_h^2 + \bar{x}_h^2) + (\sigma_m^2 + \bar{x}_m^2)}{2} \quad (3.34)$$

and entering the empirically determined Henon and mirrored-Henon values yields

$$\sigma_t^2 = \frac{(0.5210 + 0.2558^2) + (0.5210 + (-0.2558)^2)}{2} = 0.5864 \quad (3.35)$$

Transmit Sequence
$\bar{x}_t = 0$
$\sigma_t^2 = 0.5864$

These values have been proven by statistical analysis of empirical data runs. The transmit power is used to set up the proper lossless nonfading AWGN channel noise power level via an SNR specified as an input to the simulation for the bit error rate (BER) data runs.

3.2.3 Henon, Mirrored-Henon, and Transmit Sequence Power Spectral Density Plots

The PSD plots for the Henon, mirrored-Henon, and transmit sequences are presented in this section. The plots were generated using the Welch's averaged periodogram method with a Hanning window. Figure 3.16 shows the PSD of the Henon chaotic system and mirrored-Henon system, which are identical because the autocorrelation functions of negated sequences

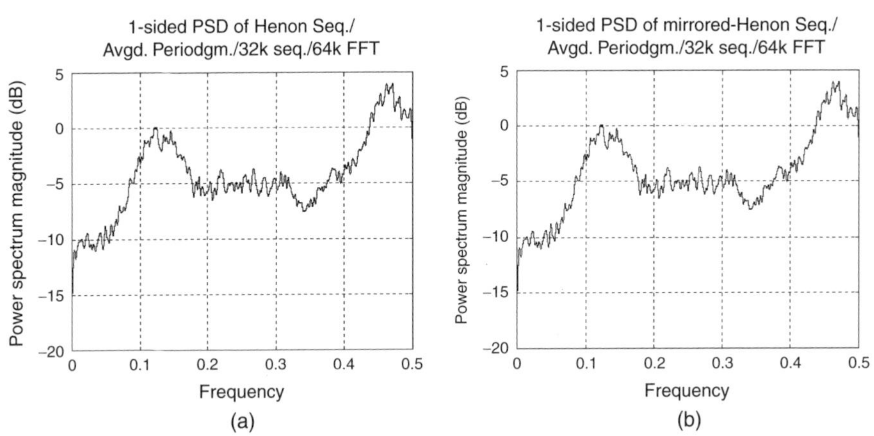

Figure 3.16 The power spectral densities for the Henon and mirrored-Henon chaotic systems. (a) Henon chaotic system. (b) Mirrored-Henon chaotic system.

are identical. They used 64-k FFTs on 32-k data strings in the spectral plotting process.

The PSD plots for various transmit sequences are depicted in Figure 3.17. It can be seen in Figure 3.17a that the spectrum is fairly flat at one iterate per bit because of the frequent changes between the Henon sequence and its negated version, the mirrored-Henon sequence. At two iterates per bit, Figure 3.17b shows that the flat PSD begins to have a slope. For 16 iterates per bit Figure 3.17c has a strong resemblance to the Henon (or mirrored-Henon) PSD because of the long dwell time on a given chaotic process, and it shows the existence of a low-frequency region of spectral content not seen in the individual chaotic system plots. Figure 3.17d illustrates the spectral content of a transmit sequence with 64 iterates per bit that further exemplifies the trend and shows a narrowing low frequency region as more

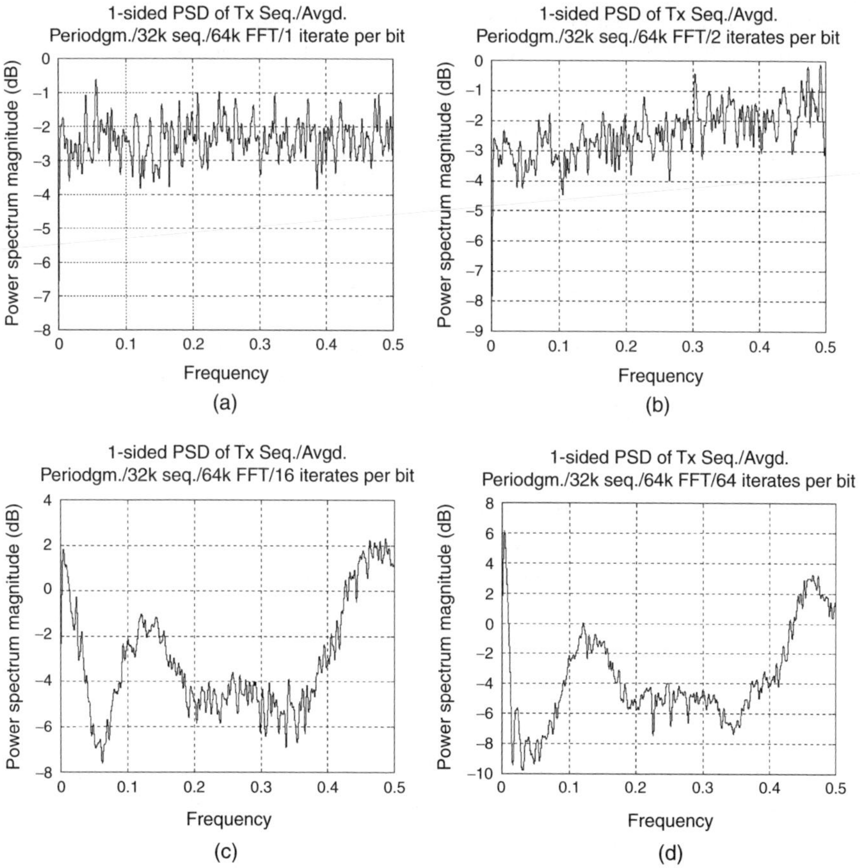

Figure 3.17 Power spectral density of the transmit sequence derived from the Henon and mirrored-Henon chaotic systems. (a) 1 iterate per bit. (b) 2 iterates per bit. (c) 16 iterates per bit. (d) 64 iterates per bit.

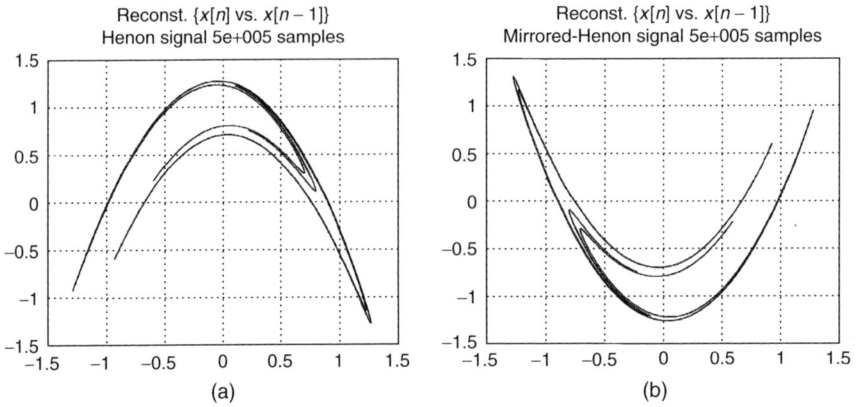

Figure 3.18 Reconstructed attractors for the Henon and mirrored-Henon chaotic systems. (a) Henon attractor reconstructed from time sequence. (b) Mirrored-Henon attractor reconstructed from time sequence.

dwell time is spent within a given chaotic system and less time at the bit duration transition points. These observations indicate that fewer iterates per bit are desirable for lowering the probability of signal interception via spectral characteristics.

3.2.4 Chaotic Reconstruction Plots

This section shows a sequence of chaotic attractor plots generated by the reconstruction method of plotting the current iterate x-value against the previous iterate x-value, which works for the two-dimensional Henon system. Figure 3.18 shows the reconstruction of the individual Henon and mirrored-Henon systems. There are differences between these attractors and the plots in Section 3.2. These attractors are in square windows with uniform axes because the y-value information is not known or calculated in these reconstructions, although this computation could be done because the Henon current y-value equals the scaled previous x-value as depicted in Equation (3.2). The orientation of the attractors is rotated 90° counterclockwise and flipped across the y-axis because the stated reconstruction method plots the x-values vs. the y-values, which are scaled versions of the previous iterate x-values. This difference interchanges the role of the axes and alters the orientation of the attractor. The discrepancy could be addressed by reversing the plot procedure, but it was not deemed important for the purposes of this investigation.

Figure 3.19 illustrates the reconstructed transmit sequence attractor, which is the same as a noiseless, lossless received data sequence attractor. Both the Henon and mirrored-Henon portions of the transmit attractor have an image flipped through the y-axis superimposed on them resulting from

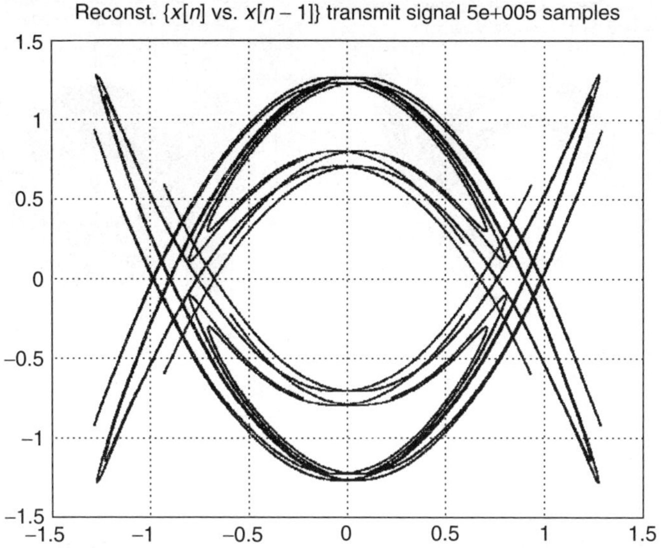

Figure 3.19 Reconstructed attractor for the transmit sequence.

the Henon and mirrored-Henon random selection process in the transmit sequence. A current Henon value plotted against a previous Henon value will produce a point on the Henon reconstructed attractor, but a current Henon value plotted against a previous mirrored-Henon value (the negative of the previous Henon value) will plot a point on the Henon attractor flipped through the *y*-axis. The same argument holds for current mirrored-Henon values, and the result is the attractor seen in Figure 3.19.

Figure 3.20 shows reconstruction plots including the effects of noise for various SNR values. The lines of the attractor have become widened by the noise at 40-dB SNR, but most detail of the reconstructed noiseless attractor is still evident. As the noise content of the received data sequence increases, the attractor lines spread and merge until at 10-dB SNR the basic crescent structure is obliterated. At 0-dB SNR the attractor is essentially circular, resembling a reconstruction plot for random Gaussian noise. Although not directly useful in this investigation, plots of this type may have an application in the adaptation of this method to a lossy channel where the receiver needs to estimate the received signal power prior to using the methods of this thesis. This investigation used a lossless channel and relied on knowledge of the transmit signal power to determine SNR. A relationship between these plots and the signal power could be derived and an algorithm developed to estimate the received signal power in a lossy channel. This result would enable the use of the SNR estimator developed in this thesis to permit successful receiver estimation engine performance in the presence of signal level attenuation between the transmitter and receiver.

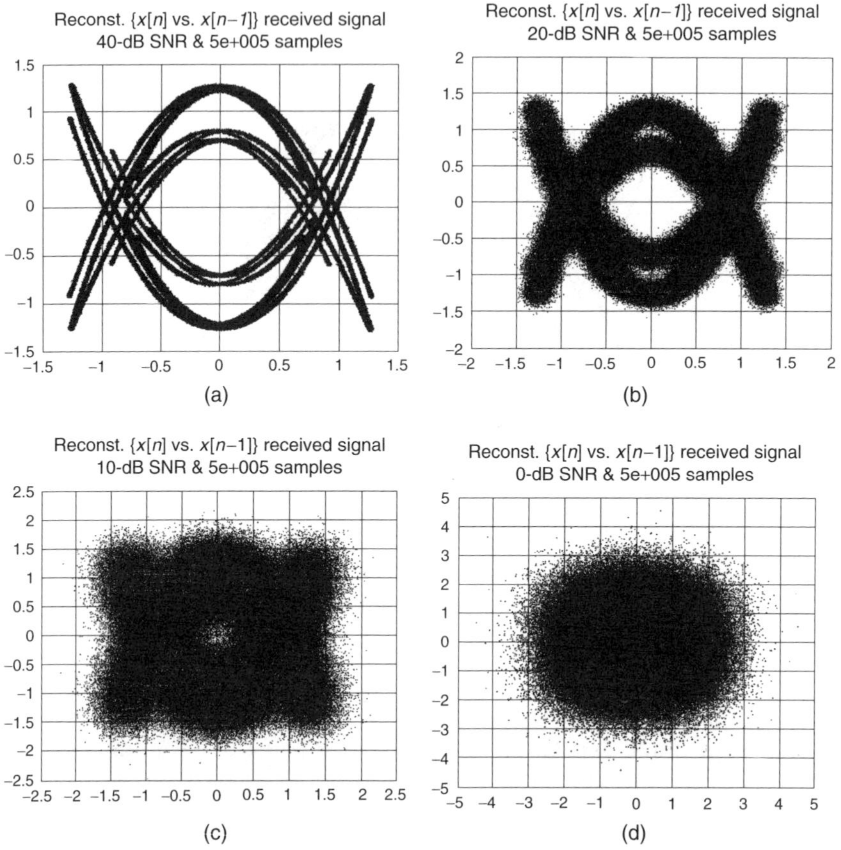

Figure 3.20 Reconstructed transmit attractors for various SNR values. (a) 40-dB SNR. (b) 20-dB SNR. (c) 10-dB SNR. (d) 0-dB SNR.

3.3 The Receiver Design

The general block diagram for the receiver is shown in Figure 3.21. Three estimates are generated for each received iterate, an initial decision is calculated from them, and a final decision is formed by mapping the initial decision onto the chaotic attractor. The mapping operation also yields the synchronization estimate. It provides chaotic synchronization without requiring the special characteristics or techniques used by other methods, such as stable and unstable chaotic subsystem separation or inverse system methods. At the appropriate times, the individual iterate decisions are combined into a bit decision. A BER is then calculated to evaluate the performance of this method.

The first estimate is the received value itself. In the absence of noise, it is an accurate indication of the transmitted value. In the presence of AWGN, the received value is a valid estimate of the transmitted iterate because

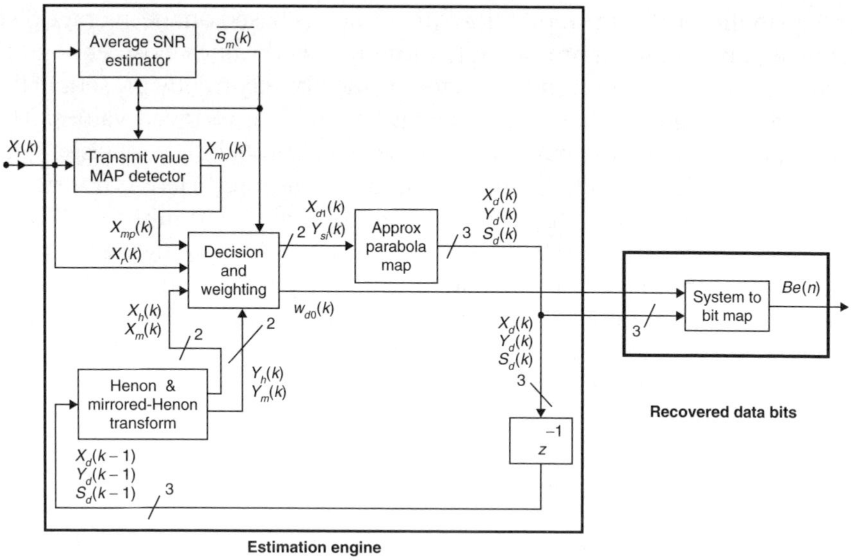

Figure 3.21 General receiver block diagram.

the most likely noise value is 0. The second estimate is the MAP transmit value, which is found for every received iterate and yields the minimal error probabilistic estimate. The third estimate is the receiver final decision, delayed by 1 iterate and processed through the Henon and mirrored-Henon equations. It provides information on the next possible received values for both chaotic systems given an accurate current decision, which aids the decision process in the presence of noise.

The decision and weighting block combines the estimates into an initial decision through a weighted average whose weights derive from probability calculations on each estimate. It is given the received iterate estimate, the MAP estimate, and two possible feedback estimates — one assuming a correct previous decision and the other assuming an erroneous previous decision. It chooses the feedback estimate closest to the received value and MAP estimates, determines the necessary probability values, and performs the weighted average calculation.

The decision and weighting block also generates a discount weight for the bit map block to use in combining individual iterate decisions into a bit decision that serves to de-emphasize iterate decisions whose received value is in close proximity to zero. Recall from Equation (3.4) that the Henon and mirrored-Henon transmit possibilities are the negatives of each other. Transmitted values in the close proximity of zero can appear to have come from the incorrect attractor with very small amounts of noise. The feedback estimate chosen as closest to the MAP and received value estimates can easily be in error, which may then create an erroneous final iterate decision.

The quantification of the term "close proximity" is based on the average SNR estimate, discounting more severely those received values close enough to zero to have been significantly affected by the noise currently present in the channel although emphasizing the larger magnitude received values. The approximation parabola map is given the initial probabilistic decision and the corresponding y-value from the Henon or mirrored-Henon transform. It uses the parabola equations that model the Henon and mirrored-Henon attractors to generate parabolic sections in the vicinity of the input point and chooses the parabola and corresponding system closest to the input point via a Euclidean distance metric. This process adjusts both the x- and y-values of the receiver initial decision to provide accurate tracking of the chaotic dynamics in the final decision.

3.3.1 The Three Estimates

As stated in the previous section, the three estimates generated for each received iterate are the received value itself, the MAP estimate, and the decision feedback estimate. The received value is used directly because it equals the transmitted value in the absence of noise and, because the AWGN has zero mean, the expected value of the received value is the transmitted value. This section explores the generation of the other two estimates.

3.3.1.1 Maximum A Posteriori (MAP) Transmit Value

The second estimate is the MAP transmit value for a given received iterate, which is the optimal probabilistic detector because it minimizes the probability of error.[3] There are differences, however, between conventional communication methods and this chaotic system in the transmission characteristics, the modulation technique, and the presence of chaotic dynamics that require additional considerations in the decision process. The MAP decision is an important contribution because it is the statistical method that yields the lowest BER. The chaotic communication system investigated here relates the binary possibilities of logical 1 and logical 0 with a constantly changing sequence drawn from a continuum of possible values, the range of which is identical for both logical states. There is not a direct relationship between a transmit value and its message content. Any given transmit value could represent either logical state depending on the attractor from which it was generated at that point in time, and the next time the value (or a value within a small neighborhood) is encountered it may have come from the opposite attractor. The MAP calculation will be performed on every received iterate, so the iterate index k will be carried explicitly in the following equations. To relate the continuum of values x_t to the PDF as empirically determined or modeled, reference is made to the formulations derived in Section 4.2.2. The continuous range of values will be considered to have a very large but finite set l of possible values as is encountered by a chaotic map processed on a computer because of the finite resolution of computer-generated numbers.

The general relationship between the probability and a specific value in the range of a PDF is

$$P[A < X \leq B] = \int_{A}^{B} p_X(x_0)dx_0 \tag{3.36}$$

The probability that the lth transmitted value possibility $x_t(l)$ falls into bin b in the quantized PDF of Section 3.2.2 is equal to

$$\begin{aligned} P_t\left[\left(x_{tb} - \frac{w}{2}\right) < x_t(l) \leq \left(x_{tb} + \frac{w}{2}\right)\right] &= \int_{x_{tb}-w/2}^{x_{tb}+w/2} {}_b p_t(x_t[l])dx_t \\ &= w[{}_b p_t(x_t[l])] \end{aligned} \tag{3.37}$$

where x_{*b} = $*$ iterate series PDF central value of bin b (i.e., $-t \Rightarrow$ Tx series, $r \Rightarrow$ Rx series, etc.)

${}_b p_*(\bullet)$ = the probability of the bin into which the value $(\bullet)$ falls, for the $*$ iterate series

l = a dummy variable of delineating the range of possible transmit values

and $b = 1, 2, \ldots, B$ = bin index.

In shorthand notation,

$$P_t(x_{tb}[l]) = w[{}_b p_t(x_t[l])] \tag{3.38}$$

The determination of a MAP transmit value seeks the maximum value of the equation

$$\max_b \{P(x_{tb}[l]|x_r[k])\} = \max_b \left\{{}_b p_{r|t}(x_r[k]|x_{tb}[l])P_t(x_{tb}[l])\right\} \tag{3.39}$$

$$= \max_b \left\{{}_b p_{r|t}(x_r[k]|x_{tb}[l])w[{}_b p_t(x_t[l])]\right\} \tag{3.40}$$

The bin width term w is the same constant value for all probability bins, and so can be dropped from the maximization process. The resulting equation to maximize is

$$\max_b \{P(x_{tb}[l]|x_r[k])\} = \max_b \left\{{}_b p_{r|t}(x_r[k]|x_{tb}[l]) {}_b p_t(x_t[l])\right\} \tag{3.41}$$

The maximization occurs across all bins in the quantized PDF range of values. The index b can be dropped, with the understanding that the entire PDF is included in the maximization range, and that the resulting values will be quantized. A further notation simplification can be made by recognizing that the transmit PDF is fixed, regardless of which time iterate index k is current. The results are

$$p_t(x_t[l]) \Rightarrow p_t(x_t) \text{ is the transmit PDF} \tag{3.42}$$

$$x_{tb}[l] \Rightarrow x_t \text{ the } l\text{th Tx value possibility that results in } x_r[k] \text{ is the Tx range} \tag{3.43}$$

$$\max_x \left\{P_{t|r}(x_t|x_r[k])\right\} = \max_x \left\{p_{r|t}(x_r[k]|x_t)p_t(x_t)\right\} \tag{3.44}$$

Equation (3.44) is the MAP technique for the estimation of continuous variables.[4]

At this point, the right hand side of the equation has a straightforward interpretation that can be implemented with a little more work. The term $p_{r|t}(x_r[k]|x_t)$ is a window that multiplies the transmit PDF $p_t(x_t)$, and the maximization of the result is the desired MAP transmit value for the kth system iterate. This window serves to define the region of the PDF over which the maximization will occur, and also to modify the shape of the transmit PDF according to the noise contained in the received sequence x_r.

For the current situation with Gaussian noise and a Gaussian approximation to the transmit PDF, it is conceptually possible to perform this multiplication and get a closed-form equation for the result. This expression could then be differentiated once or twice, depending on the iterative root approximation modification algorithm used to find the maxima, as is done for the SNR estimator in Section 3.3.1.2. However, the multiplication result will have many sharply defined peaks for the noise levels of interest for practical communications, and this approach was discarded as impractical. The multiplication result is more easily searched for its largest value with a software routine that locates the maximum value of a vector of numbers, which yields the most likely transmitted value. A *max* function is standard in many mathematical platforms such as MATLAB, and makes the search for the maximum value in a vector of numbers routine. The transmit PDF window construction is outlined below. The general result is that the noise PDF, reversed in x, is centered on the received value. This window function then multiplies the transmit PDF, and the maximum value of the result is found. A simplified decision calculation has been developed for this design to ease the computational burden of the technique by avoiding the repeated computation of Gaussian exponentials for the determination of the MAP transmitted value.

3.3.1.1.1 Transmit PDF Window

The fact that a continuum of values can be transmitted for either logical binary state complicates the determination of an a posteriori maximum likelihood transmit value. Furthermore, the result of such a calculation is not necessarily the best estimate of the transmit value because the effects of chaotic dynamics have not been taken into account. In this implementation the MAP computation uses only the current value and makes no use of past or future received values. However, it is a good contributor to the receiver decision of the transmit value, along with some other estimates that help track the deterministic random motion of the chaotic system.

A graphical proof of the window construction is presented in Figure 3.22. It draws on the fact that for additive, statistically independent noise, the received PDF is the convolution of the transmit PDF with the channel noise PDF

$$p_r(x) = p_t(x) * p_n(x) \tag{3.45}$$

It is helpful to view the succession of graphs in Figure 3.22 as depicting a sequence of transmit values that have all resulted in a received value x_r. For each of the five graphs in the sequence, the lth possible transmit value $x_t[l]$ is known, and so its PDF is a unit impulse. The convolution of the channel noise PDF with this delta function results in the noise PDF being

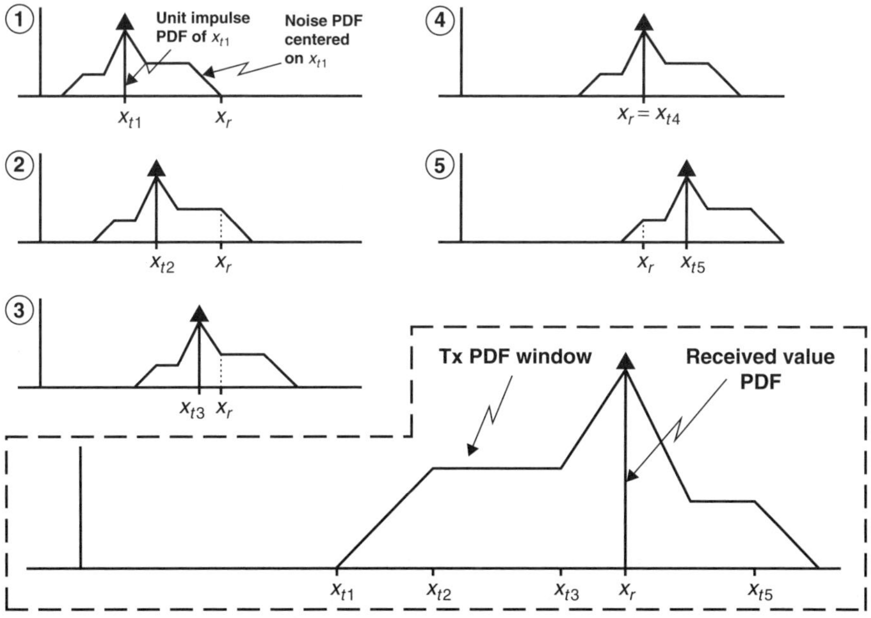

Figure 3.22 Construction of the transmit PDF window.

centered on the transmit value. So,

$$p_t(x|x_t = x_t[l]) = \delta(x - x_t[l]) \tag{3.46}$$

where $\delta(x)$ is the Dirac delta function, and

$$p_{r|t}(x|x_t[l]) = \delta(x - x_t[l]) * p_n(x) = p_n(x - x_t[l]) \tag{3.47}$$

The sequence of graphs show an asymmetrical noise PDF with its maximum peak at 0, centered on various transmit values. For all cases, the received value is the same fixed value and assumed to be the current received iterate. For the lth plot, the quantized probability that x_r is received given that $x_t[l]$ was sent is the bin width w times the value of the received PDF at x_r. However, as was discussed in Section 3.3.1.1, the bin width does not contribute to the MAP calculation and is omitted. Letting x_r be the received value on the kth iterate, $x_r[k]$:

$$P_{r|t}(x_r[k]|x_t[l]) = p_n(x_r[k] - x_t[l]) \tag{3.48}$$

The lth probability value is associated with the lth transmit value depicted in the succession of graphs in Figure 3.23, all of which are laid out for the kth received chaotic iterate. By performing this operation for all possible shifts within the range of transmit values that can result in the current received iterate, the entire noise PDF is traced out. The result is a PDF that preserves the unity area characteristic of probability density functions.

This construction is plotted in the result box labeled "Tx PDF Window." The window, W_{tx}, is seen to be the noise PDF, reversed in x, and centered on the received value.

$$W_{tx}(x_r[k]) = p_n(x_r[k] - x) \tag{3.49}$$

Note that this result has the same general form as Equation (3.47) for the convolution of the noise PDF with the current transmit value PDF, where the transmit value is considered to be known with probability 1 and consequently possesses a delta function PDF. The noise PDF is reversed in x in this case, and so the following interpretation can be adopted.

In the receiver, the current received value has been observed and is, therefore, known with probability 1. This value is not the received value probability of occurrence as determined from the received sequence PDF, it is the probability of the current observation and has a delta function PDF at the observed value for the construction of the transmit PDF window. The window is found as the convolution of the noise PDF reversed in x with the delta function PDF of the observed received value. The process of windowing the transmit PDF is performed, and the result is searched for its maximum value, yielding the corresponding MAP transmit value.

$$_{W}p_r(x_r[k]) = \delta(x - x_r[k]) \tag{3.50}$$

$$W_{tx}(x_r[k]) = p_n(-x) * {}_{W}p_r(x[k]) = p_n(x_r[k] - x) \tag{3.51}$$

$$p_{r|t}(x_r[k]|x_t) = W_{tx}(x_r[k]) \tag{3.52}$$

$$\max_{x} \left\{P_{t|r}(x_t|x_r[k])\right\} = \max_{x} \left\{W_{tx}(x_r[k])p_t(x_t)\right\} \tag{3.53}$$

3.3.1.1.2 MAP Transmit Value Calculation

The calculation of the maximum likelihood transmit value reduces to finding the maximum value of the windowed transmit PDF, the general case of which is illustrated in Figure 3.23. The convolution shown in Equation (3.51) requires the noise PDF to be reversed in x. White Gaussian noise is symmetric in x, so the reversed PDF equals the forward PDF.

$$p_n(x) = \frac{1}{\sqrt{2\pi\sigma_n}} e^{-\left(\frac{x}{\sqrt{2}\sigma_n}\right)^2} \tag{3.54}$$

where $\sigma_n^2 = s/s_{nr} =$ noise variance $=$ noise power
$s =$ signal power
$s_{nr} =$ SNR

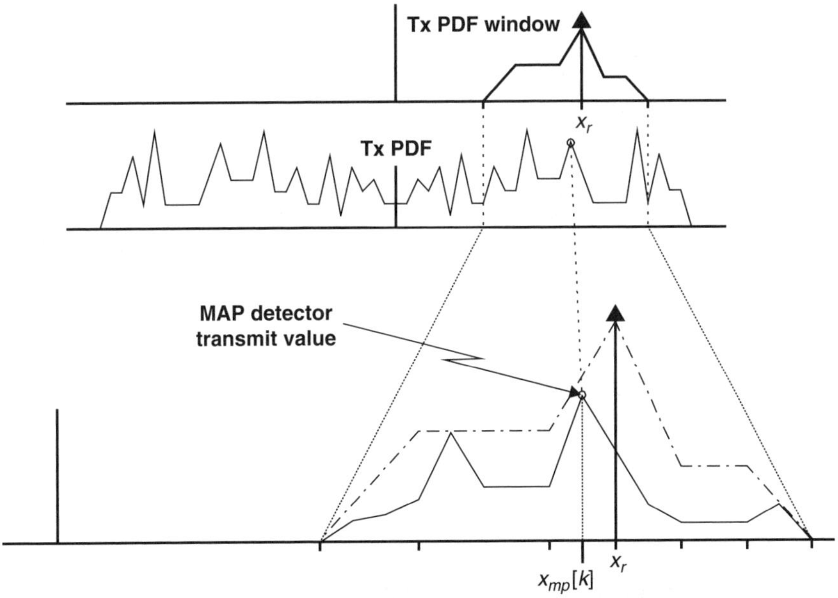

Figure 3.23 Illustration of the MAP decision windowing and max value determination.

$$p_n(-x) = \frac{1}{\sqrt{2\pi}\sigma_n} e^{-\left(\frac{-x}{\sqrt{2}\sigma_n}\right)^2} = p_n(x) \tag{3.55}$$

$$\begin{aligned} W_{tx}(x_r[k]) = p_n(x_r[k] - x) &= \frac{1}{\sqrt{2\pi}\sigma_n} e^{-\left(\frac{x_r[k]-x}{\sqrt{2}\sigma_n}\right)^2} \\ &= \frac{1}{\sqrt{2\pi}\sigma_n} e^{-\left(\frac{x-x_r[k]}{\sqrt{2}\sigma_n}\right)^2} \end{aligned} \tag{3.56}$$

$$\begin{aligned} &\max_x \left\{ W_{tx}(x_r[k]) p_t(x_t) \right\} \\ &= \max_x \left\{ \frac{1}{\sqrt{2\pi}\sigma_n} e^{-\left(\frac{x-x_r[k]}{\sqrt{2}\sigma_n}\right)^2} \left[cW_{dc} + \sum_{l=1}^{L} \frac{b_l}{\sqrt{2\pi}\sigma_l} e^{-\left(\frac{x-m_l}{\sqrt{2}\sigma_l}\right)^2} \right] \right\} \end{aligned} \tag{3.57}$$

$$\begin{aligned} \max_x \left\{ W_{tx}(x_r[k]) p_t(x_t) \right\} = \max_x \Bigg\{ \frac{1}{\sqrt{2\pi}\sigma_n} \Bigg[& cW_{dc}(x) e^{-\left(\frac{x-x_r[k]}{\sqrt{2}\sigma_n}\right)^2} \\ & + \sum_{l=1}^{L} \frac{b_l}{\sqrt{2\pi}\sigma_l} e^{-\left(\frac{\sigma_n^2(x-m_l)^2 + \sigma_l^2(x-x_r[k])^2}{2\sigma_n^2\sigma_l^2}\right)} \Bigg] \Bigg\} \end{aligned} \tag{3.58}$$

$$x_{mp}[k] = x\left(\max_x \{W_{tx}(x_r[k]) p_t(x_t)\}\right) \tag{3.59}$$

The calculations in these equations take place across the data bins that were set up in Section 3.2.2, which covered the range [−5.12, 5.12] with 4096 bins. Each new iterate produces a new received value $x_r[k]$. The noise power (or SNR) is estimated in the receiver and will also have a new value on each iterate. Every received value will have a minimum computational load of 4096 subtractions (exponent), 4096 divisions (exponent), 4096 squaring calculations (exponent), and 4096 exponentiation calculations to calculate the noise PDF, followed by a 4096 length dot product (point-by-point multiplication) of the PDF waveforms prior to performing the $\max_x$ operation to find the MAP decision. This operation count assumes that the transmit PDF was calculated and stored at receiver initialization, and accounts for the fact that the lead term $\frac{1}{\sqrt{2\pi}\sigma_n}$ contributes nothing to the $\max_x$ operation and may be ignored. A simplified MAP decision calculation, as described next, is of great benefit.

3.3.1.1.3 Simplified MAP Transmit Value Calculation

Recognizing that a squaring operation can be programmed as a multiplication, the exponentiation calculations are the major component of the computational load. The series representation of this process is

$$e^{-x^2} = 1 - x^2 + \frac{x^4}{2!} - \frac{x^6}{3!} + \frac{x^8}{4!} - \cdots \tag{3.60}$$

It can be seen that the exponentiation is an expensive calculation, and its elimination would greatly enhance the computational efficiency of the MAP calculation.

Dropping the lead term from Equation (3.57) results in the following MAP decision equation

$$\max_x \left\{ W_{tx}(x_r[k])p_t(x_t) \right\} = \max_x \left\{ e^{-\left(\frac{x-x_r[k]}{\sqrt{2}\tilde{\sigma}_n[k]}\right)^2} \left[cW_{dc}(x) + \sum_{l=1}^{L} \frac{b_l}{\sqrt{2\pi}\sigma_l} e^{-\left(\frac{x-m_l}{\sqrt{2}\sigma_l}\right)^2} \right] \right\} \tag{3.61}$$

where $\tilde{\sigma}_n[k]$ = the estimated noise power on the kth iterate. The natural logarithm is a monotonically increasing function of its argument, so the maximization of the logarithm maximizes the argument.

$$\max_x \left\{ \ln\left[W_{tx}(x_r[k])p_t(x_t) \right] \right\} = \max_x \left[\ln \left\{ e^{-\left(\frac{x-x_r[k]}{\sqrt{2}\tilde{\sigma}_n[k]}\right)^2} \left[cW_{dc}(x) + \sum_{l=1}^{L} \frac{b_l}{\sqrt{2\pi}\sigma_l} e^{-\left(\frac{x-m_l}{\sqrt{2}\sigma_l}\right)^2} \right] \right\} \right] \tag{3.62}$$

$$\max_x \left\{ \ln\left[W_{tx}(x_r[k])p_t(x_t) \right] \right\} = \max_x \left\{ -\left(\frac{x - x_r[k]}{\sqrt{2(s/\tilde{s}_{nr})[k]}} \right)^2 + \ln\left[cW_{dc}(x) + \sum_{l=1}^{L} \frac{b_l}{\sqrt{2\pi}\sigma_l} e^{-\left(\frac{x-m_l}{\sqrt{2}\sigma_l}\right)^2} \right] \right\} \tag{3.63}$$

where $\tilde{s}_{nr}[k]$ = the estimated SNR on the kth iterate.

Equation (3.63) has eliminated the recurring 4096 exponentiation operations and the 4096 length dot product of the PDF waveforms for each received iterate. Its natural logarithm function is applied only to the transmit PDF, which is calculated once and stored at the receiver initialization. Programming the squaring operation as a multiply, the MAP decision now has 4096 subtractions, 4096 divisions, 4096 multiplies, and 4096 additions for every received value.

The regions of the transmit PDF that are close to zero are of practical concern in the use of the logarithm function at receiver initialization.

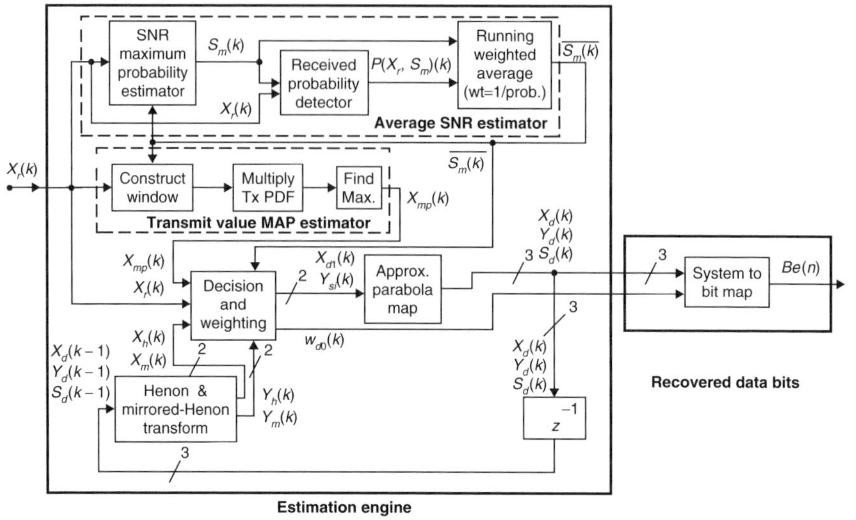

Figure 3.24 Receiver block diagram with detail for the MAP detector and SNR estimator.

The addition of a very small value (i.e., 10^{-100}) prior to taking the logarithm avoids a negative infinity result without altering the calculation or the PDF properties. The receiver block diagram showing the main processing steps for the transmit MAP detector is illustrated in Figure 3.24.

3.3.1.2 *Average SNR Estimator*

The equations of Section 3.3.1.1 and Section 3.3.1.3 clearly show that the MAP estimator requires knowledge of the noise power (or SNR). The SNR is also used to find receiver probabilities needed in the weighted averaging of the three estimates into an initial decision and, again, as the means for discounting iterate decisions that are too close to zero in the bit decision process. The receiver requires a method for estimating the noise power contained in a chaotic signal contaminated with random noise.

3.3.1.2.1 SNR Maximum Probability Estimator

The average SNR estimator is seen in Figure 3.24 to consist of an SNR maximum probability estimator followed by a running average calculation. The received PDF changes continuously as the noise content added by the channel changes. No noise results in the received PDF equaling the transmit PDF, although a noise power larger than the signal power will cause the received PDF to closely resemble the noise PDF.

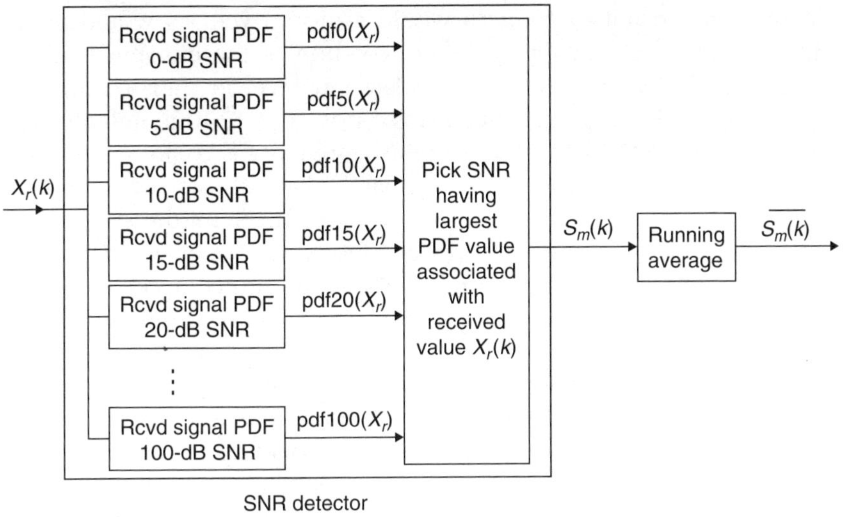

Figure 3.25 An SNR estimator based on a look-up table.

The instantaneous SNR for a given received value can be defined as the one for which a received PDF trace has the greatest probability of occurrence out of the PDF traces for all SNR values. Because the channel is lossless, the noise power is set up in the channel from knowledge of the signal power and the SNR specified to the simulation by the user. The noise power was held constant for the duration of a data run.

3.3.1.2.2 SNR Look-Up Table

A straightforward implementation of instantaneous SNR determination is a look-up table to find the maximum probability SNR for a received iterate, as shown in Figure 3.25. For this method, the received signal PDFs are stored in memory. The received value is passed into the look-up table, and the corresponding PDF value is returned. A decision block chooses the SNR with the greatest PDF value at the current received iterate value. The individual iterate decisions are smoothed through a running average calculation.

The look-up table approach is very general and can be used for any empirically generated received PDF traces and/or transmit PDFs. Empirical received PDF traces are the most general because they would be exact regardless of the transmit PDF and noise PDF for a given point in time, but are difficult to obtain because of the number of iterates required. Thus, they do not lend themselves well to changing channel conditions, where a different SNR granularity may be desired.

Empirical transmit PDF traces would perform well, as would models of them. The received PDF is the convolution of the transmit PDF with the noise PDF, which could be calculated at receiver initialization for a given noise model and predefined noise power or SNR granularity. They could be recalculated later if a different granularity is desired or if a new noise model is deemed more appropriate. The FFT-based fast convolution method works well for this method. Alternatively, if the system lends itself to description by closed-form equations, the received PDF can be calculated directly from the transmit PDF and noise PDF models via these equations. These can then be stored in the table for any desired granularity and changed at will.

In spite of its generality, this approach has some potential problems that should be recognized. The predetermined granularity could limit the precision of the average SNR estimate by providing instantaneous estimates that have too great a sample-to-sample variance to be adequately smoothed by the running average calculation, or by jumping over features such as peaks and valleys that would be captured by a continuous estimation process. The basic form of the largest-value selection block will return the globally most probable SNR value when a local solution may be more appropriate, although this concern could possibly be addressed algorithmically. This look-up table approach is also memory intensive. As an example, consider that PDF traces with 4096 bins and 1-dB granularity from 0 to 100 dB would store 400,000 numbers and occupy over 3 MB of storage space at 8 bytes/number to implement this small piece of the overall receiver functionality.

A continuous estimate algorithm based on closed-form equations for the received PDF has been developed for this investigation, so this method was never implemented. Its performance has not been evaluated, and its sensitivity to the stored waveform granularity is unknown. However, a look-up table approach remains a feasible choice for systems without a closed-form equation representation resulting in a continuous SNR estimation capability.

3.3.1.2.3 Instantaneous SNR Calculation

An algorithm has been developed for the current design based on the closed-form equations realized for the received PDF. The model chosen for the transmit PDF is based on Gaussian functions and an AWGN channel model is used. The received PDF is the convolution of the transmit PDF with the channel noise PDF, and the convolution of a Gaussian function with another Gaussian function is a Gaussian function. This closed-form representation enabled the development of a computational algorithm for the determination of the instantaneous maximum probability SNR.

Here, the decision was made to use the transmit PDF model that windows only the DC component and not the Gaussian functions (see

Section 3.2.2.4). Although both transmit PDF models result in closed-form expressions for the received PDF, the version that windowed all terms had significantly more complicated expressions involving the evaluation of Q-functions in all terms plus exponential functions in most terms. The version that windowed only the DC component had two Q-functions for the windowed DC term with the transmit PDF Gaussian terms remaining Gaussian in the received PDF. The calculation of Q-functions is computationally more expensive than the evaluation of Gaussian functions, even when a Q-function approximation formula is used. This additional complexity is easily seen by comparing the general Gaussian equation in Equation (3.66) with the integrated Gaussian structure of the Q-function definition Equation (3.70) and the extra terms preceding the Gaussian function in the Q-function approximation Equation (3.73).[6]

A few terms will be collected from previous sections to continue the discussion. As stated in Section 3.2.2.4, the equation for the transmit PDF approximation model is

$$\tilde{p}_t(x) = cW_{dc}(x) + \sum_{l=1}^{L} \frac{b_l}{\sqrt{2\pi}\sigma_l} e^{-\left(\frac{x-m_l}{\sqrt{2}\sigma_l}\right)^2} \tag{3.64}$$

with

$$\begin{aligned} W_{dc}(x) &= U(x - (-H_{\max})) - U(x - H_{\max}) \\ &= U(x - (-1.2854)) - U(x - 1.2854) \end{aligned} \tag{3.65}$$

where

$$U(x - x_0) = \begin{Bmatrix} 0 \Rightarrow x < x_0 \\ 1 \Rightarrow x \geq x_0 \end{Bmatrix} = \text{the unit step function}$$

and

$$H_{\max} = 1.2854$$

is the maximum Henon magnitude for the specified bin structure.
Channel noise PDF is as shown in Equation (3.54)

$$p_n(x) = \frac{1}{\sqrt{2\pi}\sigma_n} e^{-\left(\frac{x}{\sqrt{2}\sigma_n}\right)^2} \tag{3.66}$$

3.3.1.2.4 Received PDF Model

The model received PDF is found by the convolution of the transmit PDF model with the noise PDF model

$$\tilde{p}_r(x) = \tilde{p}_t(x) * p_n(x) \tag{3.67}$$

$$\tilde{p}_r(x) = \int_{-\infty}^{\infty} \tilde{p}_t(y) p_n(x-y) dy \tag{3.68}$$

$$\tilde{p}_r(x, \sigma_n) = \int_{-\infty}^{\infty} \left\{ \left(cW_{dc}(y) + \sum_{l=1}^{L} \frac{b_l}{\sqrt{2\pi}\sigma_l} e^{-\left(\frac{y-m_l}{\sqrt{2}\sigma_l}\right)^2} \right) \times \left(\frac{1}{\sqrt{2\pi}\sigma_n} e^{-\left(\frac{x-y}{\sqrt{2}\sigma_n}\right)^2} \right) \right\} dy \tag{3.69}$$

where

$\sigma_n^2 = s/s_{nr}$ = noise variance = noise power
s = signal power
s_{nr} = SNR
$\tilde{p}_r(x, \sigma_n) = \tilde{p}_r(x, s_{nr})$ = a function of both the received value and the noise power

Carrying the SNR variable explicitly because it is the quantity of interest yields

$$\tilde{p}_r(x, s_{nr}) = \alpha(x, s_{nr}) + \beta(x, s_{nr}) \tag{3.70}$$

where

$$\alpha(x, s_{nr}) = \int_{-\infty}^{\infty} \left\{ (cW_{dc}(y)) \left(\frac{1}{\sqrt{2\pi}\sigma_n} e^{-\left(\frac{x-y}{\sqrt{2}\sigma_n}\right)^2} \right) \right\} dy$$

$$\beta(x, s_{nr}) = \int_{-\infty}^{\infty} \left\{ \left(\sum_{l=1}^{L} \frac{b_l}{\sqrt{2\pi}\sigma_l} e^{-\left(\frac{y-m_l}{\sqrt{2}\sigma_l}\right)^2} \right) \left(\frac{1}{\sqrt{2\pi}\sigma_n} e^{-\left(\frac{x-y}{\sqrt{2}\sigma_n}\right)^2} \right) \right\} dy$$

The first term is integrated with a standard change of variables as

$$\alpha(x, s_{nr}) = c\left[Q\left(\frac{-H_{\max} - x}{\sigma_n}\right) - Q\left(\frac{H_{\max} - x}{\sigma_n}\right)\right] \quad \sigma_n^2 = s/s_{nr} \quad (3.71)$$

where (see References 6 and 7) $Q(\gamma) = \frac{1}{\sqrt{2\pi}} \int_{\gamma}^{\infty} e^{-\frac{1}{2}\gamma^2} d\gamma$ (3.72)

An approximation to the Q-function has been found in Reference 6 that reduces the computation time in MATLAB by a factor of two over its built-in error function (ERF), which is related to the Q-function by a scale factor and change of variable. The approximation is

$$Q(x) \cong \left[\frac{1}{\left(1 - \frac{1}{\pi}\right)x + \frac{1}{\pi}\sqrt{x^2 + 2\pi}}\right] \frac{1}{\sqrt{2\pi}} e^{-x^2/2} \quad (3.73)$$

and is accurate to within 1.5% across the entire range of x. Its use in the simulations helped control data run times.

The approximation has 1 to 1.5% accuracy for the argument in the range [0.3 to 1.2]. Closer to zero and greater than 1.2, yield more accurate approximations, asymptotically approaching zero error in both directions. For the windowed DC implementation here, the Q-function is used only in the first term of Equation (3.70), where the positive argument shown in Equation (3.71) results in the 1 to 1.5% error for the range

$$0.3 < \frac{H_{\max} - x}{\sigma_n} < 1.2 \quad H_{\max} - 1.2\sigma_n < x < H_{\max} - 0.3\sigma_n$$

Because $Q(-x) = 1 - Q(x)$, the greatest inaccuracy is also in the range

$$1.2\sigma_n - H_{\max} > x > 0.3\sigma_n - H_{\max}$$

The inaccuracy occurs in like manner for the other edge of the DC window. It is most pronounced just inside and outside the edges of DC window, and has greater contribution with larger channel noise power. The second term in the received PDF closed-form model Equation (3.72) is exact, and reduces the contribution of the Q-function error to the received PDF convolution result.

The second term is more difficult[10]

$$\int_{-\infty}^{\infty} e^{(-p^2x^2 \pm qx)} dx = e^{\frac{q^2}{4p^2}} \frac{\sqrt{\pi}}{p} \quad p > 0 \quad (3.74)$$

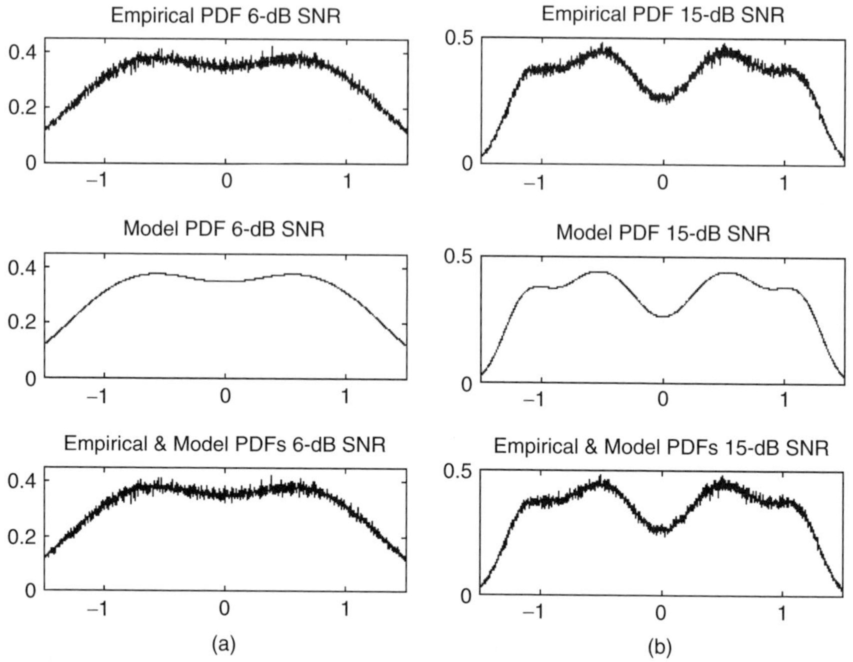

Figure 3.26 Top — Empirical PDF, Middle — $\tilde{p}_r(x, s_{nr}) = \alpha(x, s_{nr}) + \beta(x, s_{nr})$, Bottom — Overlay for (a) 6 dB and (b) 15 dB SNR.

resulting in

$$\beta(x, s_{nr}) = \sum_{l=1}^{L} \frac{b_l}{\sqrt{2\pi}\sigma_{nl}} e^{-\left(\frac{x-m_l}{\sqrt{2}\sigma_{nl}}\right)^2} \tag{3.75}$$

where $\sigma_{nl}^2 = \sigma_n^2 + \sigma_l^2$ and $\sigma_n^2 = s/s_{nr}$.

The convolution of Gaussian functions results in a Gaussian function whose variance (power) is equal to the sum of the constituent variances (powers), which is a well-known result in probability theory.

Figure 3.26 and Figure 3.27 illustrate the effectiveness of these equations in calculating PDF traces for 6- and 15-dB SNR and 30- and 60-dB SNR, respectively. In all cases, the top plot is an empirical data run of 1,000,000 iterates, the middle plot is the convolution result plotted from the equation

$$\tilde{p}_r(x, s_{nr}) = \alpha(x, s_{nr}) + \beta(x, s_{nr}) \tag{3.76}$$

and the bottom plot is the superposition of the two. This equation models the empirical data very well.

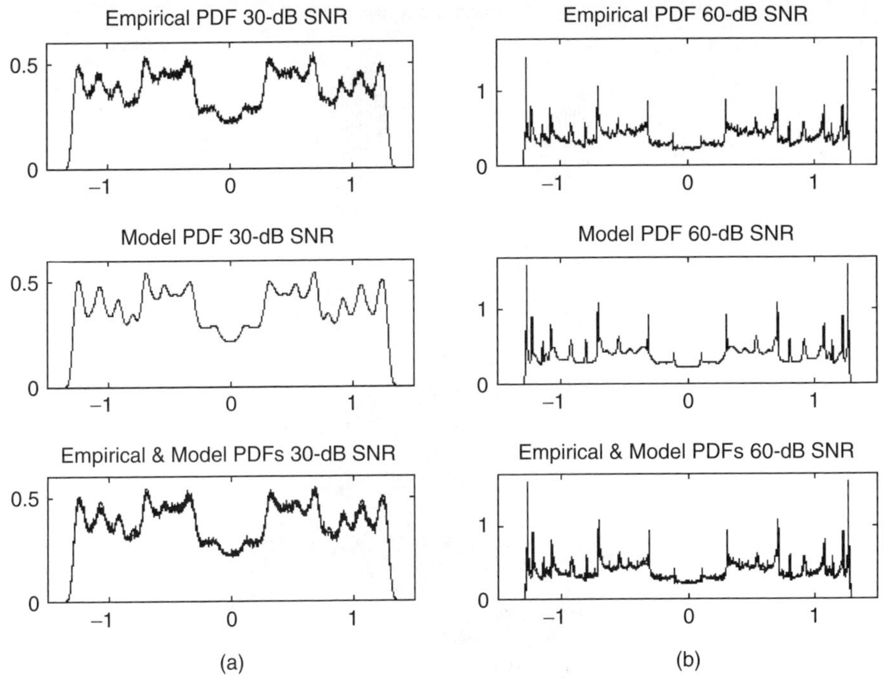

Figure 3.27 Top — Empirical PDF, Middle — $\tilde{p}_r(x, s_{nr}) = \alpha(x, s_{nr}) + \beta(x, s_{nr})$, Bottom — Overlay for (a) 30 dB and (b) 60 dB SNR.

3.3.1.2.5 Determination of Maximum Probability SNR

The basics of this method draw on the observation that the received probability $\tilde{p}_r(x, s_{nr})$ is a function of two variables, x and s_{nr}. It is not a joint probability function because only the function of x is a PDF with unity area (slices of constant s_{nr}), although the function of s_{nr} has unbounded area (slices of constant x). Figure 3.26 and Figure 3.27 in the previous section showed the x dependency, which are PDFs.

Plots of the two-dimensional function are shown in Figure 3.28 and Figure 3.29, with slightly different angles of observation to highlight the variations in the functions with s_{nr} and enable good views of the peaks and valleys of the surface topology. These peaks and valleys became an important feature in the determination of the average SNR estimate. Figure 3.30 illustrates examples of the PDF dependency on SNR for various values of the received iterate $x_r[k]$.

These shapes are continuous and vary smoothly with SNR, which opened the possibility of an algorithmic method of solving for the maximum probability point to determine the corresponding SNR as the desired instantaneous value. The preferred method will be fast and respond only

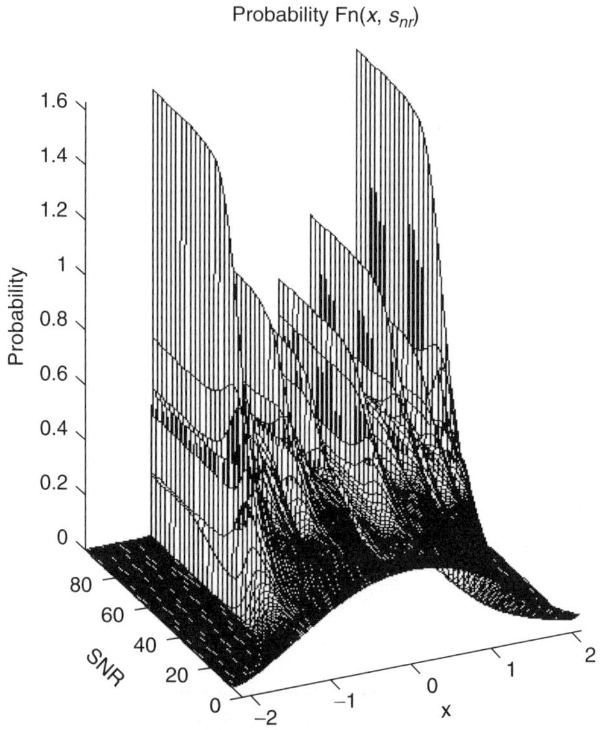

Figure 3.28 Plot of the two-dimensional probability function $\tilde{p}_r\,(x, s_{nr})$ at about 30° rotation and 20° elevation.

to maxima. In addition, the question of choosing local or global solutions arises in situations with more than one peak, as well as traces with two identical peaks that must be addressed.

The Newton–Raphson iterative root-approximation optimization technique was targeted for the algorithm because it converges quadratically to a solution, although all other possibilities such as bisection, secant, and regula-falsi converged linearly or slightly better than linearly.[5] The Newton–Raphson technique finds the zeros of a function and converges to local solutions or diverges depending on its initial estimate and the correction step size.

3.3.1.2.6 The Newton–Raphson Technique

The Newton–Raphson method[5,8] operates on functions expressed as $f(x_{sol}) = 0$, and requires the derivative of the function in its approximation

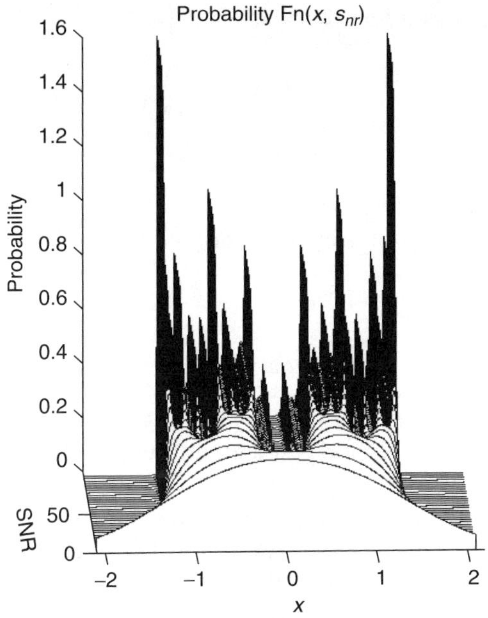

Figure 3.29 Plot of the two-dimensional probability function $\tilde{p}_r(x, s_{nr})$ at about 5° rotation and 10° elevation.

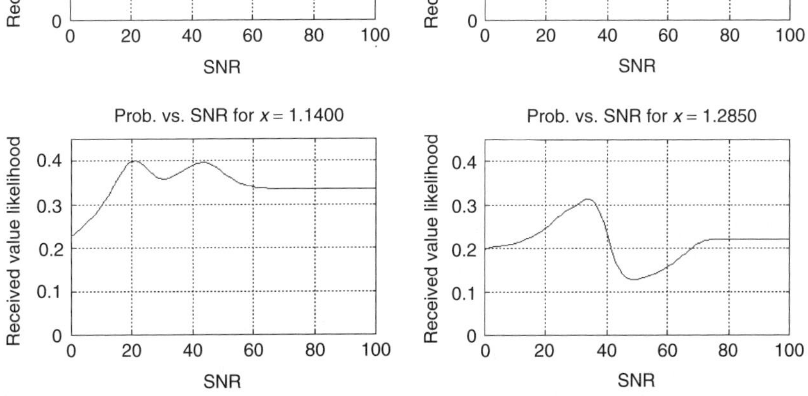

Figure 3.30 Received probability dependence on SNR (slices of constant x).

correction process. Its iterated equation is

$$x_{j+1} = x_j - \frac{f(x_j)}{f'(x_j)} \tag{3.77}$$

where

$j =$ the iterate counter
$x_j =$ the current approximation
$f(x_j) = [f(x)]_{x=x_j}$
$f'(x_j) = \left[\frac{df(x)}{dx}\right]_{x=x_j}$
$x_{j+1} =$ the next approximation

At $x = x_{sol}$, $f(x_{sol}) = 0$ and $x_{j+1} = x_j = x_{sol}$ and the iteration is complete.

Convergence of this method is not guaranteed and generally depends on the initial estimate and the step size that is calculated from the function and its derivative on each iterate. This method will find any zero of the function that is close enough to its initial value and has a well-behaved sequence of correction calculations. Although all references seen to date have similar descriptions of the technique that are not based on chaos theory, any solution point can be viewed from the standard nonlinear dynamical perspective as a fixed point of the iterative process. Correction steps that are large enough to move the next approximation out of the basin of attraction for a given solution will cause the iterative procedure to diverge from that point.

The function with zeros that the Newton–Raphson method solves for can be the derivative of another function, for example

$$f(s_{nr}) = \frac{d[\tilde{p}_r(x, s_{nr})]}{ds_{nr}} \tag{3.78}$$

The solutions $f(s_{nr_{sol}}) = 0$ in this case correspond to the maxima, minima, and points of inflection for the probability function of SNR for a given received value, where its first derivative is zero.

3.3.1.2.7 The Modified Newton–Raphson Technique

The Newton–Raphson technique is adapted to this design via three main modifications to control its functional characteristics. The first modification causes it to find only the maxima of the probability function of SNR based on the fact that the second derivative of a function identifies its concave down (maximum) or concave up (minimum) characteristic. The second change limits the maximum correction per step to avoid jumping over maxima without detecting them. And finally, the computed step size is multiplied by a factor that toggles between two values to prevent a period-two fixed point of the iterative process spanning a peak causing detection failure. Other standard programming techniques such as limiting the number of iterations allowed in each execution, hard programming upper and lower s_{nr} range limits, protecting against a zero denominator, and early loop termination via testing each new approximation for proximity to the current value round out its implementation.

The Newton–Raphson equation expressed as functions of the probability function of SNR is

$$s_{nr_{j+1}} = s_{nr_j} - \frac{\left[\frac{d[\tilde{p}_r(x, s_{nr})]}{ds_{nr}}\right]_{s_{nr_j}}}{\left[\frac{d^2[\tilde{p}_r(x, s_{nr})]}{ds_{nr}^2}\right]_{s_{nr_j}}} \tag{3.79}$$

The derivative in the numerator is carried out below, and the fact that it equals zero at the solution allows a term to be dropped that simplifies the second derivative in the denominator. The correction term actually used is, therefore, not exactly as specified in Equation (3.76), and so the notation will be changed to reflect this.

Let

$$s_{nr_{j+1}} = s_{nr_j} - \frac{f(s_{nr_j})}{f'(s_{nr_j})} \tag{3.80}$$

where

$$f(s_{nr}) \propto \frac{d[\tilde{p}_r(x, s_{nr})]}{ds_{nr}}$$

and

$$f'(s_{nr}) = \frac{df(s_{nr})}{ds_{nr}} \propto \frac{d^2[\tilde{p}_r(x, s_{nr})]}{ds_{nr}^2}$$

Differentiating Equation (3.73),

$$\frac{d[\tilde{p}_r(x, s_{nr})]}{ds_{nr}} = \frac{d[\alpha(x, s_{nr})]}{ds_{nr}} + \frac{d[\beta(x, s_{nr})]}{ds_{nr}} \tag{3.81}$$

From Equation (3.71)

$$\frac{d\alpha(x, s_{nr})}{ds_{nr}} = c\left[\frac{d}{d\sigma_n}\left\{Q\left(\frac{-H_{\max} - x}{\sigma_n}\right)\right\}\frac{d\sigma_n}{ds_{nr}} - \frac{d}{d\sigma_n}\left\{Q\left(\frac{H_{\max} - x}{\sigma_n}\right)\right\}\frac{d\sigma_n}{ds_{nr}}\right] \tag{3.82}$$

The differentiation of the Q-function was done using the Liebniz rule[11]

$$\frac{d}{dy}\left[\int_{s(y)}^{r(y)} p_X(x)dx\right] = p_X(r(y))\frac{dr(y)}{dy} - p_X(s(y))\frac{ds(y)}{dy} \tag{3.83}$$

The result was verified with an equation in Reference 7

$$\frac{d}{dx}erf(x) = \frac{2}{\sqrt{\pi}}e^{-x^2} \tag{3.84}$$

and the relationship

$$Q(x) = \frac{1}{2}\left[1 - erf\left(\frac{x}{\sqrt{2}}\right)\right] \tag{3.85}$$

as

$$\frac{d}{dx}Q(x) = -\frac{1}{\sqrt{2\pi}}e^{-\frac{x^2}{2}} \tag{3.86}$$

So

$$\begin{aligned}\frac{d\alpha(x, s_{nr})}{ds_{nr}} &= \left(\frac{1}{\sqrt{2\pi s_{nr}}}\right)\left(\frac{c}{\sqrt{2}}\right)\\ &\times\left[\left(\frac{H_{\max} - x}{\sqrt{2}\sigma_n}\right)e^{-\left(\frac{H_{\max}-x}{\sqrt{2}\sigma_n}\right)^2} + \left(\frac{H_{\max} + x}{\sqrt{2}\sigma_n}\right)e^{-\left(\frac{H_{\max}+x}{\sqrt{2}\sigma_n}\right)^2}\right]\end{aligned} \tag{3.87}$$

Carrying out the differentiation for the second term of Equation (3.81)

$$\frac{d\beta(x, s_{nr})}{ds_{nr}} = \sum_{l=1}^{L}\left\{\frac{d}{d\sigma_{nl}}\left[\frac{b_l}{\sqrt{2\pi}\sigma_{nl}}e^{-\left(\frac{x-m_l}{\sqrt{2}\sigma_{nl}}\right)^2}\right]\left(\frac{d\sigma_{nl}}{d\sigma_n}\right)\left(\frac{d\sigma_n}{ds_{nr}}\right)\right\} \tag{3.88}$$

$$\begin{aligned}&= \left(\frac{1}{\sqrt{2\pi s_{nr}}}\right)\left(\frac{\sigma_n^2}{2}\right)\\ &\times\sum_{l=1}^{L}\left\{b_l e^{-\left(\frac{x-m_l}{\sqrt{2}\sigma_{nl}}\right)^2}\left(\frac{1}{\sigma_{nl}^3}\right)\left(1 - \left(\frac{x - m_l}{\sigma_{nl}}\right)^2\right)\right\}\end{aligned} \tag{3.89}$$

Both terms of the differentiation for the Newton–Raphson correction term numerator have been found. Setting the numerator equal to zero allows the elimination of one s_{nr} term, which simplifies the next differentiation for the denominator of the correction term. It is possible to form the total correction shown in Equation (3.80) prior to simplification, but the leading term eliminated now because of numerator equality to zero contains the variable s_{nr}, which would enter into the differentiation process and result in more complicated equations. The notation change is shown below.

$$\left[\frac{d[\tilde{p}_r(x, s_{nr})]}{ds_{nr}}\right]_{s_{nr_{sol}}} = \left[\frac{d[\alpha(x, s_{nr})]}{ds_{nr}}\right]_{s_{nr_{sol}}} + \left[\frac{d[\beta(x, s_{nr})]}{ds_{nr}}\right]_{s_{nr_{sol}}} = 0 \tag{3.90}$$

Both of these terms have a common leading factor that can be eliminated.

$$\left[\frac{d[\tilde{p}_r(x,s_{nr})]}{ds_{nr}}\right]_{s_{nr_{sol}}} = \left(\frac{1}{\sqrt{2\pi}s_{nr_{sol}}}\right) f_1\left(s_{nr_{sol}}\right) + \left(\frac{1}{\sqrt{2\pi}s_{nr_{sol}}}\right) f_2\left(s_{nr_{sol}}\right) = 0 \tag{3.91}$$

Resulting in the new notation

$$f(s_{nr}) = f_1(s_{nr}) + f_2(s_{nr}) = \sqrt{2\pi}s_{nr}\left[\frac{d[\tilde{p}_r(x,s_{nr})]}{ds_{nr}}\right] \tag{3.92}$$

where

$$f_1(s_{nr}) = \left(\frac{C}{2}\right)\left[\left(\frac{H_{\max}-x}{\sigma_n}\right)e^{-\frac{1}{2}\left(\frac{H_{\max}-x}{\sigma_n}\right)^2} + \left(\frac{H_{\max}+x}{\sigma_n}\right)e^{-\frac{1}{2}\left(\frac{H_{\max}+x}{\sigma_n}\right)^2}\right] \tag{3.93}$$

$$f_2(s_{nr}) = \left(\frac{\sigma_n^2}{2}\right)\sum_{l=1}^{L}\left\{b_l e^{-\frac{1}{2}\left(\frac{x-m_l}{\sigma_{nl}}\right)^2}\left(\frac{1}{\sigma_{nl}^3}\right)\left(1-\left(\frac{x-m_l}{\sigma_{nl}}\right)^2\right)\right\} \tag{3.94}$$

The denominator in Equation (3.80) becomes

$$f'(s_{nr}) = f_1'(s_{nr}) + f_2'(s_{nr}) \tag{3.95}$$

with

$$\begin{aligned} f_1'(s_{nr}) = {} & \left(\frac{1}{2s_{nr}}\right)\left(\frac{C}{2}\right)\left[\left(\frac{H_{\max}-x}{\sigma_n}\right)e^{-\frac{1}{2}\left(\frac{H_{\max}-x}{\sigma_n}\right)^2}\left\{1-\left(\frac{H_{\max}-x}{\sigma_n}\right)^2\right\}\right. \\ & \left. + \left(\frac{H_{\max}+x}{\sigma_n}\right)e^{-\frac{1}{2}\left(\frac{H_{\max}+x}{\sigma_n}\right)^2}\left\{1-\left(\frac{H_{\max}+x}{\sigma_n}\right)^2\right\}\right] \end{aligned} \tag{3.96}$$

$$\begin{aligned} f_2'(s_{nr}) = {} & \left(\frac{\sigma_n^2}{2s_{nr}}\right)\left(\frac{\sigma_n^2}{2}\right)\sum_{l=1}^{L} b_l e^{-\frac{1}{2}\left(\frac{x-m_l}{\sigma_{nl}}\right)^2}\left(\frac{1}{\sigma_{nl}^3}\right) \\ & \times\left\{\left(\frac{1}{\sigma_{nl}^2}\right)\left[3\left(1-\left(\frac{x-m_l}{\sigma_{nl}}\right)^2\right)^2 - 2\left(\left(\frac{x-m_l}{\sigma_{nl}}\right)^2\right)^2\right]\right. \\ & \left. -\frac{2}{\sigma_n^2}\left(1-\left(\frac{x-m_l}{\sigma_{nl}}\right)^2\right)\right\} \end{aligned} \tag{3.97}$$

The equations are now fully defined, and the algorithmic modifications can be made. Equation (3.98) is repeated below as the starting point, and depicts the Newton–Raphson method with attractive fixed points at all maxima, minima, and points of inflection for the probability function $\tilde{p}_r(x, s_{nr})$.

$$s_{nr_{j+1}} = s_{nr_j} - \frac{f(s_{nr_j})}{f'(s_{nr_j})} \tag{3.98}$$

The minima have a concave-up structure with a positive second derivative of $\tilde{p}_r(x, s_{nr})$, although the maxima have a concave-down structure and a negative second derivative. The same holds true for the corresponding portions of a point of inflection in that the side leading up into the point is concave down and the side leading up away from the point is concave up. A simple test prior to the execution of Equation (3.98) turns the minima into repelling fixed points, although maintaining the maxima as attractive fixed points.

If

$$\begin{aligned} &f'(s_{nr_j}) > 0 && \text{\% Concave up — minima} \\ &f'(s_{nr_j}) = -f'(s_{nr_j}) && \text{\% Negates correction direction} \\ & && \text{\% therefore pushes away from minima} \end{aligned} \tag{3.99}$$

end

$$s_{nr_{j+1}} = s_{nr_j} - \frac{f\left(s_{nr_j}\right)}{f'\left(s_{nr_j}\right)}$$

It was found that the maximum probability algorithm performed better when working with SNR values expressed as dB rather than ratios. Although the typical calculation method is to work with power ratios and convert results to dB, the traces of Figure 3.30 have an intuitively good appearance similar to polynomial curves on a dB-SNR scale, so the various considerations are expressed in dB for this situation.

A set of maximum-step criteria were established to avoid jumping over maxima without detecting them. A sliding maximum step was used because the areas of these traces that showed little or no activity could tolerate larger steps than those areas with a lot of peaks and valleys, and maximized the efficiency of the algorithm at locating the peaks. The steps and their regions of definition are shown in Figure 3.31.

The maximum step size is multiplied at each pass by a variable that toggles between the values of 1 and $\sqrt[5]{1/2} = 0.8706$, which is an empirically determined value. The choice to use roots of ½ as candidates for a suitable step size modification was one of convenience because the various roots yielded conveniently spaced possibilities. A number of other selection methods would have worked as well. The dual-step size technique avoided the situation of a period two fixed point causing the algorithm to perpetually span the desired maximum point without ever finding it within the SNR range of a given maximum step size. Care was required at the

SNR values	Max step size
0–10 dB	5 dB
10–20 dB	5 + 1.5 dB
20–30 dB	5 + 2(1.5) dB
30–40 dB	5 + 3(1.5) dB
etc.	etc.

Figure 3.31 Sliding maximum step sizes.

crossover points of the sliding maximum steps that were used, where the wrong choice of toggle multiplier could map the larger range exactly into the smaller and cause algorithm failure due to a period two fixed point.

The detection rails were set at 0- and 100-dB SNR, which were the expected minimum SNR of performance and essentially no noise, respectively. Next approximation values that exceeded these limits were adjusted, so instantaneous SNR values are always within this range. These were chosen prior to completion of the receiver design and the subsequent evaluation of its performance, and especially the 0 dB minimum SNR may be a good subject for further investigation.

At the completion of the Newton–Raphson computation, the instantaneous maximum probability SNR estimate for iterate k has been found.

$$\hat{s}_{nr}[k] = s_{nr_{sol}} = \left[s_{nr_{j+1}}\right]_{\{|s_{nr_{j+1}} - s_{nr_j}| < \varepsilon\}} \tag{3.100}$$

where ε = desired accuracy.

3.3.1.2.8 SNR Running Weighted Average

Combining the instantaneous SNR estimates on every received iterate into an average value that accurately tracked the true SNR was rather involved and required some investigation. A straight average calculation is undesirable because it retains memory of all iterates from the start of its operation and is slow to respond to changing channel conditions. Other possibilities include a standard running average, a decaying running average where the new values count heavily and older values are discounted as a function of their age, and a running weighted average whose weights are derived from some criteria.

The standard running average method did not track the true SNR well because of the peaks and valleys in the two-dimensional received PDF seen in Figure 3.28 and Figure 3.29. The average value would move into a prominent peak of the two-dimensional probability function and get locked in. The time-discounted weighting technique did nothing to address the two-dimensional PDF topology. The situation was resolved by a running weighted average calculation whose weights are derived from probability computations so the estimated SNR tracks the true SNR closely.

The current received iterate and the corresponding instantaneous SNR estimate are processed to find their probability of occurrence. As in Equation (3.26) and Equation (3.27), the bin width w will be ignored because it is the same for all bins in the PDF model and contributes nothing to the decision process. Only the bin values of the PDF functions will be used in the weighted averaging process and will be loosely referred to as the probability.

For each received iterate, the SNR is estimated, and the probability of occurrence is found for the (received value, SNR estimate) pair. These values correspond to a point in the two-dimensional probability function shown in Figure 3.28 and Figure 3.29. Both the received value and the SNR estimate will change for every iterate, so the instantaneous probability moves with respect to both axes on every calculation.

Initially the weights were calculated as these probabilities, but it was found that this served to reinforce the entrenchment of the calculated average SNR in the peaks of the two-dimensional probability function. It became apparent that the peaks needed to be smoothed out by this averaging method in order for the estimated SNR to track the true SNR, so the averaging weights are now defined as the inverse instantaneous probability of the (received value, SNR estimate) pair

$$P_i[k] = \tilde{p}_r\left(x[k], \hat{s}_{nr}[k]\right) \tag{3.101}$$

$$w_{av}[k] = \frac{1}{P_i}[k] \tag{3.102}$$

Two considerations for calculating the weighting factors are avoiding division by zero and limiting the maximum weight. Observed probabilities of zero within the resolution of MATLAB did happen, and the probability result is now restricted to a minimum value of 10^{-10} to avoid division by zero. The maximum-weight calculation limits the amount of emphasis put on the very low probability values so that their inverses do not swamp the average, and was derived from the observed average probability and the averaging window length.

$$\bar{P}_i = 0.4444$$

$$w_{\max} = \frac{1}{\bar{P}_i}\frac{W_{ra}}{f_W} \tag{3.103}$$

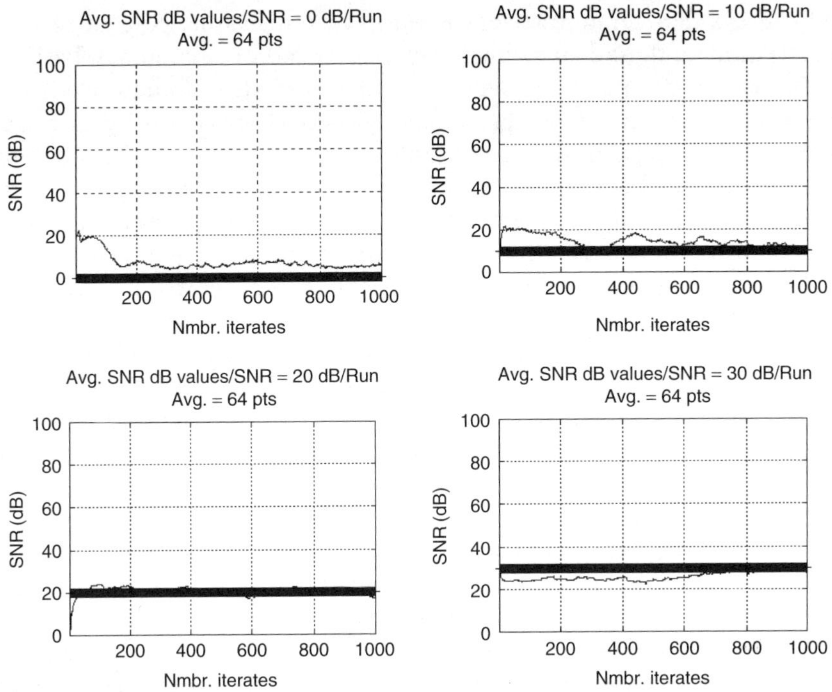

Figure 3.32 SNR estimator performance for 64-iterate window: Thick straight lines are the true SNR; curve is the receiver estimated SNR.

where W_{ra} = the running average window length and $f_W = 3$, an empirical factor.

Data was taken for various combinations of variables, and the empirical factor value of three was found to yield good results. Averaging window lengths of 8 and 16 are responsive to changes in channel conditions but have a grassy appearance with some areas of significant discrepancy between the estimated SNR and the true SNR. Windows of 32 and 64 iterates showed smoother traces with better estimate tracking. This simulation was done with a 64-iterate window, and examples of its performance are shown in Figure 3.32.

It can be seen that the running weighted-average SNR locks onto a close estimate of the true SNR within approximately 200 iterates for a 64-iterate window. The traces have some variation because they are estimates and not power measurements, however they maintain a close proximity to the true value once lock is attained. In simulations of the receiver the receiver decisions are ignored until the 200-iterate transient period has passed, after which time data bit recovery commences. In the event that the channel SNR varies, the SNR estimator will track the changes in the same manner as the receiver initialization process did with an unknown SNR. Sharp variations will cause a 200-iterate transient period to occur, as happens at receiver initialization, and gradual changes will be tracked on a continual basis.

The issues of local vs. global instantaneous SNR maximum likelihood detection and algorithm initialization are related. Global detection is desired in the absence of knowledge at estimate algorithm inception, but the average value drives the instantaneous peak detection once the averaging window has filled up with globally detected maximum likelihood instantaneous SNR values. This feedback of the current average SNR estimate is shown in Figure 3.24 and serves as the initial approximation in the Newton–Raphson root approximation optimization procedure. The average SNR feedback enables the determination of local maxima in the vicinity of the current average SNR estimate after its value has been established via global values, which tends to drive the estimated SNR toward an accurate assessment of the true SNR. Observations of the probability vs. SNR functions led to the empirical initial SNR value of 23 dB for the instantaneous SNR detector Newton–Raphson algorithm during the averaging window initialization. This value is located about midway between the lobes of the observed double-peak functions and had good mobility toward either higher or lower values for the various observed traces.

It should be pointed out that the techniques presented in this SNR estimator do not require differentiable equations for the probability function to get successful implementations. Standard derivative definition equations could be used in the Newton–Raphson iteration technique where two close values are used with the corresponding function responses, and the second derivative could be found in the same manner.

$$\frac{df(x)}{dx} \cong \frac{f(x_2) - f(x_1)}{x_2 - x_1} \quad \text{and} \quad \frac{d^2f(x)}{dx^2} \cong \frac{[(df(x_2)/dx) - (df(x_1)/dx)]}{x_2 - x_1} \tag{3.104}$$

It is necessary in the averaging process to find the probability of occurrence for the (received iterate, maximum probability SNR) pair, which could be done with empirical data rather than via closed-form equations. The difficulty with this approach is that the probability response as a function of SNR is not known and cannot be verified to be smooth and well behaved for proper Newton–Raphson iteration performance. One of the other root approximation optimization techniques with linear convergence may work well instead, but that development is beyond the scope of this investigation.

3.3.1.3 Decision Feedback

The third current estimate is derived by passing the previous estimate through the Henon and mirrored-Henon equations. The previous estimate will be exact in the absence of noise, and the result of the chaotic map operating on it will generate an exact estimate of the next received iterate.

This procedure is one of two that helps track the effects of chaotic dynamics in the decision process.

The previous estimate will have some error in the presence of noise and will be somewhat inaccurate in predicting the next nominal received iterate via the Henon and mirrored-Henon transforms. The fact that it is combined with two other estimates to make an initial decision and then mapped onto the chaotic attractor to produce a final decision enables each original estimate to impact the final decision prior to the imposition of the chaotic dynamics. This process helps to track the chaotic motion and eliminate the truly random motion in the received data stream.

The chaotic mapping provides two sets of inputs to the decision and weighting block, the Henon and the mirrored-Henon x- and y-values, $(x_h[k], y_h[k])$ and $(x_m[k], y_m[k])$. Whichever x-value is closer to the other two estimates is chosen, its corresponding Henon or mirrored-Henon system becomes the instantaneous system $s_i[k]$, and its concomitant y-value is passed with the initial decision to the parabola mapping operation.

The algorithm is outlined below. Some of the nomenclature is introduced here because it is the result of the receiver final decision, which is discussed in a later section. Referencing Equation (3.1) to Equation (3.6),

If $s_d[k-1] = h$ $\Leftarrow$ Henon previous decision

$$\left.\begin{aligned} x_h[k] &= -1.4x_d^2[k-1] + y_d[k-1] + 1 \\ y_h[k] &= 0.3x_d[k-1] \\ x_m[k] &= 1.4(-x_d)^2[k-1] + (-y_d)[k-1] - 1 \\ y_m[k] &= 0.3(-x_d)[k-1] \end{aligned}\right\} \quad (3.105)$$

else $\Leftarrow$ mirrored-Henon previous decision

$$\left.\begin{aligned} x_h[k] &= -1.4(-x_d)^2[k-1] + (-y_d)[k-1] + 1 \\ y_h[k] &= 0.3(-x_d)[k-1] \\ x_m[k] &= 1.4x_d^2[k-1] + y_d[k-1] - 1 \\ y_m[k] &= 0.3x_d[k-1] \end{aligned}\right\} \quad (3.106)$$

end

where $s_d[k]$ = receiver final system decision for the kth iterate — Henon or mirrored-Henon
$x_d[k]$ = receiver final x-value decision for the kth iterate
$y_d[k]$ = receiver final y-value decision for the kth iterate

3.3.2 The Decision and Weighting Block

The decision and weighting block performs a probabilistic combination of all three initial estimates. It decides which feedback estimate the MAP and received-value estimates are closest to, generates the weights used in a weighted-averaged computation, performs the weighted average, and calculates another factor to discount the initial decisions whose received values were in the vicinity of zero. It passes the initial decision x-value and the corresponding y-value to the approximation-parabola block and the zero proximity discount value to the system-to-bit map block, where the data bits are recovered.

The first function of the decision and weighting block is to calculate the one-dimensional Euclidean distances from the received value and the MAP decision to each of the Henon and mirrored-Henon transforms of the previous estimate. The initial current-iterate system choice corresponds to the minimum distance, and the feedback x- and y-values are taken from the chosen system.

$$d_h[k] = (x_r[k] - x_h[k])^2 + (x_{mp}[k] - x_h[k])^2 \tag{3.107}$$

$$d_m[k] = (x_r[k] - x_m[k])^2 + (x_{mp}[k] - x_m[k])^2 \tag{3.108}$$

$$s_i[k] = system(\min\{d_h[k], d_m[k]\}) \quad \text{Henon or mirrored-Henon system} \tag{3.109}$$

$$x_{si}[k] = x(\min\{d_h[k], d_m[k]\}) \quad \text{corresponding } x\text{-value} \tag{3.110}$$

$$y_{si}[k] = y(\min\{d_h[k], d_m[k]\}) \quad \text{corresponding } y\text{-value} \tag{3.111}$$

The second operation executed by the decision and weighting block is weighting factor generation based on probability calculations. Once again the bin width w is ignored as contributing nothing to the calculation, as all the required information is contained in the PDF bin value that is referred to loosely as the probability.

Each of the constituent estimates $x_r[k]$, $x_{mp}[k]$, $x_{si}[k]$ is associated with two independent probabilities. The first of these is drawn from the transmit PDF window used in determining the MAP transmission value, which is the channel noise PDF reversed in x and centered on the received value $x_r[k]$ and serves to quantify the channel effects. The second one is based on the transmitted or received PDF as appropriate and characterizes the probabilistic effects of the chaotic dynamics. These contributions are independent, and are multiplied together to produce each weighting factor, which change for every received iterate.

All estimates are of the transmitted value $x_t[k]$ and pertain to the situation that $x_r[k]$ was received on the kth iterate in the presence of noise. They are therefore associated with the noise probability window constructed on the observation $x_r[k]$ and shown in Figure 3.22 that outlines the range of possible transmit values that could result in the reception of $x_r[k]$. The difference between $x_r[k]$ and each estimate is the noise that caused $x_r[k]$ to be observed if the estimate were accurate, and the bin value of the PDF is the probability that that amount of noise occurred. The first probability contribution to each weight is expressed as

$$x_r[k] - x_*[k] = n_*[k] \tag{3.112}$$

where $* = r$, mp, or si.

$$w_{r1}[k] = p_n(x_r[k] - x_r[k]) = p_n(n_r[k]) = p_n(0) \tag{3.113}$$

$$w_{mp1}[k] = p_n(x_r[k] - x_{mp}[k]) = p_n(n_{mp}[k]) \tag{3.114}$$

$$w_{si1}[k] = p_n(x_r[k] - x_{si}[k]) = p_n(n_{si}[k]) \tag{3.115}$$

The second contribution to the weighting function comes from the transmit or received PDF. The known fact in this consideration is that $x_r[k]$ was received on iterate k. The received PDF is appropriate for the x_r estimate because the total weight reflects the probability that x_r was received and no noise occurred. The viewpoint that x_r is an estimate of the transmit value and could use the transmit PDF is not adopted here because x_r can have values outside the valid transmit range, which results in a weighting factor of zero. This weighting would nullify the estimate and its ability to contribute to the receiver decision. However, it is desirable to have the decision be based on three estimates in all cases due to the fact that no estimate is known to be correct, but $x_r[k]$ is known to be observed.

The transmit PDF is used for x_{mp} so the total weight quantifies the probability that x_r was received and the maximum probability transmitted value x_{mp} was actually transmitted. It uses the transmit PDF because this estimate is the statistically most probable transmitted value, calculated by windowing the transmit PDF, and the corresponding transmit probability is quantified in the transmit PDF bin value. It is impossible for this weighting factor to be zero because the method of finding x_{mp} inherently places it within the valid transmit range.

The feedback of the transformed instantaneous system decision uses the received PDF because the previous decision on which it is based is an estimate of the previously transmitted value. Although the previous decision had been mapped onto the attractor geometrical model, it is possible for the small discrepancies between the attractor and its model to result in a transformed iterate value slightly outside the valid region of the transmit sequence and the subsequent nullification of the estimate. This situation is avoided by use of the received PDF to generate the weighting factor second contribution.

The equations for the second set of weighting factors are

$$w_{r2}[k] = p_r(x_r[k]) \tag{3.116}$$

$$w_{mp2}[k] = p_t(x_{mp}[k]) \tag{3.117}$$

$$w_{si2}[k] = p_r(x_{si}[k]) \tag{3.118}$$

The total weights are

$$w_* = w_{*1}w_{*2} \quad * = r, mp, \text{ or } si \tag{3.119}$$

and the receiver initial decision is the weighted average of the estimates

$$x_{d1}[k] = \frac{x_r[k]w_r[k] + x_{mp}[k]w_{mp}[k] + x_{si}[k]w_{si}[k]}{w_r[k] + w_{mp}[k] + w_{si}[k]} \tag{3.120}$$

This value of $x_{d1}[k]$ is sent with $y_{si}[k]$ to the approximation parabola block.

The final action of the decision and weighting block is to calculate a discount factor in order to reduce the relative weighting of those initial decisions whose received iterate values were close to zero. Since the impact of noise is greater on small values than large, this discount process will improve the quality and accuracy of a bit decision that is formulated by combining individual iterate decisions. The term "close to" is quantified via the estimated SNR on every iterate. Figure 3.33 depicts the transmit window, which is the noise PDF reversed in x and centered on the received value. The value of this function at $x = 0$ quantifies the relative probabilistic distance of the current-received iterate from zero assuming a unimodal noise PDF, which is valid for the AWGN channel in this investigation. Received iterate values close to zero will have large probabilities and those far from zero will have small probabilities as measured by the current-average SNR estimate. The inverse of this probability is the desired discount value.

$$w_{d0}[k] = \frac{1}{p_n(x_r[k] - 0)} \tag{3.121}$$

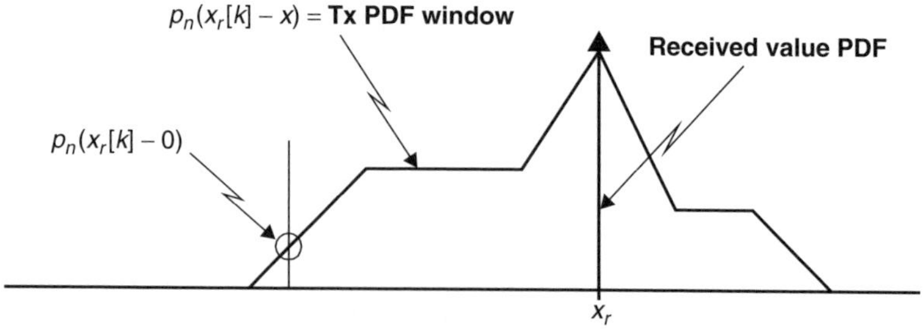

Figure 3.33 Probability of reversed noise PDF centered on received iterate and evaluated at 0.

This discount weight is sent to the "system to bit map" block for data bit recovery.

3.3.3 The Approximation Parabola Map

The approximation parabola map block accepts the initial decision $x_{d1}[k]$ and the corresponding instantaneous system y-value $y_{si}[k]$. It creates a window around $y_{si}[k]$ of 10% of the total y-range and generates the appropriate sections of the Henon and mirrored-Henon parabola models as shown below.

$$y_{\max} = 0.3862 \quad \text{absolute} \pm \text{limit of } y\text{-range on Henon attractor} \tag{3.122}$$

$$y_{tot} = 2y_{\max} \tag{3.123}$$

$$w_y = y_{tot/1000} \; y\text{-bin width — not necessary to be factor of 2} \tag{3.124}$$

$$\begin{aligned}[y_r] = [&y_{si}[k] - 500w_y, \ldots, y_{si}[k] - w_y, y_{si}[k], y_{si}[k] \\ &+ w_y, \ldots, y_{si}[k] + 500w_y]\end{aligned} \tag{3.125}$$

$$\begin{aligned}
[\tilde{x}_{h1}] &= -15.15([y_r] - 0.014)^2 + 0.710 \ldots -0.2846 < [y_r] < 0.3659 \\
[\tilde{x}_{h2}] &= -14.75([y_r] - 0.018)^2 + 0.800 \ldots -0.1911 < [y_r] < 0.2215 \\
[\tilde{x}_{h3}] &= -16.60([y_r] + 0.014)^2 + 1.245 \ldots -0.2154 < [y_r] < 0.2378 \\
[\tilde{x}_{h4}] &= -16.05([y_r] + 0.013)^2 + 1.280 \ldots -0.3862 < [y_r] < 0.3780 \\
[\tilde{x}_{m1}] &= 15.15([y_r] + 0.014)^2 - 0.710 \ldots -0.3659 < [y_r] < 0.2846 \\
[\tilde{x}_{m2}] &= 14.75([y_r] + 0.018)^2 - 0.800 \ldots -0.2215 < [y_r] < 0.1911 \\
[\tilde{x}_{m3}] &= 16.60([y_r] - 0.014)^2 - 1.245 \ldots -0.2378 < [y_r] < 0.2154 \\
[\tilde{x}_{m4}] &= 16.05([y_r] - 0.013)^2 - 1.280 \ldots -0.3780 < [y_r] < 0.3862
\end{aligned} \tag{3.126}$$

where [•] denotes a vector of numbers except when $[k]$, as in $y_{si}[k]$, is the iterate index.

The next step is the calculation of the two-dimensional Euclidean distance between its input point $(x_{d1}[k], y_{si}[k])$ and every point on all parabolas as shown in the next equation. It chooses the point on a parabola corresponding to the minimum distance, whose x-value becomes the final receiver decision for the current iterate and whose attractor is the final receiver system decision for the chaotic synchronization process. Recall from Figure 3.5 that the overlapping Henon and mirrored-Henon attractors cross in the general vicinity of $x = 0$, causing iterate decision errors at their points of crossing because the Henon and mirrored-Henon parabolas are so close that the correct attractor cannot be reliably chosen in the mapping process. This situation is exacerbated by noise, and the weighting factor that discounts decisions for received values close to zero based on the estimated SNR serves to minimize this effect.

$$
\begin{aligned}
[d_{b1}] &= [(x_d[k] - [\tilde{x}_{b1}])^2 + (y_{si}[k] - [y_r])^2] \\
[d_{b2}] &= [(x_d[k] - [\tilde{x}_{b2}])^2 + (y_{si}[k] - [y_r])^2] \\
[d_{b3}] &= [(x_d[k] - [\tilde{x}_{b3}])^2 + (y_{si}[k] - [y_r])^2] \\
[d_{b4}] &= [(x_d[k] - [\tilde{x}_{b4}])^2 + (y_{si}[k] - [y_r])^2] \\
[d_{m1}] &= [(x_d[k] - [\tilde{x}_{m1}])^2 + (y_{si}[k] - [y_r])^2] \qquad (3.127) \\
[d_{m2}] &= [(x_d[k] - [\tilde{x}_{m2}])^2 + (y_{si}[k] - [y_r])^2] \\
[d_{m3}] &= [(x_d[k] - [\tilde{x}_{m3}])^2 + (y_{si}[k] - [y_r])^2] \\
[d_{m4}] &= [(x_d[k] - [\tilde{x}_{m4}])^2 + (y_{si}[k] - [y_r])^2]
\end{aligned}
$$

$$d_{tot} = [[d_{b1}][d_{b2}][d_{b3}][d_{b4}][d_{m1}][d_{m2}][d_{m3}][d_{m4}]] \qquad (3.128)$$

$$d_{\min} = \min\{[d_{tot}]\} \qquad (3.129)$$

The receiver final iterate decision values are

$$x_d[k] = \tilde{x}(\min\{[d_{tot}]\}) \quad x\text{-value} \qquad (3.130)$$

$$y_d[k] = y_{si}(\min\{[d_{tot}]\}) \quad y\text{-value} \qquad (3.131)$$

$$s_d[k] = s_i(\min\{[d_{tot}]\}) \quad \text{chaotic system} \qquad (3.132)$$

These decisions are the approximate (x, y) attractor components of the current iterate, along with the system to which the point belongs. The current x-decision is combined with other iterate x-decisions to determine a bit decision in the system-to-bit map block, and the system decision is the synchronization output of the receiver estimation engine. The methods of the prior art have not been invoked to achieve decisions or synchronization. Special invertible chaotic systems are not required, nor is the division into subsystems that was needed for the Pecora–Carroll method. The only requirement on the chaotic system for this technique is the ability to model it with a geometric approximation.

3.3.4 The Receiver Initial Decision State

The failure of the Henon fixed point as an advantageous initial condition prompted the search for a better choice. The superimposed Henon and mirrored-Henon attractors representing the transmit sequence were observed to have a point at $(0, 0)$ that is not on either attractor but central to both. This point was tried as a possible initialization value for the receiver initial decision. The first iterate of the chaotic maps moved this common point to $(1, 0)$ and $(-1, 0)$ for the Henon and mirrored-Henon systems, respectively. These points quickly moved to their attractors with further iterations of the chaotic maps.

These chaotic dynamics of the initial condition are coupled with the decision process of combining three transmitted value estimates into a single initial decision via probabilistic computations. This step is followed by the mapping of the initial decision onto the attractors to further incorporate the chaotic dynamics into the final decision process. The net effect of combining the probability and chaotic dynamics in this manner tended to pull the initial receiver condition in the direction of the received iterates such that synchronization was quickly achieved, well within the 200-iterate transient period of the SNR estimator.

3.3.5 The System-to-Bit MAP

The system-to-bit MAP Block accepts as inputs the vector of iterate system decisions

$$[s_d] = [s_d[nN+1], s_d[nN+2], s_d[nN+3], \ldots, s_d[nN+N]] \tag{3.133}$$

and the corresponding vector of iterate discount weights

$$\begin{aligned}[w_{d0}] = [&w_{d0}[nN+1], w_{d0}[nN+2], w_{d0}[nN+3], \ldots, \\ &w_{d0}[nN+N]]\end{aligned} \tag{3.134}$$

where N is the number of iterates per bit
n is the index of bits already processed, and
iterate index k as been replaced by the bit-referenced iterate index as shown.

For example, during the second bit period $n = 1$, the iterate index range is $[N+1, 2N]$. On the final segment of a bit duration, this algorithm creates a binary antipodal vector from the system decisions and performs a weighted average to find the bit decision.

Each Henon decision $\Rightarrow$ 1 and each mirrored-Henon decision $\Rightarrow -1$. Then the vector of iterate system decisions [Equation (3.133)] in binary antipodal format is

$$[s_{ba}] = [s_{ba}[nN+1], s_{ba}[nN+2], s_{ba}[nN+3], \ldots, s_{ba}[nN+N]] \tag{3.135}$$

A weighted average performed on this vector using the discount weights for received value proximity to zero is evaluated for positive or negative quality, where a positive value is a Henon bit decision and a negative value is a mirrored-Henon bit decision.

$$B_v[n+1] = \frac{\sum_{k=nN+1}^{nN+N} s_{ba}[k] w_{d0}[k]}{\sum_{k=nN+1}^{nN+N} w_{d0}[k]} \tag{3.136}$$

$$B_s[n+1] = \begin{cases} Henon \Rightarrow B_v[n+1] \geq 0 \\ Mirrored\_Henon \Rightarrow B_v[n+1] < 0 \end{cases} \tag{3.137}$$

3.3.6 Receiver Performance

The chaotic communication system using a receiver estimation engine has been modeled and simulated in MATLAB. Data runs were taken for SNR values between 0 dB and 26 dB, and for the number of iterates per bit of 2, 3, 4, 8, 16, 32, and 64. The results are shown in Figure 3.34.

The receiver recovers data bits with as little as two iterates per bit, and performance is improved with increasing iterates per bit. As a general observation, each doubling of the number of iterates per bit sees about a 2-dB SNR advantage for roughly equivalent BERs. The performance for all traces converges to a common point at 0 dB that approaches ½, indicating that the receiver is just guessing at the low SNR values. It is possible that this is due in part to the 0 dB rail imposed on the Newton–Raphson

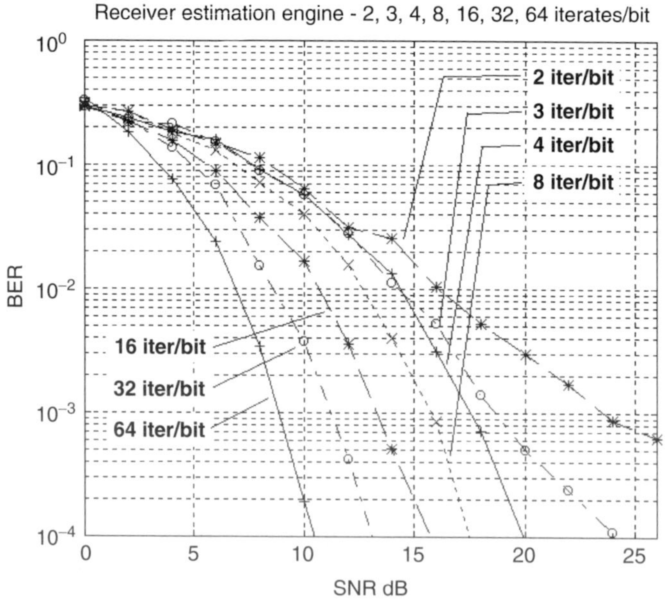

Figure 3.34 Plot of data points for the receiver estimation engine.

technique in the SNR estimator, but it seems more likely that the channel capacity limits are being reached.[9]

The data looks quite good, especially considering that the chaotic synchronization occurs on the fly in combination with the bit recovery calculations without prior lock required, and approximation models are used for the chaotic attractor and transmission PDF traces. This algorithm appears to be a generally viable communications technique requiring only that the chaotic attractor have a geometrical model to facilitate mapping the probability-based estimates onto the attractor to account for the chaotic dynamics in the receiver decision and tracking calculations.

References

1. Fleming-Dahl, A., Doctoral thesis, Georgia Institute of Technology, Atlanta, GA, 1998.
2. Peitgen, O.H., Jurgens, H., and Saupe, D., *Chaos and Fractals — New Frontiers of Science*, Springer-Verlag, New York, 1992.
3. Lee, E.J. and Messerschmitt, D.G., *Digital Communication*, 2nd ed., Kluwer Academic Publishers, Boston, 1994.
4. van Trees, H.L., *Detection, Estimation, and Modulation Theory, Part I*, John Wiley & Sons, New York, 1968.
5. Burington, R.S., *Handbook of Mathematical Tables and Formulas*, 5th ed., McGraw-Hill, New York, 1973.
6. Leon-Garcia, A., *Probability and Random Processes for Electrical Engineering*, 2nd ed., Addison-Wesley, Reading, MA, 1994.
7. Andrews, L.C., *Special Functions for Engineers and Applied Mathematicians*, Macmillan, New York, 1985.
8. van Valkenburg, M.E., Ed., *Reference Data for Engineers: Radio, Electronics, Computer, and Communication*, 8th ed., Prentice Hall Computer Publishing, Carmel, IN, 1993.
9. Drake, D.F., *Information's Role in the Estimation of Chaotic Signals*, Doctoral Thesis, Georgia Institute of Technology, Atlanta, GA, 1998.
10. Gradshteyn, I.S., Ryzhik, I.M., Alan Jeffery, Ed., *Table of Integrals, Series, and Products*, 5th ed., Academic Press, Inc., New York, 1994.
11. Sam Shanmugam, K., *Digital and Analog Communication Systems*, John Wiley & Sons, Inc., New York, 1979.

4

Chaos-Based Modulation and Demodulation Techniques

Francis C.M. Lau and Chi K. Tse

CONTENTS

4.1 Overview of Communication Systems

Communication systems are designed primarily for conveying useful information from one place to another. The two parties involved in the communications can be close to each other but they can also be very far apart. Communication systems can be used for broadcasting, in which one transmitter sends the same message signal to multiple receivers. Typical examples include amplitude modulated (AM) and frequency modulated (FM) radios, and TV broadcasting. Other commonly known communication systems include cellular mobile communications and wireless local area networks.

4.1.1 A Simple Communication System

The simplest communication system consists of only four components, namely an information source, a modulator, a channel and a demodulator, as represented by the functional block diagram shown in Figure 4.1.[35,36,47]

- **Information source:** An information source creates information to be sent via the modulator. The information can take either analog or digital forms. For example, speech and music signals broadcast by AM and FM radios are of analog nature whereas computer data are of digital form.

  In modern communication systems, analog signals can also be converted into digital form for storage, transmission, etc. This involves a process called *source coding*, in which the analog source signal is coded into digital format. Usually, the source coding

Figure 4.1 Block diagram of a simple communication system.

techniques are optimized so as to represent the analog signal by a minimum number of information bits, but without sacrificing too much fidelity. Examples such as speech coding and video coding convert analog speech and video signals, respectively, into digital forms. When source encoders are present on the transmitter side, the corresponding decoders must be installed at the receiver side to convert the recovered digital symbols back to the analog format.

In addition to the source encoder and decoder, a channel encoder and decoder may be incorporated into a digital communication system. A channel encoder adds redundant information to the digital symbols to be sent. Due to the imperfect channel conditions, errors may arise during the demodulation of the digital symbols. Based on the redundant information received, the channel decoder attempts to correct the errors, hence increasing the reliability of the communication link.

- **Modulator:** A modulator converts a given message into an appropriate form that is suitable for transmission through the communication channel. In conventional communication systems, sinusoidal carriers are employed and the message signal can be used to modulate (change) the amplitude, frequency, or phase of a sinusoidal carrier.

  Commonly known analog modulation schemes include AM and FM, which have been used in radio broadcasting for decades. As the names imply, amplitude modulation and frequency modulation change the amplitude and frequency of the carrier such that its instantaneous amplitude or frequency is linearly proportional to the amplitude of the analog message. Figure 4.2 shows an example of the AM and FM signals together with the modulating message.

  For digital modulation schemes, the amplitude of the digital signal modulating the carrier assumes only discrete values. The digital signal, when modulating the amplitude, frequency, or phase of the sinusoidal carrier, forms the basic *amplitude-shift-keying* (ASK), *frequency-shift-keying* (FSK), and *phase-shift-keying* (PSK) modulation schemes. Figure 4.3 shows an example of the ASK, FSK, and PSK signals together with the modulating message. Moreover, to make efficient use of the bandwidth, the modulation process is designed in such a way that the modulated signal occupies a bandwidth comparable to the data rate. Such communication systems are termed *narrowband systems*.
- **Channel:** The channel is the physical medium through which the modulated signal is sent to the demodulator. Communication channels can be broadly classified into *wireless* and *wired* types. Although both the atmosphere and the underwater ocean belong

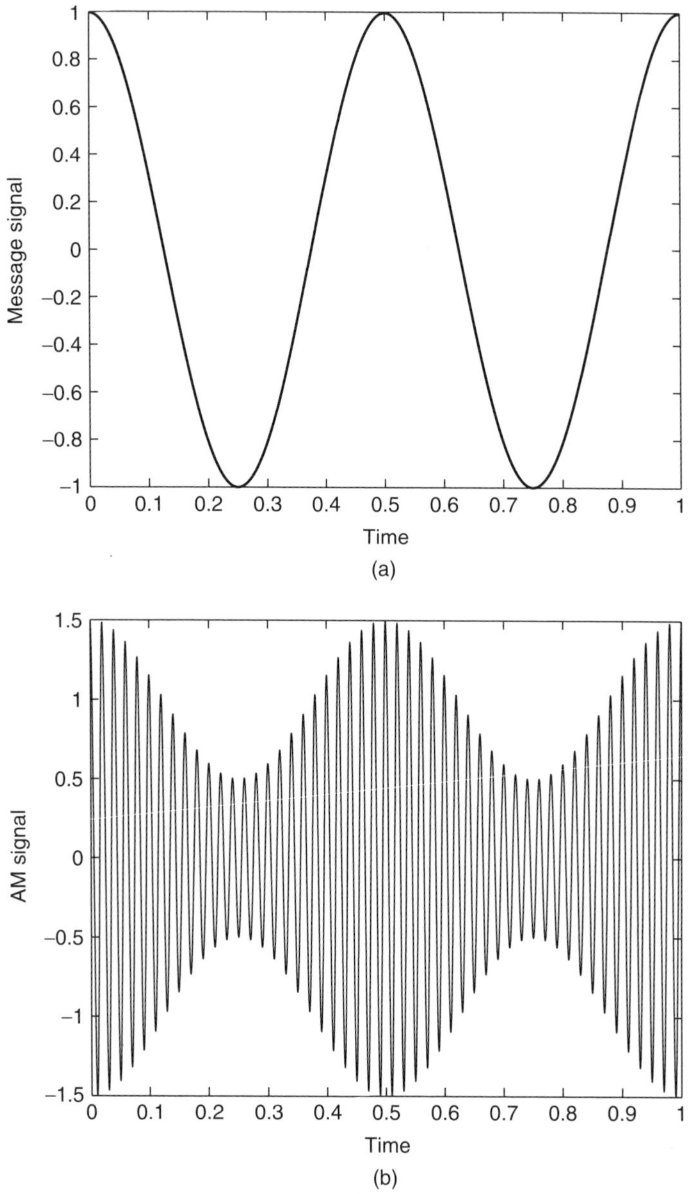

Figure 4.2 Waveforms plotted against time. (a) Message (modulating) signal to be sent; (b) amplitude-modulated signal; (c) frequency-modulated signal.

to the group of wireless channels, the former one is more commonly known and used by the public. Wired channels can take various forms such as copper wires, coaxial cables, and optical fibers. Copper wires and coaxial cables operate up to hundreds of

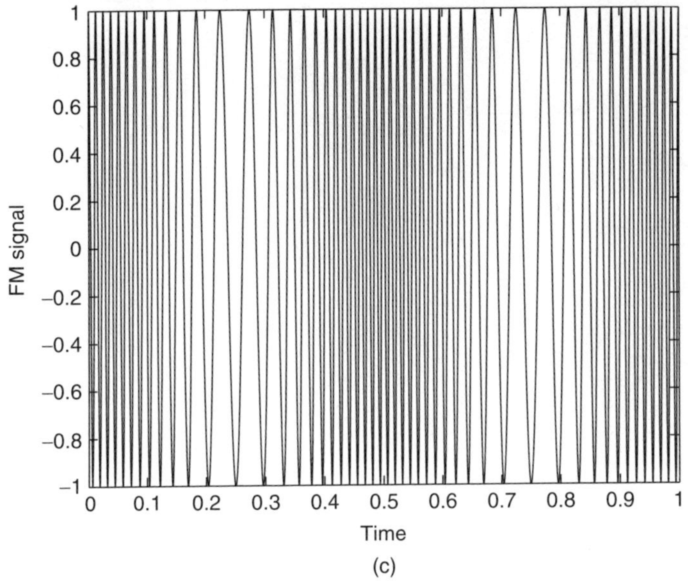

Figure 4.2 Continued.

megahertz (10^8 Hz) whereas optical fibers carry optical signals operating at hundreds of terahertz (10^{14} Hz). In general, the higher the operating frequency, the larger the bandwidth that is available and hence the higher the information transfer rate.

In a practical environment, the channel is far from perfect and the transmitted signal is always distorted in one way or another by the channel. The most popular channel that is studied by researchers and engineers is the *additive white Gaussian noise* (AWGN) channel. The AWGN, originated from the thermal processes in electrical components, is characterized by a normal distribution of its amplitude and a uniform power spectral density. Other types of additive noise include car ignitions made by humans and electrical lightning from nature. These additive noise sources are distinguished by the large noise amplitude and a very short duration. Besides adding noise to the transmitted signal, the channel attenuates the signal strength. Moreover, different frequency components travel at different velocities in a wired medium, causing delay distortion and phase distortion. In a wireless medium, usually there exists multiple transmission paths between the transmitter and the receiver. The amplitude of the resultant signal arriving at the receiver, being the vector sum of all these different components, fluctuates considerably with time. Such cause of distortion on the signal amplitude is usually called *multipath fading*. Also, in the case of a multiple

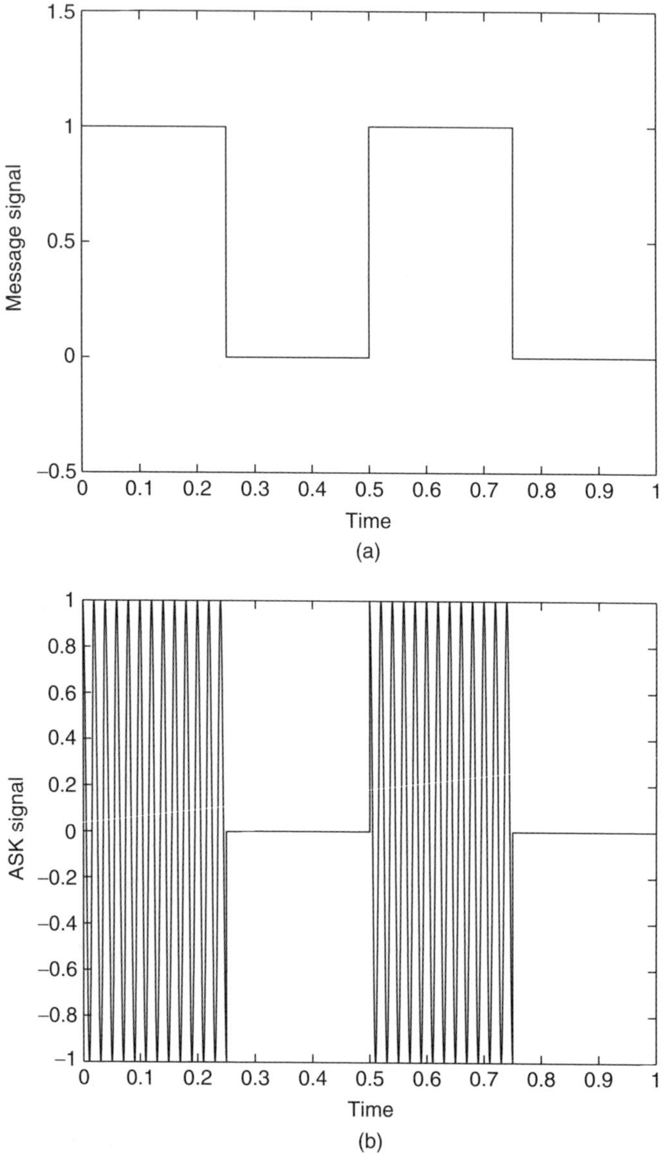

Figure 4.3 Waveforms plotted against time. (a) Digital signal to be sent; (b) amplitude-shift-keying signal; (c) frequency-shift-keying signal; (d) phase-shift-keying signal.

access environment, it is possible that more than one user is using the same channel bandwidth at the same time. The mutual interference among users will thus degrade the signal quality of the users concerned.

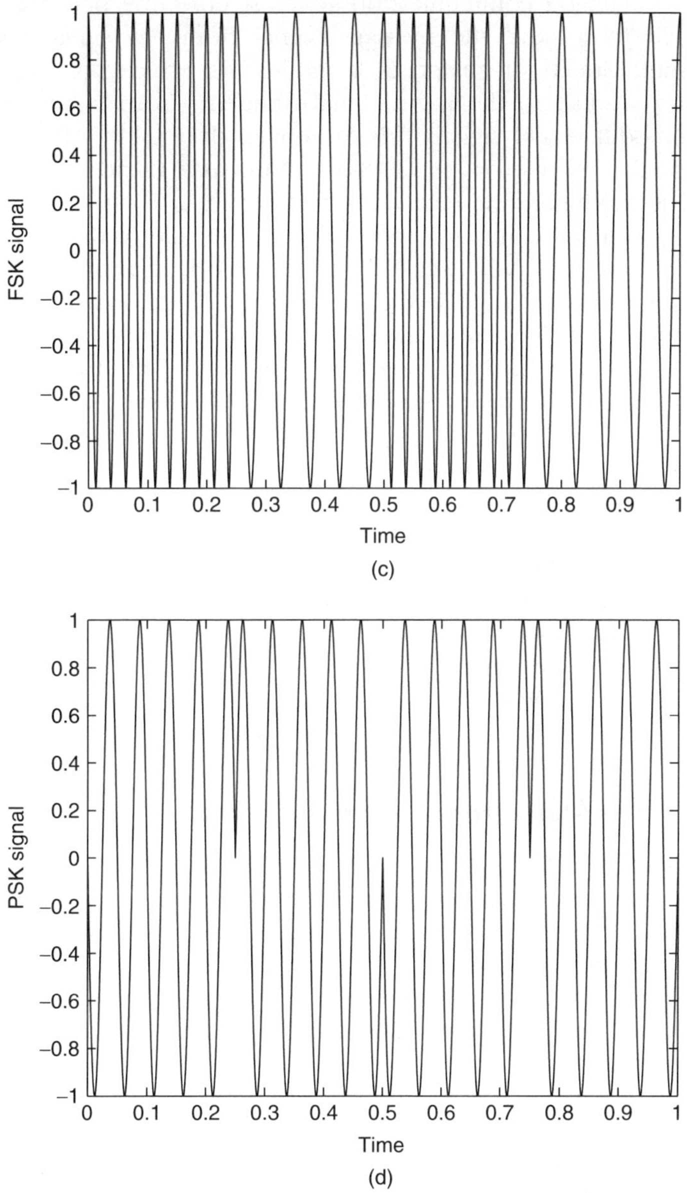

Figure 4.3 Continued.

- **Demodulator:** At the destination, the demodulator attempts to recover the original useful message. Because the transmitted signal has been distorted by various mechanisms in the channel, the recovered message cannot be perfect.

For analog communication systems, envelope detectors can be used to retrieve the message signal from AM signals, whereas tuned circuits or balanced discriminators, which produce outputs proportional to the input signal frequency, can be employed to demodulate FM signals.[36,47] Moreover, the quality of reception is usually described by the *signal-to-noise ratio* (SNR). The higher the SNR, the higher the received signal quality.

For digital communication systems, the role of the demodulator is to determine what symbol has been received within each symbol duration. For ASK, FSK, and PSK signals, correlator-type detectors or energy detectors are demodulators commonly used. For an AWGN channel, the quality of reception of the digital system is usually measured in terms of the *bit-energy-to-noise-power-spectral-density ratio* (E_b/N_0). The bit energy, usually denoted by E_b, represents the average energy transmitted per bit, and $N_0/2$ corresponds to the two-sided power spectral density of the AWGN source. Moreover, the performance of a digital communication system is typically evaluated in terms of the *bit error rate* (BER), which is defined as the probability that a bit will be decoded erroneously. Given two different digital communication systems operating at the same data rate, a common performance comparison is based on their BERs under the same E_b/N_0. Other performance evaluation criteria may include *bandwidth efficiency* and *system complexity*.

4.1.2 Spread-Spectrum Communications

Typical communication systems, such as the aforementioned ASK, FSK, and PSK systems, concentrate their transmission power over a bandwidth that is roughly equal to the data rate. To provide a reasonable BER performance, the signal strength needs to be much stronger than the background noise. In consequence, they are readily detected and picked up by anyone, even by unintended parties. Further, these signals can be decoded easily by simple correlator-type detectors or energy detectors. Hence, such narrowband systems provide little security against eavesdropping.

During the second world war, there was a constant demand from the military forces for a secure means of communications. In a military environment, communications must be protected against two types of attacks, namely, eavesdropping and jamming. Spread-spectrum technologies were then developed to provide a protected method of communications under adverse conditions.[34]

In spread-spectrum communications, the transmitted signal is "spread" over a much larger bandwidth compared with that of a narrowband signal. As depicted in Figure 4.4, when the signal spreads, the average power spectral density becomes much lower and the signal can be concealed in the

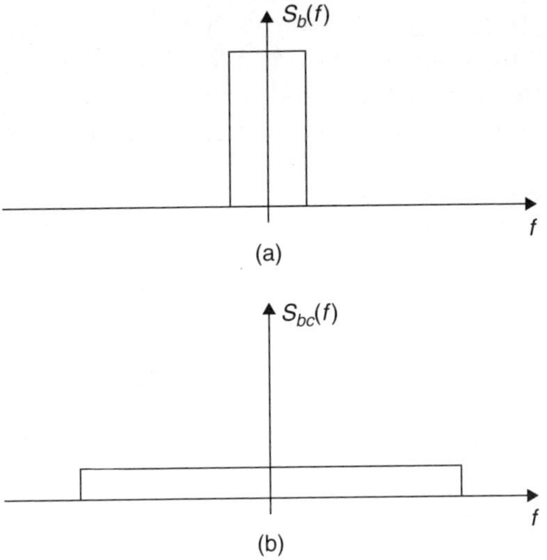

Figure 4.4 Power spectrum of a spread-spectrum signal (a) before spreading; and (b) after spreading.

background noise. Because of this feature, it is not easy to detect the presence of the signal without prior knowledge of the communication system or the use of sophisticated equipment. Furthermore, even if the presence of the signal is detected, without the appropriate decoding information, a large amount of effort will be required to recover the message. For the intended parties, the demodulation process requires the "despreading" of the incoming signal back to a narrowband signal, during which the effect of jammers like sinusoids will be substantially reduced. Thus, even in the presence of jamming signals, the message can be recovered satisfactorily in a spread-spectrum communication system. Other advantages of spread-spectrum technologies include mitigation of multipath effects in a wireless environment, reduction of frequency-planning effort and increase in system capacity in mobile cellular systems.[23]

Today, many commercial applications, e.g., wireless local area networks, mobile cellular communications and positioning systems, also make use of spread-spectrum technologies. So far, spread-spectrum techniques have developed around two main schemes: *direct-sequence* (DS) and *frequency-hopping* (FH). In addition, there are other spectrum spreading techniques in place, e.g., the time-hopping scheme and some hybrid methods that combine DS and hopping methods.[7,34,45]

4.2 Application of Chaos to Communications

Chaotic signals are nonperiodic, random-like signals derived from nonlinear dynamical systems.[1,2,3,6,12,32,38,42] In the continuous-time domain, chaotic signals are generated by ordinary differential equations whereas in the discrete-time domain, they can be produced by iterative maps. Chaotic signals are also characterized by their impulse-like autocorrelation and low cross-correlation properties, as shown in Figure 4.5. Moreover,

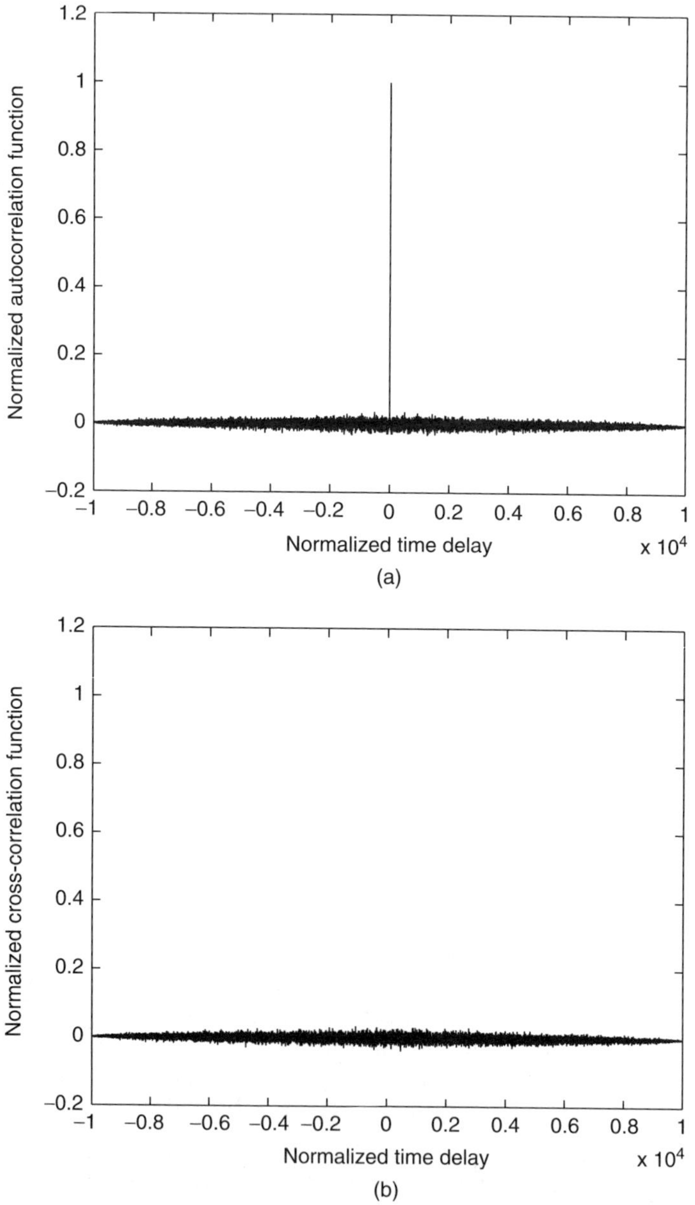

Figure 4.5 Normalized correlation functions of chaotic signals against normalized time delay. (a) Autocorrelation; (b) cross-correlation.

chaotic signals are distinguished by their inherent wideband attribute. A typical frequency spectrum of a chaotic signal is shown in Figure 4.6. Thus, when chaotic signals are used as wideband carriers to convey information, they offer potential advantages that are accomplished by conventional spread-spectrum communication systems, such as difficulty of uninformed detection and mitigation to multipath fading. Furthermore, a large number of chaotic carriers can be produced easily as a consequence of the *sensitive dependence upon initial conditions and parameter variations.*[1] Thus, chaos can provide a low-cost and versatile technique for spread-spectrum communications.

In the past decade, a number of modulation and demodulation schemes based on chaos have been proposed for use in communications. As in conventional communication systems, chaos-based communication systems can be categorized into analog and digital, as distinguished by the type of information being carried. For chaos-based analog communication systems, two main techniques have been proposed — chaotic masking[4,16] and chaotic modulation.[13,14] In chaotic masking, the analog information signal to be sent is added to the noiselike chaotic signal. At the receiver, the chaotic signal is regenerated through a *chaos synchronization* process. Then the analog information signal is recovered by subtracting the reconstructed chaotic signal from the incoming signal. The successful demodulation of the signal is subject to the availability of a robust synchronization method. Moreover, the analog information to be

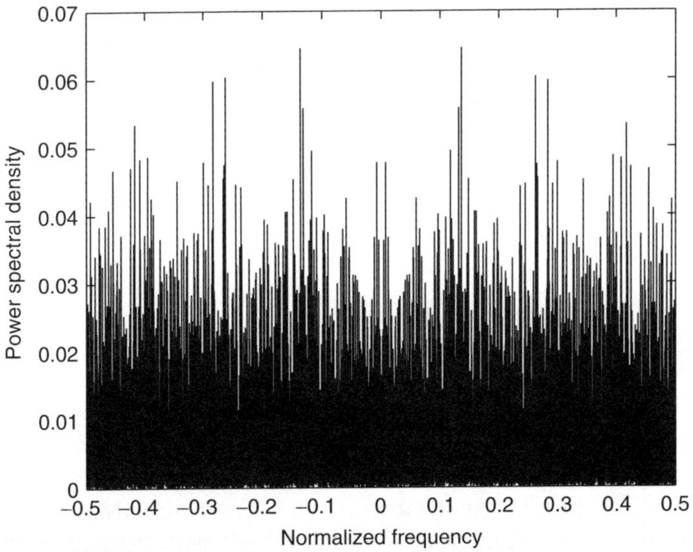

Figure 4.6 Power spectrum of a chaotic signal against normalized frequency.

sent must have a very small amplitude compared with that of the chaotic carrier. In another chaos-based analog communication system, namely chaotic modulation, the information to be sent is used to change a suitably chosen parameter of the chaos generator. The task of the receiver is then to extract the information of the same parameter based on the incoming signal. Synchronization is not required during the demodulation process and three main demodulation techniques have been proposed, namely inversion approach,[8,31] linear demodulation approach using adaptive filters,[28] and nonlinear demodulation approach using the radial-basis function neural network.[9] Although the aforementioned chaos-based analog communication systems can provide reasonable performance under a noiseless environment, they do not possess sufficient noise immunity in a practical condition. However, digital communication systems based on chaos, which is the subject of the next section, are more robust against noise.

4.3 Chaos-Based Digital Communication Systems

In conventional digital communication systems, each symbol to be sent is represented by a piece of sinusoidal signal, which is periodic by nature. In chaos-based digital communication systems, each symbol is now denoted by a segment of chaotic signal, which is aperiodic. Therefore, even if the same symbol is being sent repeatedly, the chaotic signals representing the same symbol are never the same. We can broadly categorize chaos-based digital communication systems into those that require the regeneration of chaotic carriers at the receivers, namely coherent systems, and those that do not have such a requirement, namely noncoherent systems. Although robust chaos synchronization techniques are not yet available, the study of coherent systems is important in that it can provide useful benchmark indicators for performance comparison. On the other hand, noncoherent systems are more practical. In the following, some digital communication schemes based on chaos are presented.

4.3.1 *Chaos Shift Keying*

Chaos shift keying (CSK) was first proposed by Parlitz et al.[33] and Dedieu et al.[5] Figure 4.7 shows the block diagram of a typical CSK digital communication system. The operating principle can be described as follows. Denote the bit duration by T_b. The transmitter consists of two chaos generators f and g, producing signals $\hat{c}(t)$ and $\check{c}(t)$, respectively. During the l bit duration, i.e., $[(l-1)T_b, lT_b]$, if a binary "+1" is to be sent, $\hat{c}(t)$ is transmitted, and if "−1" is to be sent, $\check{c}(t)$ is transmitted. Several demodulation algorithms have been proposed for the CSK system.

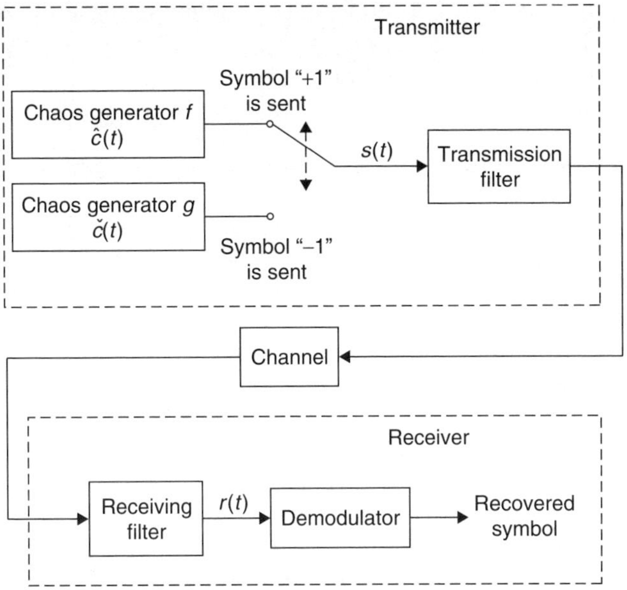

Figure 4.7 CSK digital communication system.

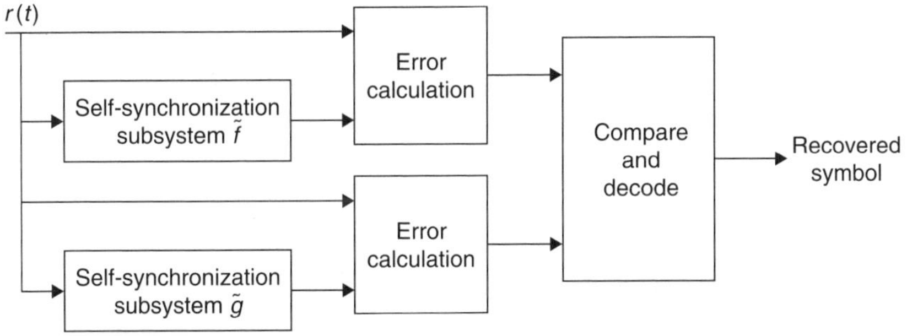

Figure 4.8 Synchronization-error-based CSK demodulator.

4.3.1.1 Coherent Demodulation Based on Synchronization Error

In the original CSK systems studied by Parlitz et al.[33] and Dedieu et al.,[5] demodulation of the received signal is based on the self-synchronizing property of chaotic systems. The structure of such a demodulator is shown in Figure 4.8. Referring to the figure, the received signal is applied to drive two self-synchronization subsystems $\tilde{f}$ and $\tilde{g}$, which are matched to f and g, respectively. Suppose that the filters at the transmitter and receiver are distortionless and the channel is perfect. If the signal $\hat{c}(t)$ has been transmitted, the subsystem $\tilde{f}$ will be synchronized with the incoming signal whereas

$\tilde{g}$ does not, and vice versa. Therefore, by evaluating the difference (error) between the incoming signal and the output of the self-synchronization subsystems, the transmitted symbol can be estimated.

4.3.1.2 Coherent Demodulation Based on Correlation

In communications, *correlation* is a generic process that is used to evaluate the "likeness" between two signals. For the CSK system, if the chaotic carriers $\hat{c}(t)$ and $\check{c}(t)$ can be recovered exactly at the demodulator, the transmitted symbol can be identified by evaluating the correlation of the transmitted signal and the regenerated chaotic carriers followed by a decision maker, as shown in Figure 4.9. In the figure, the two synchronization circuits attempt to recover the two chaotic signals $\hat{c}(t)$ and $\check{c}(t)$ from the incoming corrupted signal $r(t)$. Further, it is assumed that an acquisition time T_S is required for the synchronization blocks to lock to the incoming signal. Correlation of the reproduced chaotic signals and the received signal, implemented by a multiplier and an integrator, is then performed during the remainder of the bit duration. Afterwards, the outputs of the correlators are sampled and compared. If the input to the threshold detector, denoted by $y(lT_b)$, is positive, a "+1" is decoded for the lth symbol. Otherwise, a "−1" is decoded.

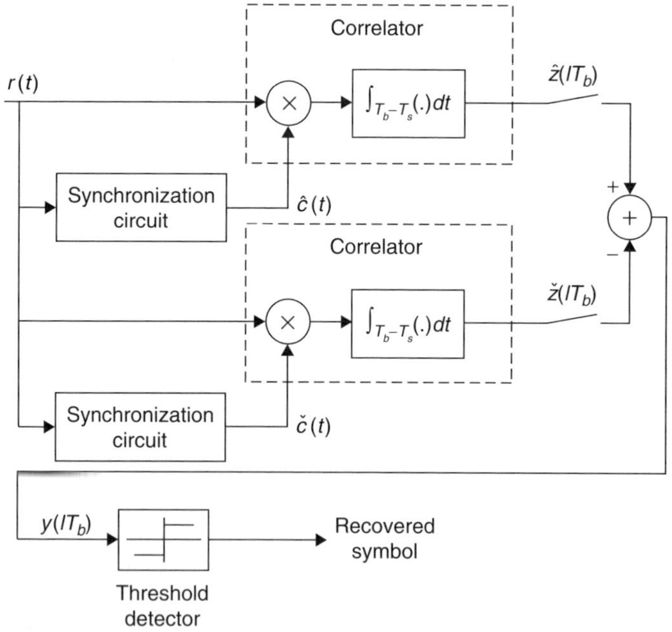

Figure 4.9 Correlator-based coherent CSK demodulator.

Assuming that the filters are distortionless and the synchronization time T_s is negligible compared to the bit period, we plot the histograms of $y(lT_b)$ for different SNRs to gain some insights into the demodulation process. Because of the nonperiodic nature of the chaotic signal, the value of $y(lT_b)$ is bound to vary even for the same symbol. As shown in Figure 4.10, when

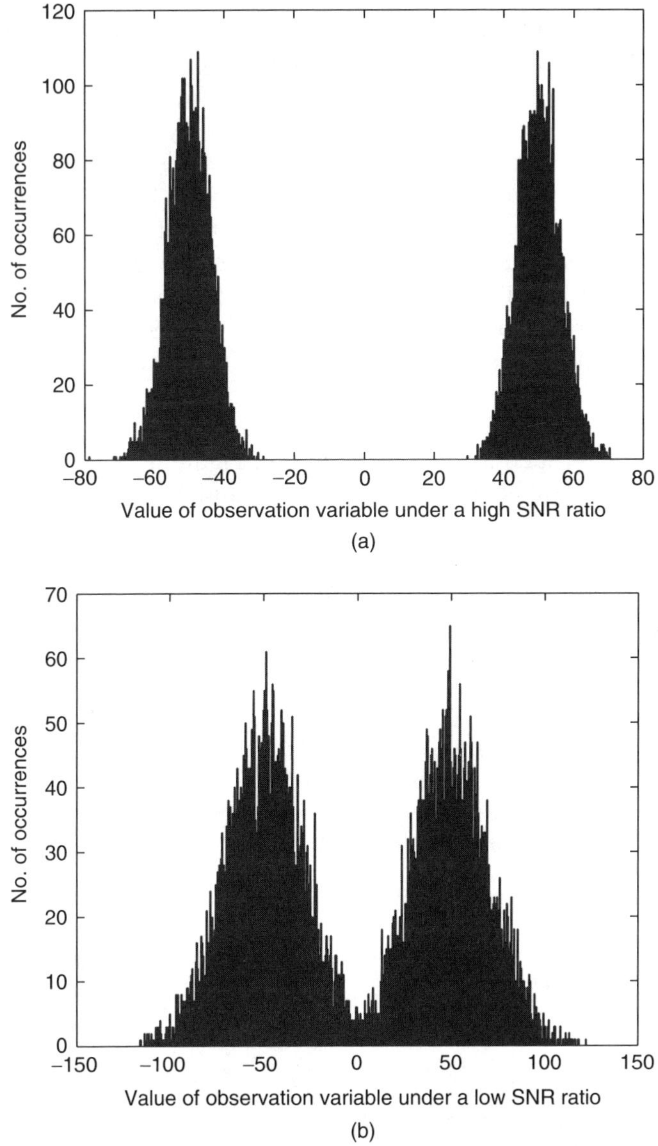

Figure 4.10 Histograms of the observation variable $y(lT_b)$ for a coherent CSK system for (a) high SNR ratio, and (b) low SNR ratio.

the SNR is high, the two distinct regions in the histogram guarantee that the transmitted symbols can be decoded correctly. However, when the SNR is low, the two regions overlap and errors become inevitable.

4.3.1.3 Noncoherent Demodulation Based on Bit-Energy Estimation

In noncoherent CSK demodulation, the chaotic carriers are not recovered at the receiver. Detection has to be done based on some distinguishable property of the transmitted signals. One such property is the bit energy, which can be deliberately made different for different symbols in the modulation process.[21] If a binary symbol "+1" is to be sent during the interval $[(l-1)T_b, lT_b]$, a chaotic signal $\hat{c}(t)$ with average bit energy $\hat{E}_c$ is transmitted, and if "−1" is to be sent, a chaotic signal $\check{c}(t)$ with average bit energy $\check{E}_c$ is transmitted. We may employ two chaos generators to produce chaotic signals with different average bit energies, or alternatively, a single chaos generator can be used to produce two chaotic signals of different bit energies with the use of two amplifiers of different gains, as shown in Figure 4.11.

At the receiving end, the bit energy can be estimated by a square-and-integrate process such as the one shown in Figure 4.12. Assume that only additive noise corrupts the transmitted signal and the noise power is limited

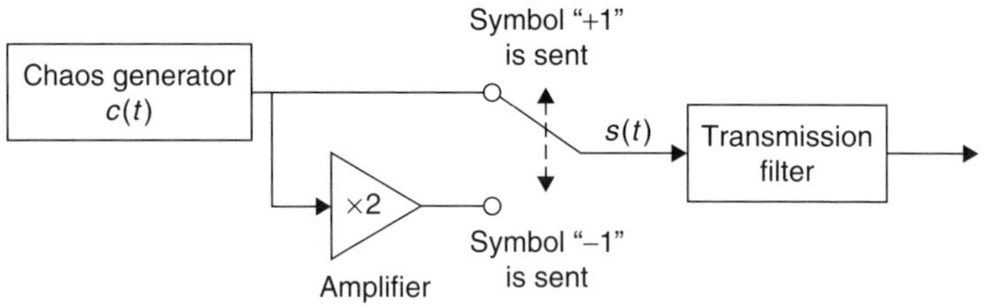

Figure 4.11 CSK transmitter sending different average bit energies for different symbols.

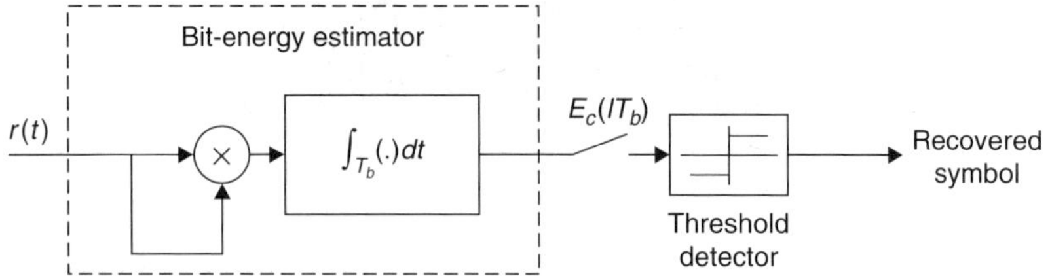

Figure 4.12 CSK receiver based on bit energy estimator.

by the receiving filter. Figure 4.13 plots the histogram of the estimated bit energy $E_c(lT_b)$. In conventional modulation schemes, the bit energy is constant for a given symbol. However, in the CSK system, due to the non-periodic nature of chaotic signals, the bit energy for the same symbol varies

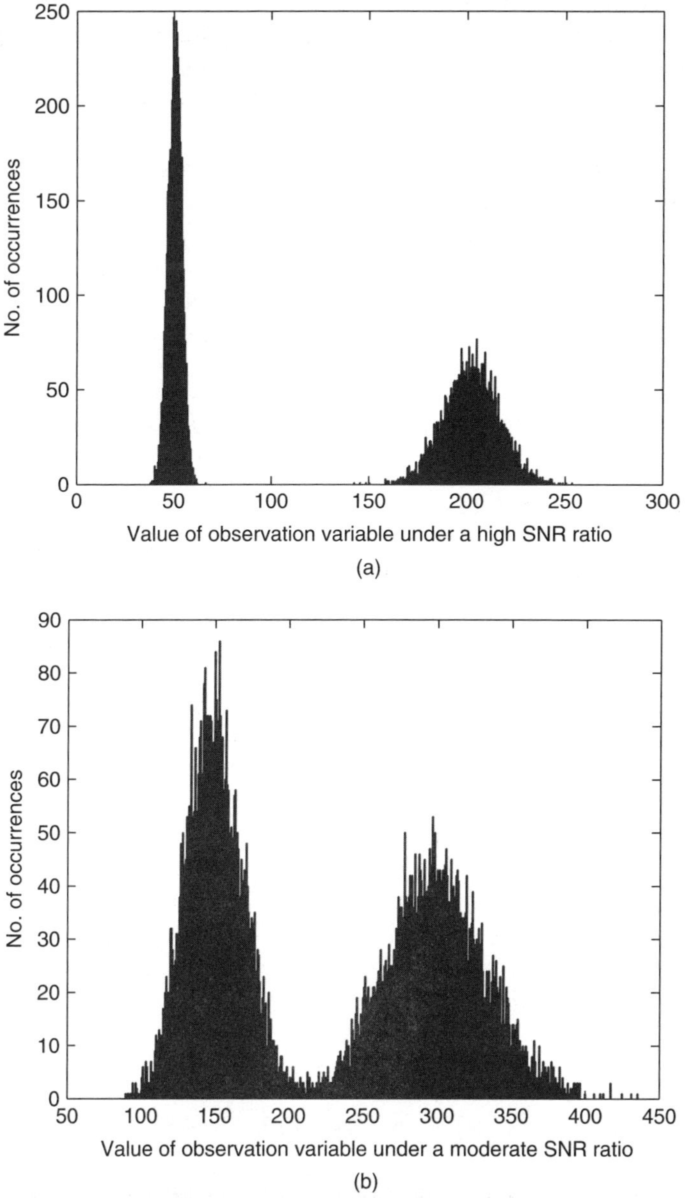

Figure 4.13 Histograms of the received bit energy for a noncoherent CSK system for (a) high SNR, and (b) moderate SNR.

with time (i.e., varies from bit to bit), as illustrated by the histogram shown in Figure 4.13a for the high SNR case. Note that $E_c(lT_b)$ does not show up as two distinct values in the histogram, but rather as clusters around the two average bit energies of the two chaotic signals with certain variances. By setting the threshold mid-way between the two average bit energies, the received symbols can be decoded correctly. For a moderate SNR environment, the calculated bit energies will generally be larger compared to the high SNR case because of the increase in noise energy. Figure 4.13b shows the histogram of $E_c(lT_b)$ under a noisy environment. In this case, the estimated bit energies are larger and the two distinct bit-energy regions in the histogram begin to widen. As the two regions start overlapping with each other, bit errors begin to emerge. Obviously, the optimal threshold is dependent upon the SNR, and this *threshold-shift* problem remains one of the drawbacks of this type of bit-energy-based noncoherent CSK system. In addition, detection based on different bit energies can be accomplished easily by intruders, jeopardizing the security of the system.

4.3.1.4 Noncoherent Demodulation Based on Return Map

In the previous section, chaotic signals with different average bit energies are applied to represent different binary symbols. However, the noncoherent demodulation technique based on the estimation of bit energy provides little security against eavesdroppers. Because chaotic signals are derived from deterministic dynamical systems, there should be some other deterministic properties of the chaotic signals, besides the average bit energy, that can be exploited for noncoherent detection. In the following, we describe how the difference in the return maps of the chaotic signals is used to distinguish the digital symbols. Two specific algorithms are discussed, namely, a *regression-based* algorithm and a *probability-based* algorithm.[24,25,43]

We consider a discrete-time CSK system, as shown in Figure 4.14. Essentially, the signal being transmitted switches itself between the two chaotic sequences, $\{a_k\}$ and $\{b_k\}$, depending upon the value of the digital message to be sent, where k is the integer index for the sequence of values generated by the chaos generators. In general the two chaotic sequences are generated by two chaotic maps f and g, such that

$$\begin{aligned} a_{k+1} &= f(a_k) \\ b_{k+1} &= g(b_k) \end{aligned} \tag{4.1}$$

Denote the number of chaotic samples sent in one bit duration by 2β, which is also known as the *spreading factor*. The output of the transmitter, x_k, during the lth bit duration, i.e., for $k = (l-1)2\beta + 1, (l-1)2\beta + 2, \ldots, 2l\beta$,

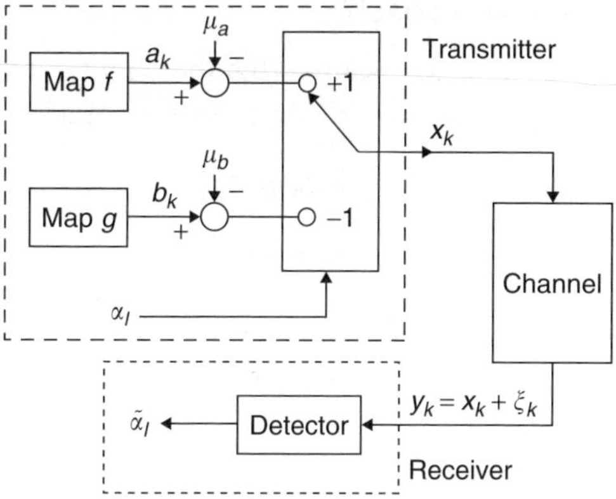

Figure 4.14 Block diagram of a discrete-time CSK system. α_l Is the digital message and $\tilde{\alpha}_l$ is the decoded message.

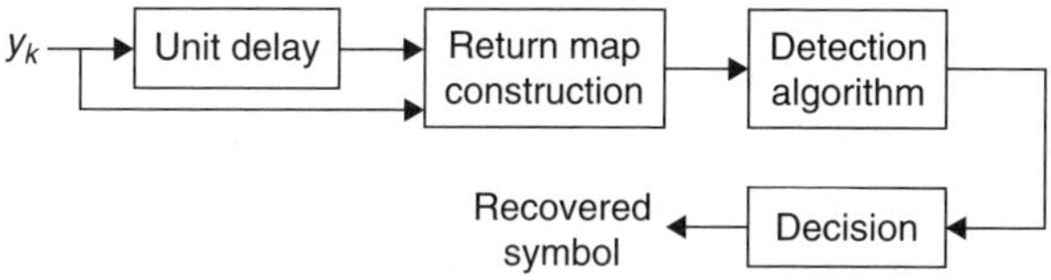

Figure 4.15 Block diagram of the return-map-based detector.

is given by

$$x_k = \begin{cases} a_k - \mu_a, & \text{if transmitted symbol is "+1"} \\ b_k - \mu_b, & \text{if transmitted symbol is "}-1\text{"} \end{cases} \tag{4.2}$$

where μ_a and μ_b are the average values of the two chaotic sequences $\{a_k\}$ and $\{b_k\}$, respectively. Therefore, the transmitter output, x_k, has a zero average. Further, assuming that the channel is subject to AWGN, the signal at the input to the receiver, y_k, is given by

$$y_k = x_k + \xi_k \tag{4.3}$$

where ξ_k is the added channel noise. At the receiving end, the objective is to recover the transmitted symbol with a minimum probability of error. Two demodulation algorithms will be given. The first one is based on regression, and the second on maximizing the *a posteriori* probability. Figure 4.15 shows a block diagram of the return-map-based detector.

4.3.1.4.1 Regression Approach[43]

In the regression approach, we assume that the chaotic maps can be written in a common form $h(.)$, with one distinguishing parameter p, i.e.,

$$\begin{aligned} \tilde{a}_{k+1} &= h(p, \tilde{a}_k) \\ \tilde{b}_{k+1} &= h(-p, \tilde{b}_k) \end{aligned} \tag{4.4}$$

where $(\tilde{a}_k, \tilde{a}_{k+1})$ and (a_k, a_{k+1}) are related by a simple transformation T, and so are $(\tilde{b}_k, \tilde{b}_{k+1})$ and (b_k, b_{k+1}). That is,

$$(\tilde{a}_k, \tilde{a}_{k+1}) = T(a_k, a_{k+1}) \tag{4.5}$$

$$(\tilde{b}_k, \tilde{b}_{k+1}) = T(b_k, b_{k+1}) \tag{4.6}$$

In particular, we consider the simple case where the parameter p appears explicitly as a multiplier to a common function $H(.)$, i.e.,

$$\begin{cases} \tilde{a}_{k+1} = pH(\tilde{a}_k) \\ \tilde{b}_{k+1} = -pH(\tilde{b}_k) \end{cases} \tag{4.7}$$

For example, the chaotic maps employing the tent maps $f(x) = 1 - 2\left|x - \frac{1}{2}\right|$ and $g(x) = 2\left|x - \frac{1}{2}\right|$ for all $x \in (0, 1)$ has a corresponding common form given by

$$H(x) = \frac{1}{2} - |2x| \tag{4.8}$$

and $p = 1$. Here, the transformed range equals $(-\frac{1}{2}, \frac{1}{2})$. Similarly, for the logistic maps $f(x) = 4x(1 - x)$ and $g(x) = 1 - 4x(1 - x)$ with $x \in (0, 1)$, the corresponding common form equals

$$H(x) = \left(x - \frac{\sqrt{2}}{4}\right)\left(x + \frac{\sqrt{2}}{4}\right) \tag{4.9}$$

and $p = 4$. Also, the range after the transformation becomes $(-\frac{1}{2}, \frac{1}{2})$. See Figure 4.16.

The detection algorithm involves first collecting the points (y_k, y_{k+1}), $k = 1, 2, \ldots, 2\beta - 1$, for each bit. Then, the aforementioned T transformation is performed on these points, such that the resulting return map resembles either the curve $y = pH(x)$ or $y = -pH(x)$ under a noise-free transmission, depending upon the transmitted bit being "+1" or "−1."

When the channel is noisy, the points on the collected and transformed return map appear dispersed, but to a certain extent (depending upon the noise level) remain close to the curves $y = \pm pH(x)$. A representative return map reconstructed under the aforementioned procedures is shown

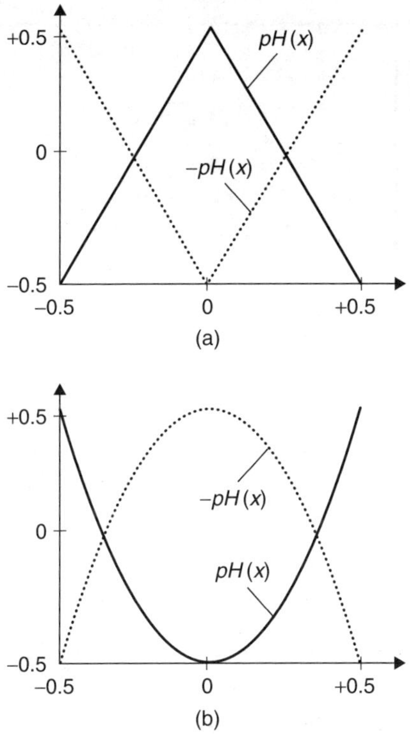

Figure 4.16 Chaotic maps. (a) $H(x) = 0.5 - |2x|$, $p = 1$; (b) $H(x) = (x - \sqrt{2}/4)\,(x + \sqrt{2}/4)$, $p = 4$.

in Figure 4.17. The detection is formulated on the basis of a regression algorithm, which aims to find the best fit of the curve

$$y = qH(x) \tag{4.10}$$

to the points of the return map for the bit. The regression method will estimate the parameter q such that the set of points is closest to the above curve in a least-square sense.[37] Afterwards, the decoded symbol, denoted by $\tilde{\alpha}$, is determined according to the following rule:

$$\tilde{\alpha} = \begin{cases} +1 & \text{if } q > 0 \\ -1 & \text{otherwise} \end{cases} \tag{4.11}$$

The essence of the transformation T is to allow the two chaotic maps to be written with *only one* parameter that "strongly" characterizes the map. If the map is written in terms of two or more parameters, then these parameters will jointly characterize the map. Unless all these parameters are estimated with a good precision, otherwise, the estimation of any one

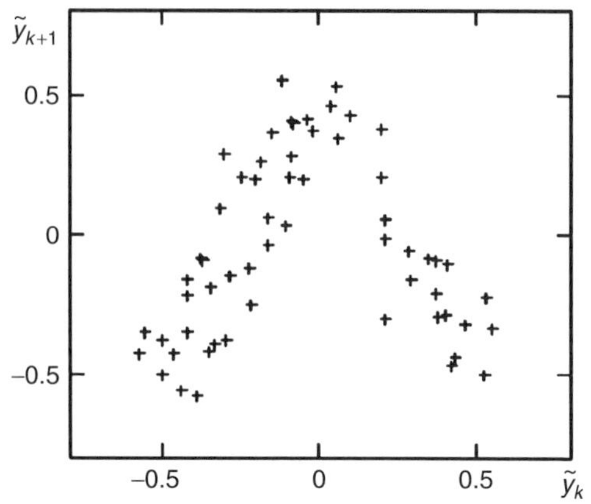

Figure 4.17 Return map of received signal for symbol "+1" at $E_b/N_0 = 25$ dB. Spreading factor is 60.

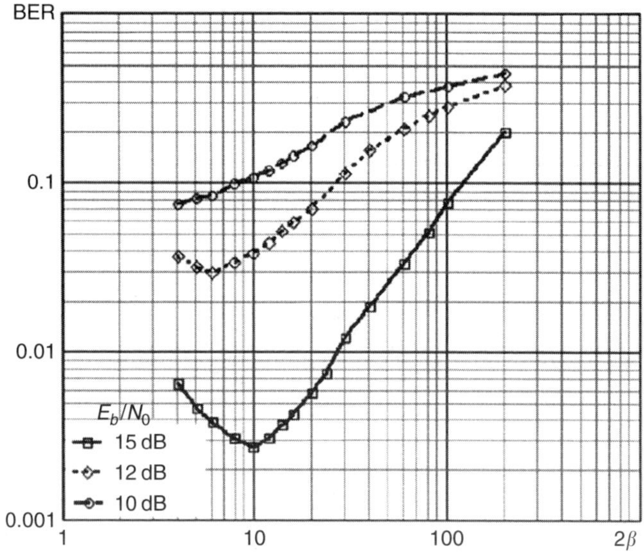

Figure 4.18 BER vs. spreading factor 2β for the regression-based detector. The tent maps are used as chaos generators.

parameter alone may not provide sufficient characterization of the particular chaotic map to allow accurate decision to be made as to which map has been sent. Hence, with only one parameter characterizing the map, the detection can be more easily done with higher accuracy.

Figure 4.18 shows a typical set of BER results. It can be observed that the BER reduces with E_b/N_0. Also, for a constant E_b/N_0, the BER reaches a minimum for a certain spreading factor, which is a usual behavior in noncoherent detection of chaos-based digital communications.

4.3.1.4.2 Probability Approach[24,25]

In the probability approach, the detection algorithm is based on maximizing the *a posteriori* probability. Denote the received signal block within a bit period by $\boldsymbol{y}=(y_1\,y_2\cdots y_{2\beta})$, and define the observation vectors as $\boldsymbol{v}_k=(y_k\,y_{k+1})$ for $k=1,2,\ldots,2\beta-1$. Define also $\boldsymbol{V}=(\boldsymbol{v}_1\,\boldsymbol{v}_2\cdots\boldsymbol{v}_{2\beta-1})$, which is simply the return map constructed for the bit. For a given $\boldsymbol{V}$, we decode the incoming chaotic signal block by selecting the symbol that would maximize the *a posteriori* probability, i.e.,

$$\tilde{\alpha} = \arg\max_{\alpha} \mathrm{Prob}(\alpha \text{ is sent} \mid \boldsymbol{V}) \tag{4.12}$$

The *a posteriori* probability is not easy to calculate, so we apply the Bayes' rule to $\mathrm{Prob}(\alpha \text{ is sent} \mid \boldsymbol{V})$ to obtain

$$\mathrm{Prob}(\alpha \text{ is sent} \mid \boldsymbol{V}) = \frac{p(\boldsymbol{V} \mid \alpha \text{ is sent})}{p(\boldsymbol{V})} \times \mathrm{Prob}(\alpha \text{ is sent}) \tag{4.13}$$

where $p(.)$ denotes the probability density function. Hence, Equation (4.12) can be rewritten as

$$\tilde{\alpha} = \arg\max_{\alpha} p(\boldsymbol{V} \mid \alpha \text{ is sent}) \tag{4.14}$$

because $\mathrm{Prob}(-1 \text{ is sent}) = \mathrm{Prob}(+1 \text{ is sent}) = 1/2$ and $p(\boldsymbol{V})$ is independent of α. In this detection scheme, we further assume that $\boldsymbol{v}_k$ and $\boldsymbol{v}_m$ are independent for $k \neq m$. Thus, $p(\boldsymbol{V} \mid \alpha \text{ is sent})$ in Equation (4.14) can be expressed as

$$p(\boldsymbol{V} \mid \alpha \text{ is sent}) = \prod_{k=1}^{2\beta-1} p(\boldsymbol{v}_k \mid \alpha \text{ is sent}) \tag{4.15}$$

With this assumption, we effectively neglect the interdependence between the observation vectors, and thus the probability of an error occurring is expected to be larger than that of the optimal case.[11]

We also assume that in each of the observation vectors $\boldsymbol{v}_k=(y_k\,y_{k+1})$ $(k=1,2,\ldots,2\beta-1)$, the initial condition of the chaotic signal x_k that gives rise to y_k is randomly selected from the chaotic range of the map according to the natural invariant probability density of the chaotic map. Now, suppose a "+1" is sent, i.e., $\alpha=+1$, and the corresponding iterative map is f. Hence, we have

$$
\begin{aligned}
& p(\boldsymbol{v}_k \,|\, {+1} \text{ is sent}) \\
&= \int_{-\infty}^{\infty} p(\boldsymbol{v}_k \,|\, ({+1} \text{ is sent}, x_k))\rho_f(x_k)dx_k \\
&= \int_{-\infty}^{\infty} \frac{1}{2\pi\sigma_n^2} \exp\left(-\frac{(y_k - x_k)^2 + (y_{k+1} - f(x_k))^2}{2\sigma_n^2}\right)\rho_f(x_k)dx_k \qquad (4.16) \\
&= \frac{1}{2\pi\sigma_n^2}\int_{-\infty}^{\infty} \rho_f(x)\exp\left(-\frac{(y_k - x)^2 + (y_{k+1} - f(x))^2}{2\sigma_n^2}\right)dx
\end{aligned}
$$

where $\rho_f(x)$ and $\sigma_n^2\,(=N_0/2)$ denote the natural invariant probability density of f and variance of noise (noise power), respectively. Similarly, when a "−1" is sent and the corresponding iterative map is g, it is readily shown that

$$
\begin{aligned}
& p(\boldsymbol{v}_k \,|\, {-1} \text{ is sent}) \\
&= \frac{1}{2\pi\sigma_n^2}\int_{-\infty}^{\infty} \rho_g(x)\exp\left(-\frac{(y_k - x)^2 + (y_{k+1} - g(x))^2}{2\sigma_n^2}\right)dx \qquad (4.17)
\end{aligned}
$$

where $\rho_g(x)$ represents the probability density of g. Note that because the transmitted signal contains no DC bias, the maps f and g should be chosen or translated appropriately such that they consistently contain no DC offset. The decision rule is simply given by

$$
\tilde{\alpha} = \begin{cases} +1 & \text{if } \prod_{k=1}^{2\beta-1} p(\boldsymbol{v}_k \,|\, {+1} \text{ is sent}) > \prod_{k=1}^{2\beta-1} p(\boldsymbol{v}_k \,|\, {-1} \text{ is sent}) \\ -1 & \text{otherwise} \end{cases} \qquad (4.18)
$$

Figure 4.19 plots the BERs vs. the spreading factor in log scale. As in the regression-based detection method, for constant E_b/N_0, the BER achieves a minimum for a certain spreading factor.

4.3.2 Differential Chaos Shift Keying

In Section 4.3.1.3, we reveal the threshold-shift problem for the noncoherent detection of a CSK system based on bit-energy estimation. In this section, we present the differential chaos-shift-keying (DCSK) modulation scheme,[17] a modulation method that is primarily designed for noncoherent detection. Moreover, the threshold level of the demodulator is fixed at zero.

Figure 4.20 shows the block diagram of a DCSK modulator. In this modulation scheme, every transmitted symbol is represented by two consecutive segments of chaotic signals. The first one serves as the reference (reference signal) whereas the second one carries the data (data-bearing signal). If a

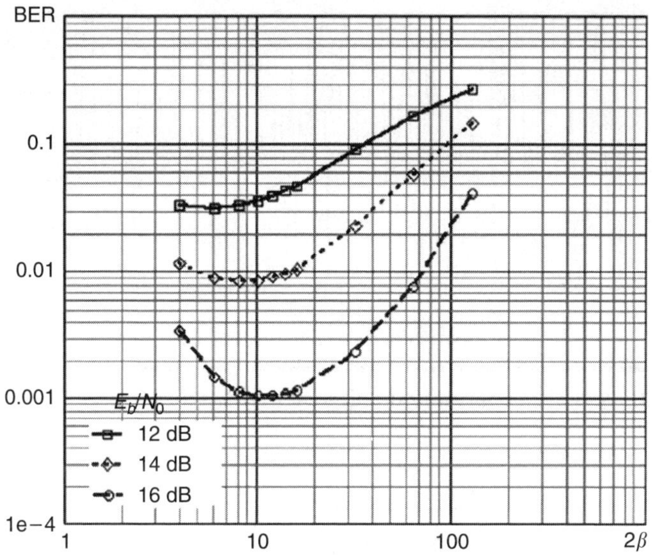

Figure 4.19 BER vs. spreading factor 2β for the probability-based detector. The tent maps are used as chaos generators.

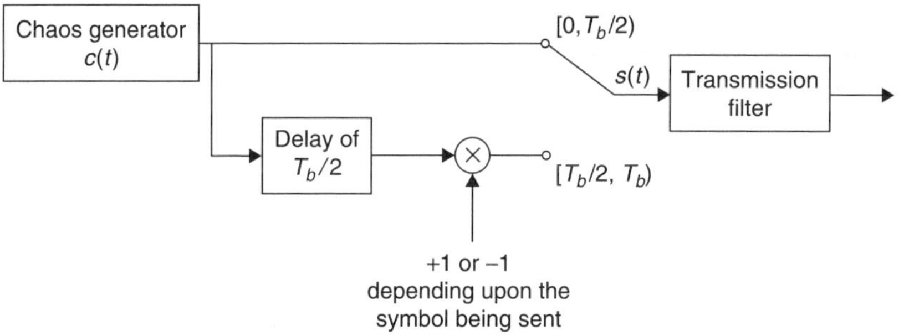

Figure 4.20 DCSK modulator.

"+1" is to be transmitted, the data-bearing signal will be identical to the reference signal, and if a "−1" is to be transmitted, an inverted version of the reference signal will be used as the data-bearing signal. Usually, the reference signal is sent in the first-half symbol period, and the data-bearing signal is sent in the second-half symbol period. Thus, for each of the symbol period, we have

$$s(t) = \begin{cases} c(t) & \text{for } 0 \le t < T_b/2 \\ c(t - T_b/2) & \text{for } T_b/2 \le t < T_b \end{cases} \tag{4.19}$$

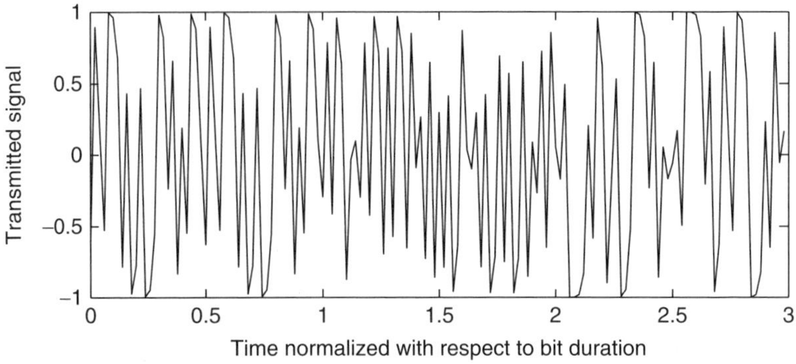

Figure 4.21 A typical transmitted DCSK signal sample.

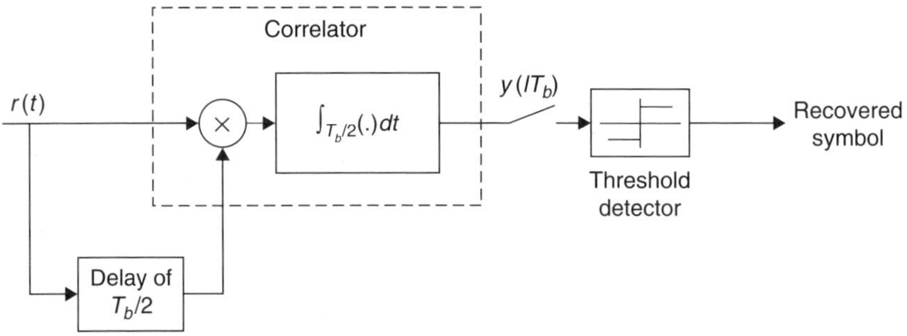

Figure 4.22 DCSK demodulator.

if "+1" is to be transmitted, and

$$s(t) = \begin{cases} c(t) & \text{for } 0 \leq t < T_b/2 \\ -c(t - T_b/2) & \text{for } T_b/2 \leq t < T_b \end{cases} \tag{4.20}$$

if "−1" is to be sent. Figure 4.21 shows a typical sample of the transmitted DCSK signal $s(t)$.

At the receiver, the correlation of the reference signal and the data-bearing signal is evaluated. This can be accomplished by correlating the incoming signal with a half-symbol-delayed version of itself, as shown in Figure 4.22. Figure 4.23a shows the statistical distribution of the correlator output under a high SNR environment. Note that the centers of the two clusters corresponding to the two symbols are located at equal distance from zero. Thus, by setting the threshold level to zero, it is sufficient to differentiate the two symbols. If the channel is noisier, however, the two clusters broaden and overlap each other, as shown in Figure 4.23b. Under such a condition, errors are inevitable but it is clear that the optimal threshold remains at zero, which is an advantage over the noncoherent CSK detection scheme based on bit-energy estimation. The main drawback

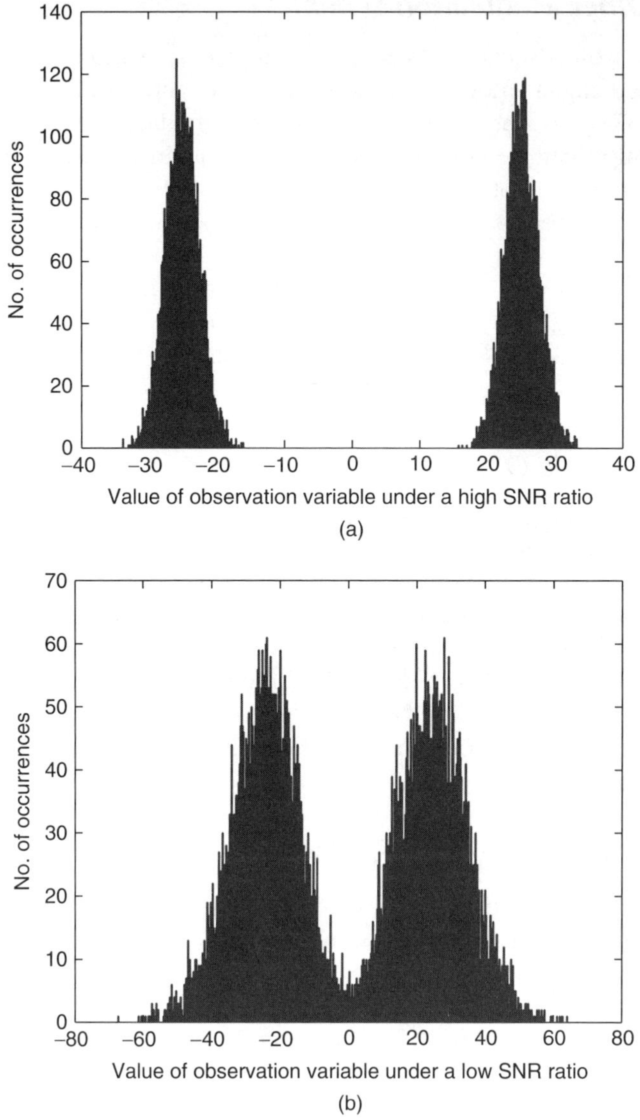

Figure 4.23 Histogram of the observation variable $y(lT_b)$ for a DCSK system. (a) High SNR ratio; (b) low SNR ratio.

of DCSK, however, is that its data rate is half of the other systems because it spends half of the time transmitting the non-information-bearing reference signals. One possible way to increase the data rate is to use a multilevel modulation and demodulation scheme.[19] The price to pay, then, is a more complicated system and a possibly degraded bit error performance when the received signal is attenuated by the channel.

4.3.3 Other Modulation Schemes

In the preceding sections, we have reviewed two most widely studied chaos-based digital modulation schemes: CSK and DCSK. The key concept in CSK is to encode digital symbols with different chaotic signals. Demodulation can be coherent or noncoherent. In DCSK, the correlation of the reference chaotic signal and the data-bearing signal is exploited to facilitate noncoherent demodulation. Since the introduction of CSK and DCSK schemes, several variations of these schemes were proposed, for example, chaotic on-off-keying, frequency-modulated DCSK (FM-DCSK), correlation delay shift keying (CDSK), symmetric CSK, quadrature CSK, and permutation-based DCSK (P-DCSK). In the following we briefly review these CSK- or DCSK-derived communication schemes.

4.3.3.1 Chaotic On-Off-Keying

Chaotic on-off-keying (COOK) scheme is a special case of noncoherent CSK method that is also based on bit-energy estimation. In this scheme, only one chaotic signal is needed. The binary symbols "+1" and "−1" are simply represented by transmission of this signal and null transmission, respectively.[18] The modulation process is thus equivalent to turning on and off a chaos generator. Clearly, the bit energy of the transmitted signal takes either a positive value or zero, depending on the symbol sent. The demodulation can be done in a noncoherent manner using a simple bit-energy estimator.

4.3.3.2 Frequency-Modulated DCSK

Because of the inherent property of chaotic signals, the energies of the chaotic signals representing the binary symbols vary with time, causing nonconstant correlator outputs in the demodulators even in a noiseless environment. To produce a wideband chaotic signal with constant power, the FM-DCSK was proposed.[20] In this scheme, a chaotic FM signal generator is needed. This chaotic FM generator can be constructed from the usual FM generator fed by a chaotic signal, as shown in Figure 4.24. The output of this chaotic FM generator is chaotic, bandlimited, and its power spectral density is determined by the chaotic signal driving the FM modulator. The FM-DCSK scheme is basically the same as the DCSK scheme, with the chaotic signals generated from chaotic FM generators. The demodulator structure

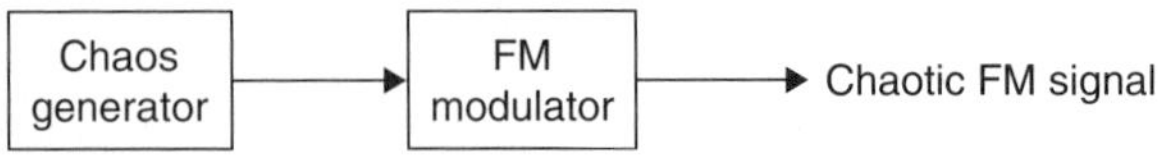

Figure 4.24 Chaotic frequency-modulated signal generator.

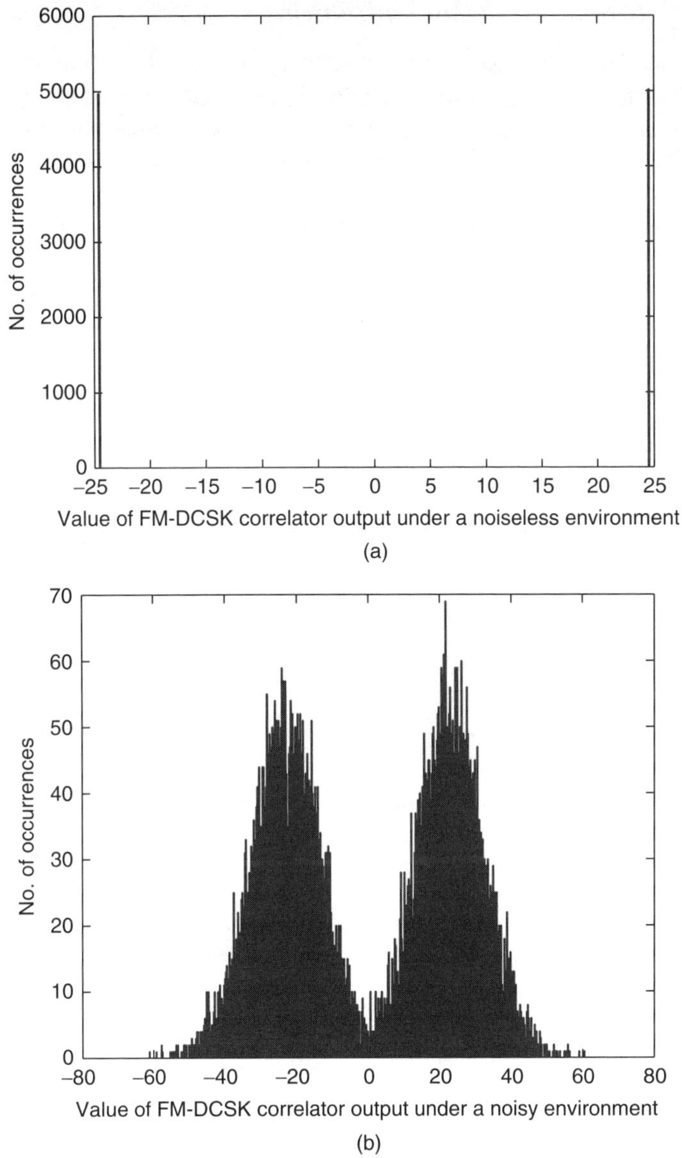

Figure 4.25 Histograms of the correlator output for an FM-DCSK system. (a) Noiseless environment; (b) noisy environment.

is the same as the DCSK demodulator. The histograms of the correlator output from an FM-DCSK demodulator are shown in Figure 4.25. Note that under a noiseless condition, the histogram shows two distinct outputs corresponding to the two symbols.

4.3.3.3 Correlation Delay Shift Keying

Correlation delay shift keying (CDSK) can be regarded as a derivative of DCSK.[39] In this scheme, the transmitted signal is the sum of two signals, namely a chaotic signal generated from a chaos generator and its delayed version multiplied by the information signal. Figure 4.26a shows the block diagram of a CDSK modulator. Compared to DCSK, the CDSK scheme does not need to perform any switching between the reference signals and the data-bearing signals. Thus, it allows a continuous operation of the transmitter. Furthermore, the delay T_D does not need to equal half of the symbol period and the transmitted signal is never repeated. This leads to a homogeneous transmitted signal and hence better tolerance to interception.

The structure of the CDSK demodulator is similar to that of the DCSK demodulator, except that the delay equals T_D, as shown in Figure 4.26b. Here, due to the nonzero correlation between different chaotic signals, the uncertainty of the correlator output generally increases and the bit error performance degrades in comparison to the DCSK scheme.

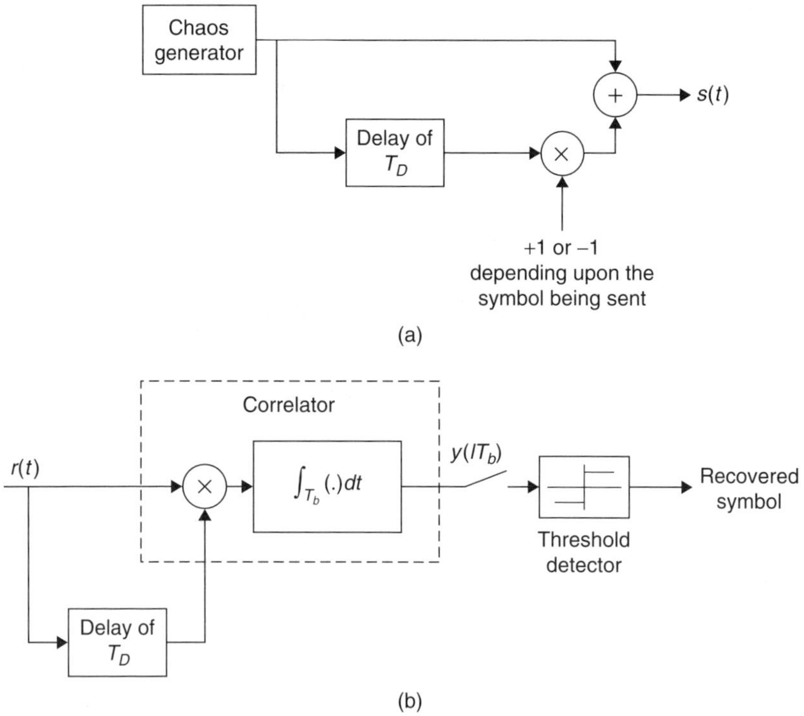

Figure 4.26 Correlation-delay-shift-keying system. (a) Modulator; (b) demodulator.

4.3.3.4 Symmetric Chaos Shift Keying

Symmetric chaos shift keying (SCSK) can be treated as a subclass of antipodal chaos shift keying.[39] The operation of the SCSK scheme is depicted in Figure 4.27. The chaotic system in the transmitter is denoted by

$$\dot{\boldsymbol{f}}(t) = \boldsymbol{F}[\boldsymbol{f}(t)] \tag{4.21}$$

where $\boldsymbol{f}(t)$ represents the internal state vector of the system at time t. The transmitted signal $s(t)$ is obtained from multiplying the first component of $\boldsymbol{f}(t)$, i.e., $f_1(t)$, by the digital symbol being sent.

At the receiver, the incoming signal is used to drive a matched chaotic system

$$\dot{\mathbf{g}}(t) = \boldsymbol{G}(|f_1(t)|, \mathbf{g}(t)) \tag{4.22}$$

Therefore, the systems $\boldsymbol{F}(.)$ and $\boldsymbol{G}(.)$ form a drive-response system with $f_1(t)$ and $g_1(t)$ being the coupling pair. In a distortion-free channel, the output of the chaotic system in the receiver is identical to that in the transmitter, i.e., $f_1(t) = g_1(t)$. The transmitted symbol can thus be decoded by observing the sign of the correlator output at the end of each symbol period.

The SCSK modulator, consisting of a chaos generator and a multiplier, is simpler compared to the DCSK and CDSK modulators. Furthermore, the SCSK signal is non-repeating, resulting a lower probability of detection.

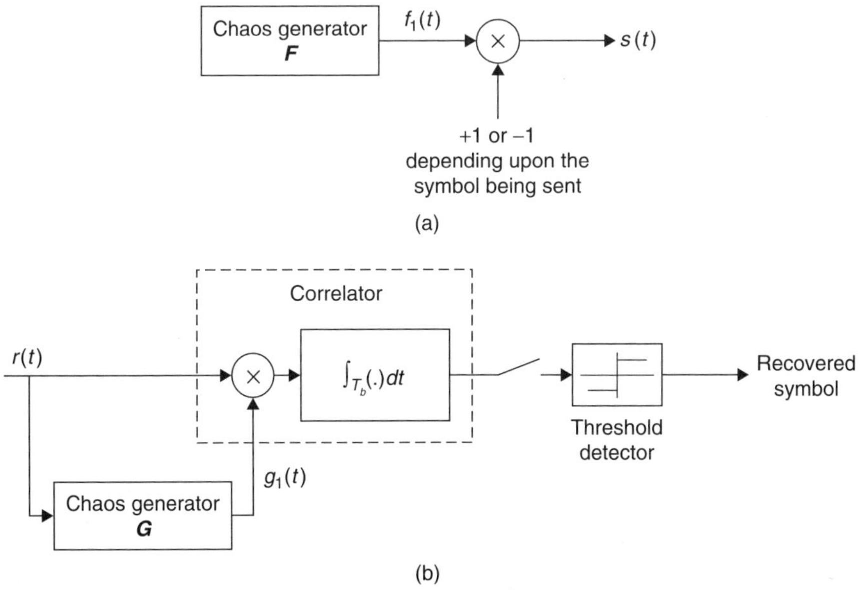

Figure 4.27 Symmetric CSK system. (a) Modulator; (b) demodulator.

Also, a higher level of security is provided against unauthorized decoding because a matched nonlinear system is essential to demodulate the SCSK signal. The disadvantages of the SCSK scheme, however, are that there is a restriction on the choice of nonlinear systems and that the bit error performance is generally worse than that of DCSK and CDSK when the channel is noisy, under which the drive-response system does not perform well.

4.3.3.5 Quadrature Chaos Shift Keying

In quadrature chaos-shift-keying (QCSK) scheme, each transmitted symbol consists of two bits of information. It can be regarded as a four-level version of DCSK which exhibits double the spectral efficiency as compared to DCSK scheme.[10]

Figure 4.28 shows the block diagram of a QCSK communication system, the operation of which is described as follows. We denote by $c(t)$ a chaotic reference signal, which is defined in $t \in [0, T_S/2)$, where T_s denotes the

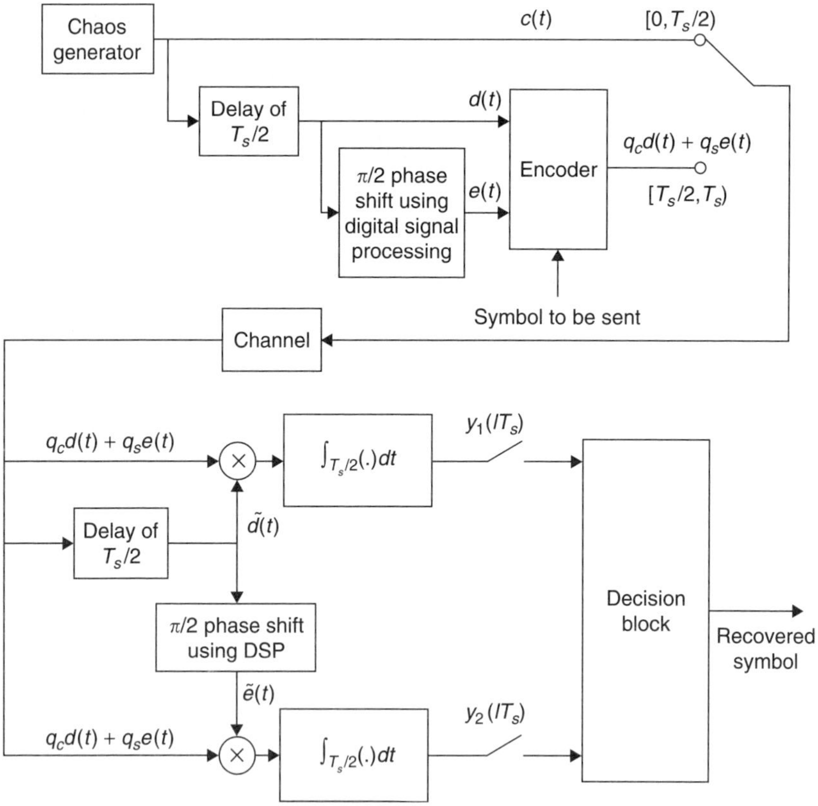

Figure 4.28 Quadrature CSK system.

symbol duration and is equal to $2T_b$. We assume that this reference signal has a zero mean. Then, we delay $c(t)$ by half the symbol duration to produce $d(t)$. Further, by making use of standard digital signal processing (DSP) techniques, the phase of all frequency components in $d(t)$ is shifted by $\pi/2$ to form a *complementary signal* $e(t)$. Because of the $\pi/2$-phase shift, the two signals $d(t)$ and $e(t)$ are orthogonal to each other. In the QCSK modulation scheme, the chaotic reference signal $c(t)$ is sent during the first half symbol period, i.e., $[0, T_s/2)$. In the second half symbol period, i.e., $[T_s/2, T_s)$, two orthogonal data-bearing signals, produced by multiplying the two information bits q_c and q_s to be sent with $d(t)$ and $e(t)$, respectively, are added together before transmission. The transmitted signal during the second half of the symbol duration, denoted by $s(t)$, is thus equal to

$$s(t) = q_c d(t) + q_s e(t) \tag{4.23}$$

At the receiving side, $d(t)$ and $e(t)$ are first estimated from the corrupted reference signal $\tilde{c}(t)$. Denote the estimated $d(t)$ and $e(t)$ by $\tilde{d}(t)$ and $\tilde{e}(t)$, respectively. Correlation of the signal received in the second half symbol period with $\tilde{d}(t)$ and $\tilde{e}(t)$ is then performed. The symbol (two bits of information) can further be decoded by an appropriate decision making algorithm based on the correlator outputs. From the construction of the transmitted signal, it can be observed that the QCSK system can be considered as the integration of two DCSK systems. The first one uses the delayed reference signal $d(t)$ and the second one uses the complementary signal $e(t)$ to carry the information bits. The QCSK scheme therefore has doubled the data rate at the expense of a higher system complexity.

4.3.3.6 Permutation-Based Differential Chaos Shift Keying

On the one hand, the regular bit structure of the DCSK signal enables a simple correlator and a fixed-threshold detector to be employed as the demodulator. On the other hand, the bit frequency can be easily determined from the transmitted signal, jeopardizing the security of the system. In Figure 4.29, the magnitude spectrum of a DCSK signal sample is plotted. It can be seen that the spectrum is white and no useful information can be retrieved. Next, we square the DCSK signal sample and plot the magnitude spectrum again. From Figure 4.30, it can be clearly observed that the spectral value goes to zero at odd multiple frequencies of the bit rate. This observation can be explained as follows. When the DCSK signal sample is squared, the resultant signals in the data-bearing slots become identical to their corresponding reference slots. Thus, no frequency component at odd multiples of the bit rate (i.e., any multiples of slot rate) exists because any contributions from the data-bearing slots will be cancelled exactly by those from the reference slots at such frequencies. This may not be very

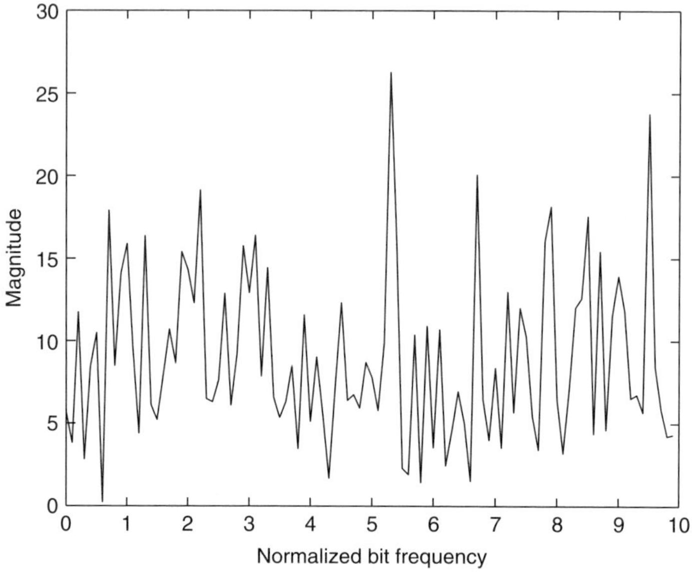

Figure 4.29 Magnitude of frequency components vs. normalized bit frequency for a conventional DCSK signal sample.

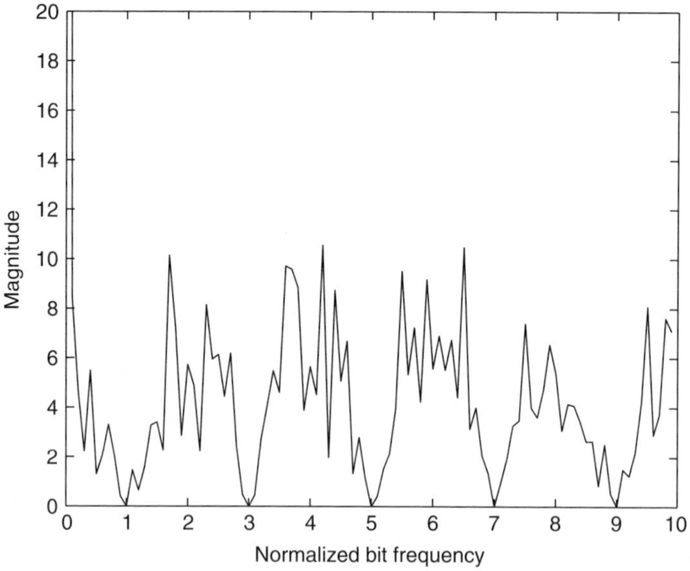

Figure 4.30 Magnitude of frequency components vs. normalized bit frequency for the square of a conventional DCSK signal sample.

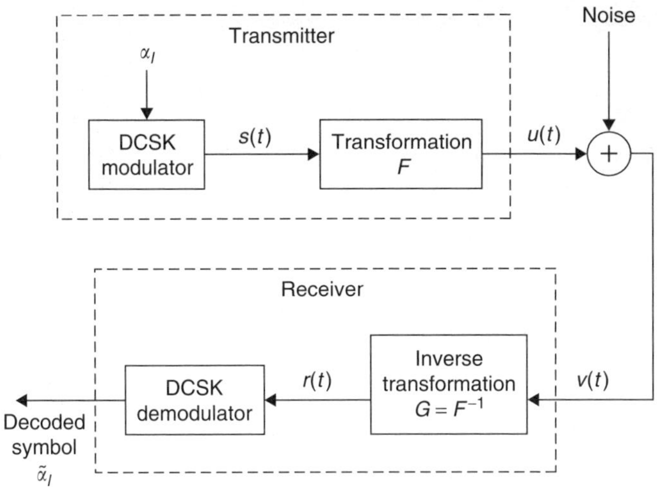

Figure 4.31 Permutation-based DCSK communication system.

desirable because anyone, intended or unintended, can retrieve the bit rate of the DCSK system easily.

Because of the aforementioned issue, the P-DCSK system was proposed.[27] Figure 4.31 depicts the block diagram of the proposed P-DCSK system. In this scheme, each signal block undergoes a transformation F before transmission. At the receiver, the incoming block will first undergo an inverse transformation $G = F^{-1}$ to retrieve the original DCSK signal. The output signal from the inverse transformation block is then passed to a conventional DCSK demodulator for decoding. Comparing with the conventional DCSK system, additional transformation and inverse transformation are needed in the transmitter and receiver, respectively. The objective of the additional transformations is to eliminate the bit rate information in the frequency spectrum. In particular, Lau et al. proposed the use of permutation in the transformation process.[27] Specifically, for each bit duration, the transformation is performed by dividing the DCSK signal into equal time intervals and then interchanging the signals among these time slots. On the receiver side, the signals in these time slots are interchanged in the reverse order as in the transmitter to recover the original DCSK signal. The frequency spectra of the P-DCSK signal sample and the square of the sample are plotted in Figure 4.32 and Figure 4.33, respectively. It can be seen that in both cases the spectrum is white and no bit rate information can be retrieved. Hence, the data security is enhanced. For the intended receiver with complete knowledge of the permutation performed, nonetheless, bit synchronization can still be easily accomplished for demodulation purposes.

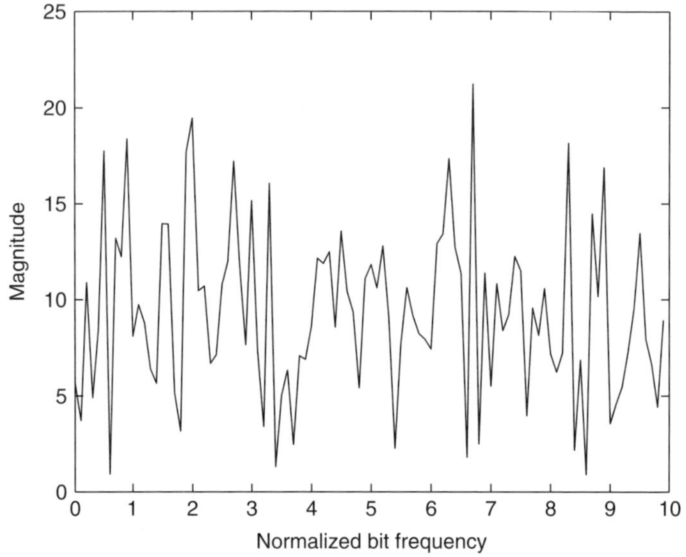

Figure 4.32 Magnitude of frequency components vs. normalized bit frequency for a permutation-based DCSK signal sample.

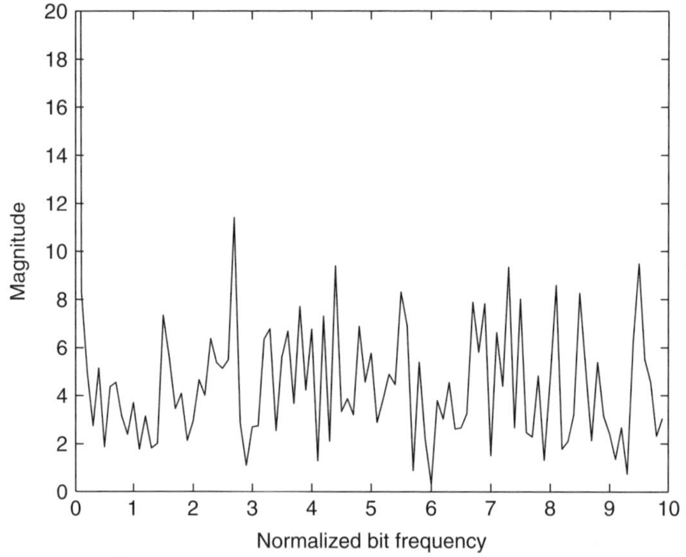

Figure 4.33 Magnitude of frequency components vs. normalized bit frequency for the square of a permutation-based DCSK signal sample.

4.4 Multiple Access Chaos-Based Digital Communication Systems

Multiple access is an important feature in spread-spectrum communications because it allows multiple users to gain access to the system simultaneously. Recently, researchers have also begun looking into the multiple access capability of chaos-based communication systems, which are inherently broadband. At the code level, spreading codes based on chaos have been applied to conventional direct-sequence code-division-multiple-access (DS-CDMA) systems.[29,30,48] Results show that using periodic quantized chaotic sequences as the spreading sequences can reduce the interuser interference in an asynchronous DS-CDMA communication system. As a consequence, the capacity of the system can be enhanced. At the signal level, coherent and noncoherent multiple access systems have been proposed and analysed.[15,19,22,23,26,40,41,44] In the coherent systems, exact replicas of the unmodulated chaotic carriers are assumed to be available at the receiver whereas no such requirement is needed for noncoherent systems. In the following, we briefly present two coherent and three noncoherent multiple access chaos-based digital communication systems.

4.4.1 Coherent Multiple Access Chaos-Based Digital Communication Systems

Among the coherent multiple access chaos-based communication systems that have been proposed, the multiple access chaotic frequency modulation (MA-CFM) and the multiple access chaos shift keying, respectively, extend the work of a single-user chaotic frequency modulation system[46] and a single-user chaos-shift-keying system.[21]

4.4.1.1 Multiple Access Chaotic Frequency Modulation

In the MA-CFM environment studied by Tsimring et al.,[44] it is assumed that there is a base station serving a number of mobile users under its coverage area. The base station will transmit a reference signal with chaotically varying frequency. Having received this reference signal, all mobile users in the system must first synchronize their own chaotic oscillators to it. Afterwards, each user can apply its own transformation to the synchronized chaotic waveform to generate its unique information-carrying CFM signal for communication purpose. Because chaos synchronization is very vulnerable to noise, the frequency band in which the mobile units are transmitting the CFM signals is separated from the band used for "synchronization" to ensure that the reference CFM signal is free from interference.

4.4.1.2 Multiple-Access Chaos Shift Keying

In the multiple-access chaos shift keying (MA-CSK) system,[23,40,41] each of the transmitters generates two chaotic signals, one used to represent the binary symbol "+1" and the other one for symbol "−1," as shown in Figure 4.34. The signals from all users are then combined before transmission. At each of the receivers, it is assumed that the corresponding synchronized versions of the chaotic carriers are available. The locally regenerated

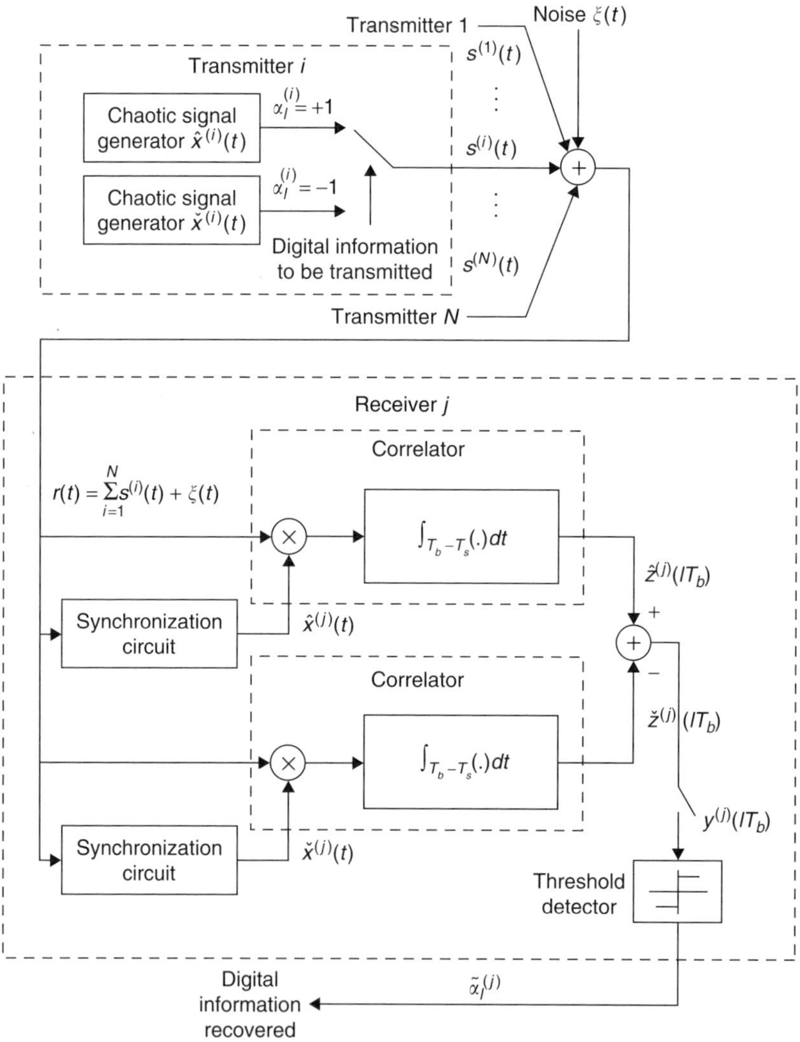

Figure 4.34 A multiple-access CSK communication system.

chaotic carriers then correlate with the incoming signal and the symbol corresponding to the larger correlator output will be decoded.

In the aforementioned multiple-access communication systems, all users are using the same bandwidth for the transmission of their chaotic carriers. Although the cross-correlation of different chaotic carriers is usually zero over an infinite period of time, the cross-correlation value becomes nonzero during a finite time interval, in this case the bit duration. Such nonzero cross-correlation values contribute to the interference among users, which degrade the bit error performance of the users. Further, at the time of writing, robust chaos synchronization techniques for the practical noise levels concerned are still lacking. Thus, the study of coherent detection schemes remains only of academic interest and the performance of coherent systems is used mainly as benchmark indicators. Noncoherent systems, which do not require the reproduction of the chaotic carriers at the receiving end, are more practical and improvements are still being made continually. Such practical systems will be the subject of the next section.

4.4.2 Noncoherent Multiple-Access Chaos-Based Digital Communication Systems

4.4.2.1 Time-Delay-Based Multiple-Access DCSK System

Recall in Section 4.3.2 that in the transmission of a DCSK signal, a reference chaotic signal is sent followed by the data-bearing signal. The operation of a time-delay-based multiple-access DCSK (TLMA-DCSK) system is built upon this principle. The TLMA-DCSK system was first introduced by Kolumbán et al.[19] For example, in a two-user environment, each bit period is divided into four time slots. For the first user, the reference signal is divided into two portions which are sent in the first and third time slots. Similarly, the data-bearing signal is also divided into two parts, which are sent in the second and fourth time slots. To reduce the interference between the transmitted signals, the order of transmission is altered for the second user. The reference signal is sent in the first two slots whereas the data-bearing signal is transmitted in the third and fourth slots. By correlating the signals in different time-slots of the incoming signal, different users can extract its own information with minimal interference from other users. Based on this principle, the TLMA-DCSK system has been generalized by Lau et al.[26] The transmission scheme of such a system is illustrated in Figure 4.35.

Referring to Figure 4.35, each bit duration is first divided into two slots for all users. Then, for the ith user, $2i$ consecutive slots are collected to form a frame. Hence, the slot duration (half of bit duration) is the same for all users but the frame periods are different for different users. Within one frame of the ith user, the first i slots (slots 1 to i) are used to transmit i chaotic reference signals whereas the remaining i slots (slots $i+1$ to $2i$)

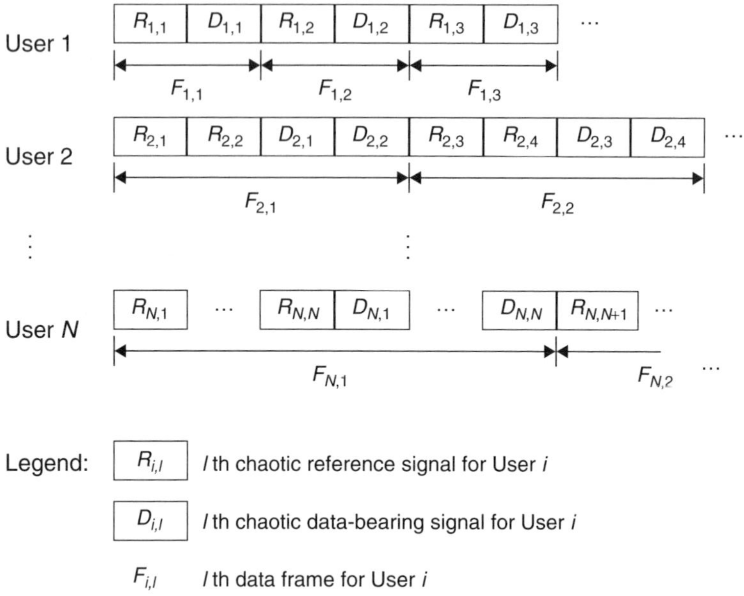

Figure 4.35 Transmission scheme for the time-delay-based multiple-access DCSK communication system.

are used to transmit i data-bearing signals. If a binary symbol "+1" is to be transmitted in slot $i+1$, the reference signal in slot 1 is repeated in slot $i+1$. If a "−1" is to be sent, an inverted version of the reference signal in slot 1 will be transmitted in slot $i+1$. Similarly, in slot $i+2$, the same or inverted copy of the reference signal in slot two is sent, and so on. As a result, the reference and data-bearing signals of the ith user will be separated by i slots, and no users will have the same separation between their corresponding reference and data-bearing signals. Moreover, an average of one bit of data is sent per bit period and the data rates of all users are thus the same.

Figure 4.36 shows the block diagram of a TLMA-DCSK communication system. In the transmitter of the ith user, a chaos generator is used to generate a chaotic signal $x^{(i)}(t)$ that has a zero mean. The chaos generators for different users are different in general. Consider the transmitted signal of the ith user during the lth time slot, i.e., for $t \in [(l-1)T_b, lT_b]$. Denote the output of the transmitter by $s^{(i)}(t)$. If the slot is a reference signal slot, then $s^{(i)}(t) = x^{(i)}(t)$. If the slot corresponds to a data-bearing signal slot sending a binary symbol "+1," we have $s^{(i)}(t) = x^{(i)}(t - iT_b/2)$. Likewise, if the slot corresponds to a data-bearing signal slot sending a binary symbol "−1," we have $s^{(i)}(t) = -x^{(i)}(t - iT_b/2)$. The overall transmitted signal is evaluated by summing the outputs of all users. For a particular user, the signal received during a reference signal slot will correlate with the signal

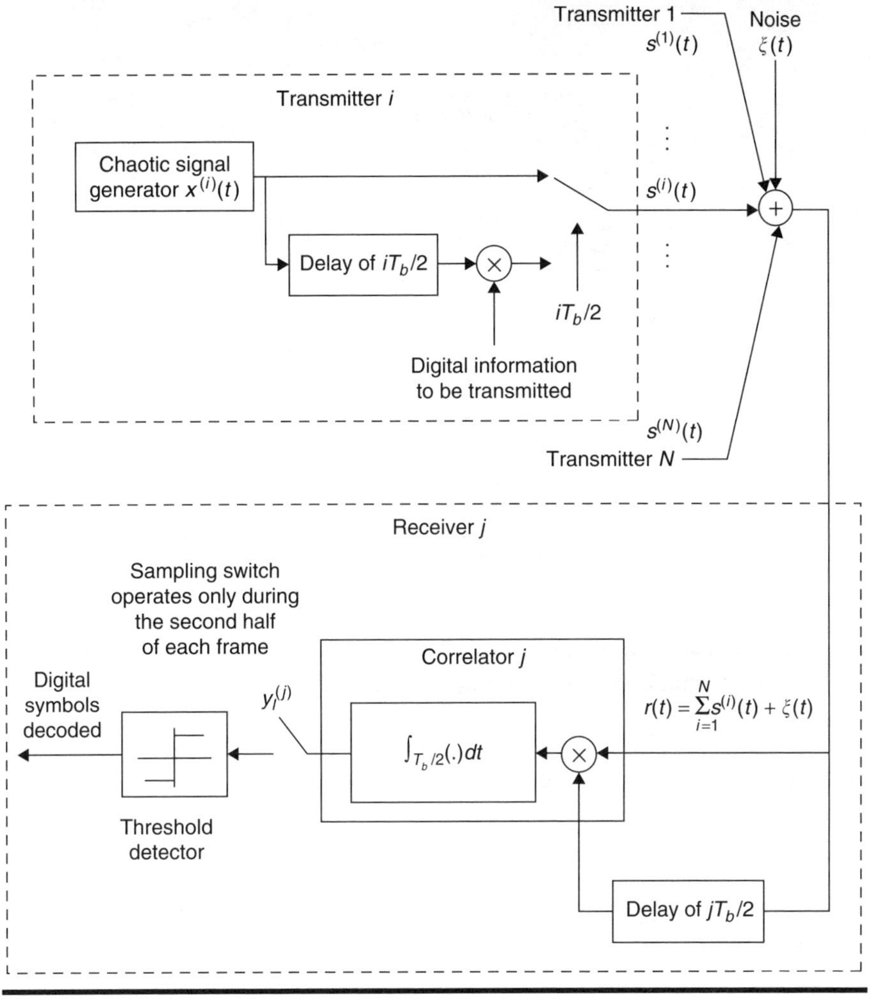

Figure 4.36 Time-delay-based multiple-access DCSK communication system.

at the corresponding data-bearing signal slot. Depending on whether the output is larger or smaller than the threshold, a "+1" or "−1" is decoded. Such a correlator-based receiver is shown in Figure 4.36. Note that the sampling switch only operates during the second half of each frame and the threshold detector produces a decoded symbol after each slot during that period of time.

4.4.2.2 Permutation-Based Multiple-Access DCSK System

An alternative approach to the implementation of multiple-access systems is to apply transformation to the original DCSK signals, with the

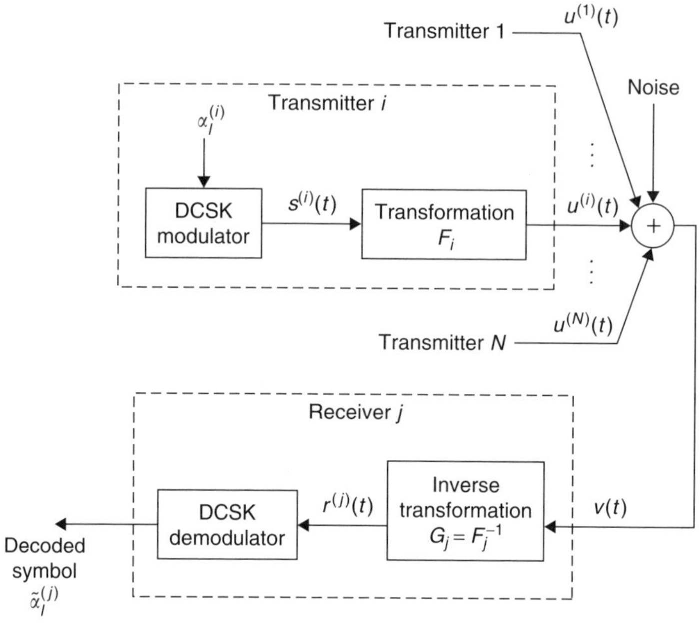

Figure 4.37 Permutation-based multiple-access DCSK communication system.

same objective of minimizing interference among users.[27] The permutation-based multiple-access DCSK (PMA-DCSK) system essentially extends the single-user P-DCSK system in Section 4.3.3.6 to a multi-user environment. Figure 4.37 shows the block diagram of a PMA-DCSK system. As in the single-user P-DCSK case, each transmitter in the PMA-DCSK system consists of a DCSK modulator followed by a transformation block. Within each bit duration, the chaotic signal generated from the DCSK modulator, consisting of the reference signal and the data-bearing signal, undergoes a transformation before it is transmitted. This transformation essentially destroys the similarity between the reference and data-bearing signals, leading to enhancement of the data security of the system. Different transformations are assigned to different users to minimize the interuser interference. The output signals from all users are added before transmission. At each of the receivers, the corresponding inverse transformation will be performed on the incoming signal and the output will be passed to a conventional DCSK demodulator for further decoding.

In particular, the PMA-DCSK system divides each DCSK signal sample into a number of small time slots and applies permutation to interchange the chaotic signals in the time slots. Also, by making use of a random permutation matrix and a "shifting" matrix, different permutation matrices can be constructed for different users. When the users interchange the chaotic signals in the small time slots based on these permutation matrices, it is shown that the interuser interference can be minimized.[27]

4.4.2.3 Walsh-Code-Based Multiple-Access System

Due to the small but finite interference between users, the performance of the TLMA-DCSK and PMA-DCSK systems degrades with the number of users. To eliminate the interuser interference completely, Kolumbán et al. proposed the introduction of Walsh codes into the multiple access FM-DCSK system.[22]

Suppose $K=2^n$ where n is a positive integer. We can construct K orthogonal Walsh codes based on the Hadamard matrix $\boldsymbol{H}_K$, which can be generated using the following recursive procedure.[36]

$$\boldsymbol{H}_1 = [-1] \quad \text{for } n=0 \tag{4.24}$$

$$\boldsymbol{H}_2 = \begin{bmatrix} \boldsymbol{H}_1 & \boldsymbol{H}_1 \\ \boldsymbol{H}_1 & -\boldsymbol{H}_1 \end{bmatrix} = \begin{bmatrix} -1 & -1 \\ -1 & +1 \end{bmatrix} \quad \text{for } n=1 \tag{4.25}$$

$$\vdots$$

$$\boldsymbol{H}_K = \begin{bmatrix} \boldsymbol{H}_{K/2} & \boldsymbol{H}_{K/2} \\ \boldsymbol{H}_{K/2} & -\boldsymbol{H}_{K/2} \end{bmatrix} \quad \text{for } n=\log_2 K \tag{4.26}$$

Each row in the Hadamard matrix represents a Walsh code. Denote the ith row of the $K \times K$ Hadamard matrix by $\boldsymbol{W}^{(i)}_{K\times K}$, $i=1,2,\ldots,K$. It is readily shown that the Walsh codes are orthogonal, i.e.,

$$\boldsymbol{W}^{(i)}_{K\times K}\left[\boldsymbol{W}^{(j)}_{K\times K}\right]^T = \begin{cases} 0 & \text{if } i \neq j \\ K & \text{if } i=j \end{cases} \tag{4.27}$$

In the multiple access scheme proposed by Kolumbán et al.,[22] two basis functions are used to represent the binary symbols for each user. Each bit duration is first divided into a number of time slots. To construct a pair of basis functions for each user, two Walsh codes are multiplied with a chaotic FM signal repeatedly, as shown in Figure 4.38 for a four-user system.

At a particular receiver, the chaotic FM signal in each time slot will multiply with the corresponding element in the Walsh code. Then an averaging process is performed at the receiver to estimate the chaotic FM signal being sent. This is done for each of the two Walsh codes assigned to each user. Finally, each of the two estimated chaotic FM signals correlates with the received signal, and the symbol corresponding to the large correlator output will be decoded. If all the transmitted signals are synchronized at the bit level, the interuser interference can be eliminated because of the orthogonal Walsh codes, and the performance is limited by the noise level only. To ensure the orthogonality between the basis functions of the users,

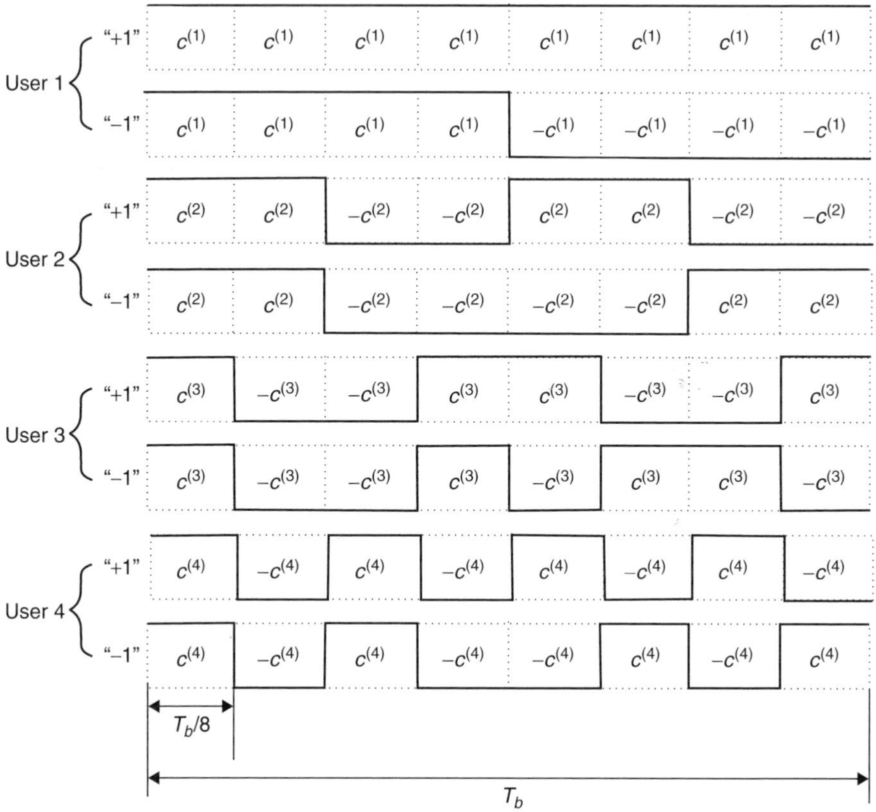

Figure 4.38 Transmission scheme of a multiuser FM-DCSK system.

longer Walsh codes may be required when the number of users increases. For example, in an N-user system, a total of $2N$ basis functions are needed. Hence, each symbol period should be divided into 2^L time slots where L is a positive integer and $2^L \geq 2N$. Note that the number of time slots equals 2^L because this is the length of the Walsh codes. Compared with the single-user FM-DCSK scheme, a larger number of time slots and hence chaotic signals are now sent for each symbol. As a consequence, the bit rate will be lowered.

4.5 Summary

In this chapter, we have briefly revealed a few selected chaos-based digital communication systems, along with some common multiple access techniques. It should be stressed that coherent systems, which require the regeneration of the chaotic carriers at the receiving end, are impractical due to the lack of robust synchronization methods. The study of coherent systems, however, can provide valuable performance benchmarks for comparison. Noncoherent systems are considered more practical because the demodulator determines the symbol being sent based only on the incoming signal. Furthermore, noncoherent methods are relatively less explored because of the vast possibilities of modulation and demodulation schemes that can potentially be used.

Finally, chaos-based digital communication systems are still regarded as immature from the practical engineering standpoint, and there are many important technical problems, such as robust synchronization and control of transmission bandwidth, that need to be tackled. Furthermore, it is the intention of the authors to present the chaos-based digital communication systems in this chapter with a minimal amount of mathematics. More in-depth study of these systems, such as the analytical techniques and the bit error performance of the systems, can be found in Kolumbán et al.[22] and Lau-Tse.[23]

References

1. Alligood, K.T., Sauer, T.D., and Yorke, J.A., *Chaos: An Introduction to Dynamical Systems*, Springer-Verlag, New York, 1996.
2. Chen, G. and Dong, X., *From Chaos to Order: Methodologies, Perspectives and Applications*, World Scientific, Singapore, 1998.
3. Chen, G. and Ueta, T., Eds., *Chaos in Circuits and Systems*, World Scientific, Singapore, 2002.
4. Cuomo, K.M. and Oppenheim, A.V., Circuit implementation of synchronized chaos with applications to communications, *Phys. Rev. Lett.*, 71, 65–68, 1993.
5. Hasler, M., Chaos shift keying: modulation and demodulation of a chaotic carrier using self-synchronizing Chua's circuit, *IEEE Trans. Circuits Systems Part II*, 40(10), 634–642, 1993.
6. Devaney, R.L., *A First Course in Chaotic Dynamical Systems*, Addison-Wesley, New York, 1992.
7. Dixon, R.C., *Spread Spectrum Systems with Commercial Applications*, 3rd ed., John Wiley & Sons, New York, 1994.
8. Elmirghani, J.M.H. and Cryan, R.A., New chaotic based communication technique with multiuser provision, *Elect. Lett.*, 30, 1206–1207, 1994.
9. Feng, J. and Tse, C.K., An on-line adaptive chaotic demodulator based on radial-basis-function neural networks, *Phys. Rev. E*, 63(2), 026202-1–10, 2001.
10. Galias, Z. and Maggio, G.M., Quadrature chaos shift keying, *Proc. IEEE Int. Symp. Circuits Syst.*, Sydney, Australia, Vol. III, 2001, 313–316.
11. Hasler, M. and Schimming, T., Chaos communication over noisy channels, *Int. J. Bifurcation Chaos*, 10, 719–735, 2000.

12. Hilborn, R.C., *Chaos and Nonlinear Dynamics: An Introduction for Scientists and Engineers*, Oxford University Press, New York, 2000.
13. Itoh, M., Murakami, H., and Chua, L.O., Communication systems via chaotic modulations, *IEICE Trans. Fund.*, E77-A(3), 1000–1006, 1994.
14. Itoh, M. and Murakami, H., New communication systems via chaotic synchronizations and modulation, *IEICE Trans. Fund. Elect. Comm. Comp. Sci.*, E78-A(3), 285–290, 1995.
15. Kolumbán, G., Multiple access capability of the FM-DCSK chaotic communications system, *Proc. Int. Workshop on Nonlinear Dynamics of Electronic Systems*, Catania, Italy, 2000, 52–55.
16. Kocarev, L., Halle, K.S., Eckert, K., Chua, L.O., and Parlitz, U., Experimental demonstration of secure communications via chaotic synchronization, *Int. J. Bifurcation Chaos*, 2, 709–713, 1992.
17. Kolumbán, G., Vizvari, G.K., Schwarz, W., and Abel A., Differential chaos shift keying: a robust coding for chaos communication, *Proc. Int. Workshop on Nonlinear Dynamics of Electronic Systems*, Seville, Spain, 1996, 97–92.
18. Kolumbán, G., Kennedy, M.P., and Kis, G., Performance improvement of chaotic communications systems, *Proc. European Conf. Circuit Theory Design*, Budapest, Hungary, 1997, 284–289.
19. Kolumbán, G., Kennedy, M.P., and Kis, G., Multilevel differential chaos shift keying, *Proc. Int. Workshop on Nonlinear Dynamics of Electronic Systems*, Moscow, Russia, 1997, 191–196.
20. Kis, G., Kennedy, M.P., and Jákó, Z., FM-DCSK: a new and robust solution to chaos communications, *Proc. International Symposium on Nonlinear Theory and Its Applications*, Hawaii, 1997, 117–120.
21. Kolumbán, G., Kennedy, M.P., and Chua, L.O., The role of synchronization in digital communications using chaos — Part II: Chaotic modulation and chaotic synchronization, *IEEE Trans. Circuits and Systems Part I*, 45(11), 1998, 1129–1140.
22. Kolumbán, G., Kennedy, M.P., Jákó, Z., and Kis, G., Chaotic communications with correlator receivers: theory and performance limits, 90(5), 2002, 711–732.
23. Lau, F.C.M. and Tse, C.K., *Chaos-Based Digital Communication Systems*, Springer-Verlag, Heidelberg, 2003.
24. Lau, F.C.M. and Tse, C.K., Approximate-optimal detector for chaos communication systems, *Int. J. Bifurcation Chaos*, 13(5), 1329–1335, 2003.
25. Lau, F.C.M. and Tse, C.K., On optimal detection of noncoherent chaos-shift-keying signals in a noisy environment, *Int. J. Bifurcation Chaos*, 13(6), 1587–1597, 2003.
26. Lau, F.C.M., Yip, M.M., Tse, C.K., and Hau, S.F., A multiple access technique for differential chaos shift keying, *IEEE Trans. Circuits and Systems Part I*, 49(1), 96–104, 2002.
27. Lau, F.C.M., Cheong, K.Y., and Tse, C.K., Permutation-based DCSK and multiple access DCSK systems, *IEEE Trans. Circuits and Systems Part I*, 50(6), 733–742, 2003.

28. Leung, H. and Lam, J., Design of demodulator for the chaotic modulation and communication systems, *IEEE Trans. Circuits and Systems Part I*, 44(3), 262–267, 1997.
29. Mazzini, G., Setti, G., and Rovatti, R., Chaotic complex spreading sequences for asynchronous DS-CDMA Part I: System modeling and results, *IEEE Trans. Circuits and Systems Part I*, 44(10), 937–947, 1997.
30. Mazzini, G., Setti, G., and Rovatti, R., Chaotic complex spreading sequences for asynchronous DS-CDMA Part II: Some theoretical performance bounds, *IEEE Trans. Circuits and Systems Part I*, 45(4), 496–506, 1998.
31. Miller, S.H., Elmirghani, J.M.H., and Cryan, R.A., Efficient chaotic-driven echo path modelling, *Elect. Lett.*, 31, 429–430, 1995.
32. Mullin, T., *The Nature of Chaos*, Oxford University Press, New York, 1993.
33. Kocarev, L., Halle, K.S., and Shang, A., Transmission of digital signals by chaotic synchronization, *Int. J. Bifurcation Chaos*, 2, 973–977, 1992.
34. Peterson, R.L., Ziemer, R.E., and Borth, D.F., *Introduction to Spread Spectrum Communications*, Prentice Hall, Englewood Cliffs, 1995.
35. Proakis, J.G., *Digital Communications*, McGraw-Hill, Singapore, 1995.
36. Proakis, J.G. and Salehi, M., *Communications Systems Engineering*, Prentice Hall, Englewood Cliffs, 1994.
37. Smith, P.J., *Into Statistics*, 2nd ed., Chapter 14, Springer, New York, 1998, 445–494.
38. Strogatz, S.H., *Nonlinear Dynamics and Chaos: With Applications to Physics, Biology, Chemistry and Engineering*, Perseus Books, New York, 2001.
39. Sushchik, M., Tsimring, L.S., and Volkovskii, A.R., Performance analysis of correlation-based communication schemes utilizing chaos, *IEEE Trans. Circuits and Systems Part I*, 47(12), 1684–1691, 2000.
40. Tam, W.M., Lau, F.C.M., Tse, C.K., and Yip, M.M., An approach to calculating the bit error probability of a coherent chaos-shift-keying digital communication system under a noisy multi-user environment, *IEEE Trans. Circuits and Systems Part I*, 49(2), 210–223, 2002.
41. Tam, W.M., Lau, F.C.M., and Tse, C.K., Analysis of bit error rates for multiple access CSK and DCSK communication systems, *IEEE Trans. Circuits and Systems Part I*, 50(5), 702–707, 2003.
42. Thompson, J.M.T. and Stewart, H.B., *Nonlinear Dynamics and Chaos*, John Wiley & Sons, New York, 2002.
43. Tse, C.K., Lau, F.C.M., Cheong, K.Y., and Hau, S.F., Return-map based approaches for noncoherent detection in chaotic digital communications, *IEEE Trans. Circuits and Systems Part I*, 49(10), 1495–1499, 2002.
44. Tsimring, L.S., Volkovskii, A.R., Young, S.C., and Rulkov, N.F., Multi-user communication using chaotic frequency modulation, *Proceedings, International Symposium on Nonlinear Theory and Its Applications*, Miyagi, Japan, Oct. 2001, 561–564.
45. Viterbi, A.J., *CDMA-Principles of Spread Spectrum Communication*, Addison-Wesley, Reading, 1995.

46. Volkovskii, A.R. and Tsimring, L.S., Synchronization and communication using chaotic frequency modulation, *Int. J. Circuit Theory Applications*, 27, 569–576, 1999.
47. Wozencraft, J.M. and Jacobs, I.M., *Principles of Communication Engineering*, Wiley, New York, 1965.
48. Yang, T. and Chua, L.O., Chaotic digital code-division multiple access communication systems, *Int. J. Bifurcation Chaos*, 7, 2789–2805, 1997.

5

A Chaos Approach to Asynchronous DS-CDMA Systems

S. Callegari, G. Mazzini, R. Rovatti, and G. Setti

CONTENTS

5.1 Introduction

Within the past decade, several research efforts have directly addressed the use of the potential of chaotic dynamics for performance improvements in several fields of electric and information engineering. The main reason for such an event is surely linked to the awareness that chaotic systems possess a nature at the borderline between deterministic and stochastic, and to the consequent adoption of tools from statistical dynamical systems theory for their analysis.[1–4]

The use of these statistical methodologies helps clarify that the application fields which are most likely to benefit from chaos-based techniques are those where the *statistical properties* of the signals are the dominant factor. Additionally, the same tools are also the key factor in developing the *quantitative models* needed to control the statistical features of the processes generated by discrete-time chaotic systems.

Applications of such chaos-related techniques in the field of information processing range from chaos-based cryptography,[6] audio and video watermarking,[7] network traffic modeling,[4,8] noise generation[9] or suppression[10] electromagnetic interference control,[11–13] and telecommunication, which has so far been the most fruitful area for applying chaos-related methodologies (see Reference 14 for a survey).

In addition to the appreciable importance of the results achieved for the mentioned applications, the existence of a plethora of different tasks to which the chaos-based approach (and the statistical techniques in particular) has been successfully applied, highlights the existence of a general scheme, which can in principle be applied to design a dynamic system able to optimize the performance for the considered application. Such a procedure follows two phases. As a first step, a model of the target application must be developed to link specifications (performance and constraints) with the statistical features of the signals entailed in the considered task. In the second phase, links will be exploited to deepen the comprehension of what improvements are achievable by the design and use, in the chosen application framework, of a chaotic system with controllable statistical features.

Among the engineering problems that benefit from this design ability, the application of chaotic dynamics to the code-level optimization of Direct Sequence-Code Division Multiple Access (DS-CDMA) systems is surely one of the most developed. The idea dates back to References 15 and 16 where a first model of the performance of a chaos-based DS-CDMA system is reported leading to encouraging analytical and numerical results. The exploitation of chaos-based techniques in this field has been further developed in References 17 to 19, where the impact of the adoption of chaos-based sequences in a standard asynchronous DS-CDMA system is studied when additive white gaussian noise (AWGN) channels are considered and multi-user interference (MAI) is the dominant cause of nonideality. For such an environment, theoretical performance bounds have been obtained, which show that significant improvement can be achieved by employing a properly designed family of chaotic systems generating real trajectories that are then quantized and periodically repeated to yield the users signature. This result has been further strengthened in Reference 20 where it is shown that chaos-based spreading permits reaching the absolute minimum of MAI and that such a lower bound is approximately 15% lower than the average interference achievable by classical pseudorandom sequences.

Few steps have also been taken in the direction of analyzing the synchronization between the transmitter and the receiver for the same user, which must be guaranteed before a reliable communication link can be established. In particular, in Reference 21 it is found that, even if synchronization can always be ensured independently of the choice of the spreading sequences, their correlation properties affect the speed and the reliability of the mechanism employed to achieve synchronization, so that chaos-based spreading may, again, offer advantages over classical ones.

Some dispersive channels, i.e., channels in which reflection, refraction, and diffraction phenomena are nonnegligible, have also been considered in connection with normal integrate-and-dump receivers.[22,23] Furthermore, the case of the more complex rake receiver has been investigated. In this setting a Finite Inpulse Response (FIR) filter is put before a classical receiver, whose taps can be adapted on-line to follow the channel dispersion characteristics, which are identified in parallel. A general method has been devised to adapt both sequences and receivers and optimize performance. Though in dispersive channels this method is approximate, theory predicts that it still results in better performance with respect to the conventional matched-filter policy usually adopted for rake-receivers.[24] Preliminary results on even more sophisticated architectures like parallel cancellers can be found in Reference 27.

Beyond theoretical results, the performance improvement exploiting chaos-based techniques have been validated in real world conditions by measurements performed using an experimental set-up using low cost Digital Signal Processor (DSP) boards.

The remainder of this chapter is organized as follows. In Section 5.2 we introduce the essential tools for computing statistical features from quantized trajectories generated by one-dimensional piecewise affine Markov (PWAM) chaotic maps. The section contains a simple and tutorially oriented bottom-up revisit of the concepts fully formalized in Reference 22: a suitable generalization of the widely employed Perron–Frobenius operator is here introduced and employed to find a way to compute high-order expectations of quantized chaotic trajectories.

In Section 5.3, the channel and receiver models are illustrated to obtain a separate expression of all the components of the correlate-and-dump output before the hard decision block. These components are random variables whose sum is matched against a zero threshold to reconstruct the symbol. Their distribution affects the bit error probability, as considered in Section 5.4. Following an established path, the standard Gaussian approximation (SGA) is adopted to tackle the cross- and self-interference terms, and, with some caution, the useful component. By exploiting this approximation, it is possible to simply express system performance as a function of the channel parameters, of the spreading sequence statistical features and of the receiver FIR taps.

In Section 5.5, such an expression is employed to state a simple constrained optimization problem in terms of the maximization of a cost function, which combines multi-index quantities depending on the previous parameters. Such a problem is analytically solved to link the two design parameters (i.e., the spreading sequences and receiver FIR taps) to the channel characteristics. As a particular but important case, it is shown that, when the channel is nondispersive, such an optimization procedure yields the same result presented in Reference 20 for the spreading sequences, allowing the minimization of MAI.

In Section 5.6, the problem of designing chaotic maps able to generate spreading sequences with different correlation behaviors, and thus with different impacts on the FIR optimization, is addressed. We show that, by properly choosing the structure of the chaotic map, we are able to generate optimal sequences when MAI is the unique cause of nonideality, and to jointly optimize sequences, statistics, and receiver FIR taps in case of selective channel.

Section 5.7 shows some details on a low-cost DSP-based DS-CDMA system, along with the measured bit error probability when classical m- and Gold and chaos-based sequences are employed for spectrum spreading. Measured performance results to be in good agreement with the theoretical analysis.

In Section 5.8 we remodel asynchronous DS-CDMA systems as a vector channel corrupted by thermal noise and we apply a well-known information theory result to numerically compute its capacity depending on spreading policies and relative delays. We then collect numerical evidence showing that chaos-based spreading is better than conventional one.

In Section 5.9 we specialize the above model for a two-user case and delve into some analytical considerations justifying the improvement noticed in the numerical simulations.

5.2 Some Tools for Computing Expectations of Quantized Chaotic Trajectories

Sensitive dependence on initial conditions, probably the best know characteristic of chaotic systems, makes it impossible to extract globally valid information from the study of single trajectories. On the contrary, considering the evolution of a nonvanishing set of points allows tracking the complex dynamic mechanism causing the loss of information and to characterizing it quantitatively.[1,2]

To follow this approach, consider a map $M : X=[0,1] \mapsto X$ and indicate with M^k its k-th iterate. Let $\mathbb{D}(X)$ be the set of probability densities defined on X and assume that the initial condition x_0 is randomly drawn according to $\rho_0 \in \mathbb{D}(X)$. Then, the probability density regulating the distribution of the random point $x_k=M(x_{k-1})=M^k(x_0)$ is $\rho_k=\mathbf{P}\rho_{k-1}=\mathbf{P}^k\rho_0$, where $\mathbf{P}$ is the Perron–Frobenius operator [Equation 4.1, Chapter 4]. Assume now that M is mixing, i.e., that a (natural) probability measure $\bar{\mu}$ exists such that $\bar{\mu}(A \cap M^{-k}(B)) \mapsto \bar{\mu}(A)\bar{\mu}(B)$ as $k \mapsto \infty$ for any set $A, B \subset X$, which roughly means that for large k the two events $\{x \in B\}$ and $\{M^k(x) \in A\}$ become statistically independent. For mixing maps, a unique invariant probability density $\bar{\rho}$ exists such that $\mathbf{P}\bar{\rho}=\bar{\rho}$, and the natural measure can be expressed as $\bar{\mu}(A)= \int_A \bar{\rho}$.

Commonly, the study of this loss of statistical dependence is carried out by considering the quantity

$$\mathbf{E}[f(x_k)g(x_{k+q})] = \int_X f(x)g(M^q(x))d\bar{\mu}(x) \tag{5.1}$$

where $f, g : X \mapsto \mathbb{R}$ are two smooth functions. If one thinks of these function as *physically observable* of the system, then Equation (5.1) quantifies the correlation between observing f at time k and g at time $k+p$.

For generic mixing maps, it is very difficult to compute $\mathbf{E}[f(x_k)g(x_{k+q})]$ and one is usually restricted to determine the rate $0 < r_{\text{mix}} < 1$ of geometric convergence of Equation (5.1) to its limit value when $q \to \infty$, which is often referred to as the *rate of mixing*.[2] More formally, it can be shown that a constant $\alpha > 0$ exists such that

$$\left|\mathbf{E}[f(x_k)g(x_{k+q})] - \int_X f(x)d\bar{\mu}(x) \int_X g(x)d\bar{\mu}(x)\right| \leq \alpha \sup|f|\sup|g|\, r_{\text{mix}}^q \tag{5.2}$$

which highlights the fact that the rate of mixing provides information on how quickly the system settles into a statistically regular behavior, and thus how quickly physical observables become uncorrelated.

Nevertheless, from an engineering point of view, the mere knowledge of r_{mix} and α cannot be considered as satisfactory. In fact, using Equation (5.2) in the model expressing the merit figure of a specific signal processing task as a function of statistical features of the chaotic sequences would simply yield to a (hopefully tight) *performance bound* (see e.g., References 17 and 19).

On the contrary, the ability of computing the exact value of Equation (5.1) would allow to analytically express the same merit figure as a function of the characteristics of M, which is surely the necessary starting point for any performance optimization procedure.

To follow this path, and to introduce the tool we need, we have to restrict the class of chaotic systems we are interested in. Consider a mixing

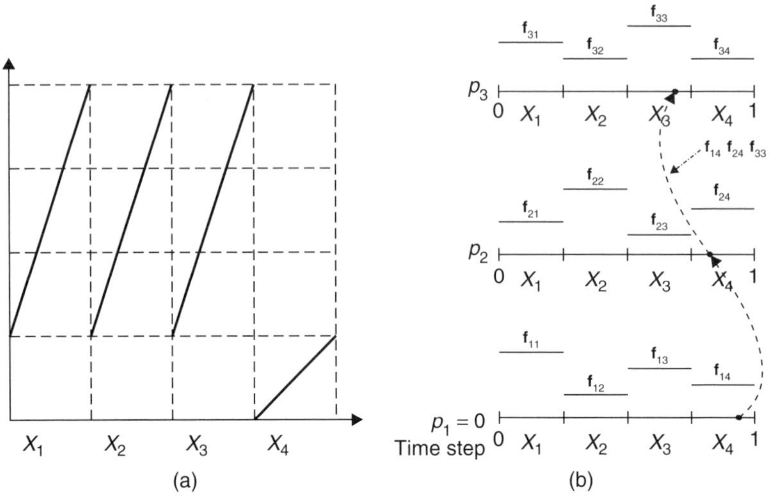

Figure 5.1 (a) Sketch of a PWAM map, and (b) quantization of the state space at three different time steps along with two possible trajectories and the values associated to each of them.

map M, and assume that a Markov partition of X exists, namely that X can be divided into n intervals $X_1, \ldots, X_n$ such that, for any pair of indexes j_1 and j_2 either $X_{j_1} \subseteq M(X_{j_2})$ or $X_{j_1} \cap M(X_{j_2}) = \emptyset$. Then M is said to be a PWAM map if it is affine in each of the intervals X_{j_i} of a Markov partition. An example of PWAM map is shown in Figure 5.1a.

For this kind of map, it is possible to give a simple matrix representation of **P**. To this aim, let us introduce the so-called *kneading matrix*,[2] which is defined componentwise as

$$\mathcal{K}_{j_1 j_2} = \frac{\mu(X_{j_1} \cap M^{-1}(X_{j_2}))}{\mu(X_{j_1})}$$

where μ is the usual interval measure, and whose entry in the j_1-th row and j_2-th column records the fraction of X_{j_1} that is mapped into X_{j_2}. From this it easily follows that the sum of the row entries of any kneading matrix is always 1. It can be proved that $\mathcal{K}$ is the restriction of the Perron–Frobenius operator to the space of functions that are constant in each X_{j_i}. As $\bar{\rho}$ for PWAM maps belong to that space, it can be expressed as $\bar{\rho} = \sum_{i=1}^{n} \boldsymbol{e}_i \chi_{X_i}$ where the vector $\boldsymbol{e} = (\boldsymbol{e}_1, \ldots, \boldsymbol{e}_n)$ is the (normalized) unique eigenvector of $\mathcal{K}$ corresponding to a unit eigenvalue, i.e., $\boldsymbol{e} = \boldsymbol{e}\mathcal{K}$, and where χ_{X_i} is the characteristic function of X_i. Obviously such a simple representation of the Perron–Frobenius operator allows one to devise a straightforward procedure to design a map with a prescribed piecewise-constant invariant probability density.

To give a closed-form representation also to quantities such as Equation (5.1), we need to introduce a suitable quantization of the chaotic map state space. Let us consider observables f_i, which quantize the state at a given time step p_i, i.e., let us assume that the function $f_i : [0,1] \mapsto \mathbb{C}$ is constant in each of n intervals of the Markov partition. Define also the n-dimensional vector $\boldsymbol{f}_i = (\boldsymbol{f}_{i1}, \ldots, \boldsymbol{f}_{in})$ so that $f_i(X_j) = \{\boldsymbol{f}_{ij}\}$; as an example, Figure 5.1b shows the functions f_1, f_2, and f_3 constant on each of the four intervals of a partition of the state space. A different quantization is adopted at different time steps and different trajectories are associated to different values depending on which intervals they visit.

As this point, we will also assume that $\bar{\rho} = 1$, so that one easily arrives at $\boldsymbol{e}_i = 1$ and also the sum of the row entries of $\mathcal{K}$ is always 1, i.e., $\mathcal{K}$ is a doubly stochastic matrix. With this we are able to compute second-order expectations of quantized chaotic trajectories of the kind in Equation (5.1). More formally we have

Property 1. *Consider a PWAM map with $\bar{\rho} = 1$ and where $\mu(X_j) = 1/n$ for $j = 1, \ldots, n$. Then, for trajectories generated with $\rho_0 = \bar{\rho}$ one has*

$$\mathbf{E}\left[f_1(x_{k+p_1}) f_2(x_{k+p_2})\right] = \frac{1}{n}\langle \boldsymbol{f}_1 \otimes \boldsymbol{f}_2, \mathcal{K}^{p_2-p_1}\rangle = \boldsymbol{f}_1 \mathcal{K}^{p_2-p_1} \boldsymbol{f}_2^T$$

where $\cdot^T$ indicates matrix transposition, and $\langle\cdot,\cdot\rangle$ and $\otimes$ denotes the usual inner and outer product among vectors.

To show this, consider first that if $\rho_0 = \bar{\rho}$ it is very easy to show that the (quantized) chaotic sequences generated by M are second-order stationary, i.e., $\mathbf{E}\left[f_1(x_{k+p_1}) f_2(x_{k+p_2})\right] = \mathbf{E}\left[f_1(x_0) f_2(x_q)\right]$, where $q = p_2 - p_1$. By its very definition one then has

$$\begin{aligned}\mathbf{E}\left[f_1(x_0) f_2(x_q)\right] &= \int_X f_1(\xi_1) f_2(M^q(\xi_1)) \bar{\rho}(\xi_1) d\xi_1 \\ &= \int_{X\times X} f_1(\xi_1) f_2(\xi_2) \delta(\xi_2 - M^q(\xi_1)) \bar{\rho}(\xi_1) d\xi_1 d\xi_2\end{aligned}$$

where the last equation holds thanks to the causality of the system and implicitly defines a two-steps Perron–Frobenius operator $\mathbf{P}^{0,q} = \bar{\mathbf{P}}^q$: $\mathbb{D}(X) \mapsto \mathbb{D}(X^2)$ such as $\bar{\mathbf{P}}^q[\bar{\rho}](\xi) = \bar{\rho}(\xi_1)\delta(\xi_2 - M^q(\xi_1))$, where $\xi = (\xi_1, \xi_2)$. This operator can be applied to the probability density $\bar{\rho}$ regulating the distribution of the points x_0 to obtain the probability density $\bar{\mathbf{P}}^q[\bar{\rho}]$ of the random vector (x_0, x_q). By exploiting the particular choice of the observables, the previous expression can be recast as

$$= \sum_{j_1=1}^{n} \sum_{j_2=1}^{n} \boldsymbol{f}_{1j_1} \boldsymbol{f}_{2j_2} \int_{X_{j_1}\times X_{j_2}} \bar{\mathbf{P}}^q[\bar{\rho}](\xi) d\xi_1 d\xi_2$$

in which the integral term is nothing but the probability that a trajectory starting in X_{j_1} falls in X_{j_2} at time step q. Because $\bar{\rho} = 1$, the previous term can be further expressed as $\int_{X_{j_1}\cap M^{-q}(X_{j_2})} \bar{\rho}(\xi_1)d\xi_1 = \mu(X_{j_1} \cap M^{-q}(X_{j_2}))$. With this, and considering that[18] $\mu(X_{j_1} \cap M^{-p}(X_{j_2}))/\mu(X_{j_1}) = (\mathcal{K}^p)_{j_1 j_2}$, we have

$$= \sum_{j_1=1}^{n} \sum_{j_2=1}^{n} \boldsymbol{f}_{1j_1} \boldsymbol{f}_{2j_2} \mathcal{K}^q_{j_1 j_2} \mu(X_{j_1})$$

from which, because $\mu(X_j) = 1/n$ and collecting the values of the quantization at the different time steps in the matrix $\boldsymbol{f}_1 \otimes \boldsymbol{f}_2$, we finally obtain the thesis.

As it is formally proven in Reference 22 and further exemplified in Reference 25, the previous result can be further generalized to obtain a closed form expression for computing the higher-order statistics of quantized chaotic processes. To this aim we have to introduce multi-index quantities that we will call *tensors* that generalize vectors and matrices. Several types of *product* can be defined among tensors and we may now define those that will be of use in the following discussion.

Given two m-order tensors $\mathcal{A}$ and $\mathcal{B}$ with identical index ranges, their *inner product* is a scalar defined as

$$\langle \mathcal{A}, \mathcal{B} \rangle = \sum_{j_1,\ldots,j_m} \mathcal{A}_{j_1,\ldots,j_m} \mathcal{B}_{j_1,\ldots,j_m}$$

where every index sweeps all its range. The inner product is a commutative bilinear operator.

Given an m'-order tensor $\mathcal{A}$ and an m''-order tensor $\mathcal{B}$, their *outer product* is an $(m' + m'')$-order tensor $\mathcal{C} = \mathcal{A} \otimes \mathcal{B}$ such that

$$\mathcal{C}_{j_1,\ldots,j_{m'+m''}} = \mathcal{A}_{j_1,\ldots,j_{m'}} \mathcal{B}_{j_{m'+1},\ldots,j_{m'+m''}}$$

where every index sweeps all its range. The outer product is a noncommutative, associative, and bilinear operator.

Given the m'-order tensor $\mathcal{A}$ and the m''-order tensor $\mathcal{B}$ such that the range of the last index of $\mathcal{A}$ is the same of the first index of $\mathcal{B}$, their *chain product* is an $(m' + m'' - 1)$-order tensor $\mathcal{C} = \mathcal{A} \circ \mathcal{B}$ such that

$$\mathcal{C}_{j_1,\ldots,j_{m'+m''-1}} = \mathcal{A}_{j_1,\ldots,j_{m'}} \mathcal{B}_{j_{m'},\ldots,j_{m'+m''-1}}$$

The chain product is a noncommutative associative bilinear operator.

The tensor products defined above enjoy few *distributive* properties, which are recalled without proof in the following

Property 2. *For any four tensors $\mathcal{A}, \mathcal{B}, \mathcal{C}$, and $\mathcal{D}$ whose index ranges make the equalities below well defined, we have*

$$\mathcal{A} \otimes (\mathcal{B} \circ \mathcal{C}) = (\mathcal{A} \otimes \mathcal{B}) \circ \mathcal{C}$$

$$\mathcal{A} \circ (\mathcal{B} \otimes \mathcal{C}) = (\mathcal{A} \circ \mathcal{B}) \otimes \mathcal{C}$$

$$\langle \mathcal{A} \otimes \mathcal{B}, \mathcal{C} \otimes \mathcal{D} \rangle = \langle \mathcal{A}, \mathcal{C} \rangle \langle \mathcal{B}, \mathcal{D} \rangle$$

The adoption of the tensor-formalism allows us to generalize the result of property 1 to the computation of m-th order expectations of the quantized trajectories.[22] In fact, we may assume $p_1=0<p_2<\cdots<p_m$ and indicate with $\bar{\mathbf{P}}^{p_2,\dots,p_m}$: $\mathbb{D}([0,1])\mapsto\mathbb{D}([0,1]^m)$ an operator that is formally expressed for any $m-1$ integers $p_2<\cdots<p_m$ as

$$\bar{\mathbf{P}}^{p_2,\dots,p_m}[\bar{\rho}](\boldsymbol{\xi})=\bar{\rho}(\boldsymbol{\xi}_1)\prod_{i=2}^{m}\delta(\boldsymbol{\xi}_i-M^{p_i-p_1}(\boldsymbol{\xi}_1))$$

which represents the joint probability density of the random vector $(x_0,x_{p_2},\dots,x_{p_m})$.

With this we obtain

$$\begin{aligned}\mathbf{E}\left[\prod_{i=1}^{m}f_i(x_{p_i})\right]&=\int_{[0,1]^m}\prod_{i=1}^{m}f_i(\boldsymbol{\xi}_i)\bar{\mathbf{P}}^{p_2,\dots,p_m}[\bar{\rho}](\boldsymbol{\xi})d\boldsymbol{\xi}\\&=\sum_{j_1=1}^{n}\cdots\sum_{j_m=1}^{n}\left[\prod_{i=1}^{m}\boldsymbol{f}_{ij_i}\int_{X_{j_1}\times\cdots\times X_{j_m}}\bar{\mathbf{P}}^{p_2,\dots,p_m}[\bar{\rho}](\boldsymbol{\xi})d\boldsymbol{\xi}\right]\end{aligned}$$

where the integral term, which represents the probability that a trajectory starting in X_{j_1} passes through X_{j_2} at time step p_2, through X_{j_3} at time step p_3, etc., can be collected in a m-order tensor $\mathcal{H}^{p_2,\dots,p_m}$

$$\mathcal{H}^{p_2,\dots,p_m}_{j_1,\dots,j_m}=\int_{X_{j_1}\times\cdots\times X_{j_m}}\bar{\mathbf{P}}^{p_2,\dots,p_m}[\bar{\rho}](\boldsymbol{\xi})d\boldsymbol{\xi}\qquad(5.3)$$

which will be indicated as the *symbolic dynamic tracking tensor* (SDTT).

Note now that the terms $\prod_{i=1}^{m}\boldsymbol{f}_{ij_i}$ for $j_1,\dots,j_m$ spanning the range $1,\dots,m$, can be also compounded in the m-order tensor $\mathcal{F}=\boldsymbol{f}_1\otimes\cdots\otimes\boldsymbol{f}_m$, so that the expression of the m-th order moment is

$$\mathbf{E}\left[\prod_{i=1}^{m}f_i(x_{p_i})\right]=\langle\mathcal{F},\mathcal{H}^{p_2,\dots,p_m}\rangle\qquad(5.4)$$

It can be proved that,[22,23] because we adopt PWAM maps, and as long $\bar{\rho}=1$ and all the intervals X_j are equal, the SDTT can be factored into smaller pieces by exploiting property 2 so that Equation (5.4) becomes

$$\mathbf{E}\left[\prod_{i=1}^{m}f_i(x_{p_i})\right]=\frac{1}{n},\langle\mathcal{F},\bigcirc_{i=2}^{m}\mathcal{K}^{p_i-p_{i-1}}\rangle\qquad(5.5)$$

where $\mathcal{K}$ is the kneading matrix defined above and the symbol $\bigcirc$ represents the iterative application of the chain-product that is associative.

Assuming the factorization of the SDTT is allowed to hold, we also know that[22,23] if $\mathcal{K}$ is primitive (i.e., a power of $\mathcal{K}$ exists in which no entry is null), a family of matrixes $\mathcal{A}(p)$ exists such that

$$\mathcal{K}^p = \Im + \mathcal{A}(p)$$

with $\mathcal{A}(p)$ exponentially vanishing as $p \to \infty$ and $\Im = (1/n)\mathbf{1} \otimes \mathbf{1}$ with $\mathbf{1} = (1, \ldots, 1)$.

With this and tensor algebra manipulations exploiting the distributivity of the tensor products we may easily derive that any inner product of the kind $\langle \mathcal{F}, \mathcal{H}^{p_2,\ldots,p_m} \rangle$ can be expressed as a weighted sum of terms of the kind $\langle \mathcal{F}, \mathcal{A}' \circ \mathcal{A}'' \circ \ldots \rangle$. For example if $\sum_{j=1}^{n} \boldsymbol{f}_{ij} = 0$ for $i = 1, \ldots, m$ then we may express the generic higher-order expectation term in Equation (5.5) as

$$\begin{aligned}
\mathbf{E}\left[f_1(x_0)\prod_{i=2}^{m} f_i(x_{p_i})\right] &= \frac{1}{n}\left\langle \bigotimes_{j=1}^{m} \boldsymbol{f}_j, \bigcirc_{j=2}^{m} \mathcal{A}(p_j - p_{j-1}) \right\rangle \\
&+ \frac{1}{n^2}\sum_{1}^{m-1}{}_{k'} \left\langle \bigotimes_{j=1}^{k'} \boldsymbol{f}_j, \bigcirc_{j=2}^{k'} \mathcal{A}(p_j - p_{j-1}) \right\rangle \left\langle \bigotimes_{j=k'+1}^{m} \boldsymbol{f}_j, \bigcirc_{j=k'+2}^{m} \mathcal{A}(p_j - p_{j-1}) \right\rangle \\
&+ \frac{1}{n^3}\sum_{1}^{m-2}{}_{k'} \sum_{k'+1}^{m-1}{}_{k''} \left\langle \bigotimes_{j=1}^{k'} \boldsymbol{f}_j, \bigcirc_{j=2}^{k'} \mathcal{A}(p_j - p_{j-1}) \right\rangle \\
&\times \left\langle \bigotimes_{j=k'+1}^{k''} \boldsymbol{f}_j, \bigcirc_{j=k'+2}^{k''} \mathcal{A}(p_j - p_{j-1}) \right\rangle \left\langle \bigotimes_{j=k''+1}^{m} \boldsymbol{f}_j, \bigcirc_{j=k''+2}^{m} \mathcal{A}(p_j - p_{j-1}) \right\rangle \\
&+ \quad \vdots \qquad\qquad \vdots \qquad\qquad \vdots
\end{aligned} \tag{5.6}$$

5.3 Asynchronous DS-CDMA Systems Model

Figure 5.2 reports the simplified baseband equivalent scheme of an asynchronous DS-CDMA system in which the carrier is common, U users are supposed, and t^u and θ^u indicate the absolute delay and carrier phase of the u-th user signal. To model the transmission from mobile terminals to a fixed base-station, we will assume that the latter quantities are independent and uniformly distributed random variables. Additionally, we will assume the v-th user as the useful one.

As far as transmitted signals are concerned, the u-th user information signal is $S^u(t) = \sum_{s=-\infty}^{\infty} S_s^u g_T(t - sT)$, where $g_T(t)$ is the rectangular pulse, which is 1 within $[0, T]$ and vanishes otherwise. We will also assume that the bipolar information symbols S_s^u are independent and produced by perfectly uncorrelated sources.

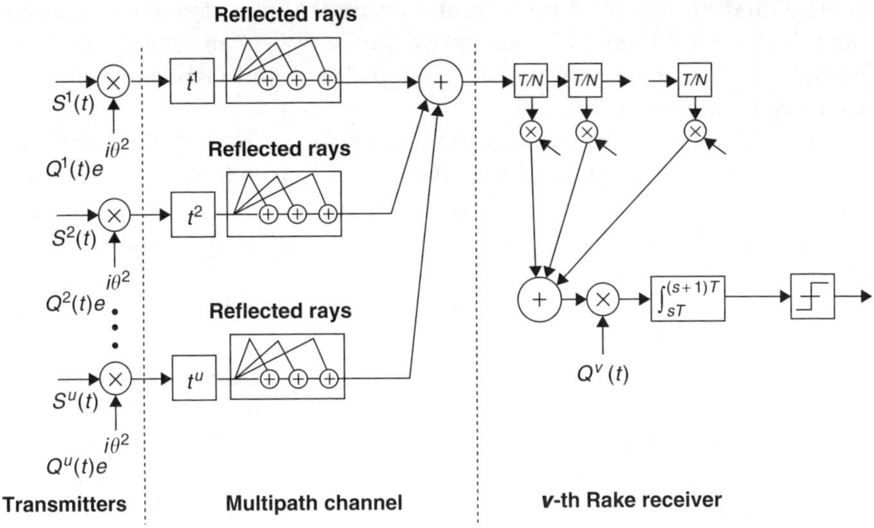

Figure 5.2 Simplified baseband equivalent of a DS-CDMA system with a multipath channel.

The spreading signal depends on sequences of symbols x_s^u from the alphabet X, which are mapped into $\{-1,+1\}$ by the function Q, and combined to form the signal $Q^u(t) = \sum_{s=-\infty}^{\infty} Q(x_s^u) g_{T/N}(t - sT/N)$. The signal $Q^u(t)$ is then multiplied by $S^u(t)$ and transmitted along the (equivalent) channel together with the spread-spectrum signals from the other users. Each user adopts a different spreading code, assigned at the connection start-up. Following the approach in References 28 and 29 we will assume that spreading sequences are periodic of a period equal to the spreading factor N.

We will model the selective fading communication channel as a linear time-varying propagation medium, where the multipath effect is due to the simultaneous presence of $\Lambda + 1$ rays for each transmitted signal. For the generic u-th user the channel is described by the following low-pass impulse response

$$h_c^u(t) = \beta_{-1}\delta(t) + \sum_{\tau=0}^{\Lambda} \beta_\tau a_\tau^u \delta\left(t - \tau\frac{T}{N}\right)$$

where $\delta(t)$ is the Dirac generalized function, β_{-1}^2 is the direct ray power attenuation, and $\beta_\tau^2 |a_\tau^u|^2$ is the instantaneous random power attenuation of the τ-th ray whose delay is $\tau T/N$. As thoroughly discussed in Reference 30, considering ray delays multiple of the chip time leads to no loss of generality.

The random nature of the first and secondary rays, assumed independent from each others, is modeled by the complex random variables a_τ^u whose real and imaginary part are independent and Gaussian with zero mean and variance of 1/2.

This channel model is the same already used in Reference 23 (with the exception of the truncation to Λ rays) and in Reference 24 because, although not the most general possible setting, it allows easing the analytical computations and to better focus the impact of the adoption of chaos-based spreading.

We indicate with K the Rice factor, i.e., the ratio between the power of the first ray and the remaining rays, such that $\beta_{-1}^2 + \beta_0^2 = K \sum_{\tau=1}^{\Lambda} \beta_\tau^2$ with $(\beta_{-1}^2 + \beta_0^2)(K+1)/K = 1$. Furthermore, to consider what fraction of the main ray power is associated to random component we refer to a second similar quantity $\alpha = \beta_{-1}^2/\beta_0^2$. The channel adds up the contributions of all the users so that the baseband receiver is presented with a signal that is the sum of the convolutions of the signal transmitted by each user and the corresponding channel impulse response. Additionally, all the random variables describing the channel contributions will be considered independent.

To cope with the selective fading effect due to the multipath propagating channel it is common practice to employ a rake filter with the aim of collecting time-scattered power before the signal undergoes a common despreading operation. More specifically, consider the v-th users rake receiver shown on the right in Figure 5.2. The presence of a further FIR filter of coefficients $\gamma_{-1}, \gamma_0, \ldots, \gamma_\Lambda$, implies that the channel is cascaded with a further linear system whose impulse response is

$$h_{\mathrm{r}}^v(t) = \gamma_{-1}\delta\left(t - \Lambda\frac{T}{N}\right) + \sum_{\tau=0}^{\Lambda} \gamma_\tau (a_\tau^v)^* \delta\left(t - (\Lambda - \tau)\frac{T}{N}\right)$$

where $\cdot^*$ stands for complex conjugation.

With this, the equivalent channel from the u-th transmitter to the v-th receiver has the impulse response

$$[h_{\mathrm{c}}^u * h_{\mathrm{r}}^v](t) = \delta\left(t - \Lambda\frac{T}{N}\right) * \sum_{\tau=-\Lambda}^{\Lambda} A_\tau^{uv} \delta\left(t - \tau\frac{T}{N}\right)$$

where the coefficient A_τ^{uv} identifies an equivalent channel model between the u-th transmitter and the v-th receiver, whose expression can be derived by direct computation of the above convolution (see e.g., Reference 24). In such a model, the main ray has amplitude A_0^{uv}, and secondary rays are delayed by $\tau T/N$ for $\tau = \pm 1, \pm 2, \ldots, \pm\Lambda$. In the following we will refer to this equivalent channel model and drop convolution with the term $\delta(t - \Lambda T/N)$, which accounts for the total transmission and filtering delay.

The joint effect of the channel and of the rake receiver on the final performance is obviously compounded in the statistics of the coefficients

A_τ^{uv}. In particular, because the features of all the random variables a_τ^v and a_τ^u are known, it is possible to compute $\mathbf{E}[A_{\tau_1}^{uv}A_{\tau_2}^{uv}]$ and $\mathbf{E}[A_{\tau_1}^{uv}(A_{\tau_2}^{uv})^*]$ with respect to all these channel parameters, and to express them in terms of the parameters β_{-1}, β_τ, γ_{-1}, γ_τ.[24]

The output of the rake filter is fed into a multiplier, which combines it with a synchronized replica of the spreading sequence of the v-th user and of an integrate-and-dump stage in charge of recovering the information symbol by correlation. Even if the unavoidable influence of thermal noise can be neglected, symbol extraction is corrupted by the presence of the delayed version of the useful signal and by the current and delayed signals transmitted by the other users, who acts as interferers for the useful one.

The key to estimate the performance of such a communication system is the knowledge of how much the nonuseful signals affect the receiver output, i.e., of how much they are correlated with the spreading sequence of the useful user.

As correlation with a fixed sequence is a linear operation, error causes add at the receiver output Υ_s^v before the hard decision block of the useful v-th user, so that it can be decomposed into three main terms

$$\Upsilon_s^v = \Omega_s^v + \Xi_s^v + \Psi_s^v$$

where Ω_s^v is the main equivalent ray carrying the information to be retrieved, Ξ_s^v represents the disturbance due to the secondary equivalent rays carrying the previous symbols of the useful user, and Ψ_s^v is the disturbance due to the primary and secondary equivalent rays carrying the information transmitted by the other users.

To compute a closed form expression for the previous quantities, one needs to consider the contribution to the correlation with the spreading sequence of the v-th user of a complex signal with envelope ϵ coming from the u-th user with relative delay Δt. Such a contribution $\Theta_s^{uv}(\epsilon, \Delta t)$ can be expressed depending on the two symbols from the u-th transmitter that are overlapping with the generic s-th symbol from the v-th one, and whose choice is linked with the integer and fractional part of $\Delta t/T$.[17]

With this we may derive[24]

$$\begin{aligned}
\Omega_s^v &= \frac{1}{2}\Theta_s^{vv}(A_0^{vv}, 0) = \frac{1}{2}S_s^v A_0^{vv} \\
\Xi_s^v &= \frac{1}{2}\sum_{\substack{\tau=-\Lambda \\ \tau\neq 0}}^{\Lambda} \Theta_s^{vv}\left(A_\tau^{vv}, -\tau\frac{T}{N}\right) \\
\Psi_s^v &= \frac{1}{2}\sum_{u\neq v}\sum_{\tau=-\Lambda}^{\Lambda} \Theta_s^{uv}\left(A_\tau^{uv}e^{\iota\theta^{uv}}, t^{uv} - \tau\frac{T}{N}\right)
\end{aligned} \tag{5.7}$$

where ι indicates the imaginary unit, θ^{uv} is the relative phase between the u-th and v-th users after demodulation and t^{uv} is the relative delay.[17]

Assuming that the receiver is synchronized with the signal it is supposed to decode, we also have $\theta^{vv}=0$ and $t^{vv}=0$.

5.4 System Performance Merit Figure

In our investigations we assume that the speed of variation of the channel parameters is of the same order as the speed of variation of the the transmitted signal. With this, expectations on both channel-related (i.e., delays and phases) and transmission-related (i.e., information symbols) random variables can be considered at the same time. These are the conditions under which the error probability $P_{\rm err}$ is the most sensible merit figure. Considering the threshold mechanism at the end of the receiver, we can expresses it as $P^v_{\rm err} = \Pr\{\Upsilon^v_s < 0|S^v_s = +1\}$.

A careful consideration of the nature of the interfering signals[24] reveals that Ξ^v_s is uncorrelated with Ω^v_s and that Ω^v_s and Ψ^{uv}_s are also uncorrelated, so that the computation of the previous quantity rely on the possibility to compute of the probability distribution of each of the three random variables composing Υ^v_s. To this aim, we mainly resort to the SGA, i.e., we assume that sums of independent random contributions are considered to be Gaussian random variables, characterized only by means of their average and variance.

The cross-interference term Ψ^v_s is made of contributions, which are independent from each other and the useful one. Hence, we may consider it as a compound zero-mean Gaussian random variable characterized by its variance $\left(\sigma^v_\Psi\right)^2 = \mathbf{E}[|\Psi^v_s|^2]$. Hence, we may use SGA again to average over all the possible pairs of users, and obtain a unique σ^2_Ψ (accounting for the average contribution to the error of the users interfering with the generic receiver) by simple multiplication by $U-1$.[23] With this we find

$$\begin{aligned}\sigma^2_\Psi &= \frac{U-1}{24N^3}\left(\beta^2_{-1}+\sum_{\tau=0}^{\Lambda}\beta^2_\tau\right)\left(\gamma^2_{-1}+\sum_{\tau=0}^{\Lambda}\gamma^2_\tau\right)\\ &\quad\times\sum_{\tau=1-N}^{N-1}\mathbf{E}_{\underline{x}^u,\underline{x}^v}\left[\Gamma^2_{N,\tau}(\underline{x}^u,\underline{x}^v)+\mathrm{Re}[\Gamma_{N,\tau}(\underline{x}^u,\underline{x}^v)\Gamma^*_{N,\tau+1}(\underline{x}^u,\underline{x}^v)]\right]\\ &= \frac{(U-1)c'}{4}\left(\beta^2_{-1}+\sum_{\tau=0}^{\Lambda}\beta^2_\tau\right)\left(\gamma^2_{-1}+\sum_{\tau=0}^{\Lambda}\gamma^2_\tau\right)\end{aligned}$$

where the quantity c' is implicitly defined by the last equality and Γ is the usual partial autocorrelation function

$$\Gamma_{N,\tau}(\underline{x}^u,\underline{x}^v) = \begin{cases}\sum_{k=0}^{N-\tau-1} Q(x^u_k)Q(x^v_{k+\tau}) & \text{if } \tau = 0,1,\ldots,N-1\\ 0 & \text{otherwise}\end{cases}$$

that is such that $\Gamma_{N,\tau}(\underline{x}^u,\underline{x}^v) = \Gamma^*_{N,-\tau}(\underline{x}^v,\underline{x}^u)$.

What we obtain for σ_Ψ^2 is nothing but the term already considered in References 17 and 18 to take co-channel interference into account, which is here multiplied by the power gain of the actual channel and of an independent rake FIR.

To characterize the self interference term we must follow a similar path first computing $\left(\sigma_\Psi^v\right)^2 = \mathbf{E}[|\Xi_s^v|^2]$ and then using again SGA to define a unique σ_Ξ^2 accounting for the average contribution of self-interference to the error on the generic receiver. By doing this, we are able to get[24]

$$\sigma_\Xi^2 = \frac{1}{4}\sum_{\tau=1}^{\Lambda}\left[c''_\tau\left(\beta_{-1}^2\gamma_\tau^2 + \beta_\tau^2\gamma_{-1}^2 + \sum_{j=0}^{\Lambda-\tau}\beta_j^2\gamma_{j+\tau}^2 + \beta_{j+\tau}^2\gamma_j^2\right)\right.$$
$$\left. + 2c'''_\tau\left(\beta_{-1}\beta_\tau\gamma_{-1}\gamma_\tau + \sum_{j=0}^{\Lambda-\tau}\beta_j\beta_{j+\tau}\gamma_j\gamma_{j+\tau}\right)\right]$$

where $c''_\tau = (\mathbf{E}_{\underline{x}^v}[\Gamma_{N,N-\tau}^2(\underline{x}^v,\underline{x}^v)] + \mathbf{E}_{\underline{x}^v}[\Gamma_{N,\tau}^2(\underline{x}^v,\underline{x}^v)])/(2N^2)$ for $\tau > 0$ and $c''_\tau = 1$ for $\tau = 0$ and $c'''_\tau = (\mathbf{E}_{\underline{x}^v}[\Gamma_{N,\tau}^2(\underline{x}^v,\underline{x}^v)])/(2N^2)$.

The treatment of the useful component Ω_s^v requires additional consideration and its approximation as a Gaussian random variable must be considered with care. A discussion of this kind of approximation can be found in Reference 24. We here exploit it to its ultimate consequences to derive that, if also the last variable can be considered Gaussian with variance σ_Ω^2, then the bit error probability can be given an expression of the kind

$$P_{\text{err}} = \frac{1}{2}\,\text{erfc}\sqrt{\rho} \tag{5.8}$$

for a signal-to-interference ratio (SIR) ρ defined as

$$\rho = \frac{\mathbf{E}[\Omega_s|S_s = +1]^2}{2(\sigma_\Omega^2 + \sigma_\Psi^2 + \sigma_\Xi^2)} \tag{5.9}$$

5.5 Performance Optimization in Dispersive and Nondispersive Channels

To optimize the final performance we may consider the expressions for $\mathbf{E}[\Omega_s|S_s = +1]^2$, σ_Ω^2, σ_Ψ^2 and σ_Ξ^2 and define the vectors $\boldsymbol{\beta} = (\boldsymbol{\beta}_1, \ldots, \boldsymbol{\beta}_{\Lambda+2}) = (\beta_{-1}, \beta_0, \ldots, \beta_\Lambda)$ and $\boldsymbol{\gamma} = (\boldsymbol{\gamma}_1, \ldots, \boldsymbol{\gamma}_{\Lambda+2}) = (\gamma_{-1}, \gamma_0, \ldots, \gamma_\Lambda)$. Then, it is convenient to define the $(\Lambda+1)\times(\Lambda+1)$ square matrix $\mathcal{A}$ as

$$\mathcal{A}_{j_1,j_2} = \begin{cases} \sum_{l=-1}^{\Lambda}\left[(U-1)c' + (1-\delta_{l,1}\delta_{j_1,1})c''_{|l-j_1|}\right](\beta_l)^2 & \text{if } j_1 = j_2 = -1,\ldots,\Lambda \\ 2c'''_{|j_1-j_2|}\beta_{j_1}\beta_{j_2} & \text{if } j_1 \neq j_2 \end{cases}$$

where $\delta_{\cdot,\cdot}$ is the usual Kronecher's symbol, and obtain

$$\rho = \frac{1}{2}\frac{\left(\boldsymbol{\beta}^T\boldsymbol{\gamma}\right)^2}{\boldsymbol{\gamma}^T\mathcal{A}\boldsymbol{\gamma}} \tag{5.10}$$

It is worthwhile to notice that in Equation (5.10) we have expressed $\rho = \rho(\mathcal{A}(\boldsymbol{\beta}), \boldsymbol{\beta}, \boldsymbol{\gamma})$ identifying the three factors affecting the performance, i.e., the statistics of the spreading sequences ($\mathcal{A}$), the channel power profile ($\boldsymbol{\beta}$), and the rake power profile ($\boldsymbol{\gamma}$).

Note also that, according to the physical intuition behind this dependence, $\rho(\mathcal{A}(\xi'\boldsymbol{\beta}), \xi'\boldsymbol{\beta}, \xi''\boldsymbol{\gamma}) = \rho(\mathcal{A}(\boldsymbol{\beta}), \boldsymbol{\beta}, \boldsymbol{\gamma})$ for any $\xi', \xi'' \neq 0$, because channel and receiver gain cannot affect the SIR.

The maximization of ρ can now be simplified exploiting this double scale-invariance of ρ and setting, without any loss of generality, $\boldsymbol{\beta}^T\boldsymbol{\beta} = 1$ and $\boldsymbol{\beta}^T\boldsymbol{\gamma} = 1$. With this, the numerator in the expression of ρ is fixed and we may maximize it solving the minimization problem

$$\min_{\boldsymbol{\gamma}} \boldsymbol{\gamma}^T\mathcal{A}\boldsymbol{\gamma} \qquad \text{s.t.} \quad \boldsymbol{\beta}^T\boldsymbol{\gamma} = 1$$

whose solution gives the optimal receiver FIR taps $\hat{\boldsymbol{\gamma}}$[31]

$$\hat{\boldsymbol{\gamma}} = \frac{\mathcal{A}^{-1}\boldsymbol{\beta}}{\boldsymbol{\beta}^T\mathcal{A}\boldsymbol{\beta}} \tag{5.11}$$

Note that the above solution relies on the possibility of extracting averages of the incoming signal and thus of estimating the transfer function of the channel. Methods for this do exist, are commonly assumed available in the classical rake receiver literature, and we assume that the resulting estimation is reliable.

Note that such an optimization depends on $\mathcal{A}$ and thus on the statistical properties of the spreading sequences and this offers room for further optimization if one may design sequence generators with prescribed statistical features, as it is possible in the chaos-based setting described in the next section. Additionally, such a remark allows us to comment about the significance of Equation (5.11) in some particular but important settings.

When information about the correlation of the spreading sequences is not available, one may assume that all the disturbances are white and thus proceed by maximization of the magnitude of the useful component in Equation (5.9) $\mathbf{E}[\Omega_s^v | S_s^v = +1] = (1/2)\boldsymbol{\beta}^T\boldsymbol{\gamma}$ only. As, under these assumptions, the power gain of the receiver acts both on the useful component and on the white disturbance, maximization can be performed for a unit power gain receiver as in

$$\max_{\boldsymbol{\gamma}} \boldsymbol{\beta}^T\boldsymbol{\gamma} \qquad \text{s.t.} \quad \boldsymbol{\beta}^T\boldsymbol{\gamma} = 1$$

which, when $\boldsymbol{\beta}^T\boldsymbol{\beta} = 1$, has the trivial solution $\hat{\boldsymbol{\gamma}} = \boldsymbol{\beta}$, as it would happen in a conventional rake receiver.

Consider now the much simpler case of a nondispersive channel, where $\beta_\tau = 0$ for $\tau \geq 0$ and $\beta_{-1} = 1$. Hence the constraint $\boldsymbol{\beta}^T \boldsymbol{\gamma} = 1$ sets $\gamma_\tau = 0$ for $\tau \geq 0$ and $\gamma_{-1} = 1$ to avoid resorting to rake-receivers for a nondispersive channel. With this, Equation (5.10) is greatly simplified and ρ is maximized when the single quantity $\mathcal{A}_{-1,-1} = (U-1)c'$ is minimized.

Details of such a minimization in the more general context of complex spreading sequences can be found in References 20 and 23. It is here enough to recall that if all the sequences can be thought as independent and if they satisfy a mild form of second-order stationarity, namely $\mathbf{E}_{\underline{x}^u}[Q(x_m^u)Q^*(x_n^u)] = \mathbf{E}_{\underline{x}^u}[Q(x_0^u)Q^*(x_{|n-m|}^u)]$ then, setting $h = 2 - \sqrt{3}$ the previous expression reaches a minimum given by

$$(U-1)\frac{\sqrt{3}}{3N}\frac{h^{-2N} - h^{2N}}{h^{-2N} + h^{2N} - 2} \tag{5.12}$$

when the sequences are characterized by a real, almost-exponential, sign-alternating autocorrelation profile given by

$$\mathbf{E}_{\underline{x}^u}[Q(x_0^u)Q^*(x_\tau^u)]^{\text{opt}} = (-1)^\tau \frac{N}{N-\tau}\frac{h^{\tau-N} - h^{N-\tau}}{h^{-N} - h^N} \overset{N \text{ large}}{\simeq} (-h)^\tau \tag{5.13}$$

It is also important to notice that, because nondispersive channels attach no randomness to the useful component of Equation (5.7) results $\Omega_s^v = 1/2S_s^v$, and the approximation that consider it as a Gaussian random variable is unnecessary. Hence, under the only assumption of Gaussian interference, we are able to identify the absolute optimum choice for spreading sequences. Because, for any practical N Equation (5.12) is approximately $(U-1)\sqrt{3}/(3N)$, i.e., 15% less than $2(U-1)/(3N)$ (the classical merit figure for ideal random sequences), a definite improvement is demonstrated.

5.6 Chaos-Based Generation of Optimized Spreading Sequences

For generating spreading sequences with chaotic maps, let us set $X = [0, 1]$ and consider a one-dimensional chaotic system (map) $x_{k+1} = M(x_k)$ which is iterated $N-1$ times starting from an initial condition x_0^u, which is randomly and independently drawn for each user according to a uniform probability density. The trajectory is quantized by a function Q: $[0,1] \mapsto \{-1, +1\}$.

Here, we assume that such a quantization is such that n is an even integer and that all the points in the intervals $X_j = [(j-1)/n, j/n]$ with $j = 1, \ldots, n/2$ or $j = n/2, \ldots, n$ are mapped by Q to the same symbol.

We also refer to a particular family of maps that has been thoroughly characterized in References 19, 22, and 23, namely the so called (n,t)-tailed shifts defined as

$$M(x) = \begin{cases} (n-t)x \,(\mathrm{mod}\ \frac{n-t}{n}) + \frac{t}{n} & \text{if } 0 \le x < \frac{n-t}{n} \\ t\left(x - \frac{n-t}{n}\right) (\mathrm{mod}\ \frac{t}{n}) & \text{otherwise} \end{cases}$$

for $t < n/2$. The family of (n,t)-tailed shifts includes some of the maps already analyzed in the chaos-based DS-CDMA framework[17] as in the n-way Bernoulli shift [which is a $(n,0)$-tailed shift] and the n-way tailed shift [which is a $(n,1)$-tailed shift]. Figure 5.1a shows the graphics of a (4,1)-tailed shift.

We are here interested in computing the entries of the matrix $\mathcal{A}$, i.e., the three quantities c', c''_τ, and c'''_τ that depend on the expectations $\mathbf{E}_{\underline{x}^v}[\Gamma^2_{N,\tau}(\underline{x}^v, \underline{x}^v)]$ and $\mathbf{E}^{u \neq v}_{\underline{x}^u, \underline{x}^v}\left[\Gamma_{N,\tau}(\underline{x}^u, \underline{x}^v)\Gamma_{N,\tau+l}(\underline{x}^u, \underline{x}^v)\right]$ for $l = 0$ and $l = 1$. To compute such expressions we may substitute the definition of $\Gamma_{N,\tau}(\underline{x}^u, \underline{x}^v)$ and perform a simple but lengthy algebraic expansion to obtain multiple sums of terms of the kind $\mathbf{E}_{\underline{x}^v}\left[Q(x^v_0)Q(x^v_{\tau_1})\right]$ and $\mathbf{E}_{\underline{x}^v,\underline{x}^u}\left[Q(x^v_0)Q(x^v_{\tau_1})Q(x^v_{\tau_2})Q(x^v_{\tau_3})\right]$.

As Q may assume only two different values, these expectations can be rewritten as a sum of the values associated to any possible product of Q multiplied by their probability, which is the joint probability that the evolution of the map state belongs to certain intervals at certain time instants.

To compute these probabilities we may exploit the detailed analysis of the statistical features of (n,t)-tailed shifts carried out in References 3, 19, and 25 and based on the tensor approach recalled in Section 5.2 and on the expression in Equation (5.6) in particular. The key property used in those works is that (n,t)-tailed shifts are PWAM maps, i.e., that the statistical features of their trajectories can be studied referring to an "equivalent" Markov chain with n states (one for each interval X_j).

When $n > 2t$, straightforward but lengthy manipulations developed in References 3, 19, and 25 yield

$$\mathbf{E}_{\underline{x}^v}[\Gamma^2_{N,\tau}(\underline{x}^v, \underline{x}^v)] = (N-\tau) + [(N-\tau)^2 - (N-\tau)]r^{2\tau}$$

and, for $l = 0$ and $l = 1$,

$$\mathbf{E}^{u \neq v}_{\underline{x}^u, \underline{x}^v}\left[\Gamma_{N,\tau}(\underline{x}^u, \underline{x}^v)\Gamma_{N,\tau+l}(\underline{x}^u, \underline{x}^v)\right] = (l+1)(N-\tau-l)r^l$$
$$+\, 2\frac{r^{2+l}}{(1-r^2)^2}\left[(1-r^2)(N-\tau-l) + r^{2(N-\tau-l)} - 1\right]$$

where the number $r = -t/(n-t)$ is a map design parameter.

Note that, as needed, these expressions are the link between n and t and the quantities c', c''_τ, and c'''_τ that are assumed known in the optimization of the receiver FIR in the previous sections.

In fact, for each r (and thus for each sequence generation mechanism) the optimization procedure sketched in Section 5.5 gives us the best receiving filter and thus the lowest achievable P_{err}. This information can be used

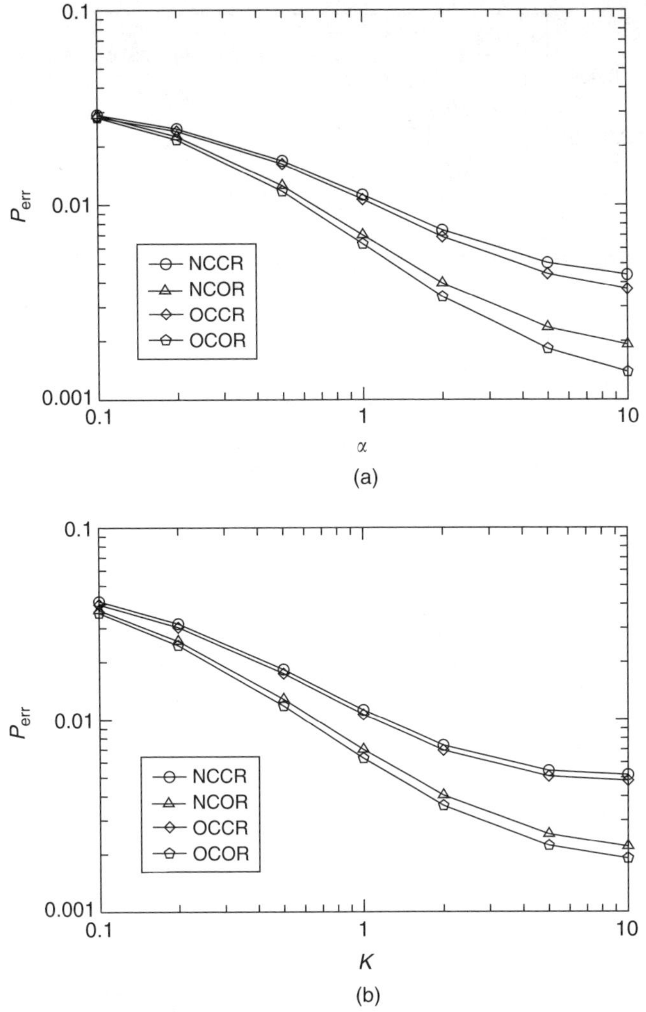

Figure 5.3 Performance of different systems for different channel configurations and different systems with $\beta_k \propto e^{-k}$ (a) $K = 1$, $\alpha \in [0.1,10]$, (b) $\alpha = 1$, $K \in [0.1,10]$; NCCR = pseudo-random with classical rake, NCOR = pseudo-random with optimized rake, OCCR = optimized chaotic with classical rake, OCOR = optimized chaotic with optimized rake.

to iteratively search for that r whose corresponding optimum filter results in the minimum P_{err} thus simultaneously optimizing sequence generation and receiver taps. We will call this system OCOR, i.e., optimal chaos with optimal rake. Among the other possible options to test, we restrict ourselves to test the most interesting ones in a multipath propagation channel, namely OCCR (optimal chaos with conventional rake) and NCOR (no chaos with

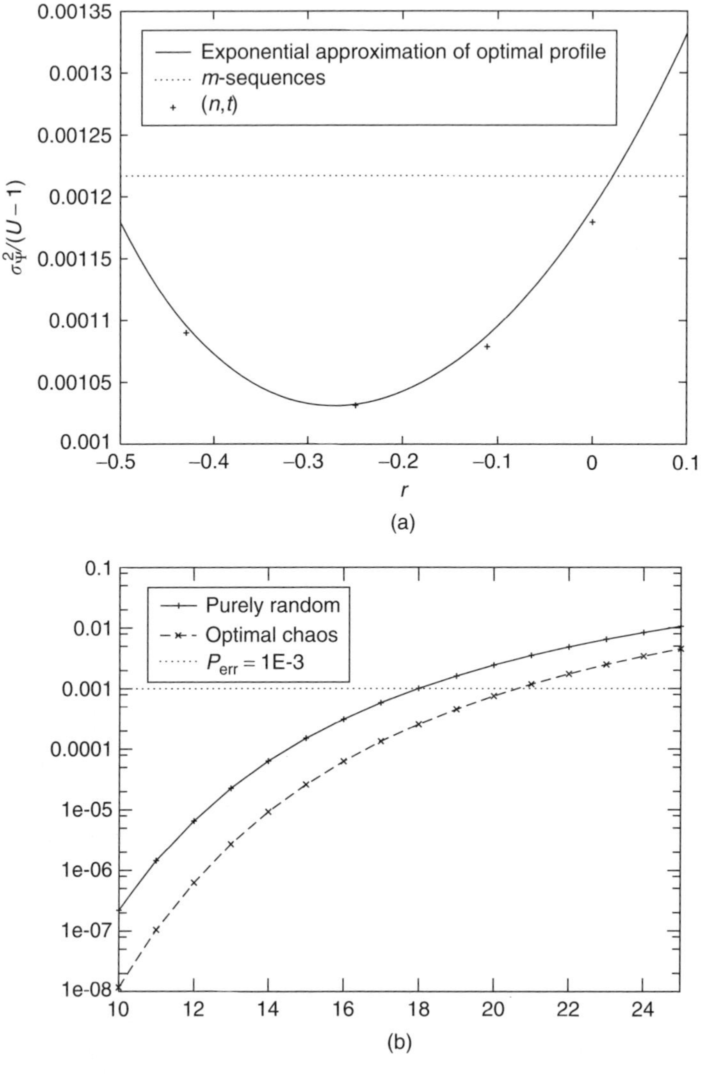

Figure 5.4 (a) Plot of $\sigma^2_\Psi/(U-1)$ as a function of r for the exponential approximation of Equation (5.13), the classical m-sequences and chaos-based ones obtained by using (10,0)-, (10,1)-, (10,2)-, and (10,3)-tailed shifts; (b) comparison of the P_{err} when classical m-sequences or the optimal ones (generated by (10,2)-tailed shifts) are adopted for spreading, as a function of the number of user U.

optimal rake) as well as NCCR (no chaos with conventional rake) where nonchaotic sequences are assumed to be a good approximation of the purely random case as, for example, m- or Gold codes (see e.g., Reference 17 and references therein).

Performance of the different systems is reported in Figure 5.3 for two different channel configurations, i.e., $\beta_k \propto e^{-k}$, $K=1$, and α spanning [0.1,10] for the first case while $\alpha=1$, and K spanning [0.1,10] for the second case. As can be easily seen, even if gains depend on the channel configuration, the relative ranking is always the same, namely $P_{\text{err}}^{\text{NCCR}} > P_{\text{err}}^{\text{OCCR}} > P_{\text{err}}^{\text{NCOR}} > P_{\text{err}}^{\text{OCOR}}$. We may therefore conclude that the adoption of chaos always benefits the overall performance even though the largest improvement is due to the availability of an optimization of the rake receiver taps.

In nondispersive channels, a no-rake receiver is needed and (n,t)-tailed shift allow the absolute optimization of performance. In fact, for any given n, we may chose t as the integer closest to $nh/(1+h)$ so that $r \simeq -h$, the accuracy being obviously improved as $n \to \infty$. With this, the autocorrelation profile of the generated sequences matches Equation (5.13) with a negligible error in practically any operation conditions. This is exemplified in Figure 5.4a which shows the trend of $\sigma_\Psi^2/(U-1)$ as a function of r; trivial visual inspection is sufficient to verify the match between the theoretical prediction and the actual data proving that the optimal profile of Equation (5.13), corresponding to the minimum of $\sigma_\Psi^2/(U-1)$, can be practically achieved by sequences generated by an (10,2)-tailed shift. The achievable advantage of chaos-based spreading is further highlighted in Figure 5.4b which shows P_{err} computed using Equation (5.8) and Equation (5.9) as a function of the number of users U. As can be seen, employing chaos-based sequences generated allows, for any fixed quality of the communication link (e.g., $P_{\text{err}} = 10^{-3}$), to increase the number of users that can be allocated in the system (e.g., from References 18 to 21) with respect to the case of conventional m- (and Gold) sequences.

5.7 Chaos-Based DS-CDMA System Prototype and Measurement Results

With the aim of obtaining a first and effective confirmation of the theoretical results reported in the previous Sections, we realized a low-cost prototype of the communication system using TMS320C542 DSP boards. A picture of the whole system is reported in Figure 5.5a.

Transmitter boards are in charge of generating baseband signals, of their BPSK modulation and of the simulation of the channel. No external RF circuit has been included in this low-cost realization. One board is devoted to the implementation of the useful transmitter, one board is devoted to the implementation of the corresponding receiver and of the estimator of the bit error rate while all the other boards can be used to account for multiple access interference.

Spreading sequences are generated off-line and stored on the transmitters. A copy of the useful sequences is also stored in the receiver. At

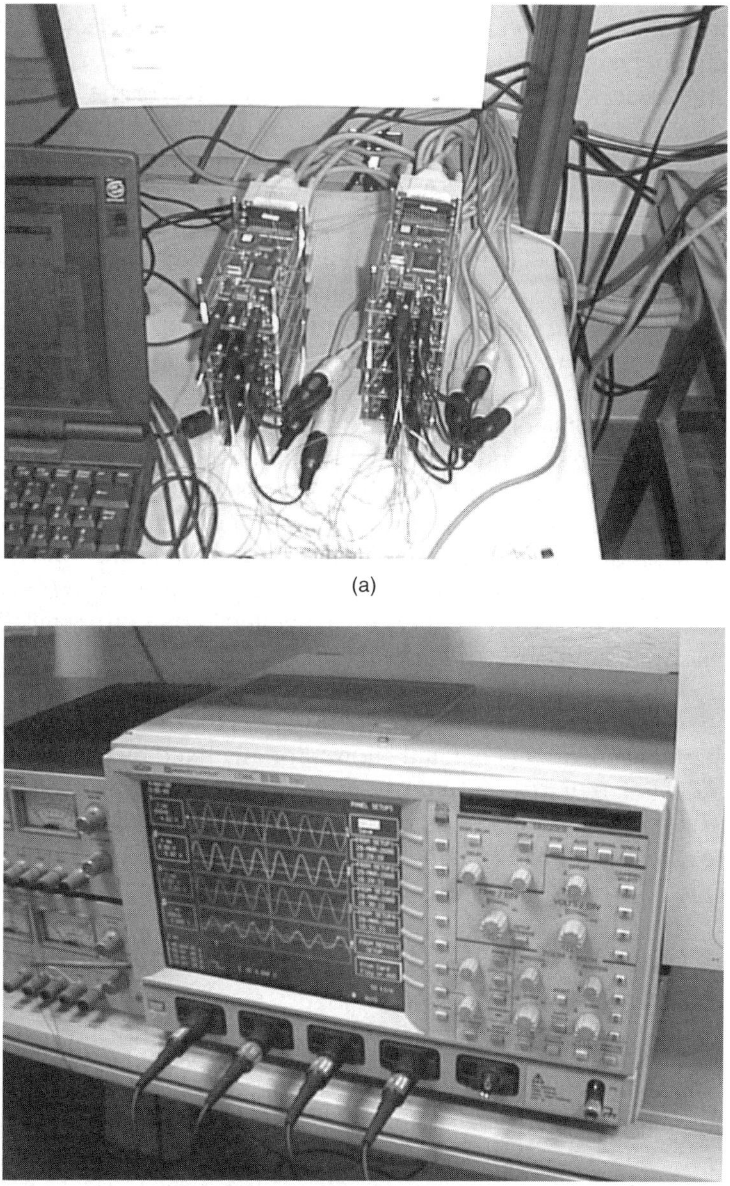

(a)

(b)

Figure 5.5 (a) Experimental set-up of the DS-CDMA communication system prototype and its schematic representation and (b) oscilloscope waveforms of (top to bottom) the BPSK transmitted signal of three different users (transmitter and two interferes) and of the received signal.

symbol rate, information bits are locally generated. They are multiplied by the spreading sequence thus generating data at chip rate that are fed into a quadrature implementation of the five-taps ($\Lambda = 4$) complex FIR simulating the channel. The two baseband components are then multiplied by stored samples of sine and cosine to generate the passband signals that are added and converted into the analog output of the transmitter. The offsets in the sine and cosine table can be adjusted to account for different time-varying delays and carrier phases in the interfering transmitters, they are kept constant in the useful transmitter.

The same structure is replicated for the interfering transmitters in which a randomly variable delay and a random carrier phase are also introduced to account for the asynchronous environment. The delay and carrier phase are obtained in the digital chain by means of the adjustable offsets in the sine and cosine table. In the transmitters, digitally generated pseudorandom variables are used to generate information bits, generate and update real and imaginary parts of the channel FIR taps once every u_c bits, as well as to possibly generate and update interferer delays and carrier phases once every u_i bits.

The signal arriving at the receiver is converted into digital and fed into a quadrature demodulator relying again on cosine and sine tables and on accumulators. This produces two streams of samples at chip-rate, which are separately processed by a quadrature implementation of the rake FIR and finally summed. The output of the rake is then fed into the correlation-based despreader and in the decision block.

To ensure correct operation of the receiver, it must be synchronized with the transmitter and the channel taps have to be perfectly identified. In addition, to ascertain if the received bit is identical to the transmitted one the information stream generated at the transmitter must be also available. In fact, the two bit streams must be compared to compute the bit error rate.

As far as all the quantities depending on digital pseudorandom variables are concerned, we use a suitable digital generator addressing the trade-off between randomness and fixed-point computation whose seed is common to both the transmitter and receiver.

As far as the time-synchronization is concerned, the Digital-to-Analog converter of the useful transmitter and the Analog-to-digital converter of the receiver share the same timing signals. In addition, at startup, a simple handshake is initiated relying on two more wires carrying interrupt signals from one board to the other.

5.7.1 Performance over Nondispersive Channel

The prototype system has been first exploited to test performance of chaos-based DS-CDMA systems when no multipath is present. In this case $\Lambda = 1$, $K = \alpha = \infty$.

Table 5.1 Theoretical and measured P_{err} with nondispersive channel

		$P_{err} \times 10^3$	
	N	*Theory*	*Measure*
Conventional Purely Random Sequences	15	13.5	16.2
Optimal Chaos-based Sequences	15	6.33	6.52
Conventional Purely Random Sequences	20	2.66	3.81
Optimal Chaos-based Sequences	20	0.98	1.46

Table 5.2 Theoretical and simulated P_{err} over a dispersive channel with exponential power profile

		$P_{err} \times 10^2$	
	Optimal r	*Theory*	*Measure*
OCOR	−0.217	2.22	2.31
NCOR	na	2.65	2.95
OCCR	−0.192	3.02	3.18
NCCR	na	3.42	3.77

Figure 5.5b shows, from top to bottom, the signal at the output of the transmitter, of two different interferes, and at the input of the receiver. Table 5.1 compares the measured bit error rate when the prototype system is operated for 2×10^6 bits for $U = 6$, $N = 15$ and $N = 20$, and $u_i = 10$. For every configuration, performance is averaged over 100 sequence sets and considering the average P_{err} experienced by every user.

Results confirm the fact that chaos-based DS-CDMA outperform classical DS-CDMA by approximately 60%. Also quantitative theoretical predictions are in good agreement with experimental results.

5.7.2 Performance over Exponential Channel

To test performance over a dispersive channel, we choose an exponential power profile characterized by $K = 1$, $\alpha = 10$, $\Lambda = 4$, and $\beta_k \propto e^{-k}$.

We used all the available interfering boards for a total of $U = 8$. We also have $N = 15$, and $u_c = u_i = 3$. The number of transmitted bits per configuration was 2×10^6. For every configuration, performance is averaged over 100 sequence sets. For every sequence set the average P_{err} experienced by all the user is considered.[26]

Again, results confirm the fact that chaos-based DS-CDMA outperforms classical DS-CDMA by up to 22% and that the adoption of nonconventional and carefully designed receiver filters also contributes nonnegligibly to the quality of the communication.

5.8 Ultimate Limits — Numerical Evidence

We here aim at taking a first step in the direction of assessing the ultimate limits of chaos-based asynchronous DS-CDMA. To do so we interpret the spreading operation as part of an AWGN vector channel and exploit the concept of channel capacity as defined by Shannon to compare chaos-based spreading with conventional shift-register based sequences and ideal random sequences.[5]

5.8.1 System Model

We will assume that each of the U users in the system is given an N-chip signature $s^u(t) = \sum_{k=0}^{N-1} y_k^u g(t-k)$ where g is a unit rectangular pulse with unit duration and $y_k^u = \pm 1/\sqrt{N}$ are symbols of the spreading sequence. The u-th bit stream b_j^u is then transmitted as $\sum_{j=-\infty}^{\infty} b_j^u s^u(t-jN)$.

In an asynchronous environment the receiver is presented with the total signal $r(t) = \sum_{u=0}^{U-1} \sum_{j=-\infty}^{\infty} b_j^u s^u(t-jN-\tau^u) + \nu(t)$ where τ^u is the random delay between the u-th transmitter and the receiver and $\nu(t)$ is the AWGN. Delay dynamics will be assumed to be much slower than bit rate to fit within the classical information theory view and delays will be considered constant for all the transmission and uniformly distributed in $[0, N]$.

Decoding is performed after acquisition of the signal in the time window $[0, BN]$, i.e., for B bit durations. Due to the asynchronicity of the transmitters, in general $B+1$ bits of each user overlap with this window.

If ϕ_i $(i=0, \ldots, \infty)$ is an orthonormal signal basis in $[0, BN]$ we may think that the decoding process depends on the projections $r_i = \langle r, \phi_i \rangle = \int_0^{BN} r(t)\phi_i(t)dt$ of $r(t)$ on such a basis.

Consider n of those projections arranged in the vector $r = [r_0, \ldots, r_{n-1}]$. They are linked to the transmitted bits $b = [b_0^0, \ldots, b_B^0, b_0^1, \ldots, b_B^1, \ldots, b_0^{U-1}, \ldots, b_B^{U-1}]$ by $r = Hb + \nu$ where ν is the vector of noise projections and the matrix H can be constructed defining $Q(x,y)$ and $R(x,y)$ to give respectively the quotient and remainder of the integer division of x by y, setting $S_j(t) = s^{Q(j,B+1)}(t - R(j, B+1)N - \tau^{Q(j,B+1)})$ for $j = 0, \ldots, U(B+1)-1$, to obtain $H_{ij} = \langle S_j, \phi_i \rangle$.

We know from Reference 32 that the Shannon capacity of a vector transmission system characterized by unit power bits, by a transfer matrix H, and subject to a noise of power σ^2 is given by

$$C = \log_2 \det\left(I_n + \frac{HH^\dagger}{\sigma^2}\right) = \log_2 \det\left(I_{U(B+1)} + \frac{H^\dagger H}{\sigma^2}\right)$$

where I_n is the $n \times n$ identity matrix, $H^\dagger$ is the transpose conjugate of H. Note that the two expressions are equivalent thanks to the fact that the nonvanishing eigenvalues of $HH^\dagger$ and $H^\dagger H$ are the same.

To check this, note that if the vector x is such that $HH^{\dagger}x = \lambda x$ then the vector $y = H^{\dagger}x$ is such that $H^{\dagger}Hy = H^{\dagger}HH^{\dagger}x = \lambda H^{\dagger}x = \lambda y$.

From our definitions, we have

$$(H^{\dagger}H)_{ij} = \sum_{k=0}^{n-1} H^*_{ki}H_{kj} = \sum_{k=0}^{n-1} \langle S_i, \phi_k\rangle^* \langle S_j, \phi_k\rangle$$

where $\cdot^*$ denotes complex conjugation. If we assume that the receiver can take advantage of the whole received signal we may assume $n \to \infty$ to get $(H^{\dagger}H)_{ij} = \int_0^{BN} S_i^*(t)S_j(t)dt$.

5.8.2 Capacity Comparison

Because the matrix $H^{\dagger}H$ is a function of both the spreading sequences and the random delays τ^u the capacity itself is a random variable whose distribution is controlled by the selection of the sequences.

A fair comparison of systems adopting different sequences should consider how each of them performs for any given set of delays. Hence it must be based on joint distribution of the capacities that, in general, cannot be assumed to be an independent random variables.

In our investigation we have chosen to compare systems based on maximum-length sequences (m-sequences, subscript m), Gold sequences (subscript g), random (i.e., of binary equivalent and independent symbols) sequences (subscript r), and chaos-based sequences (subscript c) of the kind that is known to optimize single-user correlation-based decoding.[20] For these systems we run extensive Monte-Carlo simulations to analyze the cases $N = 15, 31, 63$ and various combinations of system load U and decoding horizons B.

Let us first concentrate on $\mathbb{F}_x(c) = P\{C_x > c\}$, i.e., on the probability that the capacity of system x exceeds the threshold c.[32] Figure 5.6 shows $\mathbb{F}_m$,

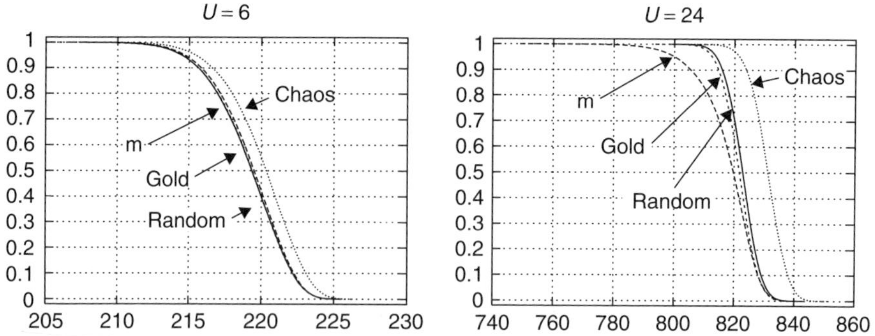

Figure 5.6 Plots of $\mathbb{F}_m$, $\mathbb{F}_g$, $\mathbb{F}_r$, $\mathbb{F}_c$ for $N = 31$ and $B = 1$.

$\mathbb{F}_g$, $\mathbb{F}_r$, $\mathbb{F}_c$ in the case $N=31$, $B=1$ for an SNR of 20 dB[32] and either $U=6$ or $U=24$. For $U=6$ note that $\mathbb{F}_m(c) \simeq \mathbb{F}_g(c) \simeq \mathbb{F}_r(c) < \mathbb{F}_c(c)$.

The indistinguishability of the nonchaotic cases is due to the substantial identity of behavior between random codes (that have been already thoroughly analyzed in the literature[34]) and deterministic codes designed to mimic them. This is further confirmed by the case $U=24$, in which the six m-sequences must be reused so often that their random-like behavior is completely lost leading to a poorer performance, i.e., to a profile $\mathbb{F}_m(c) < \mathbb{F}_g(c) \simeq \mathbb{F}_r(c) < \mathbb{F}_c(c)$.

This replicates for all the tested configurations and leads us to conclude that, once we have quantified the improvement of the chaos-based system over random sequences we have also hints on the improvement over their deterministic approximations.

To do so, in Table 5.3 we span a number of configurations for N, U, and B, each of them characterized by a system load $\lambda = U/N$ and by a decoding complexity $\gamma = (B+1)U$, i.e., the number of bits simultaneously present on the channel that we may want to capture thus discriminating among 2^γ hypotheses. For each configuration we report $P_{\alpha>0} = P\{C_c > C_r\}$; $\alpha_{10\%}$ such that $P\{C_c/C_r > 1 + \alpha_{10\%}\} = 10\%$, i.e., the least relative improvement that can be guaranteed with a 10% probability, and $\bar{\alpha} = \mathbf{E}[C_c/C_r]$, i.e., the average relative improvement.

What we get from Table 5.3 can be summarized in few points.

- $P_{\alpha>0}$ decreases as the signals tend to be "more orthogonal" i.e., for N increasing with fixed U; yet, whenever λ is kept approximately constant, $P_{\alpha>0}$ increases to almost certainty for N increasing.
- Increasing the decision horizon reduces $\alpha_{10\%}$ and $\bar{\alpha}$; yet this does not hold whenever we keep both λ and γ approximately constant (see for example the configurations U-N-B = 15-6-10, 31-12-5, 63-24-5 and U-N-B = 15-12-10, 31-24-5, 63-48-2);
- The $\alpha_{10\%}$ column reveals that relative improvements in the percent range can be always achieved with nonnegligible probability;
- The $\bar{\alpha}$ column reveals average improvements of more than 1.4% for high loads $\lambda > 0.75$ and reasonable decoding complexity $\gamma < 100$.

As a final remark, note that, though no formal link exists between capacity and performance of non-error-free realizations, even percent-range improvement may hint to significant bit error rate gain. Recall, for example, that optimum sequence selection for single-user correlation detectors[20] changes the signal-to-interference ratio from $2/(3N)$ to $\sqrt{3}/(3N)$ producing a capacity increase of less than 4% for $N > 21$ but a reduction of order of magnitudes in bit-error probabilities.

Table 5.3 Comparison between chaos-based and random spreading

N	U	B	λ	γ	$p_{\alpha\geq 0}$	$\alpha_{10\%}$	$\bar{\alpha}$
15	6	1	0.40	12	0.782	3.6%	1.3%
15	6	2	0.40	18	0.800	2.7%	1.0%
15	6	5	0.40	36	0.803	2.1%	0.8%
15	6	10	0.40	66	0.803	1.9%	0.8%
15	12	1	0.80	24	0.945	4.9%	2.7%
15	12	2	0.80	36	0.946	3.7%	2.0%
15	12	5	0.80	72	0.930	3.0%	1.6%
15	12	10	0.80	132	0.918	2.7%	1.4%
31	6	1	0.19	12	0.752	2.1%	0.8%
31	6	2	0.19	18	0.772	1.5%	0.5%
31	6	5	0.19	36	0.779	1.1%	0.4%
31	6	10	0.19	66	0.772	1.0%	0.4%
31	12	1	0.39	24	0.882	2.4%	1.1%
31	12	2	0.39	36	0.893	1.8%	0.9%
31	12	5	0.39	72	0.892	1.4%	0.7%
31	12	10	0.39	132	0.879	1.2%	0.6%
31	24	1	0.77	48	0.989	3.0%	1.9%
31	24	2	0.77	72	0.990	2.2%	1.4%
31	24	5	0.77	144	0.985	1.8%	1.1%
31	24	10	0.77	264	0.979	1.7%	1.0%
63	6	1	0.10	12	0.730	0.9%	0.3%
63	6	2	0.10	18	0.772	0.7%	0.2%
63	6	5	0.10	36	0.752	0.5%	0.2%
63	6	10	0.10	66	0.744	0.5%	0.2%
63	12	1	0.19	24	0.823	1.1%	0.5%
63	12	2	0.19	36	0.837	0.8%	0.4%
63	12	5	0.19	72	0.845	0.6%	0.3%
63	12	10	0.19	132	0.837	0.6%	0.3%
63	24	1	0.38	48	0.978	1.6%	1.0%
63	24	2	0.38	72	0.981	1.2%	0.8%
63	24	5	0.38	144	0.978	0.9%	0.6%
63	24	10	0.38	264	0.968	0.8%	0.5%
63	48	1	0.76	96	0.999	2.4%	1.8%
63	48	2	0.76	144	0.999	1.7%	1.4%
63	48	5	0.76	288	0.999	1.4%	1.0%
63	48	10	0.76	528	0.999	1.2%	0.9%

5.9 Ultimate Limits — Some Analytical Considerations

The numerical evidence in the previous section seems to confirm that the same family of chaos-based spreading sequences minimizing the bit-error[33] rate yields nonnegligible improvements in terms of Shannon capacity when compared with classical schemes.

We concentrate here on such a family that contains spreading sequences with exponentially vanishing correlations and try a further analytical step considering a performance figure related to Shannon capacity and showing how chaos-based codes can be designed to optimize it.

More formally, we will concentrate on an analytically tractable case with two users[34] ($U=2$) and assume the system model of the previous section with $B=1$.

Channel dynamics will be assumed to be much slower than bit rate to fit within the classical information theory view and delays and phases will be considered constant for all the transmission and uniformly distributed respectively in $[0,N]$.

Decoding is performed after acquisition of the signal in the time window $[0,N]$ with which, in general two bits of each users overlap.

5.9.1 Capacity Related Performance

With this and following the same approach of Reference 5 we obtain the maximum error-free transmission rate as

$$C = \frac{1}{2}\log_2 \det\left(I_4 + \rho G\right) \tag{5.14}$$

bits per independent use of the system, where $\rho = 1/\sigma^2$ is the signal-to-noise ratio, I_4 is the 4×4 identity matrix, the matrix G encodes partial autocorrelation between delayed chunks of spreading sequences in the scheme

$$G = H^{\dagger}H = \frac{1}{N}\begin{pmatrix} N-\tau_0 & 0 & a & b \\ 0 & \tau_0 & 0 & c \\ a & 0 & N-\tau_1 & 0 \\ b & c & 0 & \tau_1 \end{pmatrix}$$

where a, b, and c depend on the quantities, k_i integer and $\delta_i \in [0,1]$ such that $\tau_i = k_i + \delta_i$, on $\delta = \delta_1 - \delta_0$ and on $\left|{}^{s}_{p,q}\right| = \sum_{i=0}^{s} y^0_{i+p} y^1_{i+q}$ as

$$a = (1-\delta_1)\left|{}^{0}_{k_0,k_1}\right| + \delta\left|{}^{N-k_1-2}_{k_0,k_1+1}\right| + (1-\delta)\left|{}^{N-k_1-2}_{k_0+1,k_1+1}\right|$$

$$b = \delta\left|{}^{k_1-k_0}_{n-k_1+k_0-1,0}\right| + (1-\delta)\left|{}^{k_1-k_0-1}_{n-k_1+k_0,0}\right|$$

$$c = \delta_0\left|{}^{0}_{k_0,k_1}\right| + \delta\left|{}^{k_0-1}_{0,k_1-k_0+1}\right| + (1-\delta)\left|{}^{k_0-1}_{0,k_1-k_0}\right|$$

when $\delta \geq 0$ while

$$a = (1-\delta_0)\left|{}^{0}_{k_0,k_1}\right| - \delta\left|{}^{N-k_1-1}_{k_0+1,k_1}\right| + (1+\delta)\left|{}^{N-k_1-2}_{k_0+1,k_1+1}\right|$$

$$b = -\delta\left|{}^{k_1-k_0-1}_{n-k_1+k_0,0}\right| + (1+\delta)\left|{}^{k_1-k_0-2}_{n-k_1+k_0+1,0}\right|$$

$$c = \delta_1\left|{}^{0}_{k_0,k_1}\right| - \delta\left|{}^{k_0}_{0,k_1-k_0-1}\right| + (1+\delta)\left|{}^{k_0-1}_{0,k_1-k_0}\right|$$

when $\delta < 0$.

Note now that, from the expression in Equation (5.14) of the maximum achievable error-free rate, and from the classical construction of the coding stage achieving it (see e.g., Reference 35, Chapter 10) we get that $D(n) = \det^{\frac{n}{2}}(I_4 + \rho G)$ is nothing but the maximum number of codewords that can be reliably transmitted per n independent uses of the system, when using an ideal transmission scheme.

It can be intuitively accepted that the global performance of a system undergoing a large number of independent uses can be characterized almost equivalently by $D(n)$ for any finite n. For the sake of simple analytical tractability we here chose $n=2$ and consider the average $\mathbf{E}_{\tau y}[D(2)]$ over all the possible pairs of (τ_0, τ_1) and spreading sequences.

More in detail we have $D(2) = \sum_{i=0}^{4} \alpha_i(\rho/N)^i$ where α_i is the sum of the determinants of all the possible matrices built from i different columns of G and $4-i$ columns of I_4. With this we immediately have

- $\alpha_0 = 1$
- $\alpha_1 = \mathrm{trace}(NG) = 2N$
- $\alpha_2 = N^2 + N(\tau_0+\tau_1) - \tau_0^2 - \tau_1^2 - a^2 - b^2 - c^2$
- $\alpha_3 = N(\tau_0+\tau_1) - \tau_0^2 - \tau_1^2 - (\tau_0+\tau_1)a^2 - (N+\tau_0-\tau_1)b^2$
  $- (2N+\tau_0+\tau_1)c^2$
- $\alpha_4 = N^4\det(G) = (N-\tau_0)\tau_0(N-\tau_1)\tau_1 - \tau_0\tau_1 a^2 - \tau_0(n-\tau_1)b^2$
  $- (n-\tau_0)(n-\tau_1)c^2 + a^2c^2.$

Because delays are independent of spreading codes, for any function f we may write $\mathbf{E}_{\tau y}[f(\tau_0,\tau_1)x^2] = \mathbf{E}_{\tau}[f(\tau_0,\tau_1)\mathbf{E}_y[x^2]]$ when x is a, b, or c. Hence, we may concentrate on the $\mathbf{E}_y[x^2]$ that is the sums of terms of the kind

$$\mathbf{E}_y\left[\left|{}^{s_1}_{p_1,q_1}\right|\left|{}^{s_2}_{p_2,q_2}\right|\right] = \sum_{i_1,i_2=0}^{s_1,s_2} \mathbf{E}_y\left[y^0_{i_1+p_1}y^0_{i_2+p_2}\right]\mathbf{E}_y\left[y^1_{i_1+q_1}y^1_{i_2+q_2}\right]$$

Moreover, $\mathbf{E}_{\tau y}[a^2c^2]$ is made of terms of the kind

$$\mathbf{E}_{\tau y}\left[\left|{}^{s_1}_{p_1,q_1}\right|\left|{}^{s_2}_{p_2,q_2}\right|\left|{}^{s_3}_{p_3,q_3}\right|\left|{}^{s_4}_{p_4,q_4}\right|\right]$$

$$= \mathbf{E}_{\tau}\left[\sum_{i_1,\ldots,i_4=0}^{s_1,\ldots,s_4} \mathbf{E}_y\left[y^0_{i_1+p_1}y^0_{i_2+p_2}y^0_{i_3+p_3}y^0_{i_4+p_4}\right]\mathbf{E}_y\left[y^1_{i_1+q_1}y^1_{i_2+q_2}y^1_{i_3+q_3}y^1_{i_4+q_4}\right]\right]$$

Whenever the map producing the spreading sequence is piecewise affine[33] a systematic tensor-based computation yields analytical expressions for the above terms. In particular, we may consider the family of the so-called (n,t)-tailed shifts that are known to produce spreading sequences that are optimal in reducing multiple access interference when bit-error probability is the merit figure.

When these map are quantized to give zero-mean symbols we have $\mathbf{E}_y[y_0^u y_j^u] = r^j$ and $\mathbf{E}_y[y_0^u y_j^u y_k^u y_l^u] = r^{l-k+j}$, if $j \leq k \leq l$ and $r = -t/(n-t)$ is linked to the geometrical feature of the map.

5.9.2 Average Performance

Exploiting these expression for all the necessary correlations we arrive at a computable form for $\mathbf{E}_{\tau y}[D(2)]$ as a function of N and r. This allows us to investigate the performance of different choices in the family of the (n,t)-tailed shifts.

Note that the statistical behavior of an $(n,0)$-tailed shift ($r = 0$) is indistinguishable from that of a generator of sequences made of independent and identically distributed symbols (i.i.d.). Because i.i.d is the case that most of the common methods for sequence generation aim at approximating, we will use it as a reference.

In Figure 5.7 we set $N = 7$ and report the behavior as a function of r of the three nontrivial coefficients $\mathbf{E}_{\tau y}[\alpha_2]$, $\mathbf{E}_{\tau y}[\alpha_3]$, and $\mathbf{E}_{\tau y}[\alpha_4]$ that control the dependence of $\mathbf{E}_{\tau y}[D(2)]$ on ρ. To highlight the gain with respect to the i.i.d. case, the plots are normalized with respect to their value for $r = 0$.

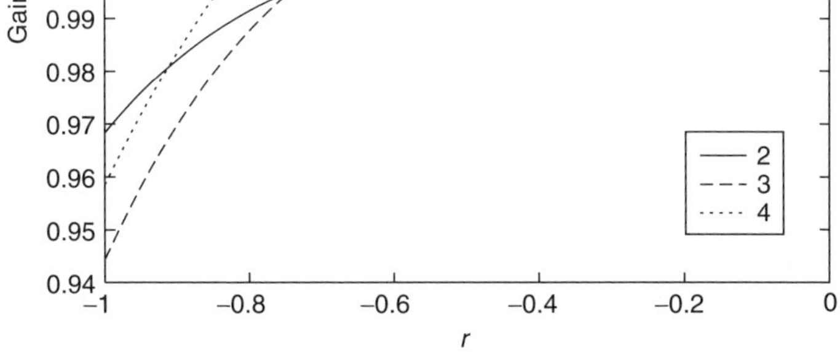

Figure 5.7 Gain in $E_{\tau y}[\alpha_2]$, $E_{\tau y}[\alpha_3]$, and $E_{\tau y}[\alpha_4]$ over i.i.d. spreading sequences as a function of the correlation decay for $N = 7$.

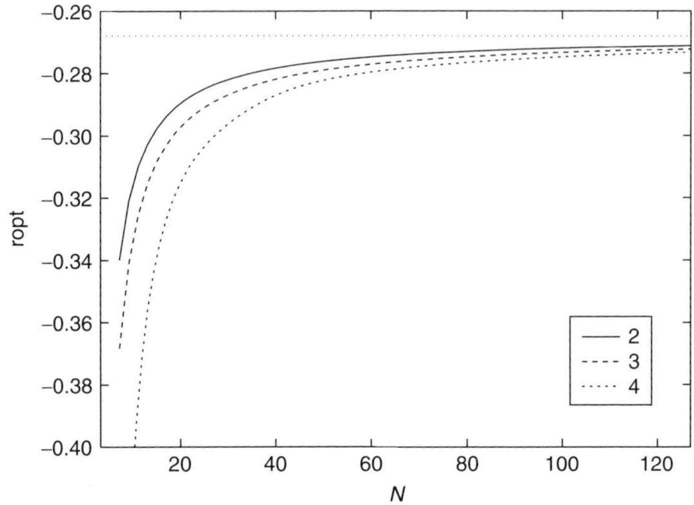

Figure 5.8 Trend of the optimum correlation decay as a function of *N*.

Note that all the coefficients feature an absolute maximum gain greater than 1 for a strictly negative r. Hence, maximum performance is always achieved for negatively correlated spreading sequences though the exact value of r for which maximum performance is achieved may vary with ρ.

It is now interesting to investigate how the optimal sequence choice behaves for large N. To this aim we plot in Figure 5.8 the values of r maximizing $\mathbf{E}_{\tau y}[\alpha_2]$, $\mathbf{E}_{\tau y}[\alpha_3]$, and $\mathbf{E}_{\tau y}[\alpha_4]$ for increasing values of N. Note how absolute maxima tend to accumulate in a neighborhood of $r = -2 + \sqrt{3} = -0.268$, which is known to optimize performance also when bit-error rate is the merit figure.[20]

This indicates that, for large N, the best choice for spreading sequences is actually independent of ρ and entails a decay regulated by $-2 + \sqrt{3}$, which therefore seems to play a special role for asynchronous DS-CDMA systems.

References

1. Lasota, A. and Mackey, M.C., *Chaos, Fractals, and Noise*, Springer-Verlag, 1994.
2. Baladi, V., *Positive Transfer Operator and Decay of Correlations*, World Scientific, Singapore, 2000.
3. Setti, G., Mazzini, G., Rovatti, R., and Callegari, S., Statistical modeling of discrete-time chaotic processes: basic finite-dimensional tools and applications, *Proc. IEEE*, 662–690, May 2002.

4. Rovatti, R., Mazzini, G., Setti, G., and Giovanardi, A., Statistical modeling of discrete-time chaotic processes: advanced finite-dimensional tools and applications, *Proc. IEEE*, 820–841, May 2002.
5. Rovatti, R., Mazzini, G., and Setti, G., Shannon capacities of chaos-based and conventional asynchronous DS-CDMA systems over AWGN channels, *IEE Elect. Lett.*, 38, 478–480, 2002.
6. Götz, M., Kelber, K., and Schwarz, W., Discrete-time chaotic encryption systems – Part I: Statistical design approach, *IEEE Trans. Circuits Syst. – Part I*, 44, 963–970, 1997.
7. Nikolaidis, N., Tsekeridou, S., Nikolaidis, A., Tefas, A., Solachidis, V., and Pitas, I., Applications of chaotic signal processing Techniques to Multimedia Watermarking, *Proc. NDES 2000*, 1–7, Catania, May 2000.
8. Erramilli, E. and Slingh, P., An application of deterministic chaotic maps to model packet traffic, *Queining Syst.*, 20, 171–206, 1995.
9. Kohda, T. and Tsuneda, A., Statistics of chaotic binary sequences, *IEEE Trans. Inform. Theory*, 43, 104–112, 1997.
10. Schimming, T., Dedieu, H., Hasler, M., and Ogorzałek, M., Noise filtering in chaos-based communication, Chap. 8 in *Chaotic Electronics for Telecommunications*, Kennedy, M.P., Rovatti, R., and Setti, G., Eds., CRC Press, Boca Raton, 2000.
11. Setti, G., Balestra, M., and Rovatti, R., Experimental verification of enhanced electromagnetic compatibility in chaotic FM clock signals, *Proc. ISCAS 2000*, III-229–III-232, Geneva, May 2000.
12. Deane, J.H.B. and Hamill, D.C., Improvement of power supply EMC by chaos, *Elect. Lett.*, 32, 1045, 1996.
13. Baranowski, A.L., Mögel, A., Schwarz, W., and Woywode, O., Statistical Analysis of a dc-dc Converter, *Proc. ISCAS 2000*, II-108–II-111, Geneva, May 2000.
14. Kennedy, M.P., Rovatti, R., and Setti, G., Eds., *Chaotic Electronics in Telecommunications*, CRC Press, Boca Raton, 2000.
15. Kohda, T., Tsuneda, A., and Sakae, T., Chaotic binary sequences by Chebyshev maps and their correlation properties, *Proc. IEEE 2nd Int. Symp. Spread Spectrum Tech. Appli.*, 63–66, 1992.
16. Kohda, T. and Tsuneda, A., Explicit evaluations of correlation functions of Chebyshev binary and bit sequences based on Perron–Frobenius Operator, *IEICE Trans. Fund.*, E77-A, 1794–1800, 1994.
17. Mazzini, G., Setti, G., and Rovatti, R., Chaotic complex spreading sequences for asynchronous CDMA – Part I: System modeling and results, *IEEE Trans. Circuits Syst.*, 44(10), 937–947, 1997.
18. Rovatti, R., Setti, G., and Mazzini, G., Chaotic complex spreading sequences for asynchronous CDMA – Part II: Some theoretical performance bounds, *IEEE Trans. Circuits Syst.*, 45(4), 496–506, 1998.
19. Rovatti, R., Mazzini, G., and Setti, G., Interference bounds for DS-CDMA systems based on chaotic piecewise-affine markov maps, *IEEE Trans. Circuits Syst.*, 47, 885–896, 2000.
20. Mazzini, G., Rovatti, R., and Setti, G., Interference minimization by autocorrelation shaping in asynchronous DS-CDMA systems: chaos-based spreading is nearly optimal, *Elect. Lett.*, 35, 1054–1055, 1999.

21. Setti, G., Mazzini, G., and Rovatti, R., Chaos-based spreading sequence optimization for DS-CDMA synchronization, *IEICE Trans. Fund.*, E82-A (9), 1737–1746, 1999.
22. Rovatti, R., Mazzini, G., and Setti, G., A tensor approach to higher-order expectations of quantized chaotic trajectories – Part I: General theory and specialization to piecewise affine Markov systems, *IEEE Trans. Circuits Syst.*, 1571–1583, 2000.
23. Mazzini, G., Rovatti, R., and Setti, G., A tensor approach to higher-order expectations of quantized chaotic trajectories – Part II: Application to chaos-based DS-CDMA in multipath environments, *IEEE Trans. Circuits Syst.*, 1584–1597, 2000.
24. Rovatti, R., Mazzini, G., and Setti, G., Enhanced rake receivers for chaos-based DS-CDMA, *IEEE Trans. Circuits Syst.*, 48, 818–829, 2001.
25. Setti, G., Rovatti, R., and Mazzini, G., Tensor-based theory for quantized piecewise-affine Markov systems: analysis of some map families, *IEICE Trans. Fund.*, 84(9), 2090–2100, 2001.
26. Mazzini, G., Rovatti, R., and Setti, G., Chaos-based asynchronous DS-CDMA systems and enhanced rake receivers: measuring the improvements, *IEEE Trans. Circuits Syst.*, 48, 1445–1453, 2001.
27. Mazzini, G., Rovatti, R., and Setti, G., The Impact of Chaos-Based Spreading on Parallel DS-CDMA Cancellers, *Proc. NOLTA 2000*, Dresden, September 2000.
28. Pursley, M.B., Performance evaluation for phase-coded spread-spectrum multiple-access communication – Part I: System analysis, *IEEE Trans. Commun.*, 25, 795–799, 1977.
29. Mazzini, G. and Tralli, V., Performance evaluation of DS-CDMA systems on multipath propagation channels with multilevel spreading sequences and rake receivers, *IEEE Trans. Commun.*, 46, 244–257, 1998.
30. Ochsner, H., Direct sequence spread-spectrum receiver for communication on frequency selective fading channels, *Proc. IEEE JSAC*, 5, 188–193, 1987.
31. Eiselt, H.A., Pederzoli, G., and Sandblom, C.L., *Continuous Optimization Models*, de Gruyter, Berlin, 1987.
32. Foschini, G.J. and Gans, M.J., On limits of wireless communications in a fading environment when using multiple antennas, *Wireless Pers. Commun.*, 6, 311–335, 1998.
33. Rovatti, R., Mazzini, G., and Setti, G., A tensor approach to higher-order correlation of quantized chaotic trajectories – Part I: General theory and specialization to piecewise affine Markov systems, *IEEE Trans. Circuits Syst.*, 47, 1571–1583, 2000.
34. Verdù, S., The capacity region of the symbol-asynchronous gaussian multiple access channel, *IEEE Trans. Inform. Theory*, 35, 733–751, 1989.
35. Cover, T.M. and Thomas, J.A., *Elements of Information Theory*, Wiley, New York, 1991.

6

Channel Equalization in Chaotic Communication Systems

Mahmut Ciftci and Douglas B. Williams

CONTENTS

6.1 Introduction

Chaos has received a great deal of attention in the past few years from a variety of researchers, including mathematicians, physicists, and engineers. Researchers in the area of signal processing and communications have largely been interested in chaos for the development of nonlinear communication techniques. A number of properties of chaotic systems, including low-power implementation, noiselike appearance, broadband spectra, and nonlinearity, are especially appealing for communications applications and can be exploited for secure communications.[1–3] A variety of approaches to chaotic communications have recently been proposed, including chaotic modulation,[4–6] masking,[7,8] and spread spectrum.[9–12] Such communications systems offer the promise of inherent security, resulting from the broadband and noiselike appearance of chaotic signals,[13] and efficiency because systems could be allowed to operate in their natural nonlinear states. Even simple one-dimensional maps can produce random-like yet deterministic signals.

A generic chaotic communications system based on chaotic modulation is shown in Figure 6.1. In such a system, the information bits to be transmitted must first be encoded in the chaotic waveform using the chaotic system's *symbolic dynamics*. Rather than using structured signals such as rectangular pulses or sinusoids, to denote "0s" and "1s" (or other information symbols), these communications systems embed the information in the time evolution of the transmitted signal. Regions of the state space formed by the chaotic system's time evolution are designated to represent different symbols. The process of mapping information bits to the state of a chaotic system through its symbolic dynamics is termed chaotic modulation. This assignment of information bits to state is not arbitrary, and the greatest efficiency is achieved when the information transmission rate matches the topological entropy of the chaotic system.[4] Here, the symbolic dynamics representation of a chaotic system is exploited to implement the chaotic modulation described later in Section 6.4.1.

Once the information sequence is encoded by the chaotic system, the modulated signal is then transmitted through the communication channel.

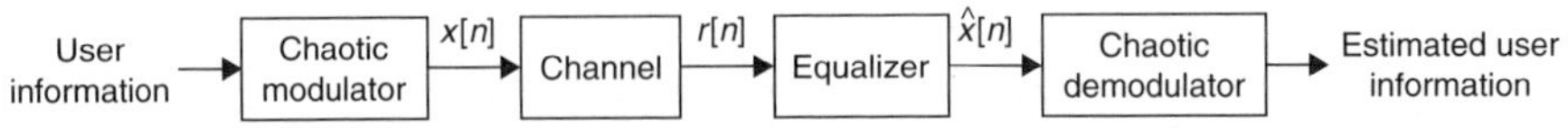

Figure 6.1 A generic chaotic communication system.

In a real communications scenario, the communications channel distorts the transmitted signal. The channel distortion can be in the form of noise or intersymbol interference (ISI) plus noise. To recover the transmitted information at the receiver, one must first undo the distortions before chaotic demodulation, which is the process of obtaining information symbols from the chaotic signal, is done.

For channels that only add noise to the transmitted sequence, such as additive white Gaussian noise (WGN) channels, the distortion can be corrected with a noise reduction algorithm. There are a number of noise-reduction techniques that take advantage of properties specific to chaotic dynamical systems. The first noise-reduction algorithms were suboptimal and ad hoc, and, thus, they would fail to remove the noise at low signal-to-noise (SNR) ratios.[14,15] In recent years, a number of optimum noise-reduction algorithms have been proposed for certain unimodal chaotic maps.[16–21] Drake has proposed an optimum estimation algorithm for the sawtooth map and related maps that exploit the alternative linear, random representation of these chaotic maps.[16] In Reference 17, an optimum noise-reduction algorithm for the tent map was presented. This approach was later generalized for one-dimensional unimodal maps on the unit interval.[18] In References 20 and 21, a probabilistic noise-reduction algorithm using the Viterbi algorithm was proposed. However, this technique does not use symbolic dynamics, relying instead on the invariant distribution of the chaotic system to approximate the transition probabilities between states.

For channel distortions where ISI as well as noise is imposed on the transmitted sequence, channel equalizers, typically implemented as linear adaptive filters, are usually employed to remove the distortion.[22] Most channel equalization development has been for conventional modulation schemes such as PAM (Pulse Amplitude Modulation) and QAM (Quadrature Amplitude Modulation).[23] Because chaotic modulation schemes are significantly different from conventional schemes, conventional equalization algorithms cannot be used, and equalization algorithms specifically designed for chaotic communication systems are needed. There has been some research into equalization algorithms for chaotic communications systems.[11,24–27] In Reference 11, a chaotic communications scheme that is robust to noise and amplitude fluctuations in the channel was presented. This scheme uses synchronizing discrete-time chaotic systems. In Reference 24, a self-synchronization-based channel equalizer was proposed. However, this approach has limited utility because most chaotic systems do not self-synchronize. In Reference 25, a technique for the blind identification of autoregressive (AR) systems driven by a chaotic signal was proposed. The algorithm is based on a minimum phase-space technique and can be applied for channel equalization of an AR system with a chaotic input. However, channel distortion is generally modeled as a finite impulse response (FIR) filter and can be time-varying.

In summary, the channel equalization algorithms that currently exist in the literature only work for certain chaotic systems and many of them perform well only for high SNR ratios or for channel distortions such as a simple time-invariant channel. Thus, they do not perform well in the presence of more realistic channel distortions such as low SNRs and for time-varying channel distortions.

In this chapter, several channel equalization algorithms specifically designed for chaotic communications systems are developed by using

- a knowledge of the system dynamics,
- linear time-invariant representations of chaotic systems, and
- symbolic dynamics representations of chaotic systems.

In addition, a multiuser communication system where a number of users share a common communication channel is presented and multiuser estimation and channel equalization algorithms are developed to remove the distortions introduced by the communication channel and separate the individual chaotic sequences.

The remainder of this chapter is organized as follows. In Section 6.2, the relevant background concerning dynamical systems and characteristics of chaotic dynamics are presented. Section 6.3 surveys applications of chaotic systems in communications. In Section 6.4, optimal estimation and sequential channel equalization algorithms based on the Viterbi algorithm are presented. In Section 6.5, a multiuser chaotic communications system is introduced. In addition, estimation and equalization algorithms for multiuser chaotic communication systems are presented.

6.2 Chaos as it Applies to Channel Equalization

This chapter reviews the basic concepts of dynamical systems and characteristics of chaotic behavior as they apply to channel equalization. More details are given in Appendix A and Appendix B. The properties of chaotic dynamical systems, as well as example chaotic systems, are presented first. Then, the symbolic dynamics representation of dynamical systems is described, in which a sequence of real numbers generated by real dynamics maps to a symbol sequence. This representation will help us to develop estimation and channel equalization algorithms for chaotic communications systems. Finally, a linear and time-invariant representation for a family of chaotic maps is described.

6.2.1 *Dynamical Systems*

Dynamical systems provide a mathematical tool for modeling physical phenomena. A dynamical system is characterized by a set of variables that changes with time in a predictable manner. The number of variables is referred to as the *order* of the system. The instantaneous description of variables is referred to as the *state*. The set of all possible states is called the *phase space* of the system. The evolution of a system refers to a sequence of states or a continuous trajectory in phase space. Simple examples of phase space for a first-order system include the real line $\Re$ or a subset of a real line; for second-order systems, examples include the plane $\Re^2$ and the complex plane.

Dynamical systems fall into two categories: continuous-time dynamical systems and discrete-time dynamical systems. Continuous-time dynamical systems are represented with ordinary differential equations as

$$\frac{d\boldsymbol{x}(t)}{dt} = f[\boldsymbol{x}(t)] \tag{6.1}$$

where $\boldsymbol{x}(t)$ is the state vector at time t and f is called the system dynamics. If the system dynamics f do not depend on time, the system is said to be *autonomous*. Otherwise, it is called nonautonomous. Discrete-time dynamical systems can be expressed with difference equations as

$$\mathbf{x}(n) = f[\boldsymbol{x}(n-1)] \tag{6.2}$$

where the mapping function f is called the system dynamics. The initial condition $\mathbf{x}[0]$ completely determines all subsequence states, often referred to as the system's orbit. Equation (6.2) can be written as

$$\mathbf{x}(n) = f^{(n)}[\mathbf{x}(0)], \quad n \geq 0 \tag{6.3}$$

where $f^{(n)}$ denotes the n-fold composition of the map f. The steady state behavior of a dynamical system is characterized by a structure in state space called the *attractor*. The types of attractors include equilibrium, where the attractor is a single point, and periodic, where the attractor is a closed path indicating that the state is periodically repeated.

The dynamical systems considered in this chapter are autonomous and deterministic. In the next section, one particular class of dynamical systems called *chaotic* is considered.

6.2.2 Chaotic Systems Definitions

Currently, there are many definitions of chaotic systems. The definitions given below will help to understand the material that follows.[28]

Definition 1 *A dynamical system $f: I \rightarrow I$ is said be chaotic if*

1. *f has a sensitive dependence on initial conditions,*
2. *f is topologically transitive, and*
3. *periodic points of f are dense in I.*

Definitions of sensitive dependence on initial conditions, topological transitivity, and dense periodic points are as follows.

Definition 2 *The map $f: I \to I$ has a sensitive dependence on initial conditions if there exists δ such that, for any $x \in I$ and any neighborhood N of x, there exists $y \in N$ and $n \geq 0$ such that $\left|f^{(n)}(x) - f^{(n)}(y)\right| > \delta$.*

Sensitive dependence on initial conditions implies that two orbits of a map with slightly different initial conditions in state-space will diverge exponentially on average.*

Definition 3 *The map $f: I \to I$ is topologically transitive on the invariant set $I = f(I)$ if for any open subsets of U and V in I, there exists a positive integer n such that $f^{(n)}(U) \cap V \neq \emptyset$.*

Intuitively, a topologically transitive map has points that eventually move under iteration from one arbitrarily small neighborhood to any other, i.e., a small part of the attractor spreads to cover the entire attractor with iteration. This property is also referred to as the *mixing property*.

Definition 4 *Periodic points of a map $f: I \to I$ are dense if for any $x \in I$ and any neighborhood N of x, there exist $y \in N$ and $n \geq 0$ such that $f^{(n)}(y) = y$.*

In short, chaos is bounded instability. In other words, chaos can exhibit strong instability locally, yet remain bounded globally. Two mechanisms contribute to this behavior: stretching, which implies instability, and folding, which provides boundedness. The amplitudes of a chaotic signal in time exhibit no specific pattern or periodicity. The trajectories of chaotic systems in state space never repeat and fill up a part of state space called the *strange attractor*. Attractors of chaotic systems are characterized as fractal, displaying a self-similarity.[29]

Definition 5 *A dynamical system $\{f_a, I\}$ is topologically semi-conjugate to the dynamical system $\{f_b, J\}$ if there exists a continuous, onto function $g: I \to J$ such that*

$$g(f_a(x)) = f_b(g(x))$$

The simplest dynamical system that exhibits chaotic behavior is called the *sawtooth map* and is given by the mapping function

$$\begin{aligned} x[n] &= f_s(x[n-1]) \\ &= 2x[n-1] \bmod 1 = \begin{cases} 2x[n-1], & x[n-1] < 0.5 \\ 2x[n-1] - 1, & x[n-1] \geq 0.5 \end{cases} \end{aligned} \tag{6.4}$$

* The average rate of change is expressed by the Lyapunov exponents.[29]

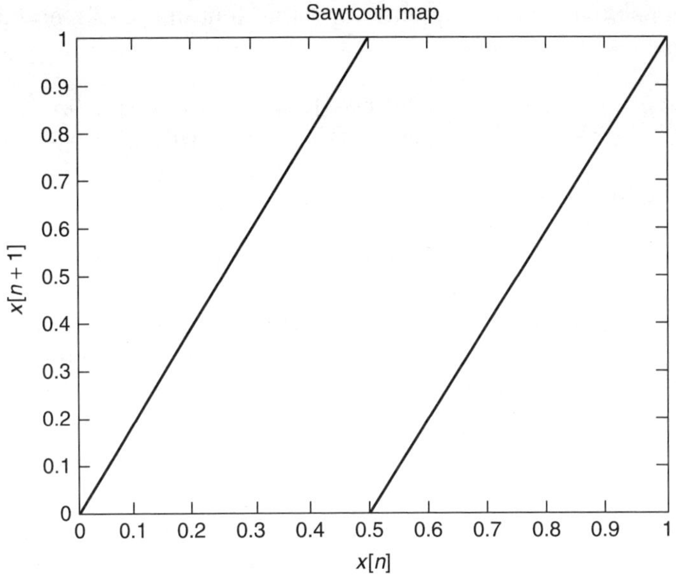

Figure 6.2 Sawtooth chaotic map.

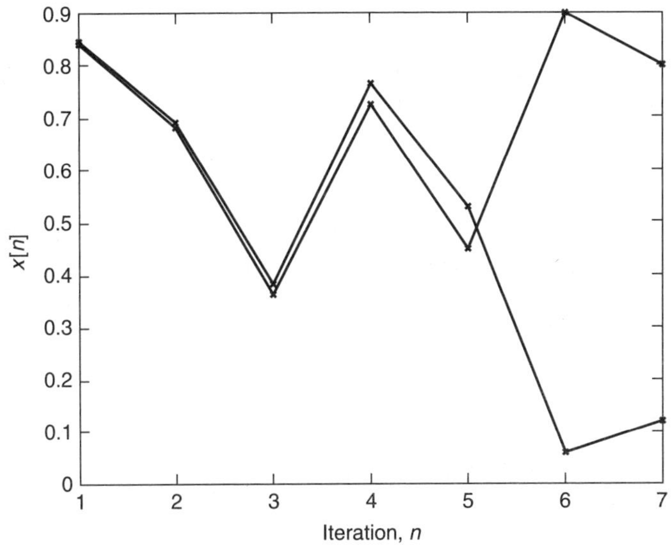

Figure 6.3 Two chaotic sequences with slightly different initial conditions.

which is defined on the unit interval, i.e., $I = [0, 1]$. The sawtooth chaotic map is shown in Figure 6.2. Figure 6.3 illustrates the sensitive dependence on the initial condition of the sawtooth map by plotting two chaotic sequences with slightly different initial conditions. Other well-known one-dimensional chaotic systems are the tent map, governed by the equation

$$x[n] = f_t(x[n-1]) = \begin{cases} 2x[n-1], & x[n-1] < 0.5 \\ 2(1 - x[n-1]), & x[n-1] \geq 0.5 \end{cases} \quad (6.5)$$

and the logistic map, given by

$$x[n] = f_l(x[n-1]) = 4x[n-1](1 - x[n-1]) \quad (6.6)$$

where both maps are defined on unit interval. Both the tent map and the logistic map inherit their chaotic behavior from the sawtooth map through the notion of topological semi-conjugacy. The definition of topological semi-conjugacy is as follows.[28]

Topological semi-conjugacy preserves properties that define chaotic behavior. Through topological semi-conjugacy, any orbit of the tent map or logistic map can be obtained from iterations of the sawtooth map dynamical systems followed by a transformation of a continuous function $g(\cdot)$. For example, the function

$$g_t(x[n]) = \begin{cases} 2x[n], & 0 \leq x[n] < 0.5 \\ 2(1 - x[n]), & 0.5 \leq x[n] < 1 \end{cases} \quad (6.7)$$

converts sawtooth map sequences to tent map sequences. Similarly, the transformation function to generate a logistic sequence from a sawtooth sequence is

$$g_l(x[n]) = \sin^2(\pi x[n]) \quad (6.8)$$

A chaotic sequence generated by the logistic map is illustrated in Figure 6.4. The semi-conjugacy relationship between different chaotic maps allows us to develop algorithms for a sawtooth map and then extend these algorithms to a large family of systems by continuous transformations.

An example of a second-order dynamical system is given by the bakers transformation:

$$\mathbf{x}[n] = f_b(\mathbf{x}[n-1]) = \begin{bmatrix} f_1(\mathbf{x}[n-1]) \\ f_2(\mathbf{x}[n-1]) \end{bmatrix} \quad (6.9)$$

where

$$f_1(\mathbf{x}[n-1]) = \begin{cases} 2x_1[n-1], & x_1[n-1] < 0.5 \\ 2x_1[n-1] - 1, & x_1[n-1] \geq 0.5 \end{cases} \quad (6.10)$$

$$f_2(\mathbf{x}[n-1]) = \begin{cases} x_2[n-1]/2, & x_2[n-1] < 0.5 \\ (x_2[n-1] + 1)/2, & x_2[n-1] \geq 0.5 \end{cases} \quad (6.11)$$

with phase space equal to the unit square.

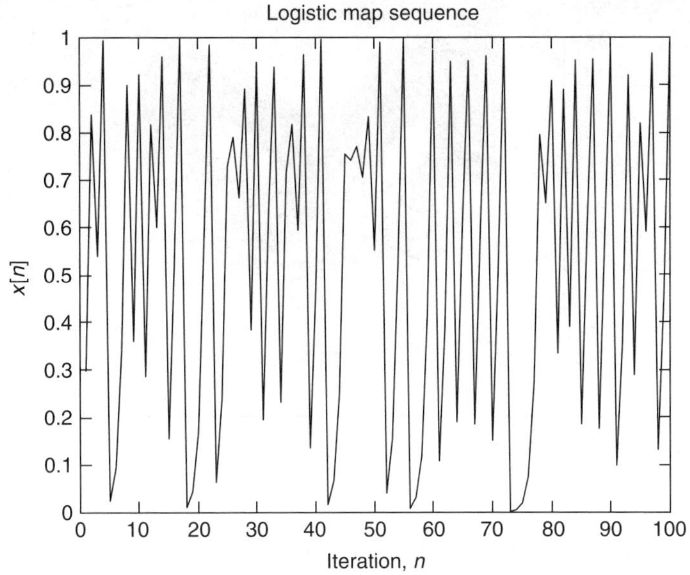

Figure 6.4 A chaotic sequence generated by Logistic map.

A continuous-time chaotic system example is given by the Lorenz system

$$\begin{aligned}\frac{dx(t)}{dt} &= -\sigma x(t) + \sigma y(t) \\ \frac{dy(t)}{dt} &= Rx(t) - y(t) - x(t)z(t) \\ \frac{dz(t)}{dt} &= -Bz(t) + x(t)y(t)\end{aligned} \tag{6.12}$$

where $\sigma = 10$, $B = 8/3$, and $R = 28$. The attractor of the Lorenz system is shown in Figure 6.5.

These chaotic systems will be used throughout this chapter to simulate the performance of equalization algorithms.

6.2.3 *Symbolic Dynamics*

Symbolic dynamics provide a means of representing and generating "finite precision" chaotic sequences. One divides the phase space into a finite number of partitions, $I = \{I_0, I_1, \ldots, I_{K-1}\}$ and labels each partition with a symbol, i.e., $v_i \in I_i$. Instead of representing trajectories with an infinite sequences of numbers, one watches the alternations of symbols. By iterating Equation (6.2), one gets the numerical sequence

$$x[0],\ x[1] = f(x[0]), \ldots, x[n] = f(x[n-1]), \ldots \tag{6.13}$$

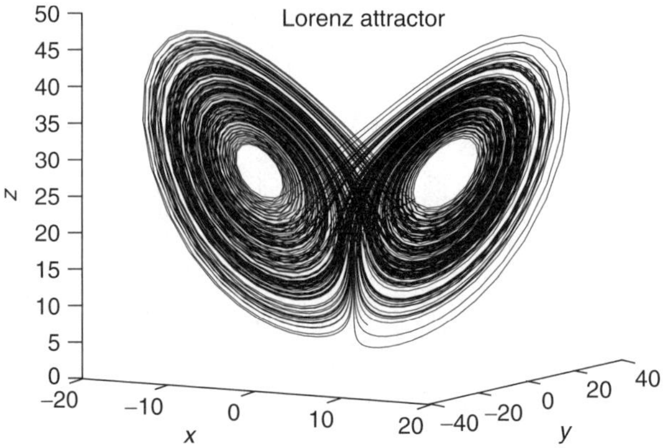

Figure 6.5 Lorenz attractor.

The corresponding symbolic sequence is

$$S(x[0]) = s_0 s_1 s_2 \dots s_n \dots \tag{6.14}$$

where s_i is one of the symbols, v_k, depending on which partition, I_k, $x[i]$ belongs to. The dynamics governing the symbolic sequence, represented by $\sigma(\cdot)$, are a left shift operation where the leftmost symbol is discarded at each iteration, i.e.,

$$\begin{aligned} S(x[n]) = \sigma(S(x[n-1])) &= \sigma(s_{n-1} s_n s_{n+1} \dots) \\ &= s_n s_{n+1} s_{n+2} \dots \end{aligned} \tag{6.15}$$

Depending on the chaotic system, there may be restrictions on the allowable symbolic sequences that generate what is called a *grammar* of sequences.[31] The chaotic systems considered in this paper are "fully developed," i.e., there are no restrictions on the symbolic sequence. However, the algorithms developed in this chapter could be applied to dynamics with restrictive grammars as well.

By reversing Equation (6.13) and Equation (6.14), we see that the initial state must be located in the set $\bigcap_{n=0}^{N-1} f^{-n}(I_{v_n})$. For a chaotic system where the orbits of the neighboring states tend to diverge as N grows large, the size of this set gets smaller with larger N. As N tends to infinity, it contains only a single initial point. If this is the case, then there is a one-to-one equivalence between the initial state $x[0]$ and the infinite sequence of symbols v_k. The mapping of symbol sequence $S(x[n])$ to state $x[n]$ is denoted by $\beta(\cdot)$. The relationship among the dynamics $f(\cdot)$, the shift $\sigma(\cdot)$, and the conversion

$\beta(\cdot)$ is illustrated in the following diagram:

$$\begin{array}{ccc} S(x[n-1]) & \xrightarrow{\sigma} & S(x[n]) \\ \beta\big\downarrow & & \big\downarrow\beta \\ x[n-1] & \xrightarrow{f} & x[n] \end{array} \tag{6.16}$$

As an example, consider the sawtooth map as given in Equation (6.4). The state space, partitions, and corresponding symbolic representations are $I=[0,1]$, $I_0=[0,1/2)$, $I_1=[1/2,1)$, and $v_0=0$, $v_1=1$, respectively. The mapping function, $\beta(\cdot)$, is

$$x[n] = \beta(S(x[n])) = \sum_{k=n}^{\infty} s_k 2^{-(k-n+1)} \tag{6.17}$$

Mapping functions for two other popular chaotic maps, the tent and logistic maps, can be obtained by using Equation (6.17) and the topological semiconjugacy properties relating the sawtooth map to these two maps. Furthermore, the chaotic maps that can be described by a linear, random representation and their generalizations as outlined in Reference 16 can easily be generated with their corresponding symbol sequence. It is also possible to numerically approximate $\beta(\cdot)$ for many chaotic maps. In this chapter, the symbolic dynamics representation of chaotic systems is exploited to develop channel equalization algorithms.

6.3 Chaos in Communication

Following the discovery of the self-synchronization property of chaotic systems,[7] numerous communications systems based on chaotic systems were proposed.[1,2,4,8,10] This chapter presents some of these proposed applications of chaotic systems for communications. It first introduces a number of properties of chaotic systems that are appealing for chaotic communications systems. Then, it illustrates some of the proposed chaotic communications system examples. Finally, the channel distortion issue is addressed.

6.3.1 *Properties of Chaotic Systems*

A number of properties of chaotic systems are attractive for communications applications. These properties include the following:

- *Low-power implementations*: The circuits that generate chaotic waveforms can be run in their natural nonlinear mode and are often implemented at the microelectronic level. It is also possible to

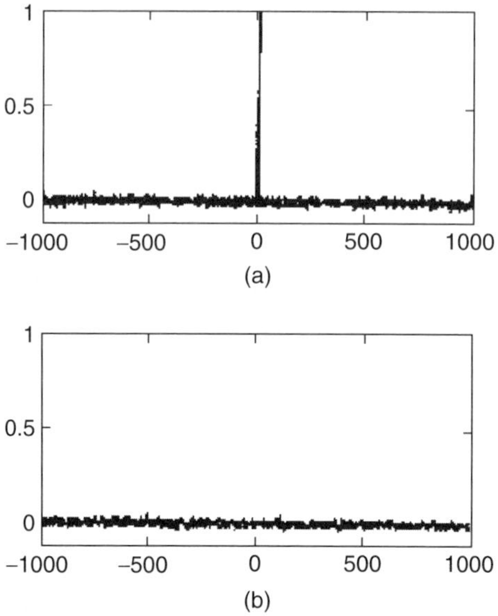

Figure 6.6 Autocorrelation and cross-correlation functions of the logistic map.[13] (a) Autocorrelation. (b) Cross-correlation.

encode information in a chaotic waveform using small perturbations to enable significant changes in the future system output with little change in its present state.[4]

- *Noiselike appearance*: Time-series representations of chaotic signals do not possess any particular pattern because the output of a chaotic system is aperiodic. Consequently, chaotic signals are often characterized as noise. Therefore, this structure makes chaotic signals more difficult to detect.
- *Broadband spectra*: The autocorrelation function of a chaotic system decreases sharply with time. The spectrum associated with such an autocorrelation function is generally broadband. Moreover, the cross-correlation function of chaotic signals corresponding to different initial conditions tends to be low because of the sensitive dependence on initial conditions. These properties make chaotic sequences attractive for spread spectrum applications. Example of autocorrelation and cross-correlation functions are illustrated in Figure 6.6, and a typical power spectrum is shown in Figure 6.7.[13]
- *Nonlinearity*: Chaotic systems are nonlinear in nature, which is also a desired feature for communications because of the possibility of increased bandwidth efficiency, which results in encoding more information than linear systems.

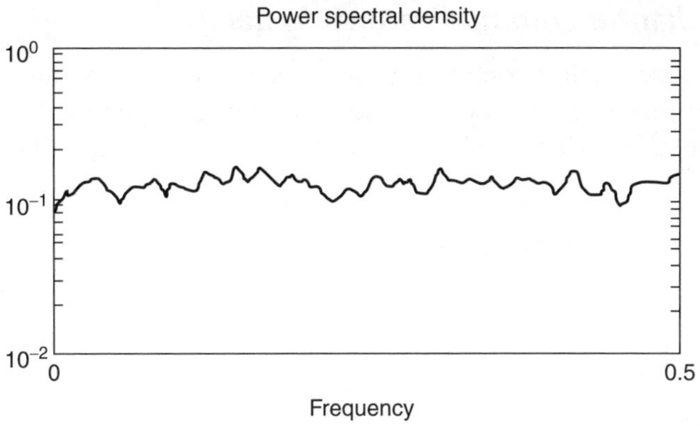

Figure 6.7 Power spectrum of the logistic map.[13]

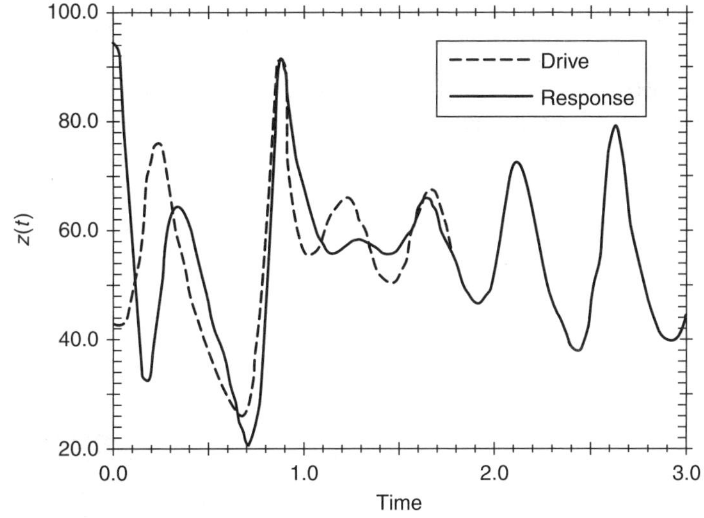

Figure 6.8 Self-synchronization of the Lorenz system.[1]

- *Self-synchronization*: It has been shown that some of the chaotic systems possess the self-synchronization property, which can be utilized in communications in various fashions[7] because the state of the receiver system tracks that of the transmitter system. Figure 6.8 illustrates synchronization of a receiver to a transmitted chaotic signal.[30]

The next section will present some of the typical approaches to using chaos in communications systems that use these properties of chaotic systems.

6.3.2 Chaotic Communications Systems

In recent years, many communications systems based on chaotic systems have been proposed, including chaotic modulation, chaotic masking, and spread spectrum. These techniques will be briefly explained here.

6.3.2.1 Chaotic Modulation

Chaotic modulation refers to the encoding of information symbols in a chaotic waveform. The input to the chaotic modulator is information symbols, whereas the output is a signal generated by a chaotic system. There have been many algorithms proposed for chaotic modulation.[3–6] In Reference 4, a symbolic dynamic-based chaotic encoding scheme was proposed.

First, a cross-sectional plane is constructed that cuts through the attractor, which is called Poincare section. The Poincare section for a Lorenz system is illustrated in Figure 6.9. The dual lobe structure of the Lorenz attractor is represented in binary terms, with one lobe corresponding to 0 and the other to 1. Prior data collection determines all the points at which the free-running trajectories cross the normalized surface of section. After each crossing point x on the Poincare section, the subsequent binary sequence is recorded. A real number is created via the coding function

$$r(x) = \sum_{n=1}^{\infty} b_n 2^{-n} \tag{6.18}$$

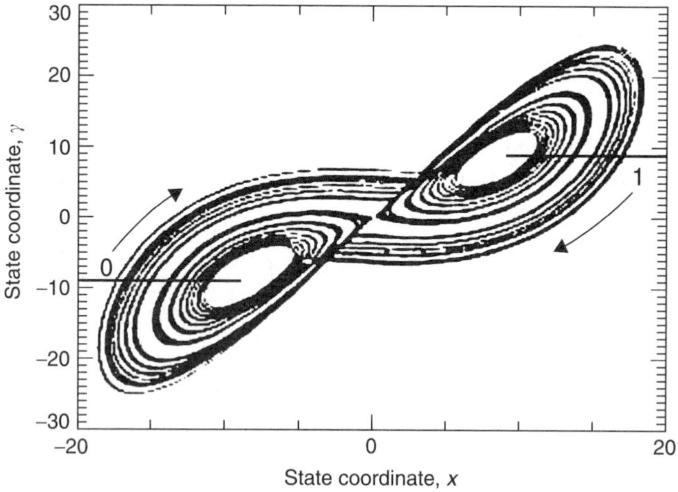

Figure 6.9 Poincare surface of section.[34]

where b_n represents the binary sequence generated after the trajectory crosses the Poincare section at point x. The inverse coding function $x(r)$ for the Lorenz system is plotted in Figure 6.10.

Once the coding function is obtained, the first N bits of the binary message sequence specifying the desired symbolic dynamics is converted to its corresponding binary fraction r_0. The corresponding surface of section coordinate is calculated as $x(r_0)$. The Lorenz trajectory is forced to this point by applying a small perturbation when the free running state coordinate passes near $x(r_0)$. Next, the message sequence is shifted to the left and the most significant bit, representing the symbol just produced by the symbolic dynamics of the system, is discarded. The next bit of the desired message sequence is placed into the least significant slot and the binary fraction, r_1 is calculated. Then, the corresponding surface of section coordinate is computed as $x(r_1)$. The Lorenz trajectory is forced to this point when it crosses the surface of section. This process is repeated until the entire message has been transmitted. The necessary perturbations are quite small as only the least significant bit needs to be compensated for and its effect on $x(r)$ is minor for the next several iterations on $r(x)$.

Chaotic demodulation, which is the process of obtaining information symbols from the chaotic signal, is a simple threshold operation for this system if there are no channel distortions. However, in practice, channel distortions must be removed before chaotic demodulation can be reliably performed.

6.3.2.2 *Chaotic Masking*

Signal masking refers to the process of embedding a message in a free-running chaotic series such that the message is hidden in the dominant

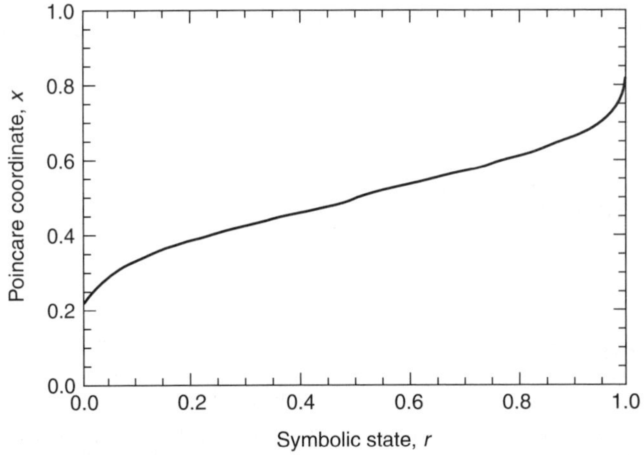

Figure 6.10 Inverse coding function.[34]

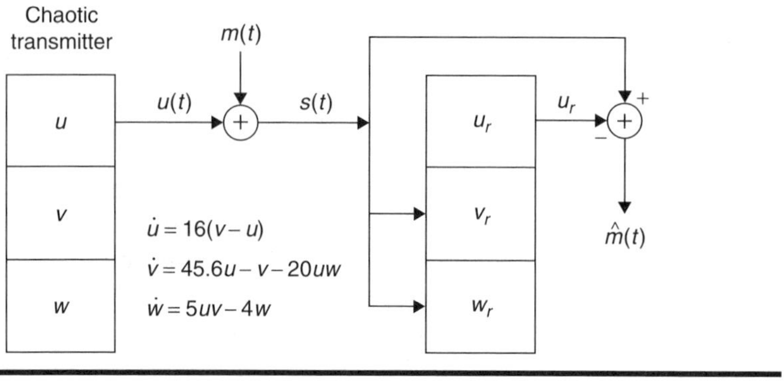

Figure 6.11 Chaotic masking system.[8]

chaotic waveform. It was the first of the proposed communications methods using chaos[8] following the discovery of the self-synchronization property of chaotic systems.

A block diagram of a chaotic masking system is shown in Figure 6.11, where the message signal is added to a chaotic signal and recovered by subtracting the synchronized chaotic signal from the received waveform. To recover the message signal, the message power should be at least 30 dB below the chaotic power. A problem with this method is that even modest channel noise results in erroneous message recovery.[5] Also, chaotic systems fail to synchronize in the presence of even mild channel distortions.

6.3.2.3 Spread Spectrum

Spread spectrum communications systems use binary pseudonoise sequences to spread the spectrum of the transmitted signal. Pseudonoise sequences, usually generated using a linear feedback shift register, are periodic and binary valued.[33] On the other hand, chaotic sequences are aperiodic and have a wideband spectrum with low autocorrelation and cross-correlation properties. Also, chaotic sequences take on a continuous range of amplitudes. Because of these properties, chaotic sequences are attractive candidates to replace binary pseudonoise sequences.[13,35,36] In recent years, a number of such chaotic spread spectrum communications systems have been proposed.[9–12]

6.3.3 Channel Effects

Many of these previously proposed chaotic communication systems disregard the distortions introduced by typical communication channels and fail to work under realistic channel conditions. Common channel distortions take the form of additive noise or ISI plus noise. A generic chaotic

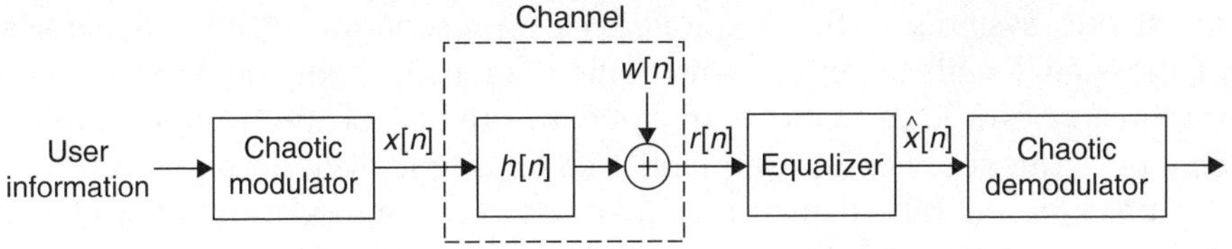

Figure 6.12 A generic chaotic communication system.

communication system based on chaotic modulation is shown in Figure 6.12. In such a system, the information bits to be transmitted must first be encoded in the chaotic waveform using the chaotic system's *symbolic dynamics*. In this chapter, it is assumed that the transmitted signal is generated by a first-order discrete-time chaotic system that satisfies the dynamical equation

$$x[n] = f[x(n-1)], \quad x[n] \in I \tag{6.19}$$

where $f(\cdot)$ is a nonlinear function mapping points from an interval I onto the same interval.

Once the information sequence is encoded by the chaotic system, the modulated signal is then transmitted through the communication channel. The distortion introduced by the channel is modeled by a linear filter with impulse response $h[n]$ and additive WGN $w[n]$ with variance of σ_w^2. The received signal, $r[n]$, can then be written as

$$r[n] = h[n] * x[n] + w[n] \tag{6.20}$$

where $*$ represents the convolution operation. To recover the transmitted information, one must undo these distortions before chaotic demodulation, which is the process of obtaining an estimate of the transmitted symbol sequence from the received signal. The goal of the equalizer is to obtain an accurate estimate of transmitted signal $x[n]$ from the observed signal $r[n]$.

In the remainder of this chapter, a number of equalization algorithms for chaotic communications systems are developed.

6.4 Optimal Estimation and Sequential Channel Equalization Algorithms Based on the Viterbi Algorithm

This section describes optimal estimation and sequential channel equalization algorithms for chaotic communications systems where information is encoded using symbolic dynamics representations. To find an optimal estimate of the transmitted signal, the symbolic dynamics representation of chaotic systems is first exploited to represent the chaotic dynamics of the signal with an equivalent trellis diagram. Then, the Viterbi algorithm can be used to achieve an optimal estimate of the noise-corrupted chaotic sequence. For the sequential channel equalization algorithm, the dynamics-based trellis diagram is expanded to accommodate a FIR model of the channel. Once an initial estimate of the channel parameters is obtained by using a training sequence, the Viterbi algorithm is then used to estimate the chaotic sequence and track changes in the channel.

6.4.1 Chaotic Modulation through Symbolic Dynamics

The symbolic dynamics representation as described in Section 6.2.3 is used here for chaotic modulation. Once the mapping function $\beta(\cdot)$ is known, one can generate chaotic sequences using the symbol sequence rather than generating the chaotic sequence directly. Such an implementation is especially important for chaotic maps where the chaotic sequence becomes irrelevant with direct iteration because of finite precision. The two-dimensional bakers transformation and two of the most popular one-dimensional chaotic maps, the sawtooth and tent maps, are examples of this problem. For example, the sequence generated by direct iteration of the sawtooth map corresponding to a finite precision initial condition, $x[0]$, will be a zero sequence after a finite number of iterations. Two different ways of generating chaotic sequences are illustrated in Figure 6.13.

Because of the limitations of digital computers, chaotic sequences can be generated only with finite precision. Thus, only a finite sequence of symbols can be considered in practice. By modifying Equation (6.17), the relationship between a finite symbol length sequence and its corresponding state can be rewritten as

$$x[n] = \beta(\hat{S}(x[n])) = \sum_{k=n}^{n+N_p-1} s_k 2^{-(k-n+1)} \tag{6.21}$$

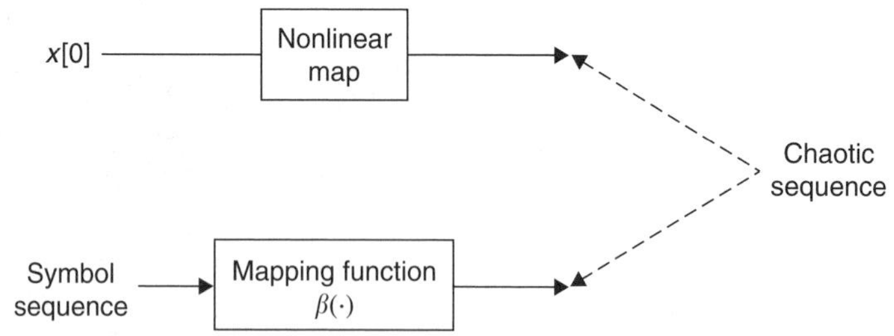

Figure 6.13 Two different ways of generating chaotic sequences.

where $\hat{S}(\cdot)$ is used to indicate that only the first N_p symbols are used. Two different implementations for generating these finite precision "pseudo-chaotic" sequences, directly and through symbolic dynamics, are illustrated in Figure 6.14.

This finite precision symbolic dynamics representation is the key to efficient chaotic modulation and demodulation. Chaotic modulation and demodulation through mapping functions are illustrated in Figure 6.15. The mapping function may not be known for certain chaotic systems and may not exist in some cases. However, it may still be possible to encode the information through symbolic dynamics. The small perturbation approach to controlling the symbolic dynamics of chaotic systems, combined with an estimate of the mapping function as described in Section 6.3.2.1, can be used to drive the system through the desired sequence of symbols. The Lorenz and double-scroll continuous-time chaotic systems are examples of chaotic systems where this scheme has been applied.[4]

6.4.2 *Trellis Diagram Representation*

The ability to use a finite number of symbols to generate chaotic sequences allows us to represent chaotic dynamics with an equivalent trellis diagram, because the symbol set is also finite. To illustrate how a trellis diagram is constructed, the binary symbol set {0, 1}, which is used for many popular unimodal chaotic maps, is used for simplicity. Figure 6.16 illustrates the trellis diagram for a precision of N_p symbols. In the trellis diagram, the states

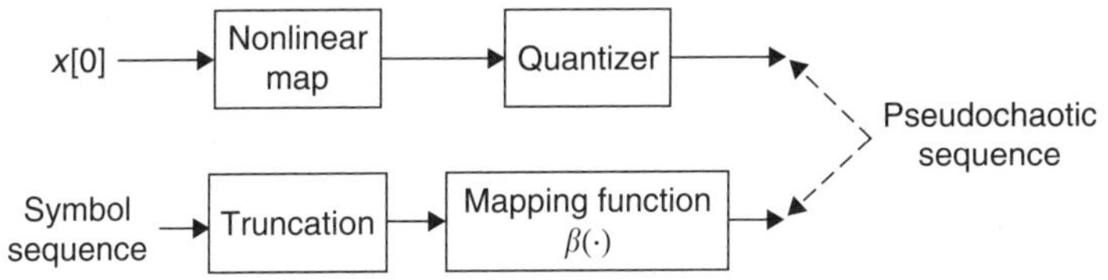

Figure 6.14 Generating pseudochaotic sequences.

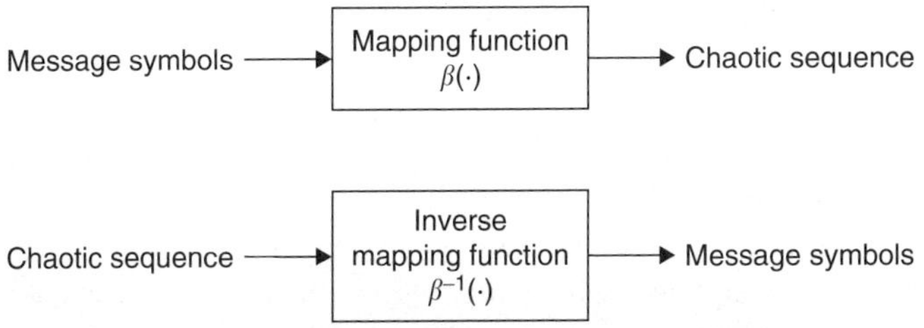

Figure 6.15 Chaotic modulation and demodulation through symbolic dynamics.

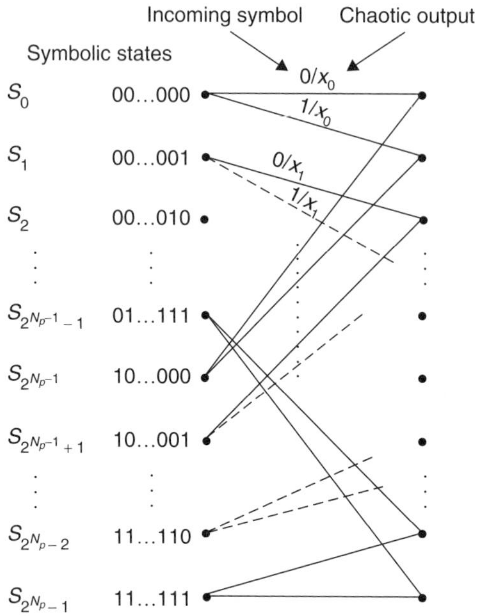

Figure 6.16 Trellis diagram representation of chaotic dynamics with a symbolic dynamics representation for the finite precision case.

are all possible combinations of N_p symbols, i.e., there are 2^{N_p} possible symbolic states. Because each incoming symbol can take on values of only 0 or 1, there are two branches leaving each state. The transition between the current state and the next state is determined by the symbolic dynamics, which is just a left-shift operation, and the incoming symbol. The branches of the trellis diagram are labeled with the branch metrics consisting of the input symbol and the output chaotic signal, which is calculated using the mapping function $\beta(\cdot)$ for each state. As an example, the trellis diagram for the sawtooth map with two-bit precision is illustrated in Figure 6.17.

The trellis diagram in Figure 6.16 is for "fully developed chaotic systems," i.e., there are no restrictions on the grammar. If there are restrictions on the grammar, the trellis diagram has to be modified accordingly. For example, some states would not exist or certain transitions between states would not be possible, depending on the grammar.

For schemes where information is encoded using symbolic dynamics through the use of small perturbations and estimated coding functions to generate the desired sequence of states,[4] a similar trellis diagram representation can be used. It is only necessary to determine which chaotic output corresponds to each symbolic state. This information can be obtained through the coding function used to generate the desired perturbations at the transmitter.

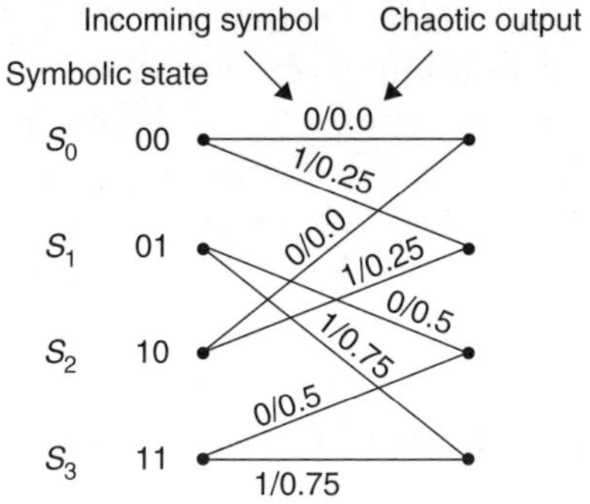

Figure 6.17 Trellis diagram for the sawtooth map with two-bit precision.

This trellis diagram representation of chaotic systems is the key for the development of the optimal estimation and channel equalization algorithms to be explained in following sections.

6.4.3 Optimal Estimation Algorithm

In this section, the development of an optimal estimator for chaotic signals is facilitated by the structure of the trellis diagram representation. An observed noisy chaotic signal can be formulated as

$$r[n] = x[n] + w[n] \tag{6.22}$$

where $x[n]$ is the chaotic signal generated by Equation (6.19) and $w[n]$ is additive noise that is typically modeled as white Gaussian with a variance of σ_w^2. The goal is to estimate $x[n]$ from the observations $r[n]$. The following cost function will be minimized to get an estimate of $x[n]$, represented by $\hat{x}[n]$

$$J(\hat{x}) = \min_{\hat{x}[n]} \sum_{n=0}^{N-1} (r[n] - \hat{x}[n])^2 \quad \text{s.t. } \hat{x}[n] = f(\hat{x}[n-1]) \tag{6.23}$$

where N is the length of the received sequence. Under the WGN assumption, the solution of this cost function yields the maximum-likelihood estimate. Because of the sensitivity of chaotic systems to initial conditions, straightforward minimization of this cost function is numerically difficult.[38]

Here, the Viterbi algorithm[22] is used to obtain the optimum estimate of $x[n]$ by exploiting the trellis diagram representation developed in the previous section. The branch metric that yields the optimal solution is given by

$$b_{Sn} = |r[n] - x_s|^2 \tag{6.24}$$

where x_s represents the chaotic output for the corresponding branch. Because any path of the trellis diagram satisfies the constraints imposed by the cost function given in Equation (6.23), the surviving path determined by the Viterbi algorithm minimizes the mean-square error. Therefore, the chaotic sequence corresponding to the surviving path is the same sequence that minimizes the cost function given by Equation (6.23).

6.4.4 A Sequential Channel Equalization Algorithm

Once the information sequence is encoded with a chaotic system, the modulated signal is then transmitted through the communications channel. The received signal can then be modeled as

$$r[n] = h[n] * x[n] + w[n] \tag{6.25}$$

where $h[n]$ and $w[n]$ represent the finite length channel impulse and WGN with variance of σ_w^2, respectively, and $*$ represents the convolution operation. To accommodate this FIR channel model, the trellis diagram representation developed in the previous section can be expanded by increasing the number of states. Previously, the number of states was 2^{N_p}. For a filter of length $L+1$, the number of states increases to 2^{N_pL}, if the consecutive states are independent of each other. However, for chaotic sequences, consecutive states are not independent of each other because of the system dynamics. Consequently, the total number of states in the trellis is just 2^{N_p+L}. The new trellis diagram is essentially the same as the previous trellis diagram with a precision of N_p+L symbols and a new branch metric. The state at time n is now given by

$$(x[n-1], x[n-2], \ldots, x[n-L]) \tag{6.26}$$

Let's assume that N symbols are transmitted over the communications channel. Then, the received sequence is $(r[N], r[N-1], \ldots, r[1])$. The maximum-likelihood (ML) receiver estimates the sequence that maximizes the likelihood function,

$$p(r[N], \ldots, r[1] \mid x[N], \ldots, x[1]) \tag{6.27}$$

or, equivalently, the log-likelihood function

$$\log p(r[N], \ldots, r[1] \mid x[N], \ldots, x[1]) \tag{6.28}$$

formed from the conditional probability density function (*pdf*). Because the noise $w[n]$ is white, $r[N]$ depends only on the L most recently transmitted symbols, and the log-likelihood function can be written as

$$\begin{aligned}\log p(r[N],\ldots,r[1] \mid x[N],\ldots,x[1]) \\ = \log p(r[N] \mid x[N],x[N-1],\ldots,x[N-L]) \\ + \log p(r[N-1],\ldots,r[1] \mid x[N-1],\ldots,x[1])\end{aligned} \tag{6.29}$$

If the second term has been calculated at time $N-1$, then only the first term, called the branch metric, has to be computed for each incoming signal $r[N]$ at time N. From the model given by Equation (6.25), the conditional pdf

$$p(r[N] \mid x[N],x[N-1],\ldots,x[N-L]) = \frac{1}{(2\pi\sigma_w^2)^{1/2}} \exp\left\{-\frac{1}{2\pi}\left|r[N] - \sum_{k=0}^{L} h[k]x[N-k]\right|^2\right\} \tag{6.30}$$

so that $\log p(r[N] \mid x[N],x[N-1],\ldots,x[N-L])$ yields the branch metric

$$b_N = \left|r[N] - \sum_{k=0}^{L} h[k]x[N-k]\right|^2 \tag{6.31}$$

The well-known Viterbi algorithm[22] can be used to implement the ML receiver by searching through an $N_s = 2^{N_p+L}$-state trellis for the most likely transmitted chaotic sequence. Because the trellis diagram is formed using symbolic dynamics, the resulting sequence satisfies the dynamics imposed by the chaotic system.

The Viterbi algorithm requires knowledge of the channel parameters $h[n]$ to compute the branch metrics in Equation (6.31), so an adaptive channel estimator is needed. An initial estimate of the channel parameters is obtained using training sequences. After the training mode, the final decisions at the output of the Viterbi algorithm can be used to update channel parameter estimates. The parameter update equation for the LMS algorithm is:

$$h_k[j] = h_{k-1}[j] + \mu \cdot e[k-D]\hat{x}[k-j-D] \tag{6.32}$$

where μ is the adaptation step size, D is the trellis depth, $\hat{x}[k-D]$ is the output of the Viterbi algorithm, and $e[k-D]$ is given by

$$e[k-D] = r[k-D] - \sum_{i=0}^{L} h_k[i]\hat{x}[k-D-i] \tag{6.33}$$

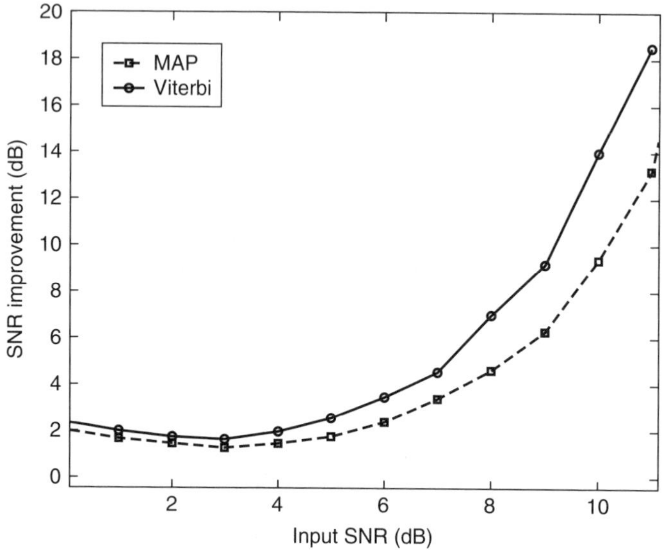

Figure 6.18 SNR improvement at the output for the sawtooth map as estimated by the MAP estimator and the Viterbi algorithm.

6.4.5 Simulations

The performance of the proposed estimation algorithm has been simulated using chaotic maps such as the sawtooth map, tent map, and logistic map, and it has been compared to some other optimum estimation algorithms. Throughout these simulations, the noise was modeled as having a white Gaussian distribution. First, the sawtooth map was used. The results are compared with the MAP estimation algorithms developed by Drake[16] in terms of SNR improvement, which is defined as

$$\text{SNR improvement} = \text{Output SNR} - \text{Input SNR} \tag{6.34}$$

SNR improvement at the output is plotted in Figure 6.18 for the proposed algorithm together with the MAP estimation algorithm. Because the sawtooth map is uniformly distributed, the cost function given by Equation (6.23), and, therefore, the Viterbi algorithm yields the MAP estimate for the Gaussian noise distribution. In these simulations the Viterbi algorithm shows slightly better performance than Drake's MAP estimator because the data was generated with finite precision and Drake's estimator does not make that assumption. SNR improvement for the logistic map and tent map are also plotted in Figure 6.19.

The estimation algorithm was also applied to a chaotic communication system where information was encoded into chaotic signal with the sawtooth map using symbolic dynamics. The communications channel was

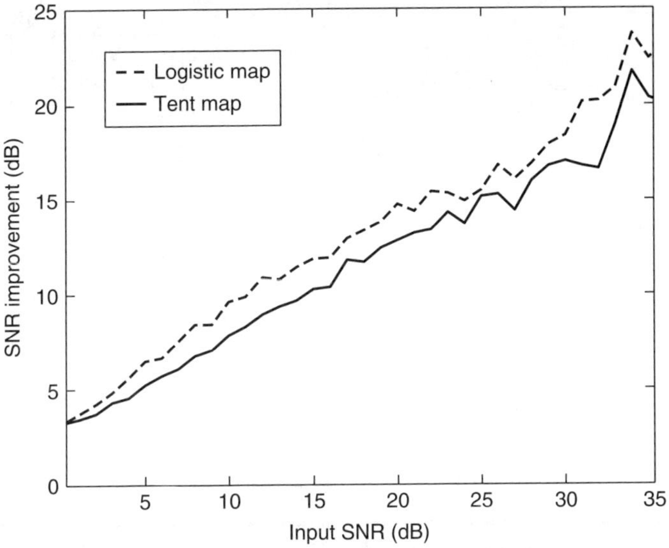

Figure 6.19 SNR improvement for the logistic map and tent map using the Viterbi algorithm.

modeled as a WGN channel. The bit error rate (BER) is plotted in Figure 6.20 for the Viterbi algorithm and the MAP estimator and compared with BPSK modulation scheme.[22] The graph shows that Viterbi-based and MAP algorithms achieve similar performance. Furthermore, the algorithms have almost identical complexity with the number of multiplications for both algorithms being $N \cdot 2^{N_p}$ where N is the length of the chaotic sequence and N_p is the amount of precision.

Because trellis size primarily determines the complexity, the effect of the trellis size on the performance of the estimation algorithm was also simulated. It was assumed that information symbols were encoded in the sawtooth map's waveform using ten-bits of precision, i.e., $N_p = 10$. Instead of using a trellis diagram with 2^{N_p}-states, a smaller trellis diagram with 2^{N_s}-states where $N_s < N_p$ was used to estimate the transmitted chaotic sequence. BER performances of the estimation algorithm corresponding to different trellis sizes, i.e., to different N_ss, are illustrated in Figure 6.21 for a WGN channel. In this case, it can be concluded that the performance of the estimation algorithm does not degrade significantly until the trellis assumes fewer than five-bits of precision. Therefore, the computational load can be reduced greatly by using a smaller trellis.

The sequential equalization algorithm was simulated for both time-invariant and time-varying channel models. In both cases, information bits were modulated by the sawtooth chaotic map. The time-invariant channel

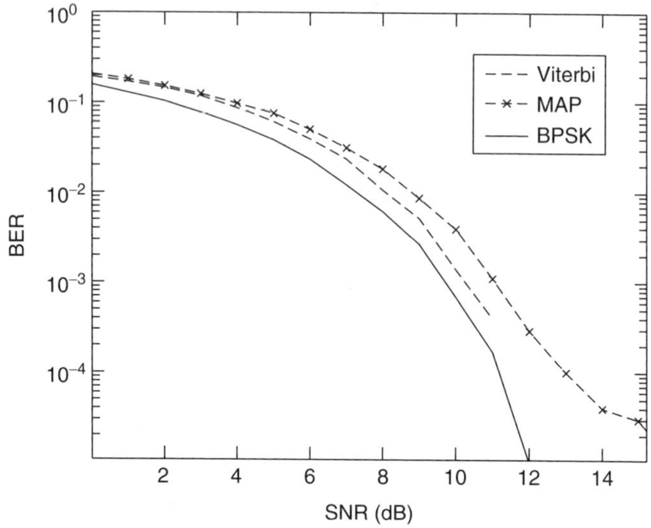

Figure 6.20 BER for a WGN channel.

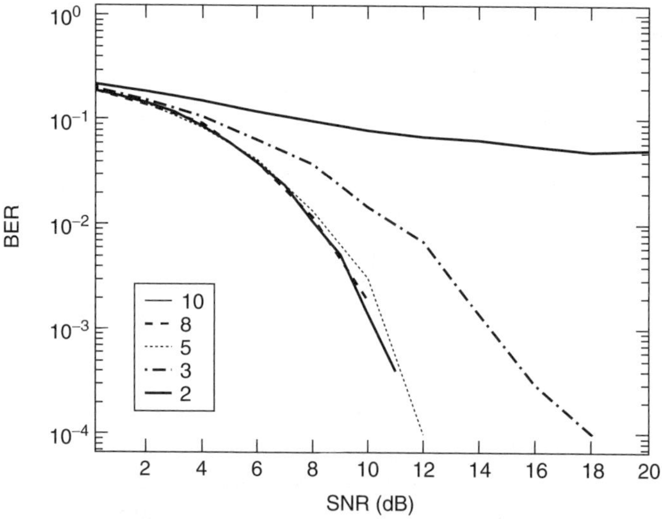

Figure 6.21 BER for a WGN channel corresponding to different trellis sizes.

was modeled as a raised cosine channel given by the impulse response

$$h[n] = \begin{cases} 1/2 + 1/2\cos[2\pi(n-2)/3] & \text{for } n = 1, 2, 3 \\ 0 & \text{otherwise} \end{cases} \tag{6.35}$$

The time-varying channel was modeled as three-tap multipath fading channel and was simulated using Jake's simulator[40] with $f_m \cdot T = 1e - 4$,

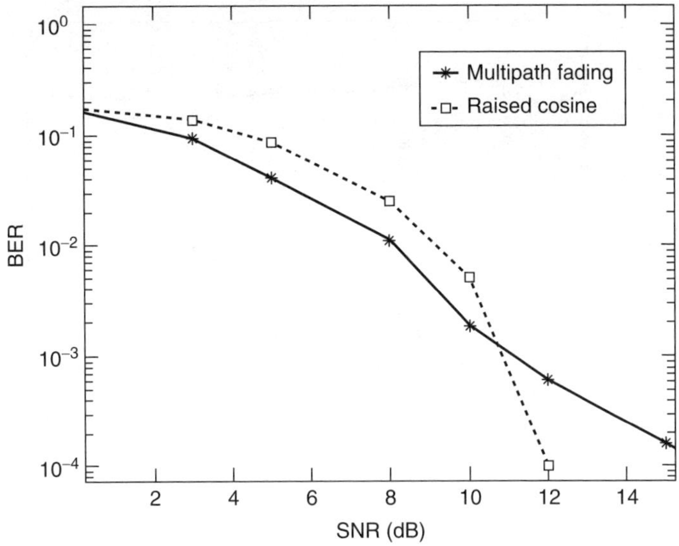

Figure 6.22 BER performances for raised-cosine and multipath fading channel models.

where f_m is the maximum Doppler frequency and T is the simulation step size. The BER performance for both channel models are illustrated in Figure 6.22.

Next, the performance of the equalization algorithm was simulated for a three-tap multipath fading channel for various channel fadings. The BER performance of the equalization algorithm corresponding to different channel fadings is plotted in Figure 6.23. It can be observed from the graph that the performance of the algorithm degrades significantly as the fading rate increases.

6.5 Estimation and Channel Equalization Algorithms for Multiuser Chaotic Communications Systems

Virtually all modern communications schemes must allow several users to share a common channel. There are currently many conventional methods for multiuser communications, such as FDMA, TDMA, and CDMA.[44] Even though many chaotic communications schemes have been proposed recently, very few of these approaches allow for multiuser communications.[9–12] Many of those multiuser chaotic systems use conventional multiuser techniques to multiplex chaotic signals. For example, in References 9 and 11, the proposed multiuser schemes based on chaotic systems are basically CDMA systems where pseudonoise sequences generated

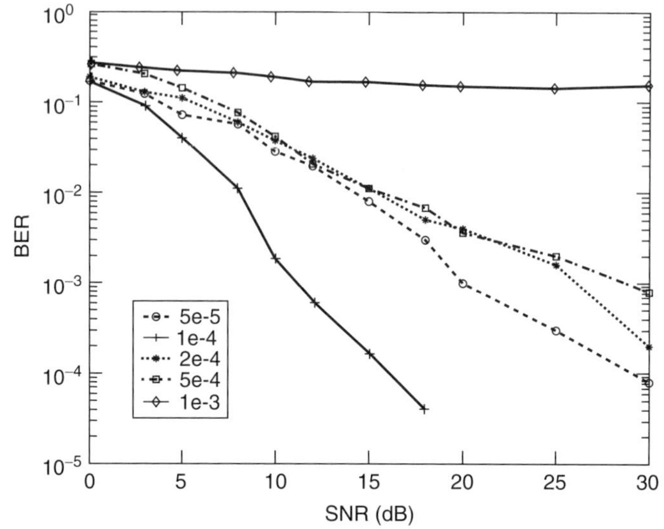

Figure 6.23 BER performances for three-tap multipath fading channel model for different fading amount expressed in terms of $f_m \cdot T$ values.

by binary linear shift registers are replaced by chaotic sequences. As such, conventional CDMA algorithms are used to separate the chaotic signals, not the nonlinear dynamic properties of the component signals.

In this section, a multiuser communications scheme based on chaotic modulation is first presented. Then, multiuser estimation and channel equalization algorithms for the proposed scheme are developed because the proposed multiuser scheme is significantly different from conventional schemes.

6.5.1 A Multiuser Chaotic Communications System

A block diagram for a multiuser chaotic communications system is shown in Figure 6.24. In such a system, each user's information bits must first be encoded through symbolic dynamics into the signal waveform generated by a chaotic system. The transmitted chaotic sequence for each user, $x_i[n]$, is generated by

$$x_i[n] = f_i(x_i[n-1]) \qquad \text{for } i = 1, 2, \ldots, K \tag{6.36}$$

where $f_i(\cdot)$ represents the dynamics of the ith chaotic modulator and K is the number of users. It is assumed here that each user's information is modulated with a different one-dimensional chaotic system, although our algorithm can easily be generalized to higher dimensional systems.

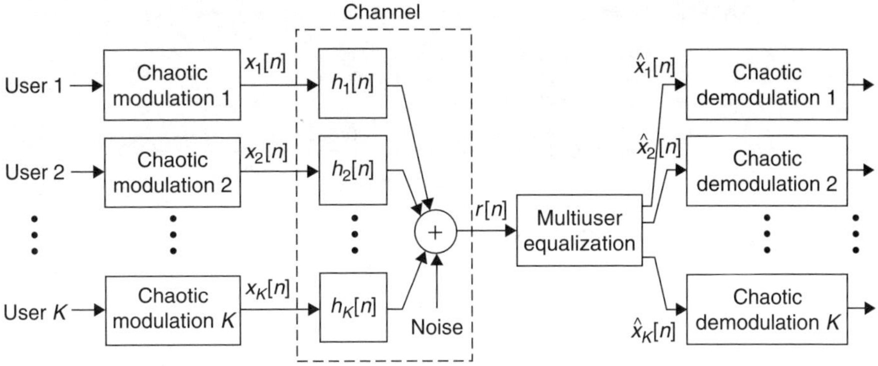

Figure 6.24 Block diagram for a multiuser chaotic communications system.

Once each user's information is encoded in a chaotic waveform, the chaotic sequences are transmitted over a common communication channel but with possibly different channel impulse responses. The communication channel distorts the transmitted signals by introducing ISI and noise, and the receiver observes the superposition of these distorted chaotic sequences. To recover each user's information, one must first undo the distortions and separate each chaotic sequence from the total received signal before performing chaotic demodulation, which is the process of obtaining estimates of each user's transmitted bit sequence from the observed signal.

In the remainder of this section, estimation and channel equalization algorithms based on the Viterbi algorithm are presented for multiuser chaotic communications. First, a symbolic dynamics representation is used to represent the multiuser chaotic system with an equivalent trellis diagram for the finite precision case. Then, the trellis diagram is expanded to include FIR channel distortions. Finally, the Viterbi algorithm is used to obtain a maximum-likelihood estimate of the component chaotic sequences from the received signal.

6.5.2 *Multiuser Trellis Diagram Representation*

The trellis diagram described in Section 6.4.2 is expanded here to accommodate multiuser chaotic systems. Figure 6.25 illustrates the trellis diagram for 2 users and for a precision of N_{p1} and N_{p2} binary symbols for user 1 and user 2, respectively. In the trellis diagram, the states consist of all possible combinations of symbols for both users, where the first N_{p1} symbols correspond to the symbolic state for user 1 and the remaining N_{p2} symbols are the symbolic state for user 2 (i.e., there are $2^{N_{p1}+N_{p2}}$ possible symbolic states). Because for binary signalling each incoming symbol for

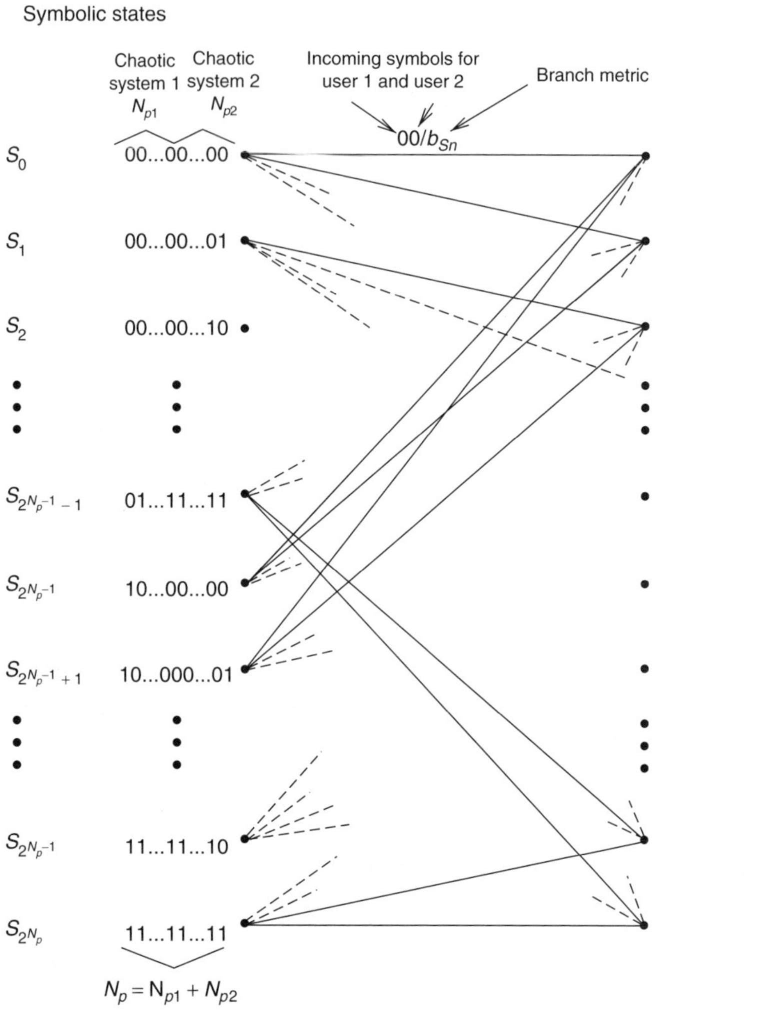

Figure 6.25 Symbolic dynamics-based trellis diagram representation of quantized chaotic dynamics for two users.

each user can only take on the value of 0 or 1, there are four branches leaving each state. The transition between the current state and the next state is determined by the symbolic dynamics, which is just the shift operation, and the incoming symbols. The branches of the trellis diagram are labeled with the input symbols and the branch metrics, which are calculated based on a minimized cost function for each state. For more than two users, the trellis diagram can be constructed by extending the states to include the symbolic state for each user. The number of states for K users is $2^{N_{p1}+N_{p2}+\cdots+N_{pK}}$ where N_{pi} is the length of the precision for the ith user.

6.5.3 A Multiuser Optimal Estimation Algorithm

Once each user's sequence has been transmitted over an additive noise channel, the receiver observes a superposition of the transmitted sequences with additive noise. The received signal can be formulated as

$$r[n] = x_1[n] + x_2[n] + \cdots + x_K[n] + w[n] \quad (6.37)$$

where $w[n]$ is typically modeled as WGN with a variance of σ_w^2.

To recover the transmitted user's information, each transmitted chaotic sequence, $x_i[n]$, must be estimated from the received signal, $r[n]$. However, one must first show if it is possible to estimate uniquely each chaotic sequence when a superposition of chaotic sequences is observed. The following theorem addresses this issue.

Theorem 1 *Under the assumptions of no noise, arbitrary initial conditions, and an infinite number of observations, each chaotic sequence, $x_i[n]$, can be uniquely estimated from the observed signal, $r[n]$, if the dynamics of each chaotic system are different from each other and do not have the same derivative at every point on the attractor.*

Proof 1 *The theorem is first proved for two chaotic sequences and then generalized for K users. For two chaotic sequences, the received signal can be written as*

$$r[n] = x_1[n] + x_2[n]$$

If there exist sequences $\tilde{x}_1[n]$ and $\tilde{x}_2[n]$ such that

$$\tilde{x}_1[n] + \tilde{x}_2[n] = r[n] \quad \text{for } n = 0, 1, \ldots, \infty$$

but $\tilde{x}_1[n] \neq x_1[n]$ and $\tilde{x}_2[n] \neq x_2[n]$, then the following is true

$$x_1[n] = \tilde{x}_1[n] - \epsilon[n] \quad \text{for } n = 0, 1, \ldots, \infty$$

$$x_2[n] = \tilde{x}_2[n] + \epsilon[n] \quad \text{for } n = 0, 1, \ldots, \infty$$

Because $x_1[n]$ and $x_2[n]$ are chaotic sequences generated by iterative functions, both dynamical functions should have the same effect for each $\epsilon[n]$. Because of the sensitivity to initial conditions and the mixing property, this is only true if both functions have the same constant derivative at every point on the attractor.

This result can be easily extended for a higher number of chaotic sequences by grouping the K users into two groups. From the above result, one can separate each group uniquely. The same process can be applied to each group separately. At the end, each chaotic sequence will be uniquely identified.

In communication applications, the received signal is observed in the presence of noise. Here, the following cost function is minimized to get an estimate of each chaotic sequence, represented by $\hat{x}_i[n]$.

$$J(\cdot) = \min_{\hat{x}_1[n],\hat{x}_2[n],\ldots,\hat{x}_K[n]} \sum_{n=0}^{N-1} (r[n] - \hat{x}_1[n] - \ldots - \hat{x}_K[n])^2$$
$$\text{s.t. } \hat{x}_1[n] = f_1(\hat{x}_1[n-1]), \ldots, \hat{x}_K[n] = f_K(\hat{x}_K[n-1]) \tag{6.38}$$

where N is the length of the received sequence. Under the Gaussian noise assumption, the solution of this cost function yields the ML estimate of the chaotic sequence. Because of the sensitivity of chaotic systems to initial conditions, straightforward minimization of this cost function is computationally difficult. Once the multiuser trellis diagram is constructed, the cost function given by Equation (6.38) can be minimized by using the Viterbi algorithm. The branch metrics at time n can be calculated as

$$b_S[n] = \left| r[n] - \sum_{k=1}^{K} x_k^S \right|^2 \tag{6.39}$$

where x_k^S represents the chaotic value for the corresponding state for user k.

6.5.4 A Multiuser Channel Equalization Algorithm

If the channel distorts the transmitted signals by introducing ISI as well as noise, then the receiver observes a superposition of distorted transmitted sequences. The received signal can be formulated as

$$r[n] = \sum_{i=0}^{K} \sum_{k=0}^{L_i} h_i[k] x_i[n-k] + w[n] \tag{6.40}$$

where $h_i[n]$ represents an FIR channel distortion of order L_i for the ith user and $w[n]$ represents WGN with a variance of σ_w^2. To recover the transmitted users' information, each transmitted chaotic sequence, $x_i[n]$, must be estimated from the received signal, $r[n]$. A channel equalization algorithm to obtain the estimate of each transmitted chaotic sequence, $\hat{x}_i[n]$, will be described here.

The equalization algorithm will be based on the multiuser trellis diagram representation of the previous section. To accommodate FIR channel distortions in multiuser chaotic communications systems, the trellis diagram representation developed above can be expanded by increasing the number of states. If the FIR channel impulse response for the ith user is of order L_i, then the number of states increases to $2^{N_{p1}+L_1+\cdots+N_{pK}+L_K}$. The new trellis diagram is similar to the previous trellis diagram with a larger precision for each user and with a new branch metric. The state at time n is now given by

$$(\overline{x}_1[n], \overline{x}_2[n], \ldots, \overline{x}_K[n]) \tag{6.41}$$

where $\overline{x}_i[n]$ represents the vector $(x_i[n-1], x_i[n-2], \ldots, x_i[n-L_i])$. Assume that each user transmits N symbols. Then, the received sequence is $(r[N], r[N-1], \ldots, r[1])$. The ML receiver estimates the sequence that maximizes the log-likelihood function,

$$\log p(r[N], \ldots, r[1] \mid \overline{x}_1^N, \ldots, \overline{x}_K^N) \tag{6.42}$$

where $\overline{x}_i^N$ represents vector $(x_i[N], x_i[N-1], \ldots, x_i[1])$. Because the noise $w[n]$ is white, $r[N]$ depends only on the L_i most recently transmitted symbols from the ith user, and the log-likelihood function can be written as

$$\begin{aligned}\log p(r[N], \ldots, r[1] \mid \overline{x}_1^N, \ldots, \overline{x}_K^N) &= \log p(r[N] \mid \overline{x}_1[N], \ldots, \overline{x}_K[N]) \\ &+ \log p(r[N-1], \ldots, r[1] \mid \overline{x}_1^{N-1}, \ldots, \overline{x}_K^{N-1})\end{aligned} \tag{6.43}$$

If the second term has been calculated at time $N-1$, then only the first term, called the branch metric, has to be computed for each incoming signal $r[N]$ at time N. From the channel model given by Equation (6.25), the conditional pdf is

$$p(r[N] \mid \overline{x}_1[N], \ldots, \overline{x}_K[N]) = \frac{1}{(2\pi\sigma_w^2)^{1/2}}$$
$$\exp\left\{-\frac{1}{2\pi}\left|r[N] - \sum_{i=1}^{K}\sum_{k=0}^{L_i} h_i[k]x_i[N-k]\right|^2\right\} \tag{6.44}$$

so that $\log p(r[N] \mid \overline{x}_1[n], \overline{x}_2[n], \ldots, \overline{x}_K[n])$ yields the branch metric

$$b_N = \left|r[N] - \sum_{i=1}^{K}\sum_{k=0}^{L_i} h_i[k]x_i[N-k]\right|^2 \tag{6.45}$$

The equalization algorithm can be easily extended to other channel distortion models with appropriate changes in the branch metric.

The well-known Viterbi algorithm can be used to implement the ML receiver by searching through the trellis diagram for the most likely transmitted chaotic sequences. Because the trellis diagram is formed using symbolic dynamics, the resulting sequence satisfies the dynamics imposed by the chaotic system.

As in the single-user chaotic communications system, the Viterbi algorithm requires knowledge of the channel parameters $h_i[n]$ to compute the branch metrics in Equation (6.45), so an adaptive channel estimator is needed. An initial estimate of the channel parameters is obtained using training sequences. After the training mode, the final decisions at the output of the Viterbi algorithm can be used to update channel parameter estimates. The parameter update equation for the LMS algorithm is:

$$h_i^n[j] = h_i^{n-1}[j] + \mu_i \cdot e[n-D]\hat{x}_i[n-j-D] \tag{6.46}$$

where μ_i is the adaptation step size for the ith user, D is the trellis depth, $\hat{x}_i[n-D]$ is the output of the Viterbi algorithm, and $e[n-D]$ is given by

$$e[n-D] = r[n-D] - \sum_{i=1}^{K}\sum_{k=0}^{L_i} h_i^{n-1}[k]\hat{x}_i[n-D-k] \tag{6.47}$$

Simulation results for this multiuser equalization algorithm are presented in the following section.

6.5.5 *Simulations*

The performance of the proposed estimation algorithm has been simulated using chaotic maps, such as the sawtooth map and the logistic map. Throughout the simulations, the noise was modeled as having a white Gaussian distribution. Simulation results are given in terms of the BER. We first simulate the results of the proposed algorithm for the noise-only case for different numbers of users with equal energy. In the one user case, the sawtooth map is used for modulation. For two users, the sawtooth map and logistic map are used for user 1 and user 2, respectively. In the three user case, the sawtooth map, logistic map, and logistic map with a constant multiplication of 0.5 are used for users 1, 2, and 3 respectively. A four-bit precision length is used for all three users. The BER performance of the first user in the cases of 1, 2, and 3 users are plotted in Figure 6.26 and compared with BPSK modulation scheme for single-user case.

The effect of the trellis size on the performance of the estimation algorithm was also simulated for the two-user case where each user's information was encoded using six-bit precision. The BER performances of the estimation algorithm corresponding to different trellis sizes are illustrated in Figure 6.27 for a WGN channel. The performance of the estimation algorithm degrades significantly for each bit lost.

Then, the equalization algorithm was applied to the two-user case with each user's channel filter first modeled as a one-tap multipath fading channel using Jake's fading simulator. Initial estimates of channel parameters

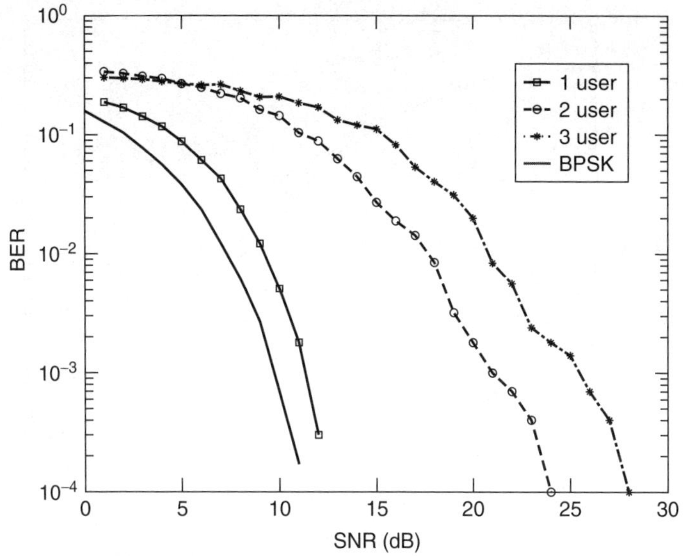

Figure 6.26 BER performance of the proposed multiuser algorithm for the noise-only case.

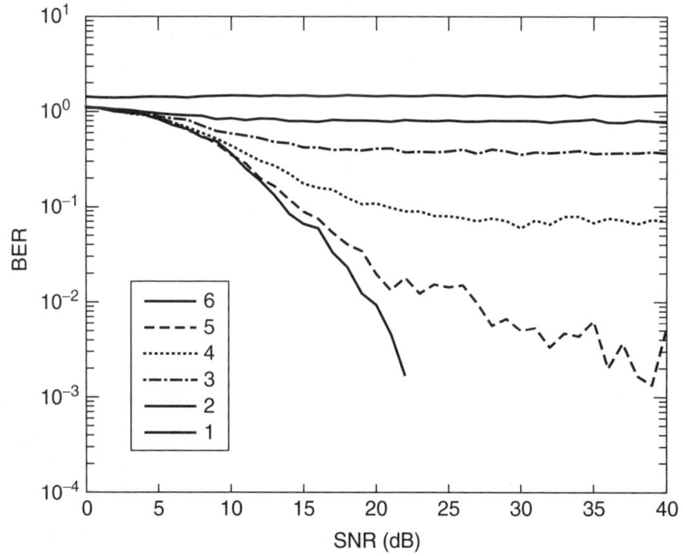

Figure 6.27 BER performance of the multiuser estimation algorithm corresponding to different trellis sizes expressed by the number of precision bits used for each state for the two-user case.

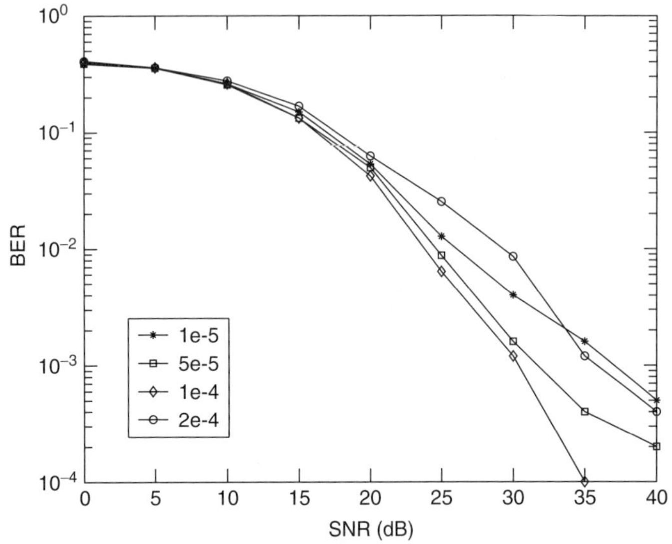

Figure 6.28 BER performance for the one-tap multipath fading channel model.

were obtained using a training sequence of length 1000 and four-bit precision length for both users. Also, the trellis depth, D, was set to 10. The BER performance of the first user is illustrated in Figure 6.28 for different $f_m \cdot T$ values, where f_m represents the maximum Doppler frequency and T is the simulation step size. Surprisingly, fast fading improves the performance of the equalization algorithm for certain amount of fading.

Next, the performance of the equalization algorithm was simulated for a three-tap multipath fading channel simulated using Jake's fading simulator for various channel fadings for the two-user scenario. Initial estimates of channel parameters were obtained using a training sequence of length 1000 and a four-bit precision length for both users. Also, the trellis depth, D, was set to 10. The BER performance of the first user is plotted in Figure 6.29 for different $f_m \cdot T$ values. It can be observed from the graph that the equalization algorithm does not perform well for fast fading channels. Similar to one-tap channel model, fast fading improves the performance of the algorithm for certain amount of fading.

The computational complexity of the proposed algorithm increases exponentially in both the length of the precision, the length of the channel filters, and the number of users. In each case, the number of states in the trellis diagram increases by an order of two with each increase in precision of one bit. On the other hand, computational complexity can be reduced by using a smaller trellis diagram, but this pruning results in performance degradation.

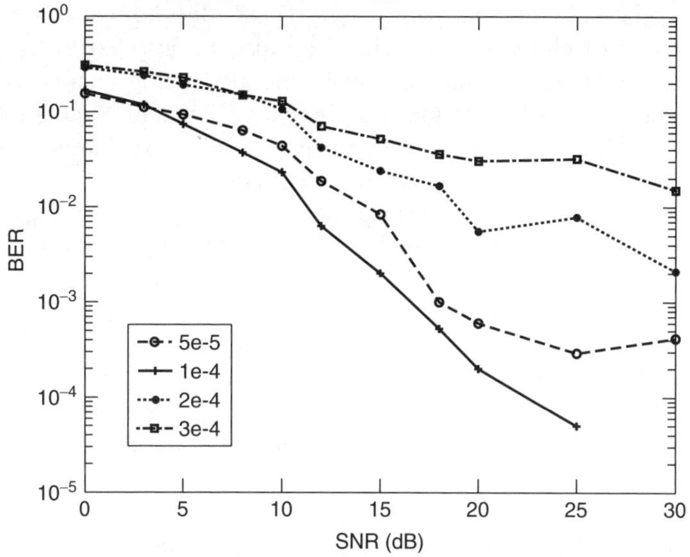

Figure 6.29 BER performance of the first user for a three-tap multipath fading channel model for different $f_m \cdot T$ values.

6.6 Conclusions

This chapter has been concerned with the development of channel equalization algorithms specifically designed for chaotic communications systems. Symbolic dynamics representations of chaotic systems have been exploited to equalize the distortions introduced by the communication channel.

First, optimal estimation and sequential equalization algorithms have been presented for chaotic communication systems based on symbolic dynamics representations. The symbolic dynamics representation of chaotic systems was the key for efficient chaotic modulation and demodulation. For optimal estimation, the chaotic system dynamics was first mapped onto a trellis diagram for finite precision implementations. Then, a dynamics-based cost function was minimized using the Viterbi algorithm to optimally estimate the transmitted chaotic sequence. Because the trellis diagram was constructed based on the system dynamics, the estimated chaotic sequence automatically satisfied the system dynamics.

For channel equalization, the dynamics-based trellis diagram was extended to accommodate FIR channel models. Then, the Viterbi algorithm was used to remove the distortion introduced by the channel and to estimate the chaotic sequence.

The channel parameters were initially estimated using a training sequence. After the training sequence, the estimated chaotic sequence was used to update the channel parameters. These algorithms can be used for a wide range of chaotic maps. They can also be applied to continuous-time systems where the information is encoded using chaotic control to determine the symbolic dynamics of the chaotic system. Simulation results show that both algorithms are able to remove realistic channel distortions and reliably determine the transmitted sequence.

Next, an iterative equalization algorithm for chaotic communications systems was developed that was similar to recent turbo equalization algorithms for conventional communications systems. In this scheme, channel equalization and chaotic demodulation were jointly implemented to improve the global performance of the algorithm. The iterative equalization algorithm had lower complexity than the sequential equalization algorithm, primarily because of the reduced trellis size. Simulation results showed that the performance of the iterative equalization algorithm improved significantly after the first iteration for time-varying channel models, while it didn't improve much for the time-invariant channel model. It was also shown that interleaving did not improve the performance.

Finally, a multiuser chaotic communications scheme where multiple users share a common communication channel was presented. Furthermore, optimal multiuser estimation and equalization algorithms were developed for chaotic systems with a symbolic dynamics representation to remove the distortions introduced by the communication channel and to separate the individual chaotic sequences. It was first proven that under the conditions of no distortion and infinite observations, it was possible to uniquely estimate each chaotic sequence with minor limitations from observed superposition of chaotic sequences. For optimal estimation, the multiuser chaotic system was first represented with an extended trellis diagram by combining the trellis diagrams of individual chaotic systems. Then, a cost function was minimized to estimate each user's chaotic sequence using the Viterbi algorithm. For equalization, the multiuser trellis diagram was further expanded to accommodate FIR channel models. Viterbi algorithms were used to remove the channel distortions and to estimate individual chaotic sequences. Simulations results showed that the algorithm was able to remove the noise at realistic SNRs. Even though the algorithm yields the maximum-likelihood sequence, the performance of the equalization algorithm degrades for fast fading environments.

References

1. Pecora, L.M., Overview of chaos and communications research, in *Chaos in Communications*, *Proc. SPIE*, 2038, 2–26, 1993.
2. Ogorzalek, M.J., Overview of electronic chaos generation, control and applications, in *Chaotic Circuits for Communications*, *Proc. SPIE*, 2612, 2–13, 1995.

3. Bollt, E., Lai, Y., and Grebogi, C., Coding, channel capacity, and noise resistance in communication with chaos, *Phys. Rev. Lett.*, 79(19), 3787–3790, 1997.
4. Hayes, S., Communicating with chaos, *Phys. Rev. Lett.*, 70(20), 3031–3034, 1993.
5. Dedieu, H., Kennedy, M.P., and Hasler, M., Chaos shift keying: Modulation and demodulation of a chaotic carrier using self-synchronization Chua's circuits, *IEEE Trans. Circuits and Systems – II: Analog and Digital Signal Processing*, 40, 634–645, 1993.
6. Heidari-Bateni, G., McGillem, C.D., and Tenorio, M.F., A novel multiple-address digital communication system using chaotic signals, in *Proc. Int. Conf. Commun.*, 1232–1236, 1992.
7. Pecora, L.M. and Carroll, T.L., Synchronization in chaotic systems, *Phys. Rev. Lett.*, 64(8), 821–824, 1990.
8. Cuomo, K.M. and Oppenheim, A.V., Synchronization of Lorenz based circuits with applications to communications, *IEEE Trans. Circuits Syst. II*, 40, 626–633, 1993.
9. Heidari-Bateni, G., and McGillem, C.D., A chaotic communication direct-sequence spread-spectrum communication system, *IEEE Trans. Commun.*, 42(2,3,4), 1524–1527, 1994.
10. Schweizer, J., Stochastic approach to spread spectrum communication using chaos, in *Chaotic Circuits for Communications*, *Proc. SPIE*, 2612, 115–125, 1995.
11. Milanovic, V., Syed, K.M., and Zaghloul, M.E., Combating noise and other channel distortions in chaotic communications, *Int. J. Bifurcation Chaos*, 7(1), 215–225, 1997.
12. Lipton, J.M. and Dabke, K.P., Spread spectrum communications based on chaotic systems, *Int. J. Bifurcation Chaos*, 6(12A), 2361–2374, 1996.
13. Heidari-Bateni, G. and McGillem, C.D., Chaotic sequences for spread spectrum: an alternative to PN-sequences, *Proc. IEEE Int. Conf. Sel. Top. Wireless Commun.*, 437–440, 1992.
14. Kostelich, E.J. and Schreiber, T., Noise reduction in chaotic time-series: A survey of common methods, *Phys. Rev. E*, 48(3), 1752–1763, 1993.
15. Abarbanel, H.D.I., *Analysis of Observed Chaotic Data*, Springer-Verlag, New York, 1996.
16. Drake, D.F., *Information's Role in the Estimation of Chaotic Signals*, PhD thesis, Georgia Institute of Technology, Atlanta, 1998.
17. Papadopoulos, H.C. and Wornell, G.W., Maximum-likelihood estimation of a class of chaotic signals, *IEEE Trans. Inform. Theory*, 41(1), 312–317, 1995.
18. Kay, S. and Nagesha, V., Methods for chaotic estimation, *IEEE Trans. Signal Proc.*, 43(8), 2013–2016, 1995.
19. Kay, S., Asymptotic maximum likelihood estimator performance for chaotic signals in noise, *IEEE Trans. Signal Proc.*, 43(4), 1009–1012, 1995.
20. Marteau, P.H. and Abarbanel, H.D.I., Noise reduction in chaotic time series using scaled probabilistic method, *J. Nonlinear Sci.*, 1, 313–343, 1991.

21. Kisel, A., Dedieu, H., and Ogorzalek, M., Noise reduction methods for chaotic communication schemes, *Proc. Int. Symp. Circuits Syst.*, 5, 446–449, 1999.
22. Proakis, J.G., *Digital Communications*, McGraw-Hill, New York, 1995.
23. Haykin, S., Ed., *Blind Deconvolution*, Prentice-Hall, New Jersey, 1994.
24. Cuomo, K.M., Oppenheim, A.V., and Barron, R.J., Channel equalization for self-synchronizing chaotic systems, *Proc. Int. Conf. Acoustics, Speech, Signal Processing*, 1996.
25. Leung, H., System identification using chaos with application to equalization of a chaotic modulation system, *IEEE Trans. Circuits and Systems-I: Fundamental Theory and Application*, 45(3), 314–320, 1998.
26. Johnson, Jr., C.R. and Thorp, J.S., An adaptive calibration algorithm for synchronized chaos, *IEEE Signal Processing Lett.*, 1(12), 194–195, 1994.
27. Dedieu, H. and Ogorzalek, M.J., Identifiability and identification of chaotic systems based on adaptive synchronization, *IEEE Trans. Circuits and Systems-I: Fundamental Theory and Application*, 44(10), 948–961, 1997.
28. Devaney, R.L., *An Introduction to Chaotic Dynamical Systems*, Addison-Wesley Publishing Company, 2nd ed., 1989.
29. Peitgen, H., Jurgens, H., and Saupe, D., *Chaos and Fractals New Frontiers of Science*, Springer-Verlag, New York, 1992.
30. Carroll, T.L. and Pecora, L.M., Synchronizing chaotic circuits, in *Proc. Am. Inst. Phys. Conf.*, 127–136, 1994.
31. Hao, B., Symbolic dynamics and characterization of complexity, *Physica D*, 51, 161–176, 1991.
32. Hao, B. and Zheng, W., *Applied Symbolic Dynamics and Chaos*, World Scientific, Singapore, 1998.
33. Gibson, J.D., Ed., *The Communication Handbook*, CRC Press, Boca Raton, FL, 1997.
34. Hayes, S. and Grebogi, C., Coding information in the natural complexity of chaos, in *Chaos in Communications, Proc. SPIE*, 2038, 153–160, 1993.
35. Rao, S.S. and Howard, S.P., Correlation performance of chaotic signals in spread spectrum systems, in *Proc. IEEE Digital Signal Processing Workshop*, 506–509, 1996.
36. Drake, D.F. and Williams, D.B., Pseudo-chaos for direct-sequence spread-spectrum communication, in *Chaotic Circuits for Communications, Proc. SPIE*, 2612, 104–114, 1995.
37. Hayes, M.H., *Statistical Digital Signal Processing and Modeling*, John Wiley and Sons, 1996.
38. Luenberger, D.G., *Linear and Nonlinear Programming*, 2nd ed., Addison-Wesley, 1989.
39. Oppenheim, A.V., Wornell, G.W., Isabelle, S.H., and Cuomo, K.M., Signal processing in the context of chaotic signals, *Proc. Int. Conf. Acoustics, Speech and Signal Processing*, 1992.
40. Stuber, G.L., *Principles of Mobile Communication*, Kluwer Academic Publishers, Boston, 1996.

41. Laot, C., Glavieux, A., and Labat, J., Turbo equalization: adaptive equalization and channel decoding jointly optimized, *IEEE J. Sel. Area. Comm.*, 19(9), 1744–1752, 2001.
42. Raphaeli, D. and Zarai, Y., Combined turbo equalization and turbo decoding, *IEEE Commun. Lett.*, 2(4), 107–109, 1998.
43. Lee, E.A. and Messerschmitt, D.G., *Digital Communications*, 2nd ed., Kluwer Academic Publishers, Boston, 1994.
44. Verdu, S., *Multiuser Detection*, Cambridge University Press, New York, 1998.
45. Ciftci, M. and Williams, D.B., A novel channel equalizer for discrete-time chaotic communications systems, *Proc. Int. Conf. Acoustics, Speech, and Signal Processing*, 1999.
46. Ciftci, M. and Williams, D.B., Optimal estimation for chaotic sequences using the Viterbi algorithm, *Proc. ASILOMAR Conf. Signals, Systems and Computers*, 2001.
47. Ciftci, M. and Williams, D.B., Optimal estimation and channel equalization for chaotic communications systems, *EURASIP J. Appl. Signal Processing, Special Issue on Nonlinear Signal and Image Processing*, 2001(4), 249–256, 2001.
48. Ciftci, M. and Williams, D.B., The sequential channel equalization algorithm for chaotic communications systems, *Proc. IEEE-EURASIP Nonlinear Signal Image Processing Workshop 2001*, 2001.
49. Ciftci, M. and Williams, D.B., Dynamics-based blind equalization for chaotic communications systems, *Proc. IEEE-EURASIP Nonlinear Signal Image Processing Workshop 2001*, 2001.
50. Ciftci, M. and Williams, D.B., Optimal estimation for multiuser chaotic communications systems, *Proc. Int. Symp. Circuits Syst.*, 2002.
51. Ciftci, M. and Williams, D.B., Channel equalization for multiuser chaotic communications systems, *Proc. Int. Conf. Acoustics, Speech, and Signal Processing*, 2002.

7

Optical Communications using Chaotic Techniques

Gregory D. VanWiggeren

CONTENTS

7.1 Introduction

The many techniques for optical communication with chaos described in this chapter share the same fundamental principle of operation. A receiver, with parameters matched to those of the transmitter, compares two time-delayed measurements of a signal to extract the message. This concept was inspired by a communication scheme developed by Volkovskii and Rulkov[1] for electronic circuits, but it has been significantly altered here to account for time-delays and the effects of optical phase, wavelength, and polarization. This general chaotic communication method can be applied to systems in which the dynamics are described by nonlinear time-delay equations, such as lasers in a ring configuration,[2–5] electro-optic loops,[6,7] or even the well-studied Mackey-Glass system.[8] It is well-known[9] that such time-delay differential equations can exhibit high-dimensional chaotic attractors.

Several other groups are currently working to develop methods for optical communication with chaos. A theoretical method for synchronizing chaotic nonlinear optical ring oscillators is proposed by Daisy and Fischer in Reference 10. They also discuss the possibility of using these synchronized ring oscillators for communication purposes. The oscillators they describe are similar in nature to erbium-doped fiber ring lasers (EDFRLs); though, the method they propose is unrelated to the methods presented in the following sections. Luo and Chu[11] have proposed a method for "secure" communications with EDFRLs that is also substantially different than the methods presented in this thesis. Actual experimental implementation of their method appears problematic, however. From an experimental perspective, the tolerances in the parameter mismatch between the two EDFRLs used in their method seem difficult to obtain and maintain. Later, the same group[12] experimentally demonstrated optical chaotic communication with a different method, which was very similar to one previously developed by VanWiggeren and Roy[13] (the method is discussed in Section 7.4).

Other groups have chosen to employ semiconductor lasers in investigations of optical communication with chaos. S. Sivaprakasam and K. A. Shore[14] at the University of Wales at Bangor have demonstrated chaotic synchronization of external cavity diode lasers. They also recently demonstrated optical communication of a square-wave at roughly 3 kHz using these chaotic external cavity diode lasers.[15] Previously, Goedgebuer et al.[4,5] used the chaotic wavelength fluctuations of tunable diode lasers to demonstrate optical chaotic communication of square- and sine-waves of a few kHz in frequency. References 4 and 5 are excellent and clearly written papers. The laser's wavelength dynamics can be clearly described by a delay-differential equation that is known to exhibit chaos, and the correspondence between the theory and experiments is very good.

7.2 Erbium-Doped Fiber Ring Lasers for Optical Chaotic Communication

Three properties of EDFRLs make them attractive for use in optical chaotic communication: (1) Existing fiber-optic communications devices and components can be conveniently incorporated into the research because EDFRLs themselves are fiber-optic devices. (2) Their very broad bandwidth suggests that they would be suitable for very high communication rates. Naturally, a chaotic optical communication system should have a large communication bandwidth to be most useful. (3) Their high-dimensional dynamics offer the potential for enhanced communication privacy.

Realizing these benefits, however, requires appropriate communication methods. In this chapter, several methods of communication using the irregular waveforms generated by EDFRLs are described. All of the methods attempt to realize the potential benefits of EDFRLs, achieving varying degrees of success.

An understanding of the nonlinear dynamics of EDFRLs is crucial to developing realizable optical chaotic communication schemes. A number of models have been developed to explain relatively slow dynamics, such as chaotic self-pulsing at kHz rates.[16–18] To take advantage of the many gigahertz bandwidth available in EDFRLs, it was important to develop a model for the faster, sub-round-trip dynamics of the EDFRL.[19] This model assumes that the classical electric field in the laser interacts with two-level quantized atoms. This assumption is called the semiclassical approximation because the electric field is treated classically, but the erbium-atoms are treated quantum mechanically. The electric-field is governed, therefore, by the Maxwell wave-equation, whereas the two-level atoms are governed by the optical Bloch equations. Combining these two sets of equations gives the Maxwell-Bloch equations. For the EDFRL, the Maxwell-Bloch equations are

$$
\begin{aligned}
\frac{1}{v_g}\frac{\partial E_m}{\partial t} &= -\frac{\partial E_m}{\partial z} - i\frac{\omega_m}{2\epsilon}P_m \\
\dot{P}_m &= -(\gamma_\perp + i(\omega_0 - \omega_m))P_m + \left(\frac{ie^2|X_{1,2}|^2}{\hbar}\right)E_m N \qquad (7.1) \\
\dot{N} &= q - \gamma_{||}N + \frac{2i}{\hbar}(E_m P_m^* - E_m^* P_m)
\end{aligned}
$$

where two orthogonal polarization are denoted by $m = 1, 2$ and a summation is performed over m (Einstein summation convention) in the equation for $\dot{N}$. Here, the group-velocity, v_g, is used because the slowly varying envelope of the electric-field, E, propagates at the group velocity of light in fiber. Equation (7.1) take into account the possibility that the angular frequency of the laser light in the EDFRL, ω, is detuned from the angular frequency corresponding to the transition between the two atomic energy levels, ω_0. This results in the $(\omega_0 - \omega_m)$ term in the equation for $\dot{P}$. The squared magnitude of the x component of the atomic displacement, X_{12} equals $|M|^2/3e^2$, where M is just the total atomic dipole moment and e is the charge of an electron. The parameter q simply describes the pumping of the erbium-atoms. As before, E is just the slowly-varying envelope of a quasi-monochromatic lightwave, $\mathcal{E} = \frac{1}{2}E(z,t)\exp(i(kz - wt) + c.c)$. Likewise, P is the slowly varying envelope of the dipole polarization per unit volume, $\mathcal{P} = \frac{1}{2}P(z,t)\exp(i(kz - wt) + c.c)$. N is the population inversion per unit volume. Specifically, it is the difference between the number of excited atoms and the number of unexcited atoms per unit volume. Consequently, it can assume both positive and negative values. The dipole decay rate and the spontaneous emission rate are represented by $\gamma_\perp$ and $\gamma_{||}$ respectively. As is customary, $\hbar$ denotes Planck's constant divided by 2π, and finally, ϵ is just the permittivity of the medium. The permittivity $\epsilon = n^2\epsilon_0$, where n is the index of refraction of the medium.

To gain tractability, the equation for $\dot{P}$ is adiabatically eliminated, as is often done when modeling class-B lasers in which the dipole decay rate, here $\gamma_\perp \approx 10^{12}$, is much faster than the decay of the population inversion, here $\gamma_{||} \approx 10^2$, and of the decay of the electric field — which in this case depends on the loss that occurs in each round trip in the output coupler. This approximation assumes that the polarization dynamics quickly evolve to a "steady-state" in which

$$P_m = i\frac{e^2|X_{12}|^2 E_m N}{\hbar(i\Delta_m + \gamma_\perp)} \tag{7.2}$$

where $\Delta_m = \omega_0 - \omega_m$. This approximation has been numerically tested[20] and is by definition appropriate for class-B lasers. A simple stability analysis in which E and N are assumed to evolve slowly is also supportive of this adiabatic elimination of $\dot{P}$.

Substituting the results of Equation (7.2) into Equation (7.1) yields two new dynamical equations,

$$\dot{E}_m = -v_g\frac{\partial E_m}{\partial z} + \frac{v_g\omega_m(i\Delta_m + \gamma_\perp)e^2|X_{12}|^2}{2\epsilon\hbar(\Delta_m^2 + \gamma_\perp^2)}E_m N$$

$$\dot{N} = Q - \gamma_{||}N - \frac{4e^2|X_{12}|^2\gamma_\perp|E_m|^2}{\hbar^2(\Delta_m^2 + \gamma_\perp^2)}N \tag{7.3}$$

When proper boundary conditions for the ring are incorporated into the model, a set of delay differential equations for the electric field, E, and inversion, N result. More detail is provided in References 19 and 21.

The model presented here is able to recreate many aspects of the experimentally observed intensity dynamics. However, it neglects the nonlinear characteristics (the Kerr effect) of the fiber medium itself. An improved delay-differential model was developed in Reference 19 concluding that the nonlinear interaction between the electric field and the population inversion, as described in Equation (7.3), is a relatively small effect. Instead, the authors assert that the evolution of the intensity waveform over many round trips through the laser is primarily due to the effect of the nonlinearity of the fiber medium.

> The analysis we provide in this paper points clearly to the role of the non-linear dependence of the polarization of the fused silica medium of the fiber as the source of the optical chaos. We test this quantitatively ... and we show that this chaos persists when we vary the difference of index of refraction between the electric field polarizations, the absorption for each polarization, and even the external injection of light into the cavity. None of these effects gives rise to chaotic oscillations in the absence of the medium nonlinearity, as expected, and each of them only slightly modifies the effect of the medium nonlinearity.

7.3 Message Injection Technique

7.3.1 Message Injection Technique using Low-Pass Filtering

7.3.1.1 Apparatus and Method

The experimental setup for the message injection technique,[13,22] is shown in Figure 7.1. As is common in conventional fiber-optic communication schemes, the message consists of an optical sequence of binary "bits." To create the message, a Hewlett-Packard model 80,000 data-generator was used to form a voltage signal consisting of a sequence of bits that drive a lithium-niobate Mach-Zehnder intensity-modulator. Continuous-wave (cw) and completely polarized light produced by an external-cavity tunable diode laser (Photonetics Tunics-BT) was intensity modulated as it passed through the modulator to form an optical sequence of pseudorandom bits. This sequence of bits served as the message to be communicated. After modulation, the message sequence was amplified in a controllable fashion by an EDFA with a 13 dBm maximum output power. This operation allowed the amplitude of the message injected into the EDFRL to be adjusted. The amplified message then passed through a polarization controller consisting of a sequence of waveplates ($\lambda/4$, $\lambda/2$, $\lambda/4$) arranged to permit complete control of the polarization state of the message as it was coupled into the ring.

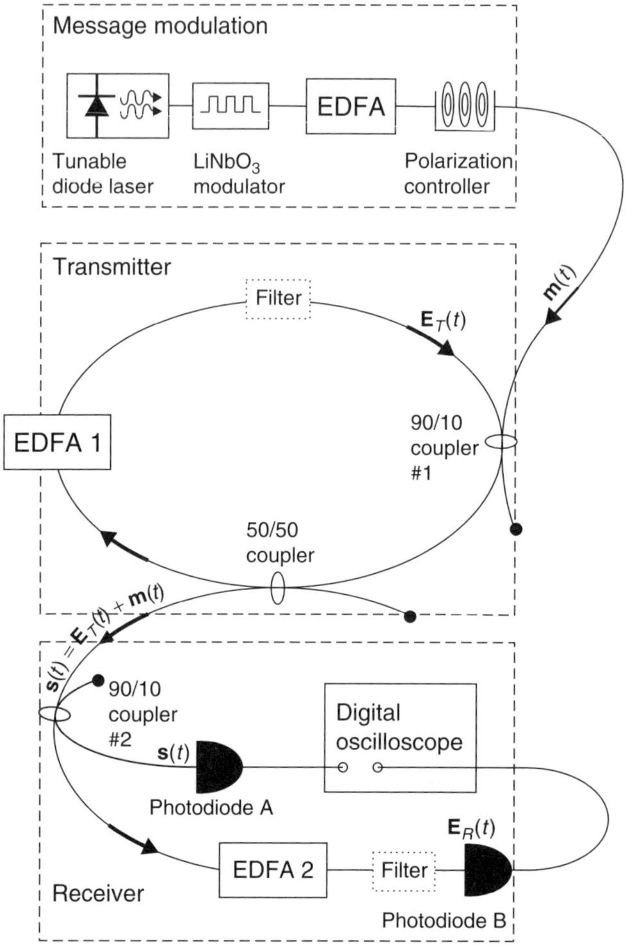

Figure 7.1 Experimental system for optical chaotic communication consisting of three parts.[13] In the message modulation unit, cw laser light is intensity modulated to produce a message signal, m(t), consisting of a sequence of binary bits. The message signal is injected into the transmitter where it is mixed with the fluctuating lightwaves produced by the EDFRL, $E_T(t)$. The optical combination of these two signals, s(t), propagates through the communication channel to the receiver where the message is recovered. During this transmission, the message is masked by the chaotic fluctuations of $E_T(t)$.

The message light was injected into the EDFRL through a 90/10 waveguide coupler, which allowed 10 percent of the message light to be injected into the ring, while retaining 90 percent of the light already in the ring. The light then propagated to a 50/50 coupler that sent half of the light in the ring to a receiver unit, while the other half passed through EDFA 1. This EDFA had 17 dBm maximum output power and 30 dB small signal gain.

The active (doped) fiber in the EDFA was 17 m long and was pumped by diode lasers with a 980-nm wavelength. Isolators in the EDFA ensured that light propagated unidirectionally in the ring, as indicated by the arrows in Figure 7.1. Unless otherwise specified in the description of subsequent experiments, the EDFA in the transmitter EDFRL is operated at pump powers more than 10 times threshold. After being amplified by EDFA 1 the light continues to circulate around the ring. The filter shown inside the EDFRL in Figure 7.1 was not used in the experiments described in this section and should be ignored until subsequent experiments are introduced. The total length of active and passive fibers in the ring was approximately 40 m, which corresponds to a round-trip time for light in the ring of about 200 ns.

Light exiting the ring through the 50/50 output coupler propagated in a fiber-optic communication channel to a receiver unit. Light entering the receiver unit was split with a 90/10 coupler. Ten percent passed through a variable attenuator to prevent photodiode A (125-MHz bandwidth) from saturating. The remaining 90 percent of the light passed through EDFA 2 and is measured by photodiode B (125-MHz bandwidth). The filter that follows EDFA 2 was not used in these experiments. The signals from photodiodes A and B are recorded by a digital oscilloscope with a 1 GS/s sampling rate for each channel and 8-bit resolution.

As indicated in the figure, the slowly-varying amplitude of the message bit-stream input into the transmitter at a particular time is denoted by $\mathbf{m}(t)$, where $\mathbf{m}$ is a vector to account for its polarization. It is added to the slowly-varying amplitude of the electric field arriving in the coupler at the same time, $\mathbf{E}_T(t)$. The amplitude of the combined electric fields, denoted by $\mathbf{s}(t)$ in anticipation of its use as a message-carrying signal, is very simply $\mathbf{s}(t) = \mathbf{E}_T(t) + \mathbf{m}(t)$. This $\mathbf{s}(t)$ propagates to the output coupler where half of the light propagates to the receiver unit while the other half continues to propagate around the ring. If the injected message power is not too large (typically on the order of a few milliwatts or less), then the message is "masked" by the larger 10–40-mW chaotic intensity fluctuations of $|\mathbf{E}(t)|^2$ as $\mathbf{s}(t)$ is transmitted through the fiber optic communication channel between the transmitter and receiver.

The fraction of $\mathbf{s}(t)$ that remains in the transmitter continues to propagate around the ring. The nonlinearities that act in the amplifier and the fiber alter the original waveform, $\mathbf{s}(t)$, to produce a new waveform $\mathbf{E}_T(t+\tau_R)$. The models presented in the previous chapter show that $\mathbf{E}_T(t+\tau_R)$ is functionally dependent on the population inversion, $N(t)$ and field-amplitude one round-trip time earlier. This relationship is written

$$\mathbf{E}_T(t+\tau_R) = f(\mathbf{E}_T(t), N(t), \mathbf{m}(t)) \tag{7.4}$$

where f is a nonlinear function of all of these variables. The new lightwave, $\mathbf{E}_T(t+\tau_R)$ is combined with a new message amplitude, $\mathbf{m}(t+\tau_R)$, in the 90/10 coupler to create $\mathbf{s}(t+\tau_R)$. As before, half of this new signal is transmitted to the receiver unit, whereas the remaining half remains in the transmitter. In the setup described here, the effect of the nonlinearities in the system were small over just one round trip. Consequently, the intensity of $|\mathbf{E}_T(t+\tau_R)|^2$ is very similar to the intensity of $|\mathbf{E}_T(t)+\mathbf{m}(t)|^2$. The communication method presented here, however, should work well even in situations in which the nonlinearities in the EDFRL are more significant and $|\mathbf{E}_T(t+\tau_R)|^2$ is not similar to $|\mathbf{E}_T(t)+\mathbf{m}(t)|^2$.

The transmitted fraction of $\mathbf{s}(t)$ arrives in the receiver unit where a coupler directs 10% of the light to photodiode A. The remaining 90% is operated on in a nonlinear way by EDFA 2 and then continues in the fiber to photodiode B. This arm of the receiver has been constructed to operate on light in exactly the same way as one round trip through the transmitter. Thus, the nonlinear function, f, which operates on $\mathbf{s}(t)$ in the transmitter as described in Equation (7.4) also operates on $\mathbf{s}(t)$ in the arm leading to photodiode B. This operation in the receiver is written

$$\mathbf{E}_R(t+\tau_R)=f(\mathbf{E}_T(t), N(t), \mathbf{m}(t)) \tag{7.5}$$

Because the function, f, is the same in both cases, $\mathbf{E}_R(t+\tau_R)=\mathbf{E}_T(t+\tau_R)$.

The lengths of the arms in the receiver have been adjusted such that photodiode B measures light with a time-delay, τ_R, relative to photodiode A. Therefore, the signal, $\mathbf{s}(t)$ is detected at photodiode A while it is still propagating toward photodiode B in the other arm of the receiver. Exactly τ_R after $\mathbf{s}(t)$ is measured at photodiode A, the nonlinear operation of EDFA 2 and the fiber on the remaining fraction of $\mathbf{s}(t)$ cause $\mathbf{E}_R(t+\tau_R)$ to be detected at photodiode B. Simultaneously, photodiode A is measuring $\mathbf{s}(t+\tau_R)=\mathbf{E}_T(t+\tau_R)+\mathbf{m}(t+\tau_R)$. Taking the difference of the two incident electric field amplitudes gives $\mathbf{m}(t+\tau_R)$ and the message would be recovered. However, photodiodes can only measure signal intensities. Therefore, the two photodiodes would measure the intensities $|\mathbf{E}_T(t+\tau_R)+\mathbf{m}(t+\tau_R)|^2$ and $|\mathbf{E}_R(t+\tau_R)|$ respectively. The difference of these two measurements, $2\text{Re}\{\mathbf{E}_T^*(t+\tau_R)\cdot\mathbf{m}(t+\tau_R)\}+|\mathbf{m}(t+\tau_R)|^2$, does not reveal the message directly.

When no message is transmitted, it is clear from the preceding discussion that both photodiodes should measure the same signal because $\mathbf{E}_R(t+\tau_R)=\mathbf{E}_T(t+\tau_R)$. In a real experiment, however, the time-delays between the photodiodes are not exactly τ_R. Consequently, phase errors between $\mathbf{E}_R(t+\tau_R)$ and $\mathbf{E}_T(t+\tau_R)$ will be observed, but $|\mathbf{E}_R(t+\tau_R)|=|\mathbf{E}_T(t+\tau_R)|$ is still valid. The results of an experiment in which no message was injected into the transmitter are shown in Figure 7.2. Figure 7.2a shows the irregular intensity fluctuations of the lightwave measured by photodiode A, in this case $|\mathbf{E}_T(t+\tau_R)|^2$. The intensity measured by photodiode A is, for the remainder of this thesis, denoted by I_A. Thus, for

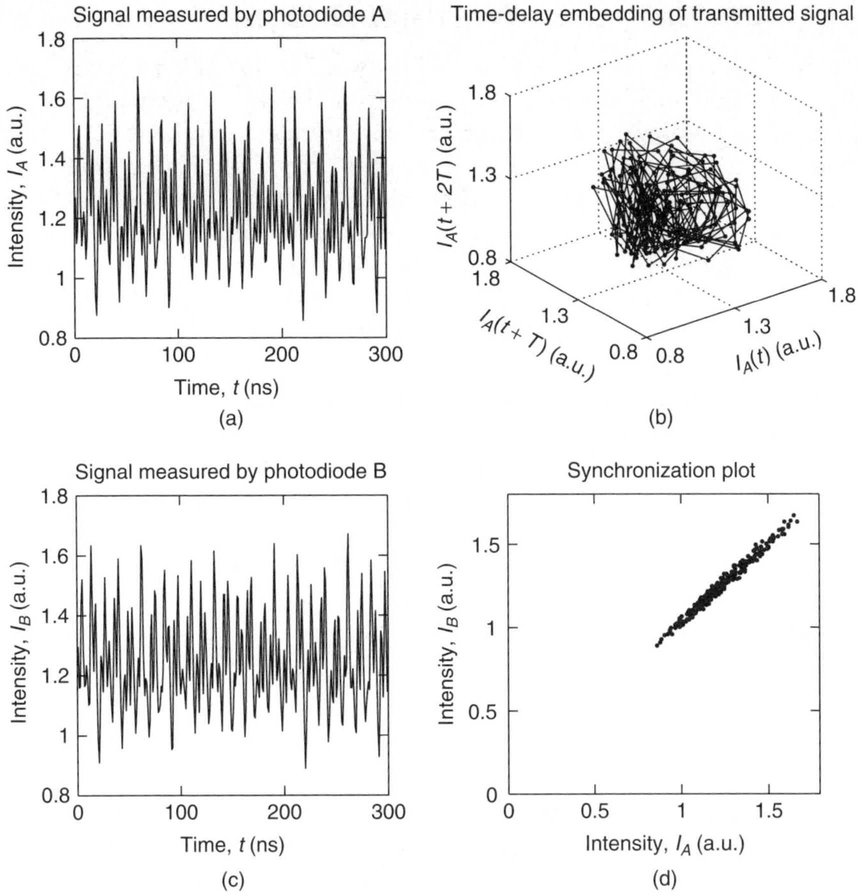

Figure 7.2 (a) Transmitted signal measured by photodiode A when there is no message component, $I_A \propto |E_T(t+\tau_R)|^2$. (b) Time-delay embedding of the intensity time trace (a) in three dimensions with a time delay $T = 4$ ns shows no low dimension structure. (c) Signal measured by photodiode B. Here, $I_B \propto |E_R(t+\tau_R)|^2$. (d) $x-y$ Plot of the signals in (a) and (c). The two signals are well synchronized, demonstrating that the receiver tracks the transmitter laser dynamics when no message is present.[22]

this figure, $I_A \propto |\mathbf{E}(t+\tau_R)|^2$. Figure 7.2c shows the irregular intensity of the lightwave measured by photodiode B, after it has passed through EDFA 2. The intensity measured by photodiode B, denoted by I_B in this thesis, is proportional to $|\mathbf{E}_R(t+\tau_R)|^2$. The two measured intensities, I_A and I_B, are quite evidently synchronized, but Figure 7.2d more clearly shows the synchronization by plotting both intensities against each other in an $x-y$ plot.

7.3.1.2 The Dimensionality of the EDFRL Dynamics

Figure 7.2b, however, is not so straightforward in its explanation. A time-delay embedding plot, such as is shown in (b), takes advantage of the embedding theorem, which Reference 23 attributes to Takens and Mañé. Reference 23 is an excellent resource on the topic of chaotic time-series analysis. Essentially, the embedding theorem states that the basic geometrical structure of a multivariable system's dynamics in phase-space can be reproduced from the observation of just one of the system variables. From one variable, say $x(t)$, of an arbitrary system, the system's phase-space geometry can be unfolded by using a set of new vectors, $y(t) = [x(t), x(t+T), x(t+2T), \ldots, x(t+NT)]$, where T is a time-delay. This new space can mathematically be related to the original multivariable phase space by smooth, differentiable transformations.

An enlightening and simple example, which may make such a theorem seem more plausible, is given here. The sine-wave, $s(t) = A\ \sin(t)$, is a solution of a two-dimensional system, in other words, a system whose phase-space geometry is two-dimensional. The two-dimensional nature is apparent if one decomposes the differential equation $\ddot{s} = s$, whose solution is a sine-wave, into the following coupled first-order differential equations:

$$\dot{s} = v \tag{7.6}$$

$$\dot{v} = s \tag{7.7}$$

This form of these equations highlights the two-dimensional nature of the dynamics — two variables are required to specify the state of the system. Plotting the dynamics of the system in two dimensions simply requires plotting the vectors $[s(t), v(t)]$, and the result is a circle. The same geometrical structure can be seen by plotting the time-delay vector $[s(t), s(t+T)]$ with time as an independent variable. In general, this method of plotting gives an ellipse, which is of the same basic geometry as the circle given by $[s(t), v(t)]$. Care must be taken, however, to avoid choosing a time-delay that is exactly equal to the period of the sine-wave, for which the ellipse appears as a line. Ensuring an appropriate choice for T is a subject that will be addressed below.

Figure 7.2b is included because of the understanding that time-delayed vectors of a single variable can reveal the basic structure of the entire multivariate dynamics of a system. The plot shown in (b) is called a time-delay embedding plot. Intensity data sampled every nanosecond were used to generate vectors, $[I(t), I(t+T), I(t+2T)]$, with a time-delay, $T = 4$ ns. These vectors were then plotted in three-dimensional space. This plot shows no structure in an embedding dimension of three, which suggests (but does not prove) that its attractor in phase space for the dynamics of the ring laser is not low-dimensional (≤ 3 dimensions).

Although the embedding theorem applies for all values of T, except for certain multiples of any periodicity of the system,[23] in real chaotic systems, such as the EDFRL, it is important to choose an appropriate time-delay, T. An appropriate time-delay is a multiple of the sampling time — any kind of interpolation scheme does not provide any more information than is already present in the sampled data and just creates extra work. Choosing a time-delay that is so short that the system has not really revealed any new information about itself is unhelpful from a numerical analysis perspective. On the other hand, the time-delay, T, can be too long. In a real chaotic system (sensitive to perturbations), small amounts of noise prevent even a very accurate measurement at one time from being dynamically correlated with a measurement taken at a time, T, that is too much later.

The procedure given in Reference 23 for determining a suitable T considers the measured chaotic dynamics from an information theory perspective. As Shannon[23] defined it, the mutual information between two measurements is just

$$\log_2\left[\frac{P_{AB}(a_i, b_j)}{P_A(a_i)P_B(b_j)}\right] \tag{7.8}$$

where $P_{AB}(a, b)$ is the joint probability density for measurements A and B giving values a and b. $P_A(a)$ and $P_B(b)$ are the individual probability densities of A and B. The average mutual information, AMI_{AB}, is

$$\mathrm{AMI}_{AB} = \sum_{a_i, b_j} P_{AB}(a_i, b_j)\log_2\left[\frac{P_{AB}(a_i, b_j)}{P_A(a_i)P_B(b_j)}\right] \tag{7.9}$$

This basic concept can be applied to time-delayed measurements of the EDFRL intensity dynamics. The calculation,

$$\mathrm{AMI}(T) = \sum_{I(t), I(t+T)} P(I(t), I(t+T))\log_2\left[\frac{P(I(t), I(t+T))}{P(I(t))P(I(t+T))}\right] \tag{7.10}$$

gives the average mutual information between two measurements of the intensity, denoted by I, at times separated by T. This AMI(T) can be considered as a kind of nonlinear autocorrelation function to determine when the values of $I(t)$ and $I(t+T)$ are independent enough of each other to be useful in a time-delay vector but not so separated in time that they are not dynamically related at all. The prescription for selecting T recommended in Reference 23 is to find the first minimum of the average mutual information and choose its corresponding T. The results of an average mutual information calculation performed on EDFRL intensity data when no message is injected is shown in Figure 7.3. A software package called

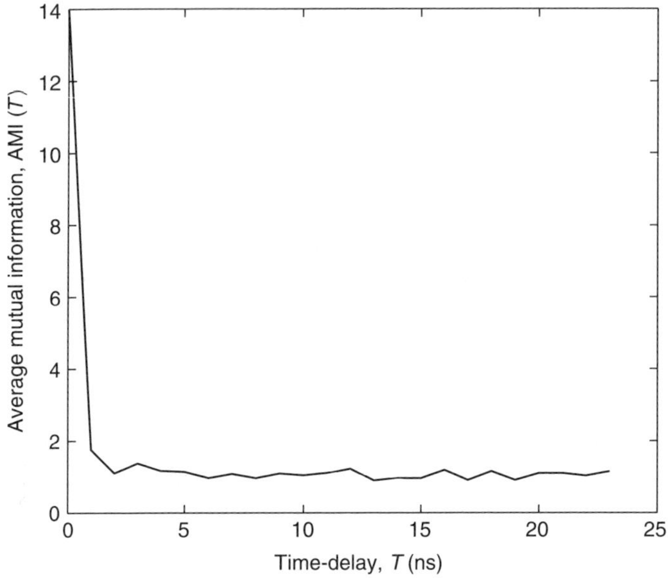

Figure 7.3 Average mutual information of EDFRL intensity dynamics plotted as a function of time-delay, *T*.

Contemporary Signal Processing for Windows version 2.1 by Applied Nonlinear Sciences/Randle Inc. was used to perform this calculation. For these data, the first minimum of the AMI occurs at $T = 2$ ns.

If the time-delay dynamics are plotted in an embedding space that is too low-dimensional, the trajectories will overlap as higher-dimensional dynamics are "folded" into the low-dimension embedding space. The uniqueness theorem states that trajectories in the phase-space of dynamical systems cannot cross. Such intersections in phase space would allow two deterministic systems with the same initial conditions to evolve differently, an occurrence that the uniqueness theorem prohibits. A related concept is employed to determine the appropriate embedding dimension. If two points in an embedding space are located near each other, their proximity is the result of two possibilities. They may be near each other because the second point has arrived through dynamical evolution at a point near the first on the system's attractor. These are proper neighboring points. Alternatively, they may be nearby in embedding space because the embedding dimension is not sufficient to "unfold" the dynamics of the system completely. The second point then has arrived near the first through projection of a higher-dimensional attractor onto a lower-dimensional space. Two such points are called *false* nearest neighbors (FNNs).

Reference 23 describes a numerical technique that uses this concept of FNNs to determine the appropriate embedding dimension, d_E. It begins with a low-dimensional space of dimension d and locates neighboring

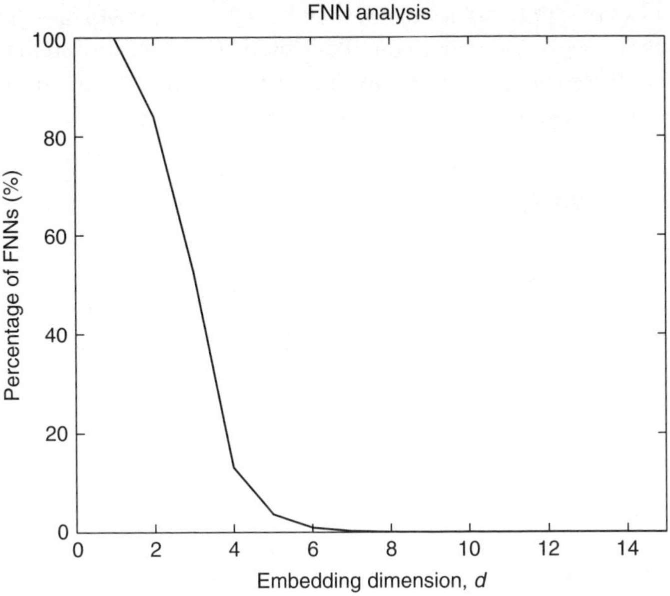

Figure 7.4 FNNs analysis of the intensity dynamics of an EDFRL. A time-series with 50,000 points was analyzed to reveal an embedding dimension of either seven or eight.

points. Then it plots the same data in a space of dimension $d+1$ and attempts to determine how many of the original neighbors are still neighbors. Those that are no longer neighbors are those that were originally FNNs. The process continues until all of the points that are neighbors remain neighbors, and at this point, one has achieved the proper embedding dimension for the data.

Figure 7.4 shows the results of numerical analysis to determine the proper embedding dimension. The numerical details were performed with the Contemporary Signal Processing Program for Windows software mentioned above on 50,000 data points. As is evident from the figure, increasing the embedding dimension decreases the percentage of neighbors that are false until this percentage reaches a value near zero. Ideally, the proper embedding dimension of the system is the smallest one for which there are no FNNs. In this particular figure, there is no such dimension at which no FNNs are detected, though by dimension seven the curve has reached a very low level. Dimension eight is the first minimum. The small residual level can be either attributed to noise in the experimental system, which is by definition infinitely dimensional, or a very small amount of high-dimensional dynamics. It can be shown[23] that low-pass filtering of time-series data can reduce the proper embedding dimension. Thus, the low-pass filtering of the EDFRL dynamics by the photodiodes may be eliminating even higher-dimensional behavior along with the higher frequencies that may be present in the real signal. The data shown in Figure 7.4 suggest that the proper embedding dimension is around seven or possibly eight. In either case, the dynamics are considered high-dimensional.

7.3.1.3 Message Recovery

The discussion above focused on the communication system when no message is being transmitted. When a message is injected into the transmitter, the two photodiodes do not measure exactly the same signal. The difference between the two photodiodes when a message is being communicated is

$$2\text{Re}\{\mathbf{E}_T^*(t+\tau_R)\cdot\mathbf{m}(t+\tau_R)\}+|\mathbf{m}(t+\tau_R)|^2. \qquad (7.11)$$

In the method described in this section, the message was a series of binary bits. Consequently, determining $|\mathbf{m}(t)|^2$ was essentially the same as determining $\mathbf{m}(t)$, the value of the bit being communicated. To isolate $|\mathbf{m}(t)|^2$, it was necessary to somehow eliminate the crossterm, $2\text{Re}\{\mathbf{E}_T^*(t+\tau_R)\cdot\mathbf{m}(t+\tau_R)\}$. The simplest way to achieve this was to choose $\mathbf{m}(t)$ to have a very low bit-rate. Low-pass filtering the difference signal could eliminate the crossterm because most of the energy in $\mathbf{E}_T(t+\tau_R)$ was typically spread among higher frequencies that were eliminated by the low-pass filter. Results of this approach are shown below in Figure 7.5 and Figure 7.6.

Figure 7.5a shows the chaotic intensity time-trace, $I_A \propto |\mathbf{s}(t+\tau_R)|^2$, output from the transmitter and measured by photodiode A. The injected message was a 10-MHz square-wave with a wavelength corresponding to one of the peaks in the EDFRL gain spectrum. Figure 7.5b shows the power-spectrum of the transmitted signal, given by $|\mathcal{F}\{I_A\}|^2$ where $\mathcal{F}$ represents the Fourier transform operator. The spectrum is very broadband; the high-frequency components are even suppressed in this figure due to the limited bandwidth of the photodiodes. The power of the message modulation light, $|\mathbf{m}(t)|^2$, was roughly 1000 times smaller than the power of the light generated by the EDFRL, $|\mathbf{E}(t)|^2$. The signal measured simultaneously by photodiode B, I_B, is shown in Figure 7.5c. In this figure, $I_B \propto |\mathbf{E}_R(t+\tau_R)|^2$. This trace is similar, though not identical, to the trace shown in Figure 7.5a. Its power spectrum is shown in Figure 7.5d.

To recover the message, the signal measured by photodiode B was subtracted from the signal measured by photodiode A. The results of this subtraction are shown in Figure 7.6a, which according to the discussion above corresponds to the signal, $2\text{Re}\{\mathbf{E}_T^*(t+\tau_R)\cdot\mathbf{m}(t+\tau_R)\}+|\mathbf{m}(t+\tau_R)|^2$. This difference signal is low-pass filtered using a first-order Butterworth filter with a corner frequency of 35 MHz. The result of this filtering is shown

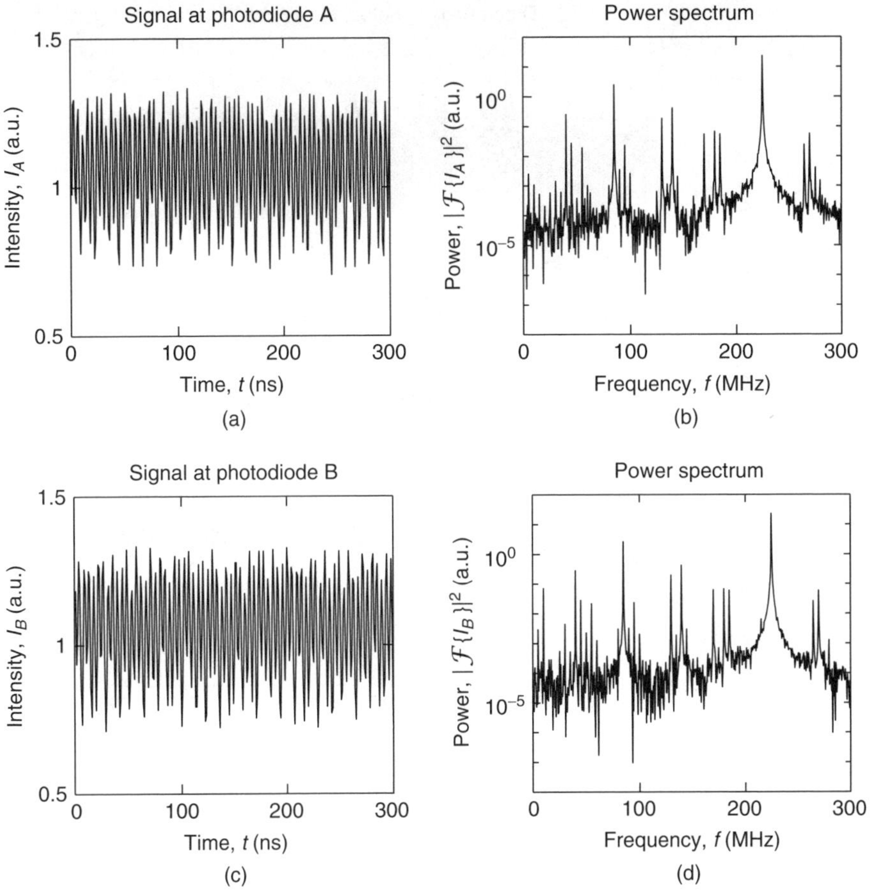

Figure 7.5 Measurements of the transmitted signal with message by photodiodes A and B.[22] (a) The transmitted signal time trace measured by photodiode A, $I_A \propto |s(t+\tau_R)|^2$, shows chaotic fluctuations. The message signal was injected into the ring laser and is present in this time-series. The message power is about 1000 times smaller than the power already present in the light output from the EDFRL. (b) The power spectrum of the signal shown in (a), given by $|\mathcal{F}\{I_A\}|^2$ where $\mathcal{F}$ represents the Fourier transform operator. (c) The time-series measured simultaneously measured by photodiode B, $I_B \propto |E_R(t+\tau_R)|^2$. (d) The power spectrum of I_B calculated in the same manner as in (b).

in Figure 7.6b. Superimposed on this low-pass filtering is a dashed line showing the signal detected by photodiode A when the transmitter EDFRL is turned off ($\mathbf{E}(t+\tau)=0$), but the message continues to be transmitted. This is the "direct-detect" signal, I_m, and its noisy nature is due to the low-power level at the detector. Interestingly, the amplitude of the recovered message, $I_A - I_B$, is slightly higher than that of the direct-detected

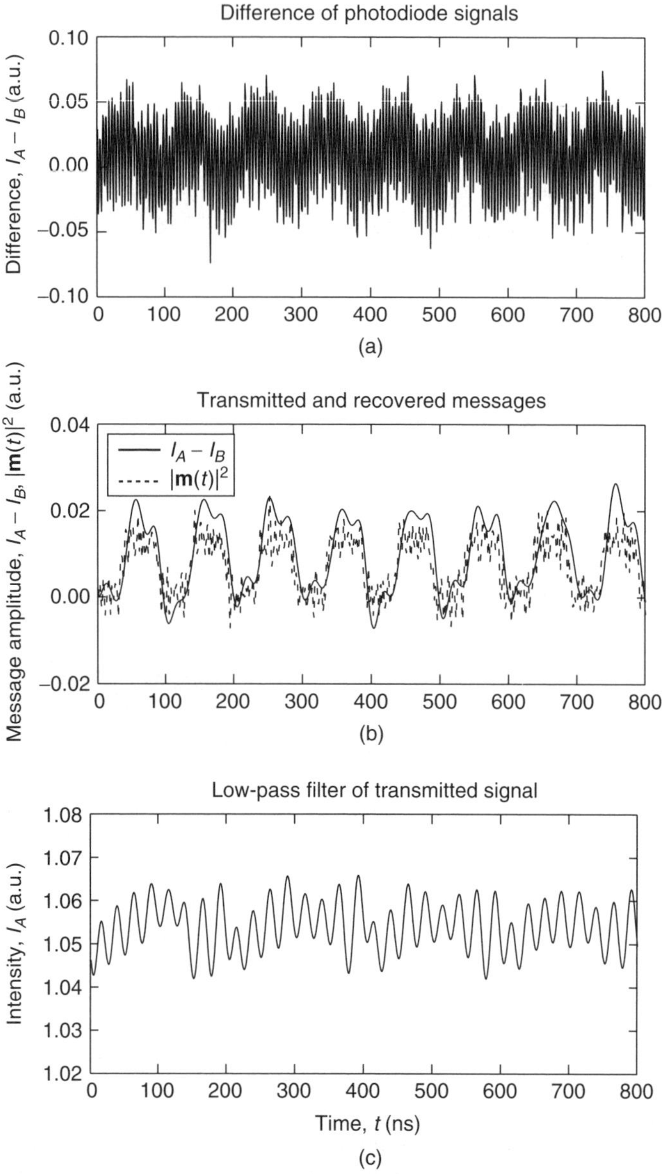

Figure 7.6 Message recovery using the message injection technique without a filter.[22] (a) The difference of the signals detected by photodiodes A and B. (b) The solid line is the recovered square-wave message after low-pass filtering of the difference signal, $I_A - I_B$. The dashed line shows the "direct-detect" message as detected by photodiode A when the transmitter EDFRL was turned off. The fluctuations of the dashed line are mainly due to noise from the photodiode amplifier and analog-to-digital converter in the oscilloscope. (c) Low-pass filtered version of the transmitted signal of Figure 7.5a, I_A, showing no trace of the square-wave message.

signal. This may be due to preferential amplification in the receiver EDFA 2, or possibly just enhancement from the low-frequencies of the crossterm, $2\mathrm{Re}\{\mathbf{E}_T^*(t+\tau_R)\cdot\mathbf{m}(t+\tau_R)\}$. To prove that it is not possible to filter out the message from the transmitted signal, Figure 7.6c shows the results of applying the same 35-MHz Butterworth filter to the transmitted signal shown in Figure 7.5a, I_A. No 10-MHz square-wave is detectable. The very subtle 5-MHz periodicity is related to the periodicity of the waveforms generated in the EDFRL.

This scheme successfully used the high-dimensional and irregular intensity fluctuations of light from an EDFRL to "mask" the transmission of a simple message. The message was also successfully recovered in the receiver unit. However, the scheme failed to take advantage of the very broad bandwidth of the EDFRL transmitter. The rapid and irregular intensity fluctuations are capable of "masking" much higher frequency signals than the 10-MHz square-wave communicated here. To increase the bit-rate of the message communication, a new technique was required.

7.3.2 Message Injection Technique using Polarization Discrimination

In the technique described in this section, the filters shown in Figure 7.1 are introduced into the ring. They consist of a polarization controller followed by a polarizer. Both of these are held within an "optical bench" manufactured by Optics for Research. A graded-index (GRIN) lens couples light out of the fiber and passes it through a polarization controller that consists of a sequence of three rotatable waveplates ($\lambda/4$, $\lambda/2$, $\lambda/4$) which enable any input polarization of light to be transformed into any desired output polarization. After passing through the polarization controller, the light is transmitted through a polarizer. The light is then coupled back into the fiber with a second GRIN lens.

Including this filter in the transmitter EDFRL causes the light in the ring to be polarized because it must pass through the polarizer. As with most lasers, the lasing modes in the EDFRL are selected through a "survival of the fittest" process. The lasing modes with least loss are those that are able, after being transmitted through the polarizer, to return to it one round-trip later perfectly aligned for maximum transmission. In this technique, the polarized message, $\mathbf{m}(t)$, is injected into the EDFRL with a polarization orthogonal to the light in the ring, $\mathbf{E}_T(t)$, but with the same wavelength. Because $\mathbf{m}(t)$ is injected orthogonal to $\mathbf{E}_T(t)$, it is blocked upon reaching the filter. $\mathbf{E}_T(t)$, however, passes through with little loss. The message is prevented from circulating in the EDFRL and mixing with chaotic light in the ring on more than one pass through the EDFA, though both are transmitted to the receiver unit.

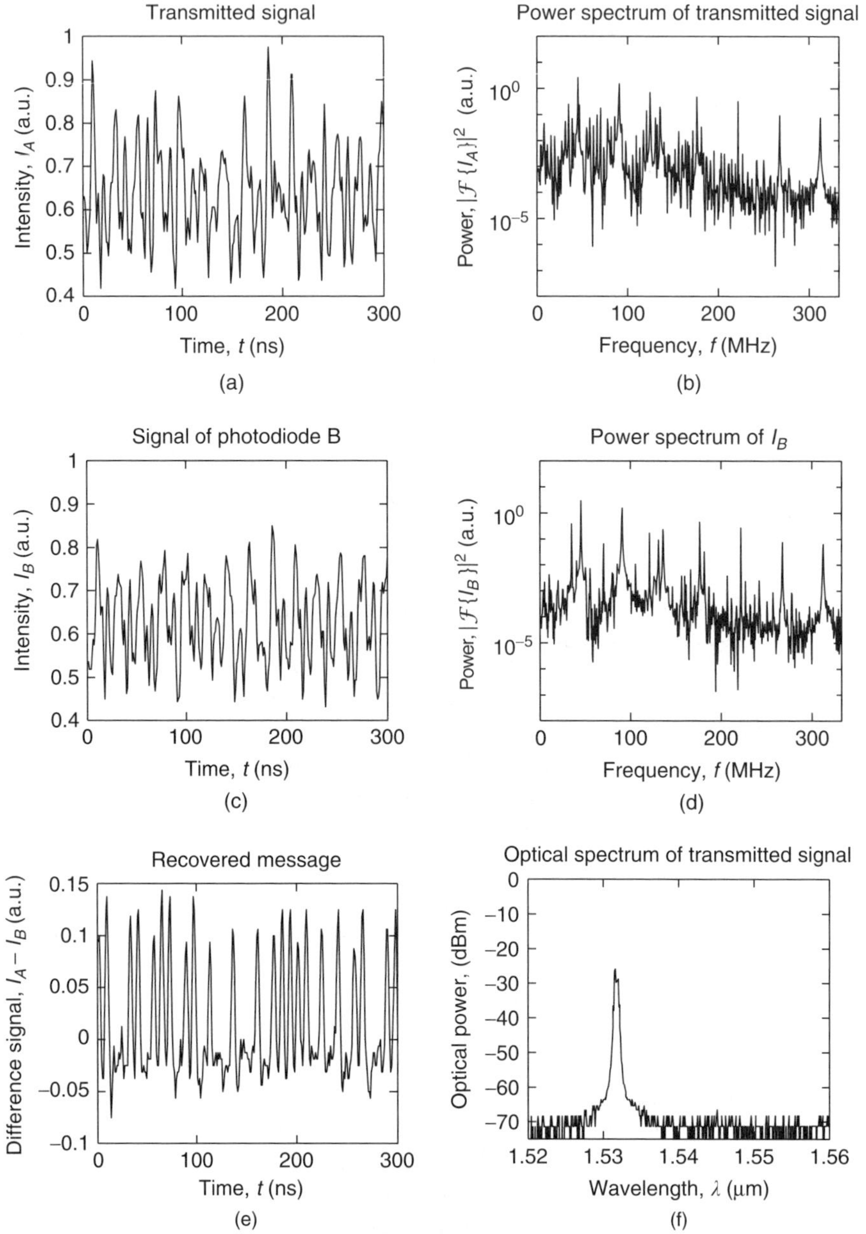

Figure 7.7 Message recovery using orthogonally polarized $m(t)$ and $E_T(t)$.[13] (a) Shows the transmitted signal measured by photodiode A, $I_A \propto |s(t+\tau_R)|^2$. (b) The power spectrum of the intensity shown in Figure 7.7a. (c) The detected signal at photodiode B, $I_B \propto |E_R(t+\tau_R)|^2$. (d) The power spectrum of I_B. (e) The difference resulting from a subtraction of the waveform in (c) from the one in (a). (f) Both the message and chaotic light share the same wavelength, $\lambda \cong 1.532\,\mu m$.

The filter in the receiver unit is also aligned to prevent light polarized in the same direction as $\mathbf{m}(t)$ from reaching photodiode B. With proper timing of the two arms of the receiver, it can be arranged so that photodiode A measures $|\mathbf{E}_T(t+\tau_R)+\mathbf{m}(t+\tau_R)|^2$ while photodiode B measures $|\mathbf{E}_R(t+\tau_R)|^2$, which is equivalent to $|\mathbf{E}_T(t+\tau_R)|^2$. As before, the difference of the two signals at the photodiodes gives $2\text{Re}\{\mathbf{E}_T^*(t+\tau_R)\cdot\mathbf{m}(t+\tau_R)\}+|\mathbf{m}(t+\tau_R)|^2$. The crossterm is eliminated in this technique because $\mathbf{E}_T(t)$ and $\mathbf{m}(t)$ are orthogonally polarized, and their dot-product is zero. This leaves just $|\mathbf{m}(t)|^2$ after subtraction, which easily reveals a binary message. No low-pass filtering is necessary to eliminate the crossterm. Consequently, the bit-rate of the message can be increased.

Figure 7.7 shows the results of an experiment using this technique. Figure 7.7a shows the time-trace of the transmitted signal, $I_A \propto |\mathbf{s}(t+\tau_R)|^2$, as measured by photodiode A. Its power spectrum is shown in Figure 7.7b. No hint of the message can be discerned in the time-trace of the signal. The broadband nature of the transmitted signal is again evident in its power spectrum. The spectrum's gradual decline with increasing frequency matches the spectral response of our photodiodes (125-MHz bandwidth) and results from this limitation. Figure 7.7c is a time-series measurement of the intensity measured by photodiode B, $I_B \propto |\mathbf{E}_R(t+\tau_R)|^2$. Its corresponding power spectrum is shown in Figure 7.7d. This power spectrum lacks the smaller, fine peaks of the transmitted signal shown in Figure 7.7b. These smaller peaks correspond to the power-spectrum of $|\mathbf{m}(t)|^2$, which, due to the polarizer in that arm of the receiver, is not detected by photodiode B. Figure 7.7e results from a subtraction of the waveform in Figure 7.7c from Figure 7.7a. The bits are clearly recovered and match the 32-bit pattern used for $\mathbf{m}(t)$ in this experiment: 11001101110110100100101111101010. The message bits were transmitted and recovered at a rate of 125 MB/s. Figure 7.7f is the optical spectrum of the transmitted signal showing that both $\mathbf{E}_T(t)$ and $\mathbf{m}(t)$ have the same wavelength, $\lambda \cong 1.532\,\mu\text{m}$.

Figure 7.8 is included as a measure of the accuracy of this technique. For comparison, Figure 7.8a shows a segment of the message sequence, $I_m \propto |\mathbf{m}(t)|^2$, as detected by photodiode A in the absence of chaos (EDFA 1 is turned off and $\mathbf{E}_T(t)=0$). Figure 7.8b is simply the power spectrum of the entire recorded message sequence, I_m. Figure 7.8c shows a portion of the message obtained through the chaotic subtraction method, $I_A - I_B$, as shown in Figure 7.7e. Clearly, the recovered message is degraded, probably due to imprecision in matching the polarization-based filters in both the transmitter and receiver or in creating the proper relative time-delays between the two arms of the receiver. The power spectrum of the recovered message, a segment of which is given in Figure 7.8c, is shown in Figure 7.8d. The peak corresponding to the 125-MHz bit-rate is evident in Figure 7.8b.

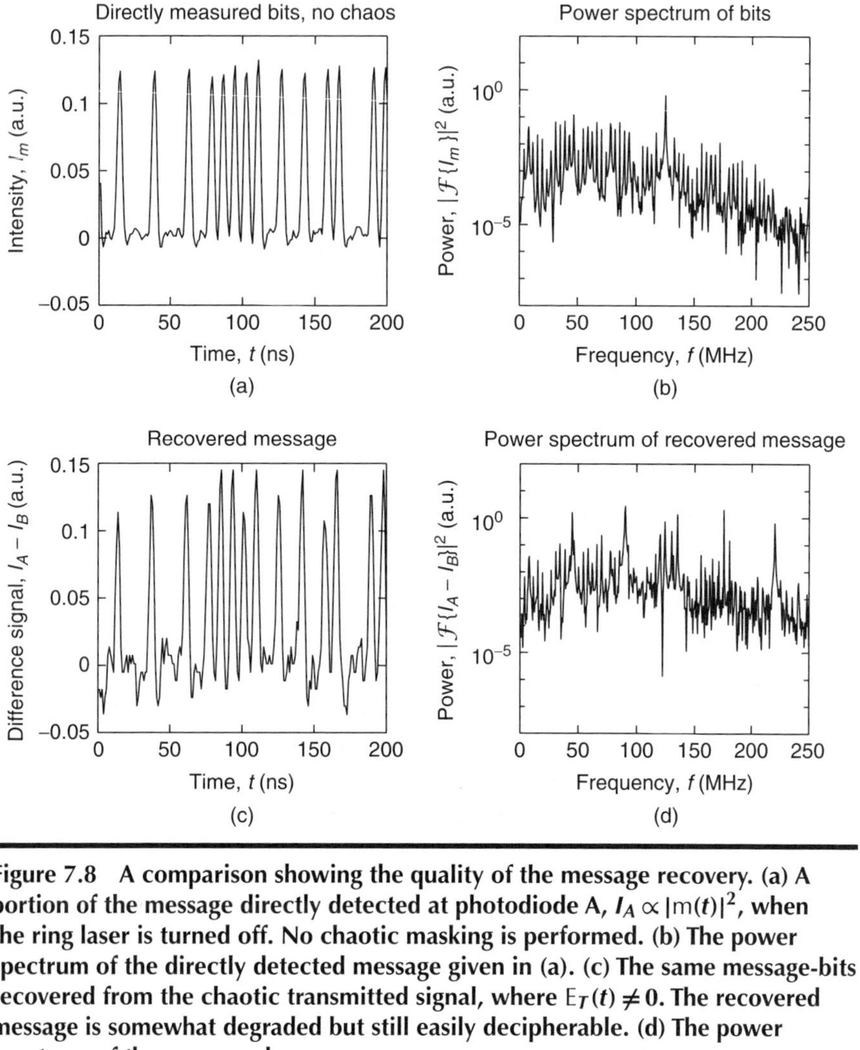

Figure 7.8 A comparison showing the quality of the message recovery. (a) A portion of the message directly detected at photodiode A, $I_A \propto |m(t)|^2$, when the ring laser is turned off. No chaotic masking is performed. (b) The power spectrum of the directly detected message given in (a). (c) The same message-bits recovered from the chaotic transmitted signal, where $E_T(t) \neq 0$. The recovered message is somewhat degraded but still easily decipherable. (d) The power spectrum of the recovered message.

7.3.3 Message Injection Technique using Wavelength Discrimination

In the preceding technique, the wavelengths of $\mathbf{m}(t)$ and $\mathbf{E}_T(t)$ were the same, but their polarizations were orthogonal. In this method, however, the wavelengths are different, but their polarization are the same. The filter in the EDFRL again consists of the polarization controller followed by a polarizer. The polarization controller in the message modulation unit is adjusted so that the polarization of $\mathbf{m}(t)$ and $\mathbf{E}_T(t)$ are parallel as they are transmitted from the EDFRL to the receiver unit. Again, the filter's purpose is to block light associated with $\mathbf{m}(t)$ from continuing to circulate in the ring.

The wavelength dependent birefringence of the fiber in the EDFRL permits the polarizer in the filter to act as a wavelength filter. The electric fields $\mathbf{m}(t)$ and $\mathbf{E}_T(t)$ have the same polarization at the coupler where the message is transmitted, but as the light propagates around the ring, the difference in wavelength, typically 20 nm, results in orthogonal polarization for $\mathbf{m}(t)$ and $\mathbf{E}_T(t)$ at the filter. The phenomenon primarily seemed to occur in the erbium-doped fiber inside the EDFA. At the filter, the message, $\mathbf{m}(t)$ is blocked by the polarizer, while the irregularly fluctuating field, $\mathbf{E}_T(t)$ continues to circulate, becoming $\mathbf{E}_T(t+\tau_R)$ when it returns to the message injection coupler. A new message, $\mathbf{m}(t+\tau_R)$ is added to this field to produce $\mathbf{s}(t+\tau_R)$. As with the other techniques, a subtraction of the signals at the two photodiodes gives $2\mathrm{Re}\{\mathbf{E}_T^*(t+\tau_R)\cdot\mathbf{m}(t+\tau_R)\}+|\mathbf{m}(t+\tau_R)|^2$. In this instance, however, the crossterm is made to vanish by using different wavelengths of light. The dot-product is a function of the phase between $\mathbf{m}(t)$ and $\mathbf{E}_T(t)$. Because they have different wavelengths, the phase-difference between them grows linearly and rapidly. On the time-scales measurable with the photodiodes, the crossterm averages to zero.

Figure 7.9 shows results obtained using this technique. Figure 7.9a shows the signal transmitted from the EDFRL to the receiver unit as detected by photodiode A. Here, the intensity detected by photodiode A, I_A, is proportional to $|\mathbf{s}(t+\tau_R)|^2$. The power spectrum of I_A is shown in Figure 7.9b, and again, the spectrum is very broadband. Figure 7.9c shows the signal recorded by photodiode B, I_B, which is proportional to $|\mathbf{E}_R(t+\tau_R)|^2$. Its corresponding power spectrum is given in Figure 7.9d. Subtracting the time-series shown in Figure 7.9c from that of Figure 7.9a recovers the message shown in Figure 7.9e, which demonstrates excellent recovery of the bits. Finally, Figure 7.9f shows the optical spectrum of the transmitted signal, revealing a narrow peak corresponding to $\mathbf{m}(t)$ at $\lambda = 1.532\,\mu\mathrm{m}$ and a broader peak corresponding to $\mathbf{E}_T(t)$ at $\lambda = 1.555\,\mu\mathrm{m}$.

7.3.4 Wavelength Discrimination with In-Ring Bandpass Filter

Another method that was investigated was the use of bandpass filters rather than polarizers in the filter shown in Figure 7.1. Once again, the purpose of the filters was to prevent light associated with $\mathbf{m}(t)$ from continuing to circulate around the ring. The bandpass filters used in these experiments were tunable with a bandwidth of roughly 1 nm. The filter in the receiver was tuned to transmit the same wavelength as the bandpass filter in the transmitter. Again, because the wavelengths of the message, $\mathbf{m}(t)$, and the electric

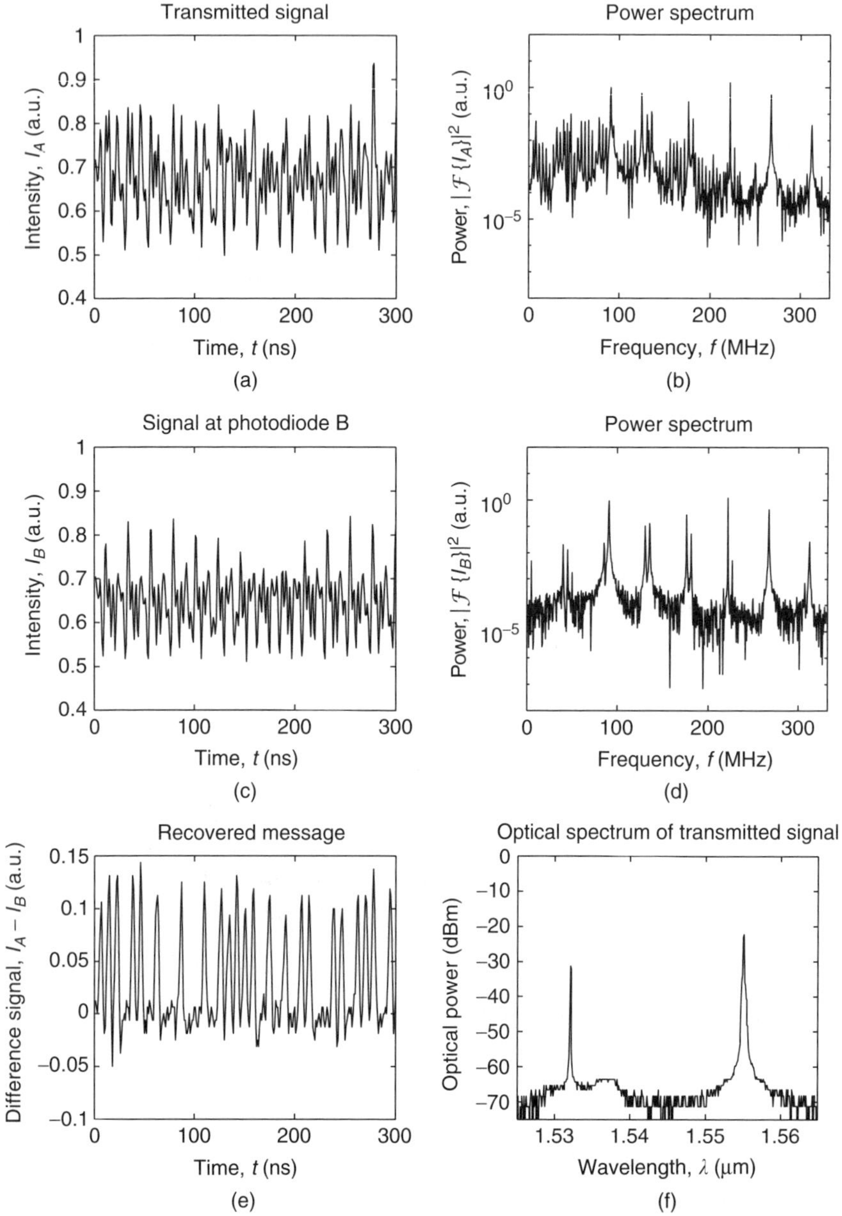

Figure 7.9 Message recovery using the wavelength discrimination message injection technique.[13] (a) The transmitted signal measured by photodiode A, $I_A \propto |s(t+\tau_R)|^2$. (b) The power spectrum of I_A. (c) The signal measured simultaneously by photodiode B, $I_B \propto |E_R(t+\tau_R)|^2$. (d) The power spectrum of I_B. (e) Again, a subtraction of the measured signal I_B from the measured signal I_A reveals the message. An optical spectrum, (f), shows that the message light and chaotic light have very different wavelengths.

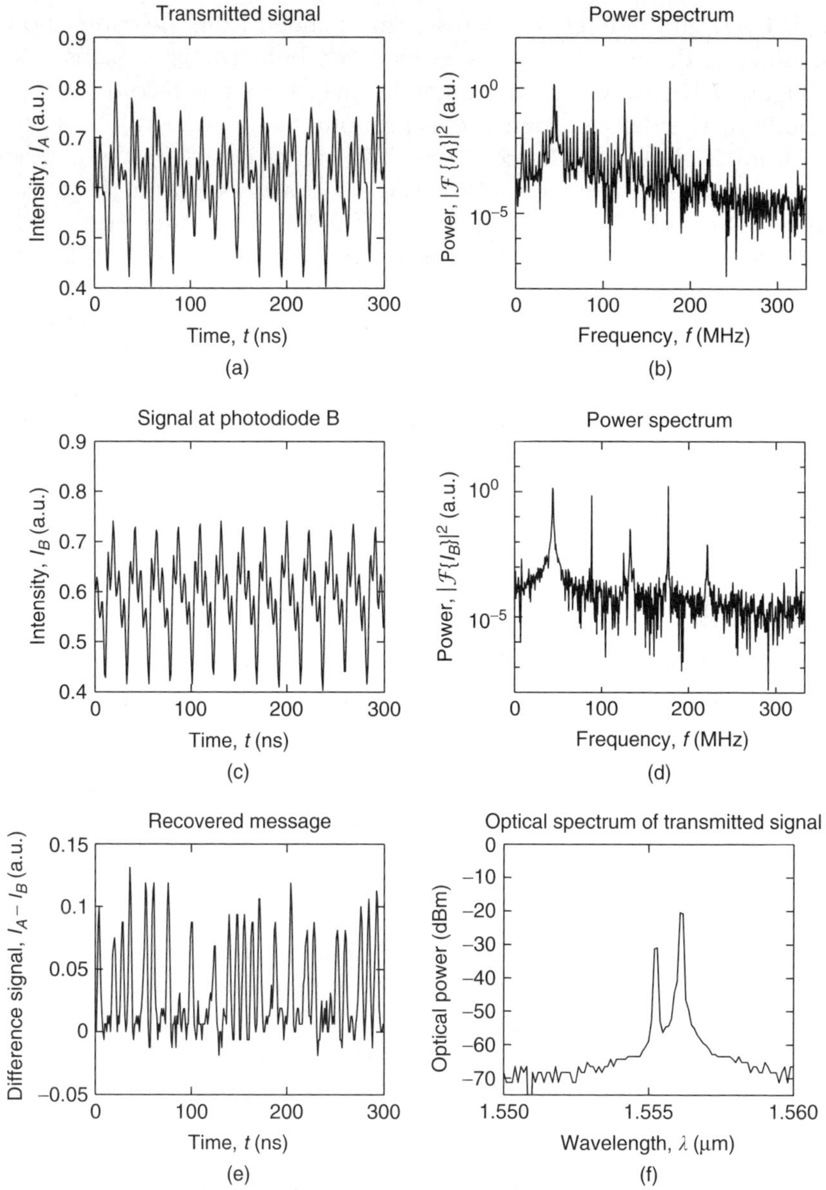

Figure 7.10 Message recovery using a bandpass filter for wavelength discrimination.[13] (a) Transmitted signal as measured by photodiode A, I_A. (b) The power spectrum of the signal detected by photodiode A. (c) The signal, I_B, measured by photodiode B. (d) The power spectrum of I_B. (e) The difference of the two signals, $I_A - I_B$, which reveal the transmitted message. (f) An optical spectrum showing that the wavelengths of the message light, m(t), and light from the EDFRL, $E_T(t)$ are different, though they are much closer together than in the experiments described in Section 7.3.3.

field, $\mathbf{E}_T(t)$, are different, the crossterm obtained from the subtraction of the photodiode signals averages to zero, and only $|\mathbf{m}(t)|^2$ remains.

Figure 7.10a shows the transmitted signal, $\mathbf{s}(t+\tau_R)$, recorded by photodiode A. The detected signal, I_A is proportional to $|\mathbf{s}(t+\tau_R)|^2$. Its power spectrum is shown in Figure 7.10b. The signal detected at photodiode B, I_B is shown in Figure 7.10c, and its power spectrum is given in Figure 7.10d. The bandpass filter in the EDFRL was observed to reduce the apparent complexity of the signals output from the EDFRL. This can be ascertained from Figure 7.10c, which shows the intensity fluctuations from the EDFRL independent of the injected message, because $I_B \propto |\mathbf{E}_R(t+\tau_R)|^2$. This decreased irregularity of the lightwaves coincides with a narrowing of the EDFRL lasing line, as can be observed in Figure 7.10f. The recovered message obtained through a subtraction of the signals from the photodiodes is displayed in Figure 7.10e. The optical spectrum of the transmitted signal is supplied in Figure 7.10f. The wavelength separation between the message light, $\mathbf{m}(t)$, and the light from the EDFRL, $\mathbf{E}_T(t)$, is $\approx$1 nm, which is much smaller than in Figure 7.9.

7.3.5 Message Injection without Filtering

7.3.5.1 Identical Wavelengths for EDFRL and Message Light

The techniques discussed in the previous sections required the use of a filter and a distinction between the wavelength or polarization of the chaotic electric field, $\mathbf{E}_T(t)$, and the message signal, $\mathbf{m}(t)$. The techniques presented in this section will demonstrate message communication without the use of filters and with the same wavelength for the message and chaotic electric field.

The time-traces from such an experiment are shown in Figure 7.11. In this example both EDFA 1 and EDFA 2 are pumped by their diode lasers at 10 mW, a level slightly less than twice the threshold pump power for the ring laser. The power level of the message in the transmitted signal is slightly less than half of the power corresponding to light generated by the EDFRL. Figure 7.11a shows data taken from photodiode A, I_A, (the thin line) and from photodiode B, I_B, (the thick line). Here, $I_A \propto |\mathbf{s}(t+\tau_R)|^2$, while $I_B \propto \mathbf{E}_R(t+\tau_R)$. As explained earlier, a subtraction of the two signals is proportional to $2\mathrm{Re}\{\mathbf{E}_T^*(t+\tau_R)\cdot\mathbf{m}(t+\tau_R)\}+|\mathbf{m}(t+\tau_R)|^2$. Figure 7.11 shows that subtraction (thin line). For comparison, the thicker line in Figure 7.11b is the message signal measured by photodiode A when the EDFRL is turned off to remove the chaotic masking. This is equivalent to measuring just $|\mathbf{m}(t)|^2$. The greater amplitude of the recovered message is simply the result of the crossterm, $2\mathrm{Re}\{\mathbf{E}_T^*(t+\tau_R)\cdot\mathbf{m}(t+\tau_R)\}$ — in the data shown in the figure, the polarizations and phases of $\mathbf{E}_T(t)$ and $\mathbf{m}(t)$ have combined to improve the message reception. The fidelity, as seen in Figure 7.11b,

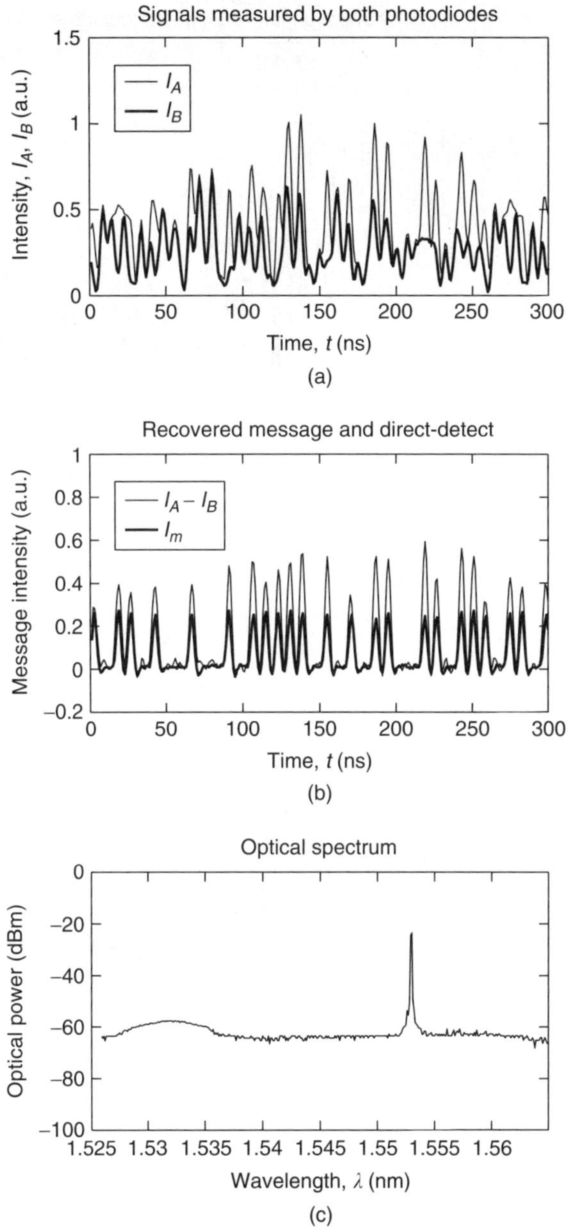

Figure 7.11 (a) The transmitted signal measured by photodiode A, I_A (thin-line) and the signal measured by photodiode B, I_B (thick-line). (b) The thick-line is the message signal, $I_m \propto |m(t)|^2$, directly detected by photodiode A when the transmitter ring laser is turned off. It is shown here to verify that the message recovered from the chaos, $I_A - I_B$, (thin-line) is indeed correct. (c) The single peak in the optical spectrum of the transmitted signal demonstrates that the message has the same wavelength as the fluctuating light from the EDFRL.[13]

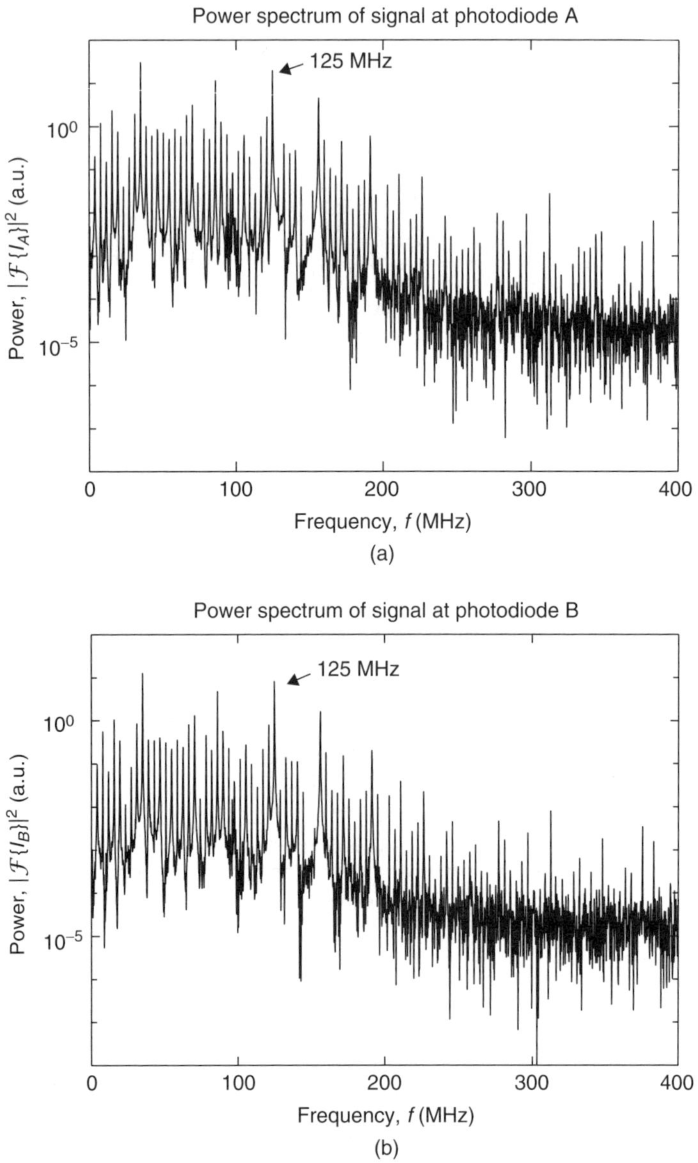

Figure 7.12 Power spectra corresponding to the signals measured in Figure 7.11.[13] (a) The power spectrum for the transmitted signal recorded at photodiode A, I_A. (b) The power spectrum for the signal measured by photodiode B, I_B. The 125-Mb/s bit-rate gives rise to a 125-MHz peak that is visible in both (a) and (b).

is quite good. The crossterm, however, changes amplitude on time-scales commensurate with the message. Consequently, message recovery is very inconsistent. An optical spectrum of the transmitted light is shown in Figure 7.11c. The message injection is strong enough that the EDFRL chooses its optical wavelength, $\lambda = 1553.01$ nm, for a lasing wavelength.

Figure 7.12 provides power spectra for the transmitted signal of Figure 7.11a and Figure 7.11b; they have a very close resemblance, as they should. The narrowly spaced discrete spikes result from the 32-bit pattern used. The bit-rate was 125 Mb/s; the corresponding peak is evident in both power spectra.

7.3.5.2 Different Wavelengths for EDFRL and Message Light

Experiments were also performed with message injection at wavelengths which were not resonant with the lasing wavelength of the EDFRL. Figure 7.13a shows signals measured by photodiodes A and B for such a case. The EDFAs were pumped at about 85 mW, many times threshold. This resulted in an optical power in the ring of ~9 dBm without any message injection. The injected message power was ~−3 dBm. As in Figure 7.11, the thin-line in Figure 7.13a corresponds to the transmitted signal as measured by photodiode A, $I_A \propto |\mathbf{s}(t + \tau_R)|^2$. The thicker line corresponds to the data measured by photodiode B, $I_B \propto |\mathbf{E}_R(t + \tau_R)|^2$. Subtracting the latter from the former results in the data shown in Figure 7.13b. Once again, the proper pattern of bits is obtained. The optical spectrum shown in Figure 7.13c shows two distinct peaks. The first of these peaks occurs at a wavelength $\lambda = 1533.01$ nm and corresponds to the message injection, whereas the second, and rather broad, peak at $\lambda = 1558$ nm corresponds to the natural lasing wavelength of the EDFRL. As was discussed in the previous chapter, the very broad linewidth is characteristic of the EDFRL when there are no polarizers or bandpass filters in the ring, and the laser is allowed to operate at pump power many times higher than threshold. The lasing line shown in Figure 7.11 is narrower due to the lower pump powers used in that experiment.

The EDFRL in this experiment did not have enough gain at 1533 nm to overcome the loss in the ring, and therefore, it did not lase. However, when the message was injected into the EDFRL at that wavelength, the EDFA was able to provide some gain to the signal. This produced circulating light in the EFDRL at 1533 nm that was constantly resupplied with energy from the message signal. This property of the EDFRL ensures that an eavesdropper using a bandpass filter would be unable to simply isolate the 1533 nm light in the transmitted signal and easily recover the message. Figure 7.14 shows the results of just such an experiment. A bandpass filter is used to isolate the transmitted light at 1533 nm and the intensity was measured. As is evident in the figure, the message remains masked. When the message is injected

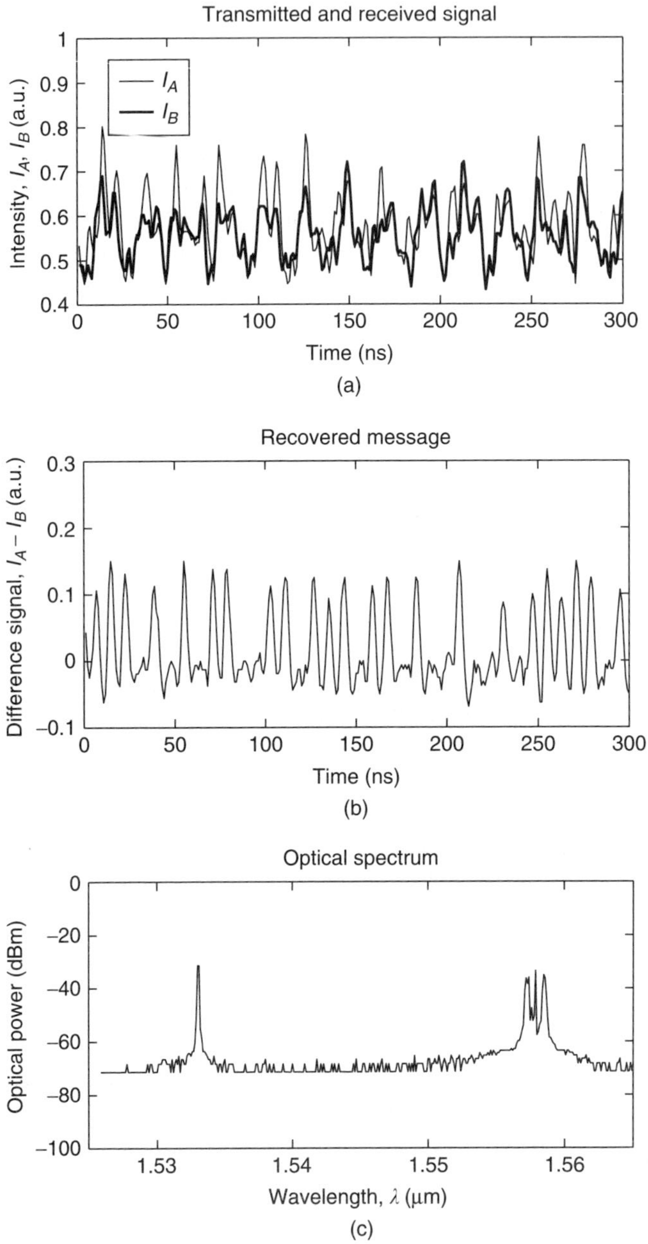

Figure 7.13 (a) The transmitted signal measured by photodiode A, I_A (thin-line) and the signal measured by photodiode B, I_B (thick-line) are superimposed. (b) The results of subtracting I_B from I_A. (c) gives an optical spectrum showing the wavelengths output by the EDFRL. The message injection is occurring at a wavelength of ~1.533 μm.[13]

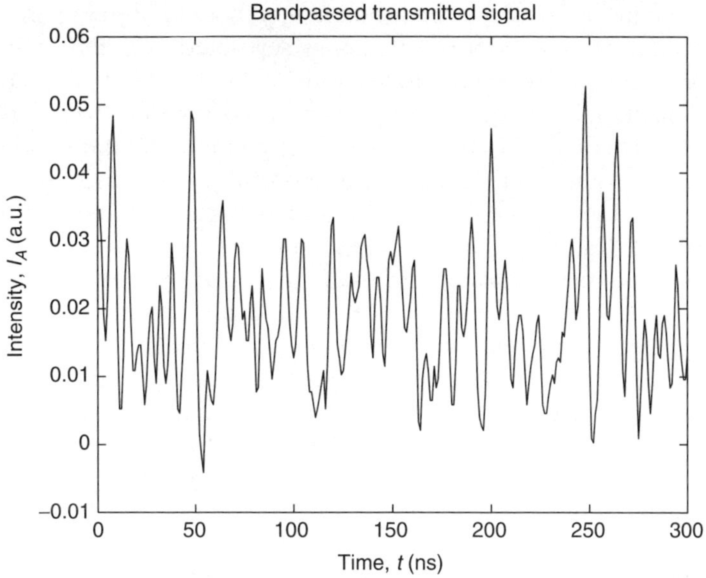

Figure 7.14 Transmitted signal after passing through a 1-nm bandpass filter at a wavelength, λ, of 1533 nm. The light intensity fluctuations still mask the message.[13]

into the EDFRL, it interferes with the already circulating light at 1533 nm. Interestingly, the light it interferes with was produced by the interference of previous message bits with yet earlier bits.

7.3.6 Discussion of Message Injection Approach

After presenting and describing the many measurements above, which are also discussed in Reference 13 by VanWiggeren and Roy, it is appropriate now to evaluate and analyze the methods that were developed. With one exception, all of the variations mentioned in Section 7.3 have demonstrated their ability to recover 125-Mb/s digital messages. Each of these variations uses a receiver in which one arm replicates the operation of one round-trip through the transmitter EDFRL, which may include a "filter." This geometry permits excellent recovery of the message from the irregular and chaotic fluctuations of intensity that are generated by the EDFRL. However, all of these methods, in one way or another, fail to realize the EDFRL's supposed potential for chaotic communication at very high-speeds with significant privacy.

In Section 7.3.1, recovering the message from the chaotic waveform that was masking it required a low-pass filtering process. This limits the frequency components that can be present in the message. If they become greater than just a few times the fundamental mode of the ring, in this case 5 MHz, the message cannot be effectively separated from the crossterm.

The experiments presented in Section 7.3.2 were able to overcome this bit-rate limitation by introducing a polarization controller and polarizer into the EDFRL. Theoretically, the bit-rate is only limited by the bandwidth of the detection equipment. The polarizations of $\mathbf{E}_T(t)$ and $\mathbf{m}(t)$ are orthogonally polarized in this method. Although dynamical message recovery was demonstrated using two time-delayed arms in the receiver, the message could also have been separated from the masking lightwave through more conventional means. Because both $\mathbf{E}_T(t)$ and $\mathbf{m}(t)$ are most likely elliptically polarized at any point in the communication channel, using a simple polarizer to separate them is not sufficient. However, using a polarization controller to transform $\mathbf{E}_T(t)$ and $\mathbf{m}(t)$ into orthogonally linearly polarized lightwaves does permit the use of a polarizer to block $\mathbf{E}_T(t)$ while allowing $\mathbf{m}(t)$ to be transmitted. An ordinary photodiode receiver could then recover the message. Thus, the high-dimensional nature of the chaotic masking lightwaves in this method is not sufficient for communication privacy.

Conceptually, the same weakness is present in the technique described in Section 7.3.3. A simple bandpass filter can be used to isolate the wavelength corresponding to $\mathbf{m}(t)$ while removing the masking lightwave. The same reasoning applies to the method discussed in Section 7.3.4, though the relative proximity in wavelength of the two signals, $\mathbf{E}_T(t)$ and $\mathbf{m}(t)$, might make such filtering somewhat more challenging. Certain techniques based on the use of bandpass filters could be developed to make extracting the message in these conventional ways more difficult. One could imagine varying randomly the wavelength of $\mathbf{m}(t)$ in time. For example, the message could be transmitted with a wavelength shorter than the EDFRL lasing line for some time and then switched to a wavelength longer than the EDFRL lasing line. It would be difficult to construct a bandpass filter to keep up with these wavelength changes of $\mathbf{m}(t)$. However, the dynamical message recovery method presented in Section 7.3.4 would be unaffected by these changes in the wavelength of the message.

Rather than using a bandpass filter, the eavesdropper might choose to use a bandstop filter which would eliminate just the wavelength corresponding to the EDFRL lasing line. Varying the wavelength of $\mathbf{m}(t)$ would not prevent the eavesdropper from obtaining the message in this case. As a countermeasure, one could imagine transmitting the light from several EDFRL's, all with equal round-trip times but different lasing wavelengths, to mask the message. A filter designed simply to block one wavelength would not be able to stop all of the masking light wavelengths. Again, the receiver described in Section 7.3.4 would still be able to recover the message.

The techniques above to combat eavesdropping seem inelegant, and certainly, they are not good examples of the communication privacy that is potentially available using the EDFRL as a transmitter. From a communication privacy point of view, the techniques of Section 7.3.5 are probably to be preferred. The message and the masking light are both of the same wavelength and components of the masking light have the same wavelength as the message. Yet, this method, too, has its limitations. Principally, it was not possible to control the phase relationship of $\mathbf{E}_T(t)$ and $\mathbf{m}(t)$. Consequently, the crossterm that appears in the difference signal, $2\mathrm{Re}\{\mathbf{E}_T^*(t+\tau_R)\cdot\mathbf{m}(t+\tau_R)\}$, prevented consistent recovery of the message. Minimizing the magnitude of $\mathbf{E}_T(t)$ helped to minimize the value of the crossterm in the experiments described in Section 7.3.5.

In spite of the flaws, the experiments of Section 7.3.5 do point out an interesting possibility. The experiment suggests wavelength-division-multiplexing (WDM) might be possible using this type of communication method. If multiple wavelengths of light are injected into the EDFRL, the EDFRL would generate light with irregular intensity fluctuations at each of these wavelengths, effectively masking all of them. Another very elegant way of defeating the crossterm that afflicts this message injection technique is to use incoherent light in the message beam. This type of experiment has been performed, and the results are very good, though a more formal investigation remains to be performed. These two concepts could be investigated simultaneously to demonstrate chaotic WDM with incoherent message light.

7.4 The Intraring Intensity Modulator Technique for Message Encryption

7.4.1 Single Ring Technique

In the previously described experiments, consistent message recovery is difficult to achieve because of interference effects between the chaotic light from the EDFRL and the injected optical message. Without a filter, the interference term $2\mathrm{Re}\{\mathbf{E}_T^*(t+\tau_R)\cdot\mathbf{m}(t+\tau_R)\}$, fluctuates due to changing relative phase and polarization between $\mathbf{E}_T(t)$ and $\mathbf{m}(t)$. A new approach is taken to overcome this difficulty.[13] In this approach, the message is applied directly to the chaotic light in the EDFRL using an electro-optic modulator located with the EDFRL. The message, in this approach, is a modulation rather than an injected lightwave; consequently, the message does not interfere with the chaotic lightwaves generated by the EDFRL.

Figure 7.15 shows a transmitter consisting of an EDFRL and a $LiNbO_3$ intensity modulator. The EDFA has a small signal gain of ~30 dB and a maximum output power of 17 dBm. The modulator in the ring also acts as a polarizer because its waveguides only allow one polarization to propagate. A 90/10 output coupler directs 10 percent of the light out of the ring and

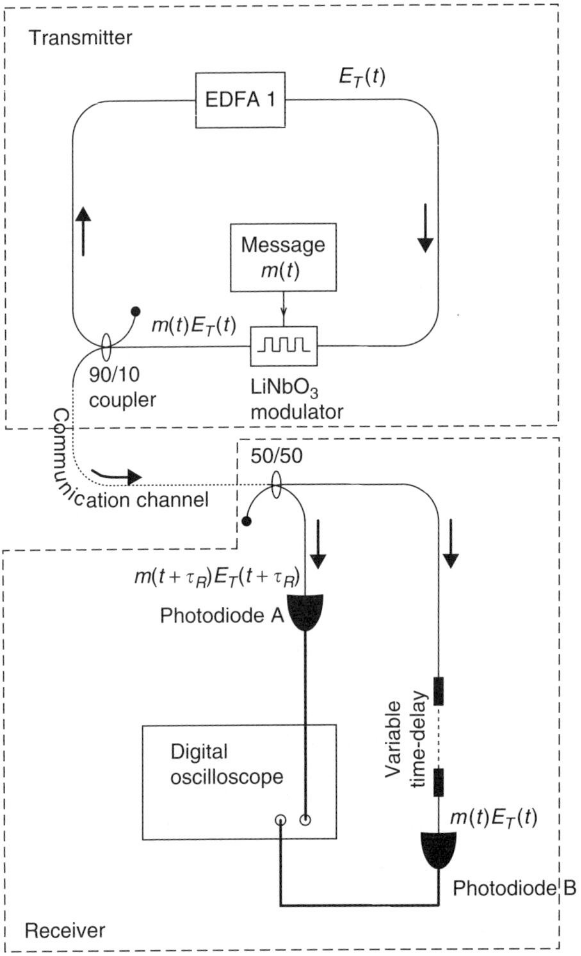

Figure 7.15 Configuration for the intraring modulator approach.[13] An intensity modulator is used to encode a digital message onto chaotic lightwaves produced by the EDFRL. Light from the ring travels to a receiver where a precise time-delay between the photodiodes allows for the message to be recovered from the chaos.

into the communication channel. The remaining 90 percent of the light continues to circulate around the ring.

In the receiver, the light is split again. The transmitted signal is directly measured at photodiode A. The remaining light passes through a variable time-delay device. A precision time-delay is achieved using a GRIN lens to couple light from a fiber into free-space. The light traverses a distance, the source of the variable time-delay, before it is coupled back into fiber using a second GRIN lens. By controlling the separation between the two

GRIN lens couplers, the time-delay between photodiodes A and B can be precisely controlled. For this experiment, photodiode B measures the same signal as photodiode A, but with a time-delay matched to within 0.1 ns of the round-trip time of the EDFRL, τ_R.

The slowly varying envelope of the irregular lightwave just after the EDFA in the receiver can be represented as $E_T(t)$. The lightwave propagates through the ring and is amplitude modulated as it passes through the modulator to create $m(t)E_T(t)$, where $m(t)$ is the message signal. Note that in this case, $m(t)$ is a scalar modulation rather than a vector amplitude as in the previous experiments. Likewise, $E_T(t)$ can have only one polarization, and so its vector nature is neglected. Ninety percent of this light continues circulating and passes through EDFA 1 to create $E_T(t+\tau_R)\cong m(t)E_T(t)$ while the remaining fraction of $m(t)E_T(t)$ is output to the receiver. To focus the discussion on the shape of the waveform, rather than the relatively unimportant amplitude, the amplification of the lightwave in passing through the amplifier is not written explicitly. The shape of the waveform is only slightly distorted in passing through the amplifier due to nonlinearities in the fiber and amplification. $E_T(t+\tau_R)$ also circulates through the ring and is modulated to produce $m(t+\tau_R)E_T(t+\tau_R)$, and a fraction of this light is also output to the receiver.

In the receiver, photodiode B is delayed relative to photodiode A by one round-trip time, τ_R, to within an accuracy of 0.1 ns. Consequently, when photodiode A is measuring $m(t+\tau_R)E_T(t+\tau_R)$, photodiode B is measuring $m(t)E_T(t)$. Because $E_T(t+\tau_R)\cong m(t)E_T(t)$, a division of the signal measured at photodiode A by the signal measured by photodiode B should give the message $m(t+\tau_R)$, thereby recovering the message from the chaotic carrier wave. Unlike the methods described in Section 7.3, there is no EDFA in the receiver arm leading to photodiode B to replicate the dynamics of EDFA 1 in the transmitter. The EDFA in the receiver could be omitted because, although the nonlinearities in the EDFA and its fiber are sufficient to induce the chaotic intensity fluctuations in the EDFRL, the lightwave is not significantly altered (except in amplitude) in passing once through the EDFA. This same reasoning holds for the methods presented in Section 7.3 as well.

Figure 7.16 gives the results of such an experiment. A repeating digital message with a 32-bit length was applied to the modulator in the ring. The message sequence was as follows: 01111101010110011011101101001001. The recovered message shown in Figure 7.16e replicates this pattern. Figure 7.16a shows the signal measured by photodiode A, I_A, and Figure 7.16c shows the signal simultaneously measured by photodiode B, I_B. No sign of the message is discernible in either I_A or I_B. The power spectra of the two measured signals, I_A and I_B, are given in Figure 7.16b and Figure 7.16d, respectively. The 125-MHz peak, corresponding to the bit-rate of the message, is evident in the power spectrum of the recovered message given

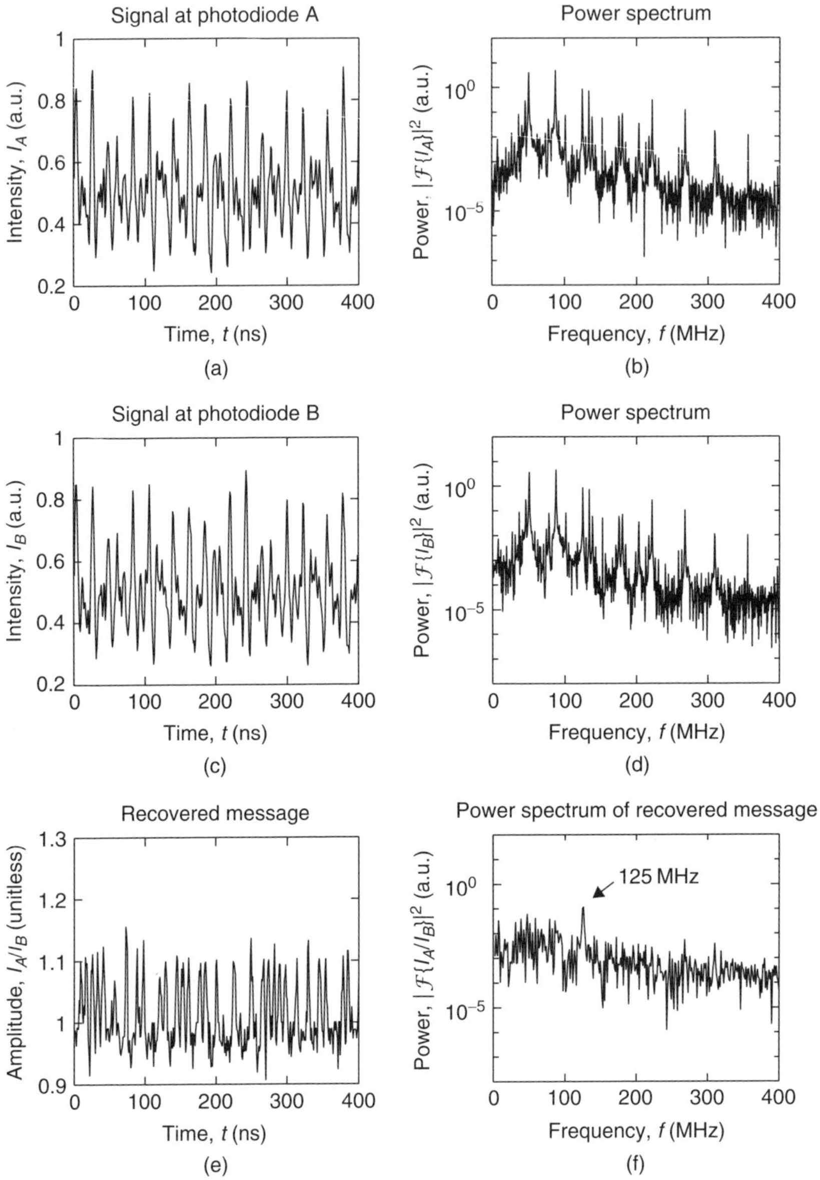

Figure 7.16 Message recovery using an intraring modulator to encode the message.[13] (a) The recorded signal from photodiode A, $I_A \propto |m(t+\tau_R)E_T(t+\tau_R)|^2$. (b) The power spectrum of I_A. (c) The signal recorded by photodiode B, $I_B \propto |m(t)E_T(t)|^2$. (d) The power spectrum of I_B. (e) Because $m(t)E_T(t) \cong E_T(t+\tau_R)$, a division of I_A by I_B recovers the message. The recovered message is shown in (e). (f) Power spectrum of the recovered message, with a prominent peak at the bit-rate, 125 MHz. The 125-MHz peaks are not conspicuous in either (b) or (d).

Figure 7.16f. However, it is not particularly conspicuous in Figure 7.16b and Figure 7.16d. This suggests that even the bit-rate of the message could not be easily discerned by an eavesdropper.

7.4.1.1 Discussion of Single-Ring Technique

Only a small message modulation was used to obtain the data in Figure 7.16; the depth of message modulation that could be used is limited by the dynamics of the ring laser. If the message modulation is very small, bit recovery is impaired because noise in the system may be commensurate with the amplitude of the message. Alternatively, if the message modulation is too large, it drives the laser into an unstable spiking regime. In that spiking regime, the transmitted intensities approach zero sometimes, though this may just be overshoot in the photodiodes. Because the message is recovered through a mathematical division of two signals, any noise present in the signal detected by photodiode B when it is near zero intensity has very detrimental effects.

Once the appropriate adjustment has been made, the message recovery is very consistent because no interference term exists to distort the message recovery. The intraring modulator method also had additional privacy benefits when compared with the techniques described in Section 7.3. In this method, the message is incorporated as a part of the very irregular carrier lightwave, $m(t)E(t)$. Consequently, the message cannot be separated from the chaotic carrier using optical devices such as polarizers or bandpass filters.

Two other factors are also important to privacy considerations: (1) the dimensionality of the transmitted light, and (2) the number of parameters that must be matched to build a proper receiver. The signal from the transmitter EDFRL, while being driven by a message, has been analyzed using an FNNs algorithm.[23] Using 100,000 data points, the analysis indicates that the embedding dimension is very high, of order 10 or higher.

The number of parameters that must be matched to build a receiver, however, is very small. Precise knowledge of just one parameter, τ_R, is necessary to recover the message. As shown in this experiment, delaying the transmitted signal by τ_R and dividing the original signal by this new signal enables message recovery. An eavesdropper could presumably scan over many time-delay values until the message is observed.

7.4.2 Two Ring Technique

A new configuration (presented in References 13 and 24) that requires multiple-matched parameters in the receiver and allows consistent and clear message communication is shown in Figure 7.17. An additional outer-loop is added to the solitary EDFRL in the previous system. The outer-loop

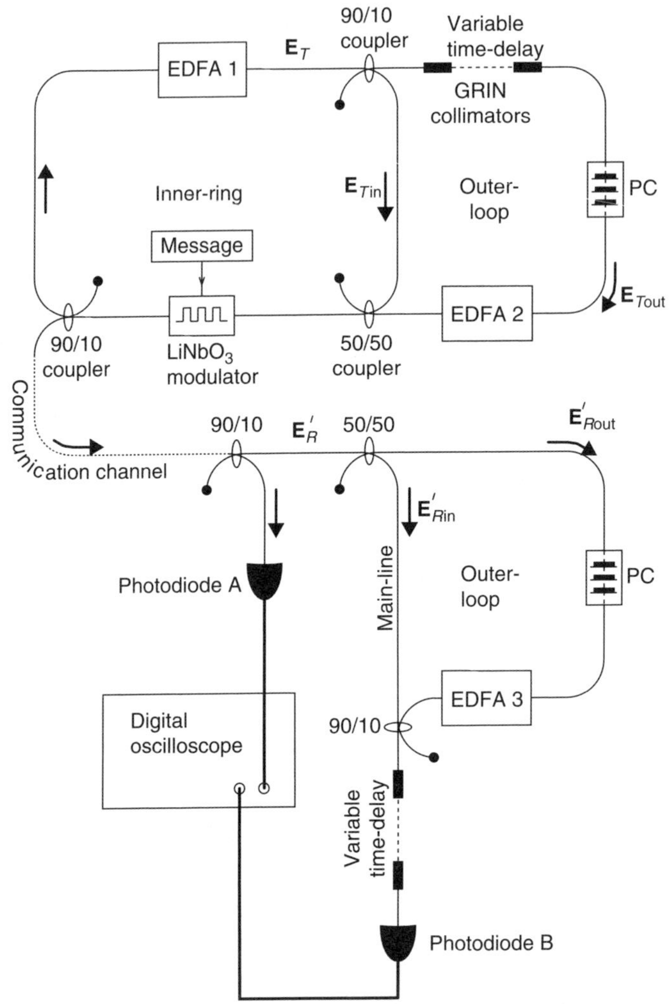

Figure 7.17 Experimental setup of the improved intraring modulator approach.[13] An EDFRL with an additional outer-loop is used as a transmitter. Again, an intensity modulator is used to encode a digital message onto the chaotic optical carrier. The carrier and message propagate to a receiver constructed to reproduce the dynamics of the transmitter. Proper configuration, time-delays, and power levels in the receiver allow recovery of the message.

extracts a portion of the light in the inner-ring, delays it relative to the light in the inner-ring, and reinjects it. With just a small amount of reinjected light, the spiking behavior observed with larger modulation amplitudes in the previous experiment is eliminated, and the transmitter becomes more stable.

The outer-loop in the transmitter consists of a variable time-delay, EDFA 2, and a polarization controller. In the communication experiments shown below, it is approximately 34 m in length, whereas the inner-ring is approximately 42 m long. The variable time-delay is created using a pair of GRIN collimating lenses separated by an adjustable free-space distance. EDFA 2 in the outer-loop is used to control the relative amplitude of the light that is sent back into the inner-ring. The polarization controller is used to change the relative polarization and phase between light in the outer-loop and light in the inner-ring.

The outer-loop in the receiver system has the same length as the outer-loop in the transmitter system. Its EDFA, EDFA 3, and polarization controller have the same functions as their counterparts in the transmitter. An additional variable time-delay has been introduced into the receiver's main-line to match the time-delay between photodiodes A and B to the round-trip time of the inner-ring of the transmitter.

The slowly-varying envelope of the lightwave in the transmitter, located as shown in Figure 7.17 just after EDFA 1, is denoted by $\mathbf{E}_T(t)$. The lightwave is split upon reaching the 90/10 coupler with ten percent sent into the outer-loop and 90 percent remaining in the inner ring. At the 50/50 coupler, the light in the outer-loop is added to the light in the inner-ring, giving $\mathbf{E}_{T\text{in}} + \mathbf{E}_{T\text{out}}$. The sum passes through the modulator and exits as $m(t)(\mathbf{E}_{T\text{in}} + \mathbf{E}_{T\text{out}})$. Ten percent of this signal is output to the communication channel. The remaining 90 percent continues to circulate in the transmitter. It passes through EDFA 1 and is amplified to produce $\mathbf{E}'_T \cong m(t)(\mathbf{E}_{T\text{in}} + \mathbf{E}_{T\text{out}})$. The amount of amplification is disregarded both for simplicity and to focus the discussion on the shape of the envelope. $\mathbf{E}'_T$ propagates as above through the transmitter, ultimately sending $m(t+\tau)(\mathbf{E}'_{T\text{in}} + \mathbf{E}'_{T\text{out}})$ to the communication channel, where τ is the round-trip time for light in the inner-ring.

In the receiver, the transmitted signal is divided as before between two arms leading to photodiodes A and B. Photodiode A measures the transmitted signal directly. Light sent toward photodiode B is split and recombined in such a way as to recreate the dynamics of the transmitter. The transmitted lightwave $\mathbf{E}'_R = m(t)(\mathbf{E}_{T\text{in}} + \mathbf{E}_{T\text{out}})$ is split and recombined to become $\mathbf{E}'_{R\text{in}} + \mathbf{E}'_{R\text{out}}$. Experimentally, the length of the outer-loop in the receiver is matched with an accuracy of ± 3 cm to the length of the outer-loop of the transmitter. At the frequencies detectable with the photodiodes used in this experiment, $|\mathbf{E}'_{R\text{in}}| \cong |\mathbf{E}'_{T\text{in}}|$ and $|\mathbf{E}'_{R\text{out}}| \cong |\mathbf{E}'_{T\text{out}}|$, but the relative optical phase between $\mathbf{E}'_{R\text{in}}$ and $\mathbf{E}'_{R\text{out}}$ is uncorrelated to the relative phase between $\mathbf{E}'_{T\text{in}}$ and $\mathbf{E}'_{T\text{out}}$. The time-delay between the photodiodes A and B has also been matched to an accuracy of 0.1 ns.

Recovering the message is accomplished by dividing the signal measured at photodiode A by the signal measured at photodiode B. The time-delay between photodiodes A and B has been adjusted to ensure that photodiode A measures the intensity

$$|m(t+\tau)(\mathbf{E}'_{T\text{in}} + \mathbf{E}'_{T\text{out}})|^2 = |m(t+\tau)|^2 \left(|\mathbf{E}'_{T\text{in}}|^2 + |\mathbf{E}'_{T\text{out}}|^2 + 2\,|\mathbf{E}'_{T\text{in}}||\mathbf{E}'_{T\text{out}}| \cos\theta_T \cos\phi_T\right) \quad (7.12)$$

while at the same time, photodiode B measures

$$|(\mathbf{E}'_{R\text{in}} + \mathbf{E}'_{R\text{out}})|^2 = |m(t+\tau)|^2 \left(|\mathbf{E}'_{R\text{in}}|^2 + |\mathbf{E}'_{R\text{out}}|^2 + 2\,|\mathbf{E}'_{R\text{in}}||\mathbf{E}'_{R\text{out}}| \cos\theta_R \cos\phi_R\right) \quad (7.13)$$

The quantities $\theta_{R,T}$ and $\phi_{R,T}$ represent the relative polarization angle and relative phase angle between $\mathbf{E}_{R,T\text{in}}$ and $\mathbf{E}_{R,T\text{out}}$ respectively. If $\cos\theta_T \cos\phi_T = \cos\theta_R \cos\phi_R$, a simple division obviously gives $|m(t+\tau)|^2$. However, as mentioned above, the relative phase in the receiver, ϕ_R is not correlated with the relative phase in the transmitter, ϕ_T. To ensure that these phases were the same, the lengths of the outer-loops in both the transmitter and receiver would need to be matched to within small fractions of a micron. Because the optical path length in fiber is sensitive to small changes in temperature and stress, matched lengths at one time does not imply matched lengths at later times.

A property intrinsic to lasers allows this difficulty in matching the lengths precisely to be overcome. In the description of the apparatus, the transmitter was described as consisting of an inner-ring and an outer loop. From a dynamical perspective, however, the transmitter consists of two coupled EDFRLs. One of the ring lasers is the inner-ring. The second ring laser is formed by the perimeter of the transmitter including the outer-loop. The two lasers have limited flexibility to adjust their lasing modes to maximize the intensity that passes through the modulator. The optimized laser modes possess a fixed value for $\cos\theta_T \cos\phi_T$ when they are recombined. Although $\cos\theta_T \cos\phi_T = 1$ maximizes the intensity when the lightwave form the inner-ring and outer-loop are recombined, that particular polarization and phase relationship may not be optimal for transmission through the modulator, which also acts as a polarizer. The fixed value $\cos\theta_T \cos\phi_T$ that results from this optimization can be altered by adjustment of the polarization controller in the outer-loop. Using the polarization controller in this way, initial observations suggest rough values for $\cos\theta_T \cos\phi_T$ can vary from $\sim$0.3 to $\sim$1, with a value near one being typical.

Although the value of $\cos\theta_T \cos\phi_T$ is relatively stable, experiments suggest that the transmitter's lasing wavelength fluctuates on a time scale of $\sim$5 ms. In the receiver unit, the wavelength fluctuation ensures that $\cos\theta_R \cos\phi_R$ also fluctuates. In the transmitter, the EDFRL can keep its value of $\cos\theta_T \cos\phi_T$ relatively optimized over long times, but the receiver has no feedback mechanism to fix the value of ϕ_R. A change in the wavelength of the transmitted signal changes the phase, ϕ_R, with which the light in the two arms of the receiver recombines.

Figure 7.18 shows two 100 ms time-traces of signals detected by photodiodes A and B. It should be noted that in the two-ring experiment, back-reflections from the interfaces of fiber-connectors seemed to stabilize the dynamics of the two rings. The two ring setup shown here required many of these connectors to be used in the ring. Thus, the unmodulated transmitter often emitted simply cw light. The output of cw light from the transmitter also implies that $\cos\theta_T \cos\phi_T$ has a stable fixed value. Otherwise, the intensity output would fluctuate much more. This cw output is shown in Figure 7.18 as the almost constant intensity trace, I_A. In contrast, the other trace, I_B, measured by photodiode B makes significant jumps. These jumps are attributed to the changing wavelength output from the transmitter. The changing wavelength leads to fluctuations in ϕ_R and therefore fluctuations in the intensity measured by photodiode B. A close examination of the figure shows that the amplitude of the transmitted signal also changes slightly when the amplitude of the signal measured by photodiode B changes.

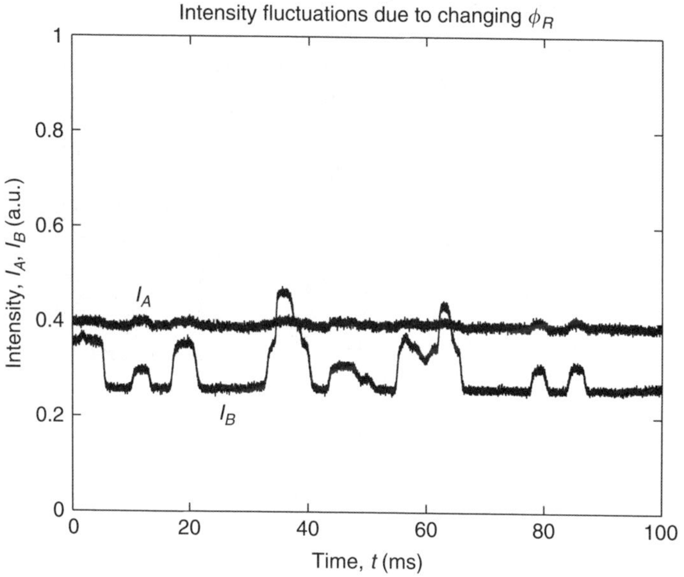

Figure 7.18 An illustration of fluctuations induced in ϕ_R by changes in the wavelength of the light emitted by the transmitter.[13] The more constant signal, I_A, is the transmitted signal detected by photodiode A when there is no intensity modulation in the transmitter. The fluctuating signal, I_B, was detected by photodiode B. The fluctuations show the very large effect that the fluctuating phase angle ϕ_R has on the signal measured by photodiode B. In the experiments, however, the polarization angle θ_R was adjusted to eliminate the effect of the ϕ_R fluctuations.

When angled connectors (FC-APC), rather than the standard flat FC-PC connectors were used in the transmitter to reduce back-reflection, the transmitter dynamics spontaneously (no modulation required) produced irregular lightwaves. The lightwaves appeared to be quasiperiodic even on sub-round-trip times, but this turned out to be because the two coupled lasers in the transmitter could only lase at frequencies that were resonant with both ring-laser resonators of the transmitter. Thus, the intensity fluctuations resulting from interference between longitudinal modes had fewer frequencies interfering below 125 MHz, which caused the intensity to appear more quasi-periodic on sub-round-trip times. Figure 7.19 is an attempt to represent the intensity dynamics of this setup. The figure was created from a vector of one million intensity values sampled at 1 GS/s.

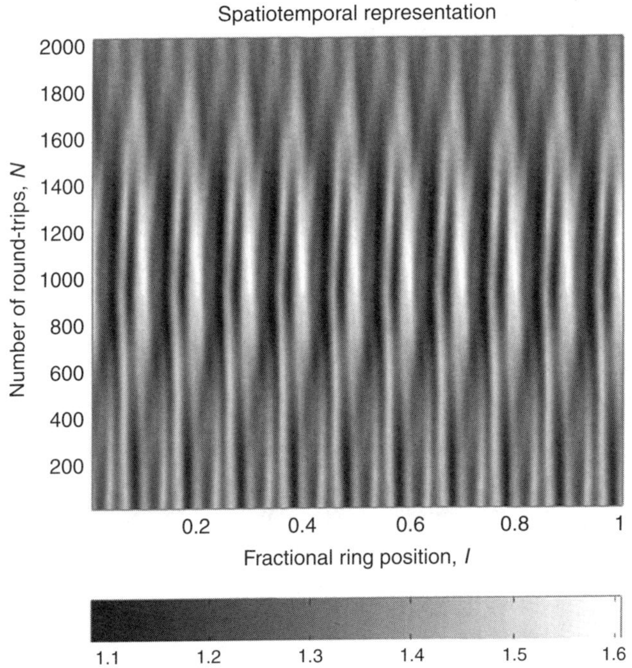

Figure 7.19 The evolution of the sub-round-trip time structure of the intensity waveform generated by the two-ring transmitter. Each fractional position, l, along the x-axis represents a position in the inner-ring. To create the figure, the intensity, at a particular time, of each point is plotted horizontally along one line for round-trip, N. The intensities are indicated according to the gray-scale given below the figure — high-intensities are white and low-intensities are black. Time is advanced by τ, and the intensities are plotted again for $N + 1$ and the process is repeated to show the evolution of the intensity dynamics over many round-trips through the inner-ring.

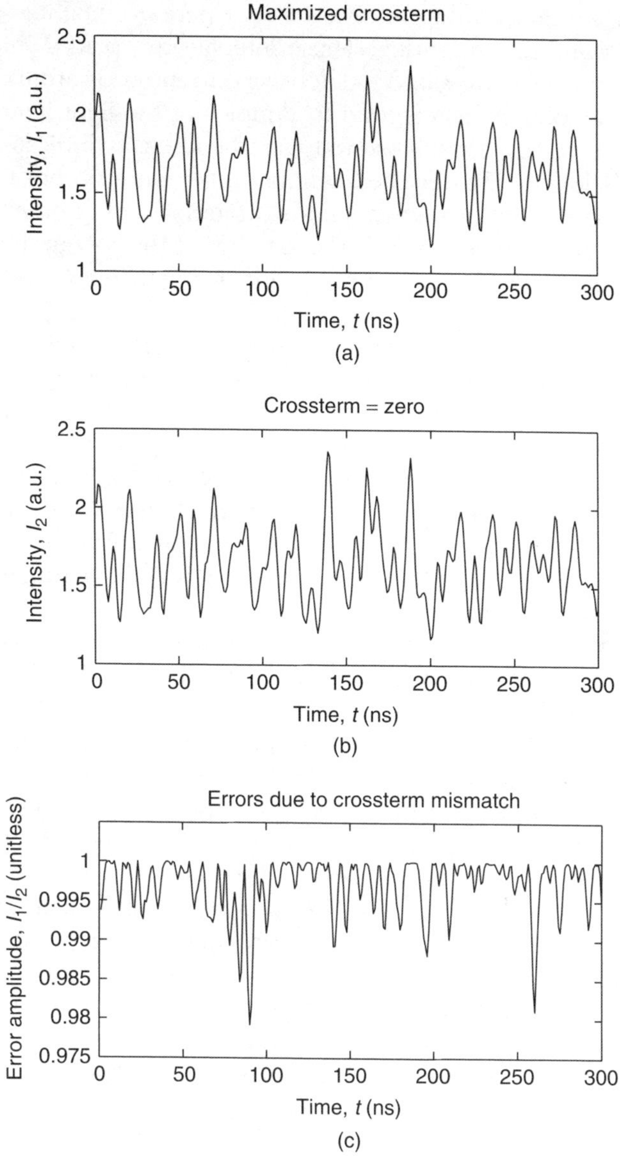

Figure 7.20 The error resulting from setting $\cos\theta_R = 0$ in Equation (7.13) when the crossterm in Equation (7.12) is maximized is shown to be small enough that it does not interfere with message recovery.[13] (a) The intensity time-series, I_1, which is proportional to $|E'_{Rin}|^2 + |E'_{Rout}|^2 + 2|E'_{Rin}||E'_{Rout}|$. (b) The intensity time-series, I_2 corresponding to $2\left[|E'_{Rin}|^2 + |E'_{Rout}|^2\right]$. (c) Dividing the data in (a) by the data in (b) gives an estimate for the magnitude of the errors that result from eliminating the crossterm in Equation (7.13) by setting $\cos\theta_R = 0$. The result of that division is shown in (c). Comparing (c) with Figure 7.21e shows that such small errors do not hinder message recovery.

These data are divided into intervals with a period equal to τ, the round trip time around the inner-ring. These intervals are placed into an array, such that the rows correspond to successive round-trips around the inner-ring, and the columns correspond to samples of the light intensity within each round-trip. This array is plotted with the intensity indicated using the gray scale below the figure. The fractional sub-round-trip position is given by l, which is just the position from the arbitrary starting point divided by the total length of the inner-ring. The number of the row in the matrix, N, is the number of round-trips around the inner-ring that have occurred since the starting point of the data.

Naturally, consistent message recovery is impaired by the fluctuations of ϕ_R described previously because in Equation (7.13) $\cos\theta_R \cos\phi_R$ will be fluctuating while the corresponding interference term in Equation (7.12) will not. However, if $\cos\theta_R$ is set to zero using the polarization controller, $\cos\theta_R \cos\phi_R = 0$ regardless of fluctuations in ϕ_R. With this setting for $\cos\theta_R$, message recovery is consistent and clear even when $\cos\theta_T \cos\phi_T \neq 0$. This surprising result is possible because, for the types of waveforms that are involved here, $|\mathbf{E}_{T\text{in}}||\mathbf{E}_{T\text{out}}|$ evolves in time in a way that strongly resembles $|\mathbf{E}_{T\text{in}}|^2 + |\mathbf{E}_{T\text{out}}|^2$.

To observe this using experimental data, light from the main-line and outer-loop in the receiver were directed into photodiodes directly, rather than recombined in a coupler. The relative lengths of the arms were the same as for the experiment. One photodiode (main-line) measured a signal proportional to $|\mathbf{E}'_{R\text{in}}|^2$ while the other photodiode (outer-loop) measured a signal proportional to $|\mathbf{E}'_{R\text{out}}|^2$. These data allows us to calculate an intensity, I_1, proportional to $|\mathbf{E}'_{R\text{in}}|^2 + |\mathbf{E}'_{R\text{out}}|^2 + 2|\mathbf{E}'_{R\text{in}}||\mathbf{E}'_{R\text{out}}|$ and display its time-series in Figure 7.20a. It also allows calculation of a signal, I_2, that is proportional to $2\left[|\mathbf{E}'_{R\text{in}}|^2 + |\mathbf{E}'_{R\text{out}}|^2\right]$. Figure 7.20b shows I_2. A comparison of I_1 and I_2 shows that they are quite similar. To obtain an estimate for the magnitude of the message recovery error in the experiment when $m(t+\tau) = 1$ and $\cos\theta_R$ is set equal to zero, the time-series shown in Figure 7.20a, I_1, is divided by the time-series shown in Figure 7.20b, I_2. The results are shown in Figure 7.20c. The values shown in (c) are very nearly one, which suggests that the two calculated intensities, I_1 and I_2, evolve very similarly.

This procedure is intended to simulate what occurs in the process of message recovery when the signal measured at photodiode A, given in Equation (7.12), is divided by the signal measured by photodiode B, given by Equation (7.13). Any deviation from a value of 1 is a message recovery error resulting from setting $\cos\theta_R = 0$ in the receiver while the crossterm is not zero in Equation (7.12). A comparison of the magnitude of these errors to the amplitude of a typical binary message bit that is experimentally recovered, as shown in Figure 7.21, shows that they are small enough that message reception is not impaired.

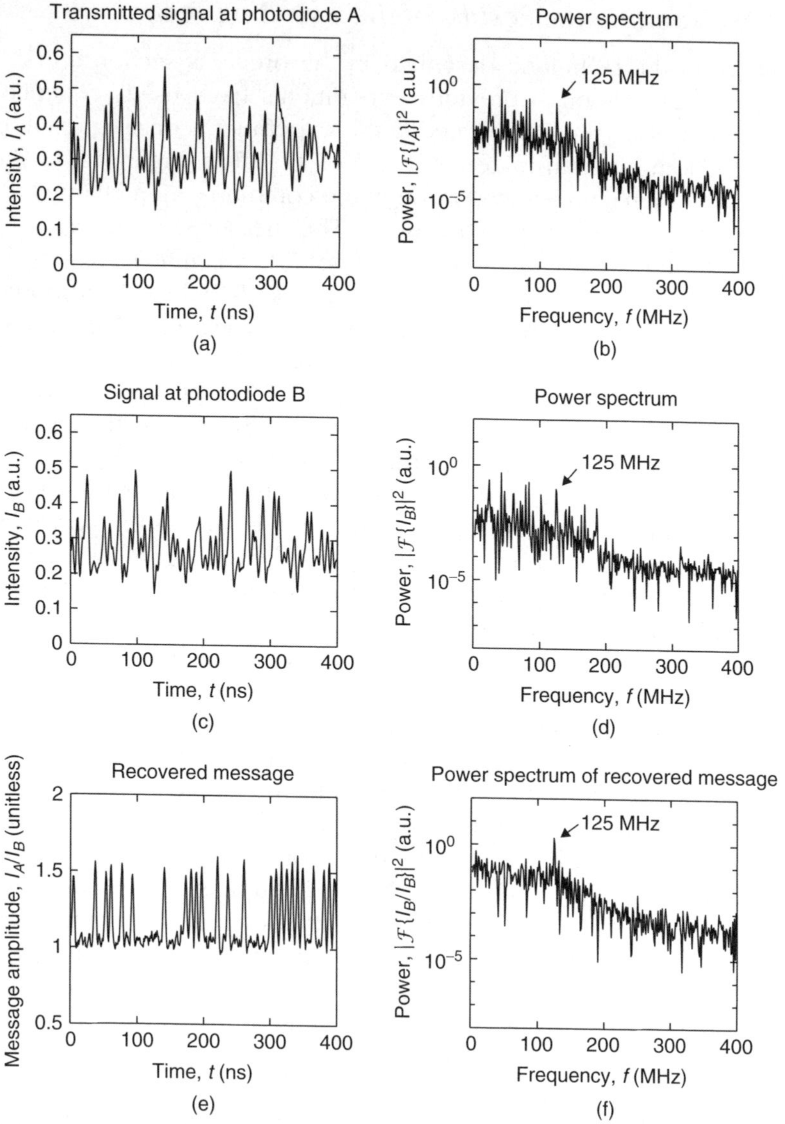

Figure 7.21 Successful recovery of a message signal from its irregular carrier.[13] The message signal consisted of a sequence of 100,000 pseudorandom bits transmitted at a rate of 125 Mb/s. (a) The transmitted signal detected at photodiode A, I_A. (b) The power spectrum of I_A. Note that a small, though inconspicuous, peak is visible at 125 MHz corresponding to the bit-rate. (c) The signal detected at photodiode B, I_B. (d) The power spectrum of I_B. Again, a small peak is visible at 125 MHz. (e) The recovered message formed by a division of I_A by I_B. (f) The power spectrum of the recovered bit sequence, with a clear peak at 125 MHz. The message itself is not discernible in either of the signals shown in (a) or (c).

7.4.2.1 *Experimental Results of the Two-Ring Technique*

Results from the technique described in the previous section are shown in Figure 7.21. Although previous experiments have used sequences of 32 bits, the message communicated in this experiment consisted of 100,000 pseudorandom bits transmitted at a rate of 125 Mb/s. Figure 7.21a shows 400 ns of the signal transmitted through the communication channel to the receiver and detected at photodiode A. The measured signal, I_A, is proportional to the intensity given by Equation (7.12). Figure 7.21c shows the signal simultaneously detected at photodiode B, I_B, which is proportional to the intensity values given by Equation (7.13). Figure 7.21e shows the division of I_A by I_B. The recovered bits are very clear in Figure 7.21e, but they are not discernible in either Figure 7.21a or Figure 7.21c. The power spectra of the signals are shown in the panels on the right. Figure 7.21b shows the power spectrum of the transmitted signal, I_A. The 125-MHz frequency peak, corresponding to the bit-rate, appears as just one of many small peaks in the spectrum. The same is true for the power spectra of the signal recorded by photodiode B, shown in Figure 7.21d. Finally, Figure 7.21f shows the power spectrum of the recovered bits. Only one peak, at 125 MHz, is significant.

Initial tests show very consistent recovery of the data. Bit-error-rates (BERs) of $<10^{-5}$ are routinely achieved. The BER may be significantly lower, but memory limitations of the digital oscilloscope limit the number of message bits that can be acquired for analysis at one time. However, when using nonangled connectors in the ring for the experiments, the transmitter occasionally "bursts" into an unstable, high-frequency chaos for approximately 1 ms. The amplitude and frequency of these bursts can be greatly diminished and even eliminated with angled-couplers or other methods to reduce back-reflection in couplers and connectors. Environmental perturbations, such as tapping on the table under the experiment, seem to trigger the bursts as well. During these intermittent bursts, the BER can be much higher than described above, though in most instances, the error produced is small enough that message recovery is not impaired at all.

To recover the communicated message, the receiver must be "tuned" to the dynamics of the transmitter. This requires matching certain transmitter parameters in the receiver. In this system, the parameters that must be matched in the receiver include the configuration, time-delays in each arm, and the relative amplitudes for the lightwaves in each arm. Figure 7.22 shows the detrimental effects of various parameter mismatches.

Figure 7.22a shows successful recovery of a repeating pseudorandom 40-bit message sequence at 125 Mb/s using the same experimental parameters as used to obtain the data in Figure 7.21. The proper sequence is clearly recovered by a division of the two signals measured by the photodiodes, I_A/I_B. The other panels represent attempted recovery of this message with

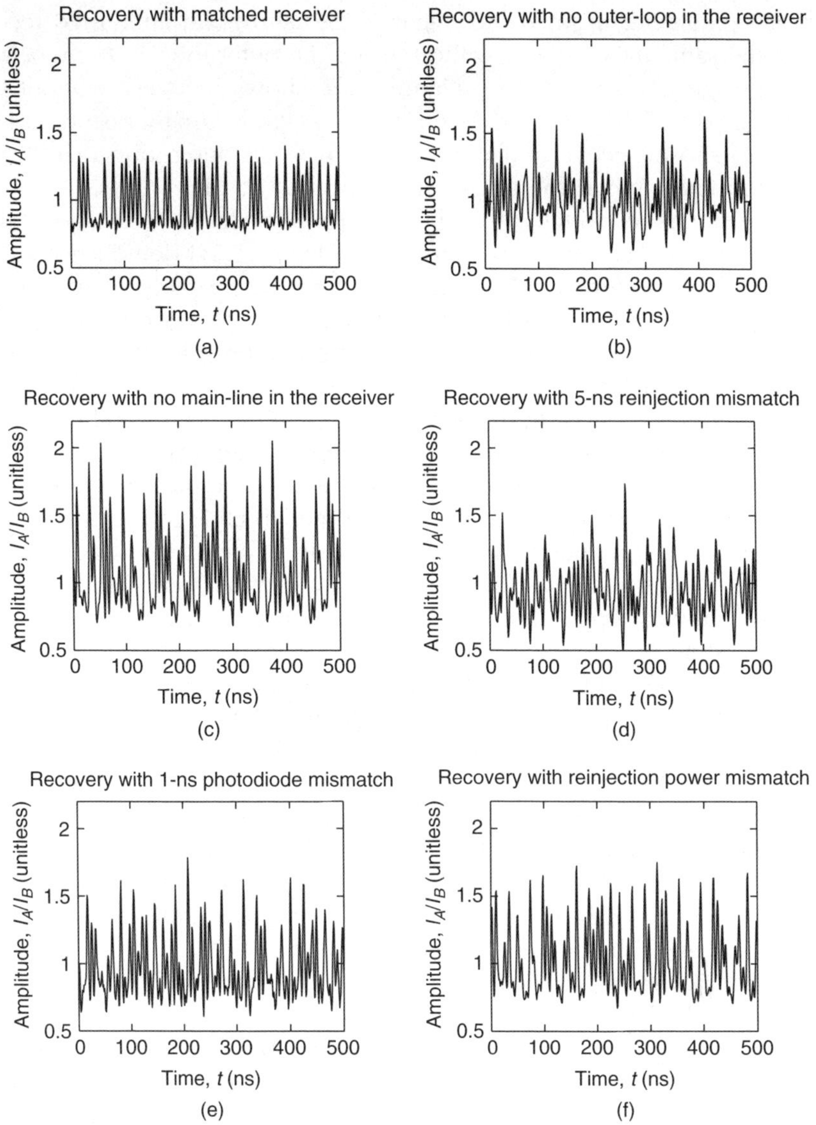

Figure 7.22 Recovery of the message requires that certain parameters in the transmitter be matched in the receiver.[13] (a) Recovery of the message with all appropriate parameters matched. The other panels show attempted recovery with just one mismatched parameter. (b) Attempted recovery with a geometrical configuration in the receiver that lacks the outer-loop. (c) Recovery is attempted without the main-line part of the receiver. (d) An extra meter of optical fiber is used in the outer-loop and results in distortion of the message. (e) Attempted recovery when the time-delay between photodiodes A and B is mismatched by just 1 ns (±20 cm of optical fiber). (f) Recovery when the amplitude of E'_{Rout} is too large relative to E'_{Rin}.

just one parameter mismatched. Figure 7.22b shows an attempted recovery of the same message but without using the outer-loop in the receiver. The message is very distorted. Figure 7.22c shows recovery without the main-line part of the receiver. Again, the message is greatly distorted. Figure 7.22d shows recovery of the message with an outer-loop that is one meter too long. When the time-delay between the photodiodes is incorrect by just 1 ns, or roughly 8 inches of optical fiber, the message is also severely distorted as shown in Figure 7.22e. Figure 7.22f demonstrates that the relative power levels of $\mathbf{E}'_{R\text{in}}$ and $\mathbf{E}'_{R\text{out}}$ must be properly set to recover the message accurately. For Figure 7.22f, the pump power on EDFA 3 was adjusted from its optimal setting of 21 mW (used to produce the data shown in Figure 7.21) to 100 mW.

To determine this method's viability as a communication scheme, it is also important to investigate the effect of long communication channels on the technique. After all, most people want to send encrypted messages over long distances rather than from one end of the lab table to the other. With this in mind, an experiment was performed using a communication channel consisting of another EDFA (to compensate for channel losses) followed by 35 km of ordinary single-mode fiber. As in Figure 7.21, a repeating message consisting of a sequence of 100,000 pseudorandom bits at 125 Mb/s is communicated. Figure 7.23a shows the transmitted signal as recorded at photodiode A after passage through the longer channel. Its power spectrum is shown in Figure 7.23b. Figure 7.23c shows the signal measured by photodiode B. Its power spectrum is shown in Figure 7.23d. The power spectra of the signals measured by the photodiodes show peaks at 125 MHz corresponding to the 125-Mb/s bit-rate, but these peaks are not conspicuous. Figure 7.23e shows the recovered message. The message does not reveal any significant distortion caused by the long communication channel. Figure 7.23f gives the power spectrum of the message.

The bit-rate in all of these experiments is limited by the bandwidth of the photodiodes (125 MHz 3-dB roll-off). Faster bit-rates require higher frequency components in the signals that simply cannot be accurately observed. However, the bit-rate can be increased without increasing the frequency components in the message signal by changing communication formats. Throughout this chapter, a return-to-zero (RZ) format has been used. In this format, the amplitude of a one bit always falls to zero before the next bit. By using a non-return-to-zero (NRZ) format, the bit-rate can be doubled without increasing the frequency content. In an NRZ format, the amplitude of the signal does not return to zero between bits.

Figure 7.24 shows results using the NRZ format. The message was a pseudorandom sequence of 100,000 bits communicated at a rate of 250 Mb/s. The communication channel for this experiment consisted of an EDFA postamplifier and 38 km of dispersion shifted (zero dispersion at 1550 nm) fiber. Figure 7.24a shows the transmitted signal measured by

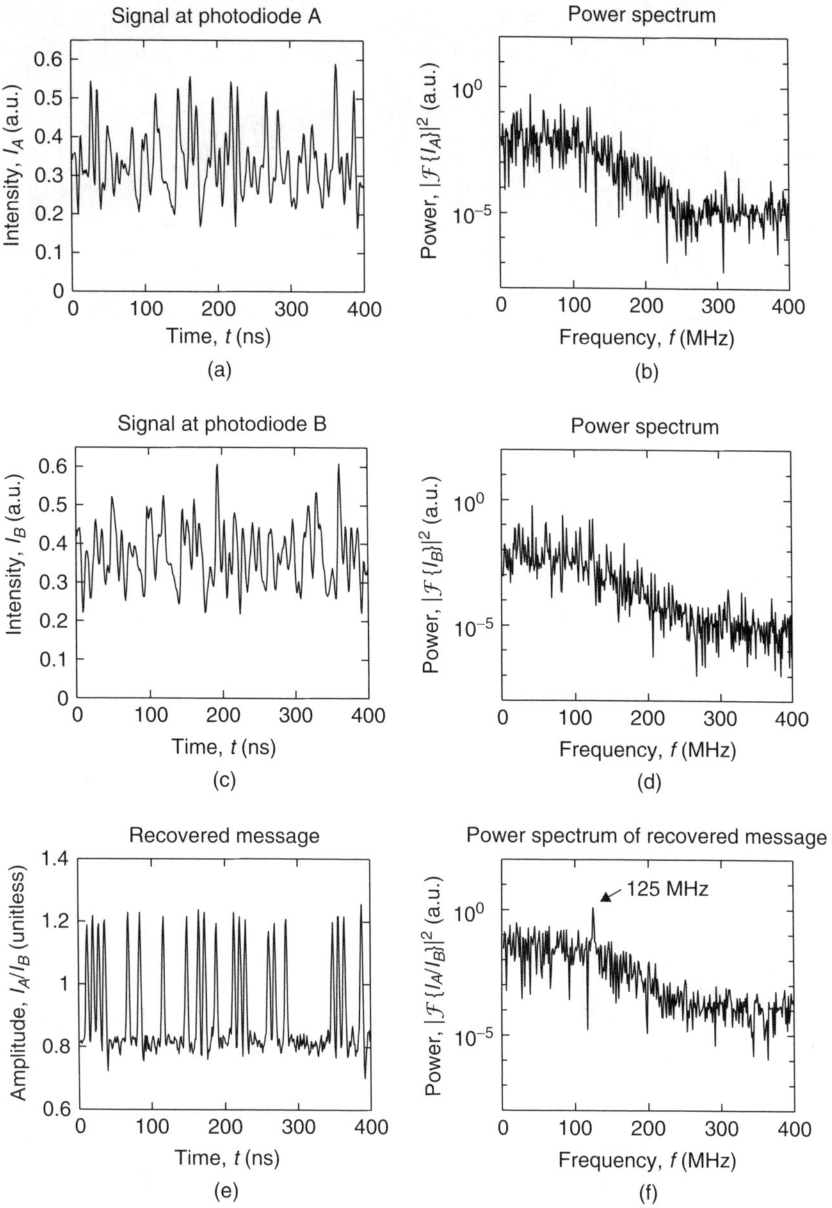

Figure 7.23 Successful recovery of a message after propagation through 35 km of ordinary single-mode optical fiber.[13] (a) The transmitted signal detected at photodiode A, I_A. (b) The power spectrum of the signal detected by photodiode A. A small peak at 125 MHz corresponding to the 125-Mb/s bit-rate is visible. (c) The signal measured by photodiode B, I_B. (d) The power spectrum of I_B. (e) The clearly recovered message after propagation through 35 km of fiber. (f) The power spectrum of the recovered message.

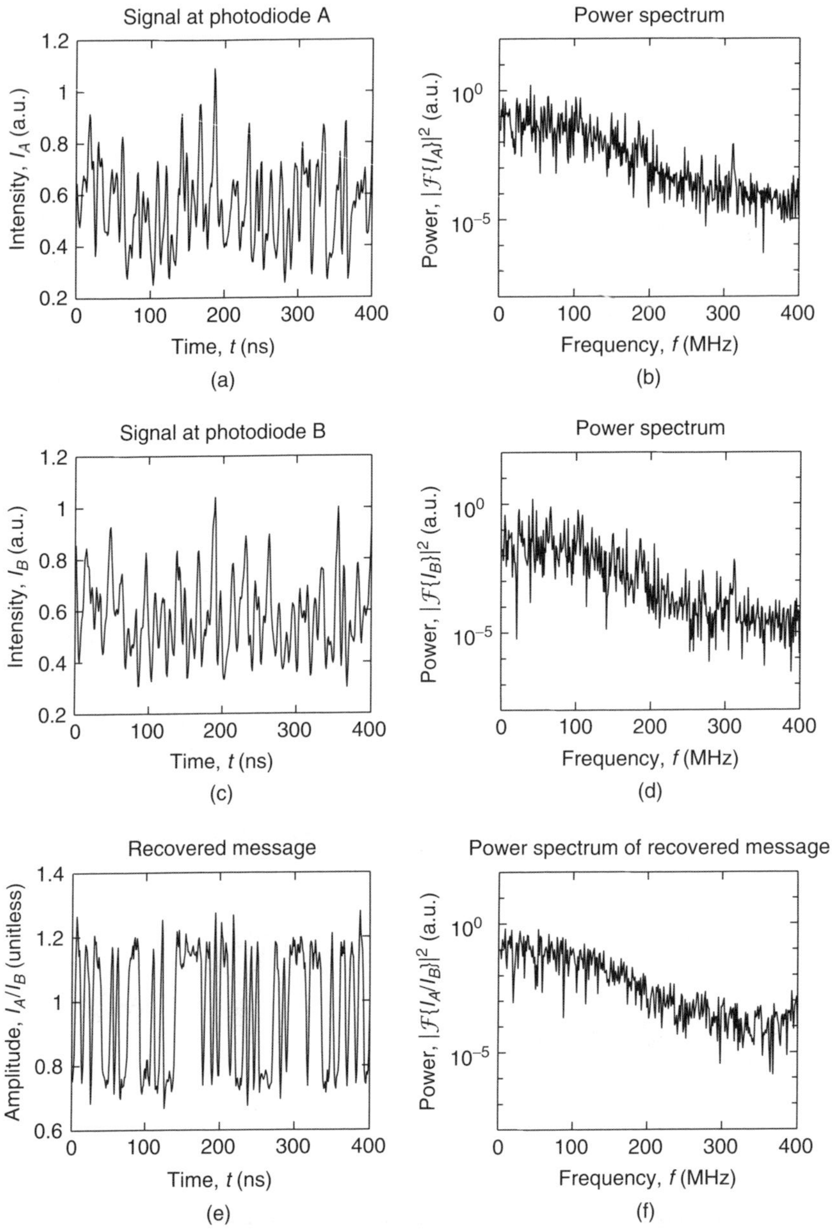

Figure 7.24 Successful recovery of an NRZ message at 250 Mb/s after propagation through a 38 km fiber with its dispersion-zero at 1550 nm.[13] (a) The transmitted signal detected by photodiode A. (b) The power spectrum of the transmitted signal. The bit-rate of the message is not conspicuous in the power spectrum of the transmitted signal. (c) The signal recorded by photodiode B, I_B. (d) The power spectrum of I_B shown in (c). (e) The message bits recovered by dividing I_A by I_B. (f) The power spectrum of the recovered message.

photodiode A, I_A. Figure 7.24b shows its power spectrum. Figure 7.24c shows the signal measured by photodiode B, I_B, and Figure 7.24d shows its power spectrum. Figure 7.24e shows the recovery of the 250 MB/s-NRZ message through the division of I_A by I_B. The message's power spectrum is shown in Figure 7.24f.

The same data for I_A/I_B, given in Figure 7.24e is also used to generate an eye-diagram for the recovered bits, shown in Figure 7.25. Eye-diagrams are often used to illustrate the quality of message recovery in binary communication systems. An "open" eye in the eye-diagram suggests good message recovery, whereas a closed eye represents a message in which distinguishing a one from a zero might be difficult. Clearly the eye is quite open in Figure 7.25. The slope of the lines corresponding to switches from a zero level to a one level gives the slew-rate of the photodiodes. As can be seen from the figure, any faster transitions between the levels would require faster photodiodes to observe properly.

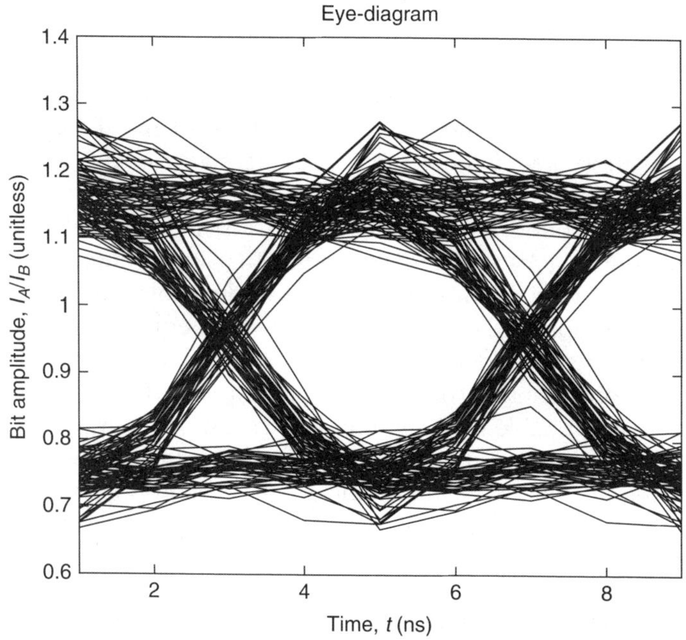

Figure 7.25 An eye-diagram showing the quality of the recovery of NRZ digital bits after propagation through 38 km of dispersion-shifted fiber.[13] The eye-diagram is open, indicating good message recovery. The slew-rate of the photodiodes is evident in the slope of the intersecting lines. The bit-rate is limited by the bandwidth of the detection equipment.

7.4.3 Discussion of the Intraring Modulator Technique

The methods presented in this section (Section 7.4) are conceptually different than those presented in Section 7.3. In the methods presented in Section 7.3, the chaotic lightwaves from the EDFRL were used to mask the message being sent. A receiver was able to recover the message from the chaotic masking. Although good privacy was not always provided by these masking techniques because conventional optical methods could also recover the message, the techniques did demonstrate that the message could be recovered from the optical chaotic masking using dynamical principles.

The messages communicated in Section 7.4 were not simply masked by chaotic lightwaves. The lightwaves transmitted from the EDFRL carried the messages on them. The message could not be separated from the carrier with conventional optical techniques. In fact, the message itself was used to drive the dynamics of the EDFRL. Consequently, the fluctuations in which the message bits were encrypted were nonlinear and time-delayed functions of previous message bits.

The first method presented in Section 7.4 involved a single-ring. The method was able to demonstrate consistent and reasonably clear message recovery using the irregular (and probably chaotic) lightwaves generated by the EDFRL transmitter. The method, however, did not realize good communication privacy because an eavesdropper needed only to determine one parameter, τ_R, to recover the message. An almost identical communication scheme was presented at the Australian Conference on Optical Fibre Technology by Luo et al.[12] after the research presented here had been performed and accepted for publication. The conference paper was entitled "Optical secure communications with 1-GHz bandwidth using erbium-doped chaotic fiber lasers." The method presented there was identical to the one presented in this thesis except for the use of faster detection equipment. Because the methods are otherwise identical, the title, which claims "secure" communication, is probably not appropriate. As in our experiment, knowledge of τ_R is sufficient to enable message recovery by an eavesdropper.

The two-ring method described in Section 7.4.2 increased the number of parameters that must be matched by an eavesdropper to recover the message. Even small mismatches of some of these parameters significantly distorts the recovered message. At the same time, the method demonstrated consistent and clear optical communication at bit-rates of 125 Mb/s and 250 Mb/s. The method works well even over long communication channels (~35 km), and in both ordinary and dispersion shifted fibers. These bit-rates were only limited by the speed of the detection electronics; though, it should be clarified that the message is not easily recovered in real time. The point-by-point division operation performed using the two signals from the photodiodes is not easily achieved in electronic circuits at high speeds, such as 125 MHz. The signals must be stored and deciphered by computation at a later time.

Another privacy consideration is dimensionality of the system attractor. As mentioned earlier in this thesis, it has proved possible to recover a message from certain lower-dimensional chaotic communication systems.[25–27] A communication method using higher-dimensional chaos is likely to provide enhanced privacy. The transmitted signals produced in this experiment have been analyzed using a FNNs algorithm.[23] Because the dynamics are driven with a pseudorandom bit-sequence, it seems reasonable to expect high-dimensional behavior, and the data plotted in Figure 7.26 supports this. Using 100,000 points acquired at 1 GS/s, the algorithm indicated that the dimensionality of the attractor is of order 10 or higher, as seen in Figure 7.26.

Still, the measure of security provided by the two-ring technique or any other chaotic communication method often is difficult to ascertain clearly. No well established or formalized cryptographic principles for dealing with chaotic communication schemes seem to have been developed. It was shown in Reference 28 that, for the two-ring method of Section 7.4.2, the dynamics of the transmitter was dominated by the message convolved with

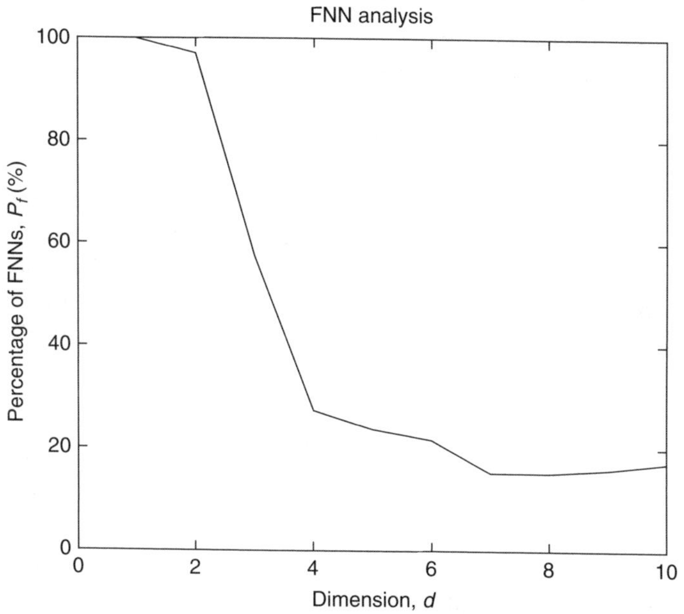

Figure 7.26 FNNs analysis showing the transmitted signal to be high-dimensional, of order ten or greater.[13] The high-dimensional nature of the chaos can be attributed to the fact that the pseudorandom message modulation drives the chaotic dynamics of the transmitter.

the response function of the two-ring laser. The authors demonstrated that four parameters were necessary to recover the message from the transmitted signal, the time delays of the two rings and the gain (or loss) in each arm. Furthermore, they were able to extract these parameters from the transmitted signal itself using mathematical techniques such as partial autocorrelation functions. Finally, in Reference 28 the following is concluded:

> The laser communication scheme suffers from a fundamental weakness — any chaos which may be present seems to be too slow to affect our method of message recovery. While the laser echo chamber effectively wraps the message into a "high-dimensional" space, this can be inverted once the laser parameters are recovered from the transmitted signal.

The methods described here in Section 7.4 all demonstrate that irregular and chaotic lightwaves can be used to carry information, and that a properly matched receiver is able to recover the message from the chaotic light.

7.5 Communication using Chaotic Polarization Fluctuations to Carry Information

From a message privacy perspective, the previously discussed techniques for communication using the optical chaos generated by EDFRL's all have an Achilles' heel: the irregular lightwaves evolve on a slow time-scale relative to the round-trip time of the ring. One possible solution is to lengthen the round-trip time to be 10 km or more, but matching this delay in the receiver would have been difficult in a laboratory setting. Another potential solution is suggested in Reference 22. Although the total intensity almost repeats on subsequent round-trips, the polarization components contributing to the total intensity do not. The Achilles' heel present in the other techniques might not apply to a technique that employs the chaotic polarization fluctuations of an EDFRL to carry messages.

Several methods of communication using modulations of a lightwave's state-of-polarization (SOP) have been proposed previously.[29–31] In fiber-optic system, these methods are referred to as polarization shift-keying (POLSK) to relate them to more familiar methods, such as amplitude shift-keying (ASK) and phase shift-keying (PSK). The simplest POLSK technique involves a binary message in which a particular SOP is used to represent a "1" bit and its orthogonal SOP represents a "0" bit. Another promising method, referred to as 8-POLSK,[31] communicates an 8-bit digital signal using eight different SOPs located at the vertices of a cube in Stokes space. In all of the proposals for modulating SOPs to carry information, a one-to-one correspondence relates a particular SOP to a particular value for the recovered signal.

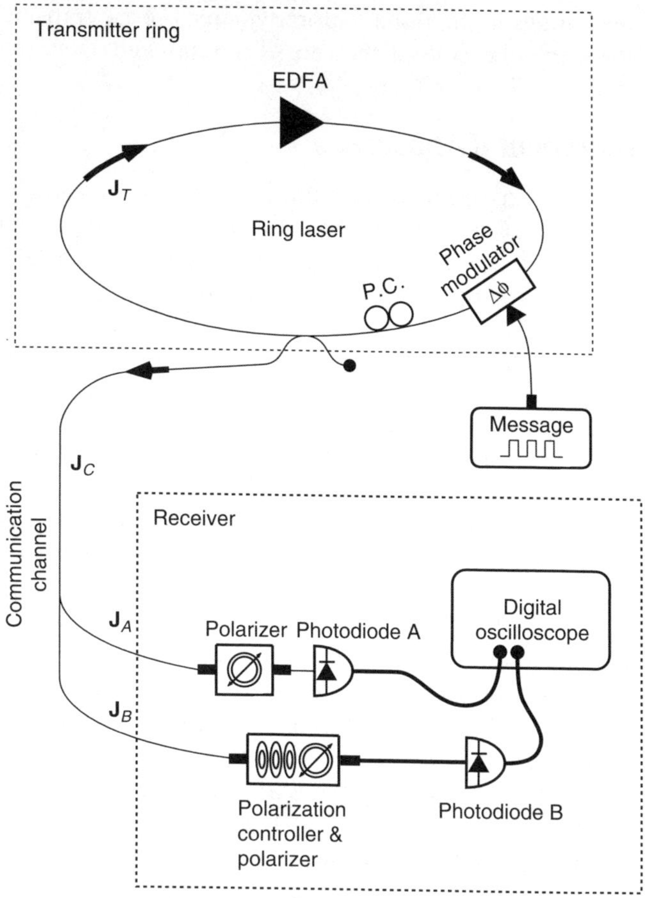

Figure 7.27 Experimental apparatus for the communication technique. A unidirectional EDFRL including both a polarization controller and a phase modulator. The phase modulator is used to vary the polarization of light in the ring. Light coupled out of the ring is transmitted to a receiver with two arms. In both arms, the light is transmitted through a polarizer before it is measured by the photodiodes. In the arm leading to photodiode B, however, the light first passes through a sequence of waveplates acting as a polarization controller. The light measured by photodiode B at a time, *t*, was output from the transmitter exactly one round-trip time, τ_R, before the light measured by photodiode A at time, *t*.

The method described in this section is different. No one-to-one correspondence exists between the message carried by the light and its SOP at the receiver. In the transmitter, the message modulates the SOP of light, but it also modulates parameters that affect the polarization dynamics of the transmitter. The receiver recovers the message from the transmitted light by detecting changes in the polarization dynamics of the transmitter, rather than from the particular SOP of the received lightwave.

7.5.1 Experimental Apparatus

A diagram of the experimental apparatus is shown in Figure 7.27. The transmitter consists of a unidirectional EDFRL with both a mandrel-type polarization controller and a phase modulator within the ring. The polarization controller is simply loops of fiber that can be twisted to alter their birefringence. Light in the phase modulator propagates in a titanium in-diffused $LiNbO_3$ strip waveguide. Electrodes on either side of the waveguide can apply a voltage, which in turn changes the index of refraction along the TE and TM modes of the waveguide, but with almost all of the change occurring for only one of these modes. One round-trip in the ring takes 239 ns, which corresponds to a length in the ring of ~48 m. A 70/30 output coupler directs a fraction (~30%) of the light in the ring into a fiber-optic communication channel while the remainder continues to circulate in the ring. The communication channel leads to a receiver comprising two branches. Light in the first branch of the receiver passes through a polarizer before being detected by photodiode A (125-MHz bandwidth). Light in the other branch passes through a polarization controller before it is incident on a polarizer. The polarization controller consists of a sequence of wave-plates ($\lambda/4$, $\lambda/2$, $\lambda/4$) that allow the controller to convert any input SOP into any other output SOP. After passing through the polarizer, the light is measured by photodiode B (also 125-MHz bandwidth). Signals from these photodiodes are recorded by a digital oscilloscope at a 1-GS/s rate.

7.5.2 Theory of Operation

Previously in this chapter, the chaotic polarization dynamics of the EDFRL were described. In the communication method presented in this section, these dynamics are harnessed to carry information. The explanation begins, using the complex amplitude convention, with the electric field of a lightwave located just before the output coupler at time t, given by $\mathbf{E}(t)$. The vector nature of the amplitude can be written more explicitly as $\mathbf{E}(t) = \begin{pmatrix} E_x(t) \\ E_y(t) \end{pmatrix}$, where x and y are arbitrarily defined orthogonal directions and both elements are complex. Birefringence in the optical fiber of the EDFRL operates on the electric field polarization in such a way that one round-trip later, the electric field can be well-approximated by $\mathbf{E}(t+\tau_R) = \mathbf{J}_T\mathbf{E}(t)$, where τ_R is the time required for light to make one circuit in the transmitter ring. The matrix $\mathbf{J}_T$ is a unitary, 2×2 Jones matrix representing the effect of birefringence on the polarization of the electric field.

Changing the voltage applied to the phase modulator in the EDFRL causes the elements of $\mathbf{J}_T$ to be altered. In the experiments performed in this paper, a data generator converts a binary message into a two-level voltage signal that is applied to the phase modulator. Thus, the Jones matrix for representing propagation around the ring has one of two values, $\mathbf{J}_T = \mathbf{J}_0$ or $\mathbf{J}_T = \mathbf{J}_1$, depending on whether a "0" or a "1" bit is to be transmitted. The phase modulator, therefore, is able to alter the polarization dynamics of the EDFRL transmitter.

Light that is coupled out of the ring propagates via a communication channel that could be either free-space or fiber-optic. In the experiments that are presented in this section, the channel was fiber-optic. Particularly in fiber-optic channels, the channel itself operates on the SOP of the light. The net effect of its birefringence on the SOP can be represented by a Jones matrix, $\mathbf{J}_C$. Like the Jones Matrices introduced before, $\mathbf{J}_C$ is a unitary 2×2 matrix with complex elements.

In the receiver, half of the light is directed toward photodiode B. Before it is measured by the photodiode, it propagates through fiber to a polarizer, where only one component is transmitted. The effect of the fiber in that arm of the receiver can be represented by the unitary Jones matrix, $\mathbf{J}_B$. The effect of the polarizer on the polarization of the light can be described by another Jones matrix, $\mathbf{P}$. However, this Jones matrix is nonunitary. For example, a polarizer transmitting only the x polarization component of the light would be represented as $\mathbf{P} = \left(\begin{smallmatrix} 1 & 0 \\ 0 & 0 \end{smallmatrix}\right)$. Thus, the light actually measured by photodiode B can be written as a concatenation of Jones matrices,

$$\mathbf{P}\mathbf{J}_B\mathbf{J}_C\mathbf{E}(t) \tag{7.14}$$

A similar process occurs in the arm of the receiver leading to photodiode A. However, the length of this arm is shorter to ensure that photodiode A measures a component of $\mathbf{E}(t + \tau_R)$ at the same time that photodiode B measures $\mathbf{E}(t)$. Said another way, the light measured by photodiode A is output from the transmitter one round-trip time, τ_R, after the light that is measured at photodiode B. The light that is measured at photodiode A, therefore, is written as

$$\mathbf{P}\mathbf{J}_A\mathbf{J}_C\mathbf{E}(t + \tau_R) \tag{7.15}$$

which can be rewritten as

$$\mathbf{P}\mathbf{J}_A\mathbf{J}_C\mathbf{J}_T\mathbf{E}(t) \tag{7.16}$$

because $\mathbf{E}(t + \tau_R) = \mathbf{J}_T\mathbf{E}(t)$. For the sake of simplicity, the polarization component selected by the polarizers in each arm has been assumed to be the same, though this is not actually necessary for the successful operation of this method.

In the technique presented here, synchronization of the signals from both photodiodes is interpreted as a "0" bit. Lack of synchronization is interpreted as a "1" bit. Therefore, to recover the proper message, the two photodiodes must measure the same signal when $\mathbf{J}_T = \mathbf{J}_0$. As is clear from Equation (7.14) and Equation (7.16), this requirement is satisfied when

$$\mathbf{J}_B\mathbf{J}_C = \mathbf{J}_A\mathbf{J}_C\mathbf{J}_0 \tag{7.17}$$

The polarization controller in the arm leading to photodiode B allows complete control over $\mathbf{J}_B$. It can be shown mathematically that a proper adjustment of $\mathbf{J}_B$ always exists, which can satisfy Equation (7.17). If this adjustment is made, the signals at the two photodiodes are synchronized. However, if $\mathbf{J}_T$ is set equal to $\mathbf{J}_1$, this synchronization will be lost because Equation (7.17) is no longer satisfied. Thus, by applying a binary voltage signal to the phase modulator in the transmitter, the signals measured by the photodiodes either synchronize or lose synchronization depending on whether a "1" bit or a "0" bit is transmitted.

7.5.3 Interpretation of Results

In the manner described above, the rapid and chaotic polarization dynamics of an EDFRL are driven by a message signal. As mentioned earlier, no one-to-one correspondence exists between the message being communicated and the light communicating it. Instead, the message is *dynamically* encoded in the SOP of the lightwave, and a properly adjusted receiver is able to recover the message using two photodiodes.

The phase modulator used in the EDFRL was chosen to perform the polarization modulation because phase modulators are relatively inexpensive and easily obtainable devices. However, the phase modulator cannot change the polarization state of all input SOPs. Some input polarizations of light are affected to a larger extent than others. A phase modulator induces a polarization change by delaying the TE mode of the waveguide with respect to the TM mode (or vice-versa), but light polarized primarily in just one or the other of these two modes is not affected as significantly as light with equal components in each mode. For this reason, a more general "polarization rotator" could be preferred, but none could be obtained without impractical costs and effort. Fortunately, the polarization in the EDFRL is not static, but fluctuates widely. This fluctuation ensures that a phase modulator is able to modulate the polarization dynamics in the ring, even if for very brief times, the SOP of the input light is such that it is not affected during transmission. A different SOP will be input into the modulator 1 ns later.

The argument above is given in terms of the electric-field amplitude. In fact, the intensity of the light that passes through the polarizers is detected, rather than the electric-fields. Equation (7.17) does satisfy the requirements for synchronization of the two photodiodes, but

$$|\mathbf{J}_B\mathbf{J}_C\mathbf{E}(t)|^2 = |\mathbf{J}_A\mathbf{J}_C\mathbf{J}_0\mathbf{E}(t)|^2 \tag{7.18}$$

satisfies the requirement as well. There are a multiplicity of solutions for $\mathbf{J}_B$ that satisfy Equation (7.18) for any given polarization of $\mathbf{E}(t)$. Although for any particular polarization of $\mathbf{E}(t)$ there are multiple solutions for $\mathbf{J}_B$, only one setting for $\mathbf{J}_B$ satisfies Equation (7.18) for all polarizations of $\mathbf{E}(t)$. Consequently, the fluctuations in the SOP of light output from the EDFRL ensure that a signal carrying a "1" bit will not falsely appear to synchronize.

Communicating using this technique provides another potential benefit that was not anticipated when the technique was developed. In fiber-optic communication channels (installed links), variations in temperature or stress in the channel leads to evolution of the elements of $\mathbf{J}_C$ on time scales of tenths of seconds.[31] This evolution makes communication via POLSK difficult. Complicated electronic algorithms have been developed[29,31] to account for this channel evolution. In the technique presented here, this evolution of the channel birefringence can have a deleterious effect on the message recovery, as indicated by Equation (7.17). However, if the transmitter Jones matrix, $\mathbf{J}_T$ is set equal to the identity matrix, $\begin{pmatrix} 1 & 0 \\ 0 & 1 \end{pmatrix}$, then fluctuations of $\mathbf{J}_C$ do not affect message recovery at all. This result is in some ways counterintuitive. It suggests that the SOP of a lightwave can carry messages through a channel whose effective birefringence, $\mathbf{J}_C$, is evolving and that the messages can be recovered without accounting for the evolving birefringence. This property results from the fact that the actual SOP of the light does not carry the message; instead, the receiver is able to recover the message by detecting changes in the polarization dynamics of the transmitter.

7.5.4 Experimental Results

The preceding discussion assumed that there is no polarization-dependent loss in the phase modulator. A significant polarization-dependent loss prevents the chaotic polarization dynamics in the EDFRL and leads to polarized output from the EDFRL. The commercially available phase modulator used in the experiment had a very significant polarization-dependent loss of 0.5 dB (12%). To compensate for this loss, a second almost identical phase modulator was placed in the ring. An additional polarization controller was also introduced between the two phase modulators. With the proper adjustment of the second polarization controller, the polarization-dependent loss in the first phase modulator could be compensated using the second phase modulator (also with ~0.5-dB polarization-dependent loss). The second phase modulator played no role in sending a message; its sole purpose was to compensate for the polarization-dependent loss of the first phase modulator. This compensation was never perfect, and consequently, the light in the ring tended to become polarized unless an electronic message signal, V, was driving the phase modulator, in which case the light fluctuated rapidly.

Figure 7.28 shows experimental results demonstrating the communication technique described in this section. The actual electronic signal, V,

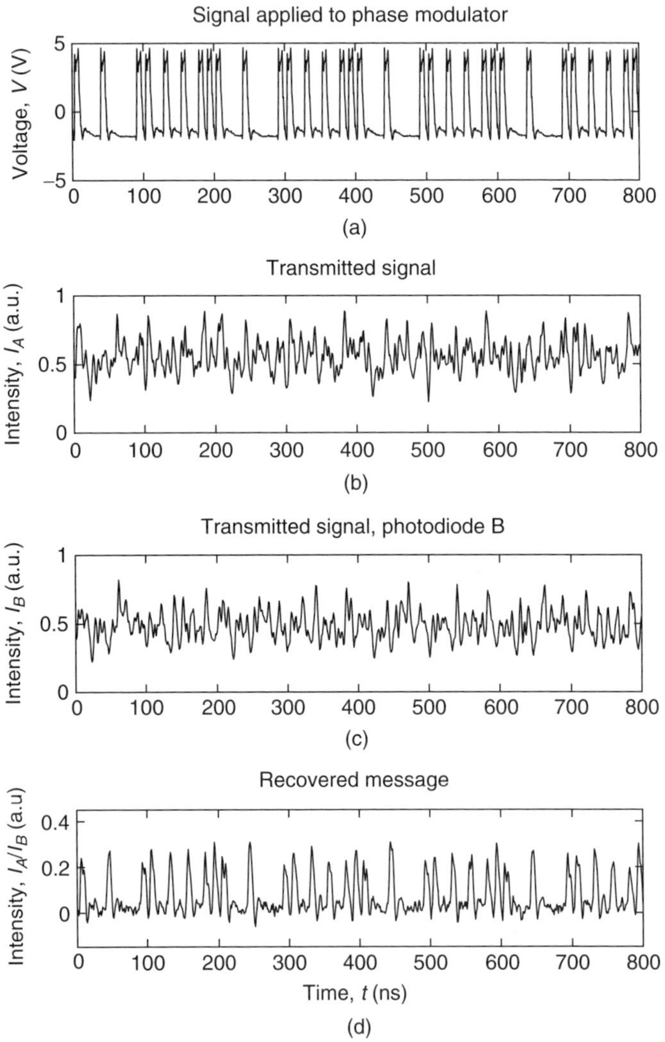

Figure 7.28 Experimental demonstration of message recovery using the irregular SOP fluctuations to carry messages. (a) The actual voltage signal applied to the phase modulator in the transmitter. (b) The transmitted signal as measured by photodiode A. (c) The signal as measured by photodiode B. (d) The result of a subtraction of (c) from (b). Loss of synchronization (nonzero values) are interpreted as "1" bits, whereas synchronization (zero values) are interpreted as "0" bits.

used to drive the phase modulator is shown in Figure 7.28a. It is a repeating sequence of 16 RZ bits at 80 Mb/s. Roughly nineteen bits can be transmitted at this rate during one round trip of light in the ring. A completely random sequence of bits can also be successfully transmitted, but the repetitive message signal provides some insights into the dynamics of the transmitter that are discussed below. Figure 7.28b shows the transmitted signal as measured by photodiode A. Despite the repeating nature of the message, the measured signal, I_A, does not show the same repetition. Figure 7.28c shows the signal measured by photodiode B, I_B. As mentioned earlier, photodiode B measures, with a time-delay, a different polarization component of the same signal measured by photodiode A. A subtraction of the data in Figure 7.28c from the data in Figure 7.28b gives the recovered message, shown in Figure 7.28d. As described earlier, a "0" bit is interpreted when the subtracted signal is roughly "0" (synchronized signals) and a "1" bit is indicated when the subtracted signal is nonzero. The consistently positive difference signals for the "1" bits results from the uncompensated polarization-dependent losses in the transmitter. If the polarizers had been set differently in the receiver, the "1" bits could have been primarily negative, or even both positive and negative. Finally, it should be noted that the message recovered on Figure 7.28d cannot be observed when the polarizers are removed from in front of the photodiodes. This proves that the message is being carried by the polarization of the light rather than by the total intensity.

The data shown in Figure 7.29 show the polarization dynamics of the EDFRL during the time this bit-pattern was being transmitted. Figure 7.29a shows that the SOP of the light was localized in one region of the Poincaré sphere. This localization is probably due to the polarization-dependent loss in the EDFRL. Figure 7.29b shows the degree of polarization (DOP) of the light from the transmitter. Figure 7.29c shows the total intensity of the light output from the transmitter. In the EDFRL, the total intensity fluctuates even when no message is being sent. Figure 7.29d–f show the normalized Stokes parameters that were measured; these were plotted in Figure 7.29a. Unlike the total intensity, the Stokes parameters, upon close inspection, reveal a semiperiodicity of 200 ns. This 16-bit message repeats exactly every 200 ns. Although the Stokes parameters reveal very imperfect periodicity that corresponds with the message repetition rate, the message itself is not obviously disclosed. The experimental techniques involved in making the measurements in Figure 7.29 are described in Reference 21. Information about the rough magnitude of the experimental errors present in the measurements.

Figure 7.30 illustrates one of the unique aspects of this communication technique. Previous discussions have explained that in conventional POLSK, a one-to-one correspondence exists between the polarization state of the lightwave and the value of the message bit carried by the lightwave.

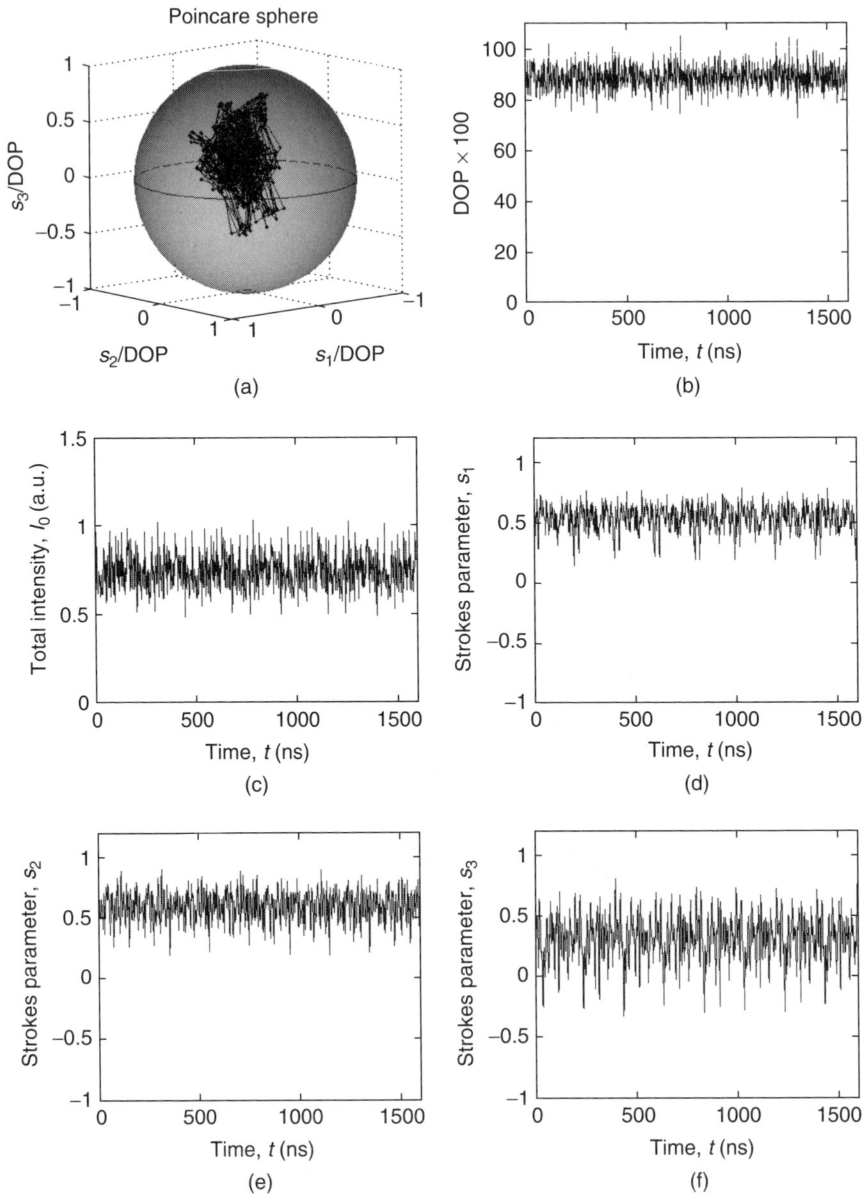

Figure 7.29 Polarization dynamics during message transmission. (a) The polarization dynamics of the transmitter are plotted on a Poincaré sphere. (b) The DOP of the light output from the transmitter. (d–f). The normalized Stokes parameters showing fluctuations in the SOP of the transmitted light.

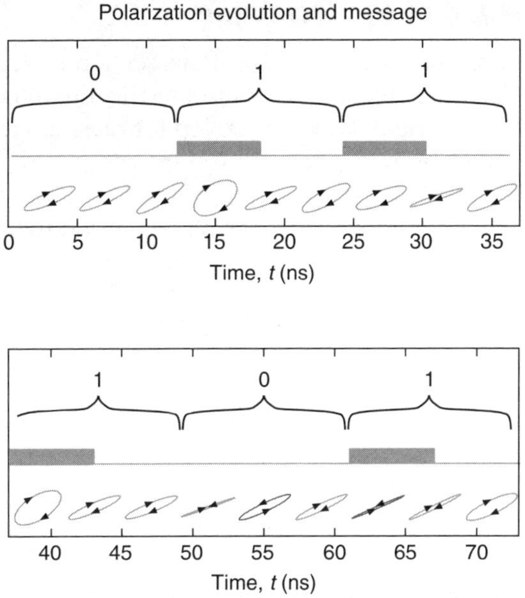

Figure 7.30 A comparison of the time-scales for the message-bits and for the evolution of the transmitted lightwave's SOP. There is no one-to-one correspondence between message bit value and the SOP of the lightwave. The handedness of the ellipses is indicated both by the arrow-heads placed on the ellipses and by the two different shades of ellipses.

Figure 7.30 shows that this is not true for the technique discussed here. It shows that the SOP of the lightwave, indicated by the ellipses, evolves over just a few nanoseconds. An RZ bit-pattern (0, 1, 1, 1, 0, 1) is shown as the gray square-wave in the figure. It is included only to indicate the time-scale on which bits are transmitted; clearly, the SOP of the lightwave evolves on time-scales shorter even than one bit.

The possibility of overcoming a fluctuating $\mathbf{J}_C$ by setting $\mathbf{J}_T$ equal to the identity matrix has also been investigated experimentally. The fiber-optic channel was manipulated (causing significant changes in $\mathbf{J}_C$), but the signal could always be recovered, with the exception of a small minority of settings. The polarization-dependent loss in the transmitter tends to cause the difference signal between photodiode A and B to fluctuate between roughly two levels. At particular settings for $\mathbf{J}_C$, the difference signal fluctuates between zero for a "0" bit and nearly zero for a "1" bit. The fluctuations were not large enough (due to the polarization-dependent loss) to discern with ease a difference between synchronization ("0" bit) and loss of synchronization ("1" bit).

7.5.5 Discussion of the Technique

The communication method presented in this section is both similar to and different from the other techniques presented in this chapter. As in the other experiments, a time-delay is employed for message recovery and the measurements made by two photodiodes are compared. The time-delay between the photodiodes is equal to the round-trip time of the transmitter ring. As in Section 7.4, a modulator was placed inside the ring.

Unlike the previous techniques, neither receiver arm attempts to replicate the dynamical behavior of the transmitter — both arms in the receiver have polarizers, whereas the light in the EDFRL transmitter was not polarized. The bits are not recovered, in this technique, in the same way as in the experiments described in the other sections. In those experiments, the message bit was recovered by subtracting or dividing the signals from the two photodiodes; essentially, the inverse of the message modulation was performed. Multi-level digital messages could be sent in this way if the recovery was sufficiently accurate. However, the technique presented in this section is limited to binary messages because the difference signal obtained from the photodiodes only permits a discrimination between synchronization and unsynchronization of the photodiode signals.

Finally, the transmitted signal from in the experiment might not be chaotic in the strictest sense. The polarization-dependent loss in the experiment caused the light to tend toward a particular polarization. One could imagine that the polarization is conceptually similar to a globally stable fixed point of the system. The modulation of the polarization drives the system away from that fixed point to produce complicated SOP fluctuations in the light output from the EDFRL. The nonlinear evolution of the polarization dynamics, however, probably occurs on time-scales longer than the time required for the system to return to the minimum loss polarization after the modulation has stopped. The significant factor for the dynamics, therefore, is simply the driving with the phase modulator. When driven with a periodic message (repetitive pattern of sixteen bits), the Stokes parameters, although they do not repeat exactly, do display a periodicity equal to the periodicity of the message.

References

1. Volkovskii, A.R. and Rulkov, N., Synchronous chaotic response of a nonlinear oscillator system as a principle for the detection of the information component of chaos, *Tech. Phys. Lett.*, 19, 97–99, 1993.
2. Ikeda, K., Multiple-valued stationary state and its instability of the transmitted light by a ring cavity system, *Opt. Commun.*, 30, 257–261, 1979.
3. Ikeda, K., Kondo, K., and Akimoto, O., Successive higher-harmonic bifurcations in systems with delayed feedback, *Phys. Rev. Lett.*, 49, 1467–1470, 1982.

4. Goedgebuer, J.P., Larger, L., and Porte, H., Optical cryptosystem based on synchronization of hyperchaos generated by a delayed feedback tunable laser diode, *Phys. Rev. Lett.*, 80, 2249–2252, 1998.
5. Larger, L. and Delorme, J.P.G.F., Optical encryption system using hyperchaos generated by an optoelectronic wavelength oscillator, *Phys. Rev. E*, 57, 1–7, 1998.
6. Aida, T. and Davis, P., Oscillation modes of laser diode pumped hybrid bistable system with large delay and application do dynamical memory, *IEEE J. Quantum Electron.*, 28, 3–15, 1992.
7. Aida, T. and Davis, P., Oscillation mode selection using bifurcation of chaotic mode transitions in a nonlinear ring resonator, *IEEE Trans. on Quantum Electron.*, 30, 2986–2997, 1994.
8. Mackey, M.C. and Glass, L., Oscillation and chaos in physiological control systems, *Science*, 197, 287–289, 1977.
9. Bünner, M.J., Meyer, T., Kittel, A., and Parisi, J., Recovery of the time-evolution equation of time-delay systems from time series, *Phys. Rev. E*, 56, 5083–5089, 1997.
10. Daisy, R. and Fischer, B., Synchronization of chaotic nonlinear optical ring oscillators, *Opt. Commun.*, 133, 282–286, 1997.
11. Luo, L.G., and Chu, P.L., Optical secure communications with chaotic erbium-doped fiber lasers, *J. Opt. Soc. Am. B.*, 15, 2524–2530, 1998.
12. Luo, L.G., Chu, P.L., and Liu, H.F., Optical secure communications with 1 ghz bandwidth using erbium doped chaotic fiber lasers, in *Proc. 24th Australian Conference on Optical Fibre Technology, ACOFT '99*, 4, 1999.
13. VanWiggeren, G.D. and Roy, R., Chaotic communication using time-delayed optical systems, *Int. J. Bifurcation and Chaos*, 9(11), 2129–2156, 1999.
14. Sivaprakasam, S. and Shore, K.A., Demonstration of optical synchronization of chaotic external-cavity laser diodes, *Opt. Lett.*, 24, 466–468, 1999.
15. Sivaprakasam, S. and Shore, K.A., Signal masking for chaotic optical communication using external-cavity diode lasers, *Opt. Lett.*, 24, 1200–1202, 1999.
16. Boudec, P.L., Jaouen, C., François, P.L., Bayon, J.-F., Sanchez, F., Besnard, P., and Stéphan, G., Antiphase dynamics and chaos in self-pulsing erbium-doped fiber lasers, *Opt. Lett.*, 18, 1890–1892, 1993.
17. Lacot, E., Stoeckel, F., and Chenevier, M., Dynamics of an erbium-doped fiber laser, *Phys. Rev. A*, 49, 3997–4007, 1994.
18. Luo, L., Lee, T.J., and Chu, P.L., Chaotic behavior in erbium-doped fiber ring lasers, *J. Opt. Soc. Am. B*, 15, 972–978, 1998.
19. Abarbanel, H.D.I., Kennel, M.B., and Lewis, C.T., Chaotic dynamics in erbium doped fiber ring lasers, *Phys. Rev. A*, 60, 2360–2374, 1999.
20. Arecchi, F.T., Lippi, G.L., Puccioni, G.P., and Tredicce, J.R., Coherence and phase dynamics of spatially coupled solid-state lasers, *Opt. Commun.*, 51, 308–314, 1984.
21. VanWiggeren, G.D., *Chaotic Communication with Erbium-Doped Fiber Ring Lasers*. PhD thesis, Georgia Institute of Technology, Atlanta, GA 30332, 2000.

22. VanWiggeren, G.D. and Roy, R., Communication with chaotic lasers, *Science*, 279, 1198–1200, 1998.
23. Abarbanel, H.D.I., *Analysis of Observed Chaotic Data*. New York: Springer, 1996.
24. VanWiggeren, G.D. and Roy, R., Optical communication with chaotic waveforms, *Phys. Rev. Lett.*, 81, 3547–3550, 1998.
25. Perez, G. and Cerdeira, H.A., Extracting messages masked by chaos, *Phys. Rev. Lett.*, 74, 1970–1973, 1995.
26. Short, K.M., Steps toward unmasking chaotic communication, *Int. J. Bifurcation and Chaos*, 4, 959–977, 1994.
27. Short, K.M., Unmasking a modulated chaotic communication scheme, *Int. J. Bifurcation and Chaos*, 6, 367–375, 1994.
28. Geddes, J.B., Short, K.M., and Black, K., Extraction of signals from chaotic laser data, *Phys. Rev. Lett.*, 83, 5389–5391, 1999.
29. Betti, S., Marchis, G.D., and Iannone, E., Polarization modulated direct detection optical transmission systems, *J. Lightwave Technol.*, 10, 1985–1997, 1992.
30. Benedetto, S., Djupsjöbacka, A., Hui, R.Q., Lagerström, B., Poggiolini, P.T., and Gaudino, R., Linbo$_3$ modulator for binary and multilevel polarization modulation, in *1994 Conference on Optical Fiber Communication, OFC '94*, 4 (San Jose, CA), 286–287, IEEE, Feb. 1994.
31. Benedetto, S., Gaudino, R., and Poggiolini, P., Polarization recovery in optical polarization shift-keying systems, *IEEE Trans. Commun.*, 45, 1269–1278, 1997.

Appendix A

Fundamental Concepts of the Theory of Chaos and Fractals

Tassos Bountis

CONTENTS

A.1 Introduction

In the early 1980s a new scientific discipline, called "chaos," was launched worldwide by a group of physicists, mathematicians, and biologists, with an impact that shook the "quiet waters" in many areas of scientific research. Chaos was heralded as no less than a revolution in science, as its followers, disciples, and innovators around the world increased at an explosive rate: Chemists, engineers, economists, artists, and philosophers joined the ranks and by the end of the decade, chaos had become a cult, a subject that one could read about in newspapers and popular magazines.

In the 1990s, the scientific world began to take a closer, more pragmatic look at the promises and achievements of chaos. As the flow of significant new theoretical advances began to subside, attention was focused on applications and the ability of this new science to answer questions of practical concern.

In this Appendix, we shall first present an introduction to the main concepts of classical chaos and bifurcation theory. We then focus on their contribution to the understanding of different transitions to the chaotic state and describe some of their applications to realistic physical systems.

When we speak of *classical* chaos we mean the irregular, aperiodic, or "chaotic" evolution of a dynamical (physical, chemical, biological, or socio-economic) system, over long periods of *time*. By the word *classical*, we distinguish the concepts discussed here from their quantum manifestations, which the interested reader can find elsewhere in the literature.[12,52]

Time, therefore, is essential in the scientific field of chaos—also called chaotic dynamics—in which one is interested in the movement, variation in the position, or *orbits* of *points*, $\boldsymbol{x}(t)$, in an n-dimensional Euclidian space $\mathfrak{R}^n$. The basic question in dynamics is therefore the characterization of the motion of $\boldsymbol{x}(t)$ in $\mathfrak{R}^n$ for arbitrary long times, $t \to \infty$. Will it approach an equilibrium point (which may be at infinity), will it execute periodic or quasiperiodic oscillations, or will it move chaotically forever in a bounded subspace of $\mathfrak{R}^n$? The answer to this question is given in any case by the analytical or numerical study of the solutions of a system of ordinary differential or difference equations describing the dynamics of the problem at hand.[1,17,51]

One very effective way to perform this analysis, as suggested already by Poincaré at the end of the 19th century, is to study qualitatively the *geometrical* properties of orbits as *curves* in $\mathfrak{R}^n$, thereby obtaining some very useful information about the dynamics of the system in time: closed curves correspond to periodic orbits, quasiperiodic motion (described by k rationally independent frequencies) occurs on k-dimensional tori, while chaotic orbits can be found near the intersections of invariant manifolds of saddle fixed points.[13,32,41]

Space, of course, need not have a discrete set of dimensions. If the medium under consideration is a continuum, as, e.g., in the case of a vibrating membrane or a turbulent fluid, one needs to study the dynamics of a system of *partial* differential equations describing the properties of variables $u_i(\boldsymbol{x}, t)$, $i=1,2,\ldots$, as functions of $\boldsymbol{x} \in \mathfrak{R}^n$ and time t. This extends the realm of chaos to a broader scientific field, which may be called *complexity* as it examines and discusses the evolution of nonlinear systems in both time and space.[11,47]

Complexity, however, does not occur only in time but also in space. As first suggested by Poincaré and later observed by E. Lorenz numerically[42] and supported theoretically by D. Ruelle and F. Takens,[53] chaotic orbits are often attracted by spatial structures of extreme geometric complexity called *strange attractors*.[4] These structures are characterized by the property of *self-similarity under scaling*, as they exhibit detailed features, similar to their original form, at all scales of magnification.

It becomes evident, therefore, that *geometry* plays an important role in the study of *dynamics*, even though one might suspect, at first, that these two notions of space and time, or kinematics and statics, are mutually exclusive (one thought about general relativity, of course, would suffice to shake this belief). Nowhere, however, in the study of classical chaos is the complementarity between geometry and dynamics more tangibly manifest than in the concept of *fractals*, as we shall also explain in this Appendix (see Section A.5).

"Fractal" is a word coined by B.B. Mandelbrot[46] to describe static objects, or collections of points in $\mathfrak{R}^n$, which are self-similar down to their finest detail and have dimension $D \leq n$, which often takes *fractional*, i.e., *non-integer* values. Remarkably, this rather nonintuitive property is shared by most inanimate or living forms around us: from galaxies and clouds to porous rocks and sponges, from tree foliages and water crystals to blood vessel networks and clusters of bacteria, nature abounds with complex structures that mock our attempts to describe them in terms of the familiar (integer-dimensional) shapes of elementary geometry.

It is important to point out, however, that besides their impressive numerical manifestations and wide-ranging practical applications, chaos and fractals rest on a firm theoretical foundation, encompassing several mathematical topics: analysis, differential equations, topology, differential geometry, number theory, measure theory, discrete mathematics, and the geometry of "fractal" sets.

It is in this sense that we speak of *complexity* as a new scientific discipline[11], aiming to combine a large body of experimental knowledge with a unified methodology in order to better analyze, predict, and control *complex systems* studied so far by many other sciences.

After introducing the basic concepts of nonlinear dynamics and chaos in Section A.1, we proceed in Section A.2 to describe in detail four types of local *bifurcations*—or qualitative change of the dynamics—connected with the destabilization of equilibria or periodic states. We focus on discrete-time dynamical models, since they are easier to analyze and contain all the important features of systems working in continuous time. Two of these local bifurcations are related to two of the three "routes" (or, transitions) to chaos described in Section A.3. Thus, the occurrence of one or more local changes of stability can bring about global changes in the dynamics.

In Section A.4, we discuss the fundamental notions of *renormalization* and *universality* which characterize all three routes to chaos discussed in this Appendix. They are closely related to the property of self-similarity and the presence of scaling (power) laws encountered in the theory of fractals described in Section A.5. Together with the theory, we shall also discuss some practical applications of generalized dimensions in analyzing and improving the short-term prediction of chaotic time series data. The contents of this Appendix are based on two review articles of this author[6,7], which were aimed at new students of this field and are, therefore, of an introductory and pedagogical nature.

A.1.1 Nonlinearity and Dynamical Systems

There are few times in the history of science that a revolutionary development takes place: When Kepler realized that planetary motion could only be explained by ellipses and Newton discovered his law of gravitational attraction, when Planck postulated that on the microscopic level energy is quantized, and Einstein explained how gravity is related to the geometry of space-time, science made big steps forward. In all these cases, of course, the importance of a scientific achievement was recognized long after the discovery was made.

Nowadays, every new scientific discovery is rapidly communicated around the world and immediately put to test, scrutinized, reproduced, and verified in laboratories, research centers, and universities everywhere. As examples, we need only recall the high T_C superconductivity results of the mid-1980s or, somewhat later, the experiments claiming to have discovered "cold fusion." Ultimately, of course, only time will tell if real scientific progress has been made.

Chaos has also taken some time to develop and become recognized as a major field in its own right. By some, it has been hailed as a revolution, comparable to the discoveries of quantum mechanics and general relativity. By others, it has been called a new scientific discipline, which promises to resolve several long-standing issues and provide fresh insight into many physical phenomena, too complicated to explain by more traditional approaches.

So, what is chaos? Interestingly enough, there is no simple answer to this question. Most researchers would hasten to reply that it is the "extremely sensitive dependence of dynamics on initial conditions." This is, however, half of the story, since it refers only to the *dynamical* manifestation of chaos, as complicated behavior in *time*.

What about chaos in *space*? How are we to understand certain highly complicated spatial patterns, like crystal growth, the structure of tumors, the complexity of blood vessel networks, the formation of trees, the irregularity of fractures, or the shape of a thunderbolt? Do we need perhaps a new kind of geometry to be able to describe such objects?

And, finally, what about the relationship of chaos to unpredictability and loss of information? Is there any connection between chaotic behavior and what most people understand as "randomness" or "noise"? But, let's take things from the start.

In the beginning, there was *nonlinearity*. Virtually no object occurring in nature has straight line edges or perfectly planar sides. The earth is not flat, mountain slopes are not constant, rivers do not flow in fixed directions, and birds don't fly in straight paths. There are curves and anomalous surfaces everywhere, and yet man has persistently tried to overlook them and study nature by linear approximations.

The harmonic oscillator, for example, consisting of a mass m attached to a spring, whose force F increases linearly in the displacement $x(t)$, i.e.,

$$m\frac{d^2x}{dt^2} = F = -kx \tag{A.1.1}$$

($k > 0$ is the spring constant), has been for centuries the classical model for studying mechanical oscillations and lattice vibrations. Population studies, however, have sought for years to explain the exponential growth of a number of species $N(t)$ by the linear equation

$$\frac{dN}{dt} = \alpha N \tag{A.1.2}$$

$\alpha > 0$ being a rate constant.

These models may be simple and fairly accurate in some situations. They do suffer, however, from very serious drawbacks: Equation (A.1.1) assures us that, if we wish, we can stretch the spring to infinity without breaking it or altering its frequency of oscillation, whereas Equation (A.1.2) implies indefinite population growth, excluding any decrease due to deaths or limited food supply.

In order to make these models more realistic, we need to include nonlinearities in their equations. Let us, therefore consider instead of Equations (A.1.1) and (A.1.2) the modified equations

$$\frac{d^2x}{dt^2} = -\omega_0^2 x + x^2 \tag{A.1.3}$$

$$\frac{dN}{dt} = \alpha N - N^2 \tag{A.1.4}$$

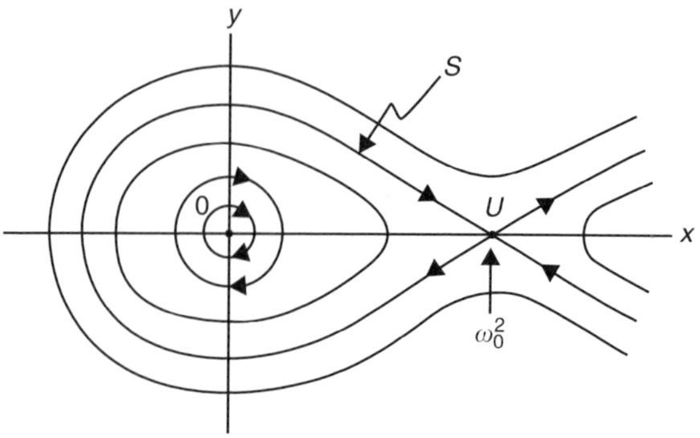

Figure A.1 The phase portrait of Equation (A.1.7), cf. Equation (A.1.6).

where $\omega_o^2 \equiv k/m$. Note that we have not added any extra parameters before the nonlinear terms, as these could easily have been removed by scaling x, or N, by a constant.

In what way are the solutions of Equations (A.1.3) and (A.1.4) different than those of Equations (A.1.1) and (A.1.2)? The most important difference is that Equations (A.1.3) and (A.1.4) possess additional *equilibrium* or *fixed point solutions* (where the derivatives of x and N vanish) beyond the trivial ones at $x=0$ and $N=0$. Indeed, if we set the righthand sides (rhs) of Equations (A.1.3) and (A.1.4) equal to zero, we find two stationary points for each system:

$$\begin{array}{lll} \text{(i) } \bar{x}=0, & \text{(ii) } \bar{x}=\omega_o^2 & \text{for Equation (A.1.3)} \\ \text{(i) } \bar{N}=0, & \text{(ii) } \bar{N}=\alpha & \text{for Equation (A.1.4)} \end{array} \tag{A.1.5}$$

In order to determine the importance of these fixed points, we need to take a closer look at the dynamics near them.

Multiplying Equation (A.1.3) throughout by dx/dt and integrating once we derive an expression that remains constant along every solution,

$$\frac{1}{2}y^2 + \frac{1}{2}\omega_o^2 x^2 - \frac{1}{3}x^3 = E = \text{const.} \tag{A.1.6}$$

with $y \equiv dx/dt$. This expression is called an *integral* of the motion and E is referred to as the total energy. Let us also write Equation (A.1.3) as a system of first-order ordinary differential equations (ODEs)

$$\begin{aligned} \dot{x} &= y \\ \dot{y} &= -\omega_o^2 x + x^2 \end{aligned} \tag{A.1.7}$$

where $(\dot{\ }) \equiv d()/dt$. We may now use Equation (A.1.6) and plot in the (x, y) *phase plane* the solutions (or *orbits*) of the system for different values of E. When this is done one obtains a picture similar to that of Figure A.1.

There is one particularly important curve S in Figure A.1, corresponding to $E = E_s = \omega_o^2/6$. It is called a *separatrix* since it separates bounded, oscillatory motion in its interior from orbits outside S, which escape to infinity. We linearize Equation (A.1.7) about the fixed point U, substituting in Equation (A.1.7)

$$x(t) = \bar{x} + \xi(t), \quad y(t) = \bar{y} + \eta(t) \tag{A.1.8}$$

with $(\bar{x}, \bar{y}) = (\omega_o^2, 0)$ to obtain

$$\begin{bmatrix} \dot{\xi} \\ \dot{\eta} \end{bmatrix} = \begin{bmatrix} 0 & 1 \\ \omega_o^2 & 0 \end{bmatrix} \begin{bmatrix} \xi \\ \eta \end{bmatrix} = L \begin{bmatrix} \xi \\ \eta \end{bmatrix} \tag{A.1.9}$$

The linear system [Equation (A.1.9)] can be solved after obtaining the eigenvalues and eigenvectors of L:

$$\lambda_1 = -\omega_o, \quad \lambda_2 = \omega_o, \quad e_1 = \begin{bmatrix} 1 \\ -\omega_o \end{bmatrix}, \quad e_2 = \begin{bmatrix} 1 \\ \omega_o \end{bmatrix} \tag{A.1.10}$$

The general solution near U, expressed in the $\{e_1, e_2\}$ basis now reads

$$c_1 e^{-\omega_0 t} e_1 + c_2 e^{\omega_0 t} e_2$$

where $c_1 = \frac{1}{2}\left[\xi(0) - \omega_0^{-1}\eta(0)\right]$ and $c_2 = \frac{1}{2}\left[\xi(0) + \omega_0^{-1}\eta(0)\right]$. This explains the qualitative picture of the dynamics near U in Figure A.1. U is called *saddle node* and is obviously unstable at a fixed point. A similar linearization analysis near the origin $O = (0, 0)$ shows that it is stable under small perturbations. O is called a *center point.*

The equilibrium points of Equation (A.1.4) are $\bar{N}_1 = 0$ and $\bar{N}_2 = \alpha$. Linearizing Equation (A.1.4) about these points with $N = \bar{N} + \xi$ one obtains

$$\begin{array}{lll} \text{(i) near } \bar{N}_1, & \xi(t) = \xi(0)e^{\alpha t} & \text{(unstable node)} \\ \text{(ii) near } \bar{N}_2, & \xi(t) = \xi(0)e^{-\alpha t} & \text{(stable node)} \end{array}$$

This can now be verified by the analysis of the general solution of Equation (A.1.4), which has the form

$$N(t) = \frac{\alpha N(0)}{N(0) + (\alpha - N(0))e^{-\alpha t}}$$

For any $N(0) \neq 0$ we have $N(t) \to \alpha$ with $t \to \infty$, hence $N = \bar{N}_2 = \alpha$ is (asymptotically) stable while $N = \bar{N}_1 = 0$ is unstable. More specifically, for $0 < N(0) < \alpha$ the solution $N(t)$ approaches α from below, and for $N(0) > \alpha$ – from above. The unphysical initial condition $N(0) < 0$ yields a singular solution since the denominator in $N(t)$ becomes 0 for $t = t_0 = \frac{1}{\alpha}\ln\frac{\alpha + c}{c}$, where $c = |N(0)|$.

Now let's make our original nonlinear oscillator problem in Equation (A.1.3) even more realistic by adding a *damping* force proportional to the velocity:

$$\ddot{x} = -\omega_o^2 x - b\dot{x} + x^2, \quad b > 0 \tag{A.1.11}$$

Observe that we cannot integrate Equation (A.1.11), as we did with Equation (A.1.7) because the system no longer possesses an integral of the form of Equation (A.1.6). In fact, it has now become a *nonconservative* system, which may be called *dissipative*, as elementary "areas" $\Delta A = \Delta x \Delta y$, everywhere in (x, y) phase space, *dissipate away* under time evolution:

$$\frac{d}{dt}(\Delta A) = (\Delta x \Delta y)\left[\frac{1}{\Delta x}\frac{d\Delta x}{dt} + \frac{1}{\Delta y}\frac{d\Delta y}{dt}\right] = (\Delta A)\left[\frac{\Delta \dot{x}}{\Delta x} + \frac{\Delta \dot{y}}{\Delta y}\right]$$

$$\underset{\substack{\Delta x \to 0 \\ \Delta y \to 0}}{\longrightarrow} (\Delta A)\left[\frac{\partial \dot{x}}{\partial x} + \frac{\partial \dot{y}}{\partial y}\right] = -b(\Delta A) \qquad \text{(A.1.12)}$$

i.e., $(\Delta A) \propto \exp(-bt)$. The last step in Equation (A.1.12) follows from writing Equation (A.1.11) as a system

$$\begin{aligned} \dot{x} &= y &&= f_1(x, y) \\ \dot{y} &= y - \omega_0^2 x - by + x^2 &&= f_2(x, y) \end{aligned} \qquad \text{(A.1.13)}$$

More generally, we can say that the rate of change of elementary areas in a dynamical system of the form Equation (A.1.13) is determined by the quantity

$$\nabla \cdot f = Tr[Df(x, y)] = \frac{\partial f_1}{\partial x} + \frac{\partial f_2}{\partial y} \qquad \text{(A.1.14)}$$

where Tr denotes the trace and Df denotes the Jacobian matrix of $f \equiv (f_1, f_2)$ evaluated at (x, y),

$$Df(x, y) = \begin{bmatrix} \partial f_1/\partial x & \partial f_1/\partial y \\ \partial f_2/\partial x & \partial f_2/\partial y \end{bmatrix}(x, y) \qquad \text{(A.1.15)}$$

Note that this matrix is exactly what is needed for the study of local dynamics near fixed points, since $L = Df(\bar{x}, \bar{y})$.

The linear analysis near the fixed points of Equation (A.1.13) reveals the behavior of orbits in the (x, y) phase plane for $b < 2\omega_0$, cf. Figure A.2. Geometric arguments and the analysis of the direction field $dy/dx = (-\omega^2 x - by + x^2)/y$ at various points (x, y) show that the separatrix S of Figure A.1 has split into two curves, W^s and W^u, called the *stable* and *unstable manifold*, respectively, of the saddle fixed point at U. The shaded area in Figure A.2 is called the *basin of attraction* of the stable *focal* point at the origin 0. This basin inflates gradually as b increases (stronger damping) and, respectively, shrinks for smaller, positive b. It disappears completely for $b = 0$, as the origin, though stable, is no longer an attractor and the basin around O does not exist for $b < 0$, as the origin is an unstable fixed point then.

The equations we have considered above, are examples of *dynamical systems*, which, in the case of Equations (A.1.1), (A.1.2), and (A.1.9)

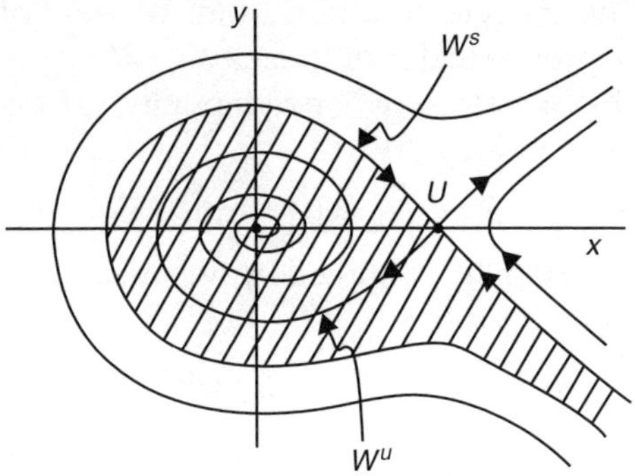

Figure A.2 Phase portrait of Equation (A.1.13). The shaded area shows the basin of attraction of the origin for $b > 0$.

are called *linear*, while in Equations (A.1.3), (A.1.4), (A.1.7), and (A.1.13) they are called *nonlinear*. The population models of Equations (A.1.2) and (A.1.4) describe dynamics in one dimension while Equations (A.1.3), (A.1.7), and (A.1.9) to (A.1.11) are said to be two-dimensional.

More generally, a system of m first-order ODEs

$$\frac{dx_i}{dt} = \dot{x}_i = f_i(x_1, x_2, \ldots, x_m), \quad i = 1, 2, \ldots, m \tag{A.1.16}$$

is said to describe an m-dimensional *flow* in $\Re^m$. Its *phase space* is a subset $D \subset \Re^m$, where the functions f_i are defined, continuous and at least once continuously differentiable. These conditions are sufficient to guarantee the *existence* and *uniqueness* of solutions. If some of the f_i's are nonlinear functions of the x_i's, of Equation (A.1.16) is said to describe a nonlinear dynamical system.

But time need not always be continuous. It may also be taken in discrete steps, corresponding to different instances $t_1, t_2, \ldots, t_n \ldots$ $(t_j < t_{j+1})$ at which interactions occur or measurements are taken. In such cases, the differential equations of Equation (A.1.16) can be replaced by *difference* equations:

$$x_{i,n+1} = g_i(x_{1,n}, \ldots, x_{m,n}), \quad i = 1, 2, \ldots, m \tag{A.1.17}$$

where $x_{i,n} \equiv x_i(t_n)$. Now the dynamics is followed not by solving a set of differential equations, but by iterating a system of *algebraic equations,* which is much simpler to do on a computer.

Let us consider the apparently simple example of a discrete dynamical system in one dimension:

$$x_{n+1} = ax_n - x_n^2, \quad n = 0, 1, 2, \ldots \tag{A.1.18}$$

with a parameter $a \in \Re$. At first sight, one might think of comparing it with the one-dimensional system of Equation (A.1.4), represented by a first-order ODE. And indeed, from the point of view of fixed points and their stability, there are no major differences:

Equation (A.1.18) also has two fixed points, given by $x_n \equiv \bar{x}$, for all n, i.e.,

$$\bar{x} = a\bar{x} - \bar{x}^2, \quad \bar{x}_1 = 0 \quad \text{and} \quad \bar{x}_2 = a - 1 \qquad \text{(A.1.19)}$$

cf. Equation (A.1.15). Stability is studied by linearization, setting $x_n = \bar{x} + \xi_n$ in Equation (A.1.18) and gives:

(i) Near $\bar{x}_1$: $\xi_{n+1} = a\xi_n$, i.e., $\xi_n = a^n\xi_o$, whence

$$|a| < 1 \Rightarrow \xi_n \to 0, \quad n \to \infty, \quad \bar{x}_1 \text{ is stable}$$

$$|a| > 1 \Rightarrow |\xi_n| \to \infty, \quad n \to \infty, \quad \bar{x}_1 \text{ is unstable}$$

(ii) Near $\bar{x}_2$: $\xi_{n+1} = (2-a)\xi_n$, $\xi_n = (2-a)^n\xi_o$ and

$$|2 - a| < 1 \Rightarrow \xi_n \to 0, \quad n \to 0, \quad \bar{x}_2 \text{ is stable}$$

$$|2 - a| > 1 \Rightarrow |\xi_n| \to \infty, \quad n \to \infty, \quad \bar{x}_2 \text{ is unstable}$$

Phase space is also taken here to be the (one-dimensional) x-axis. The observing reader, however, will notice that an interesting situation arises from the above analysis: First of all, both fixed points cannot be stable for the same a, as the stability of $\bar{x}_1$ requires $-1 < a < 1$, while that of $\bar{x}_2$, $1 < a < 3$. They can, however, be simultaneously *unstable*, e.g., for $a > 3$. What happens then to the orbits of Equation (A.1.18)? Notice that this cannot occur with Equation (A.1.4), where the stability of one of the fixed points implies necessarily the instability of the other.

The answer, of course, lies in the fact that the solutions of the ODE of Equation (A.1.4) move always on the x-axis (t being a continuous variable), where they are not allowed to cross each other (or themselves), as that would violate uniqueness. However, the orbits of Equation (A.1.18) are free to "jump" from one point of the x-axis to another, as time is treated now as a discrete variable. The only constraint the solutions of Equation (A.1.18) have to obey is that, if they return to their starting value x_o, after m iterations, say, they must then repeat the same sequence of points $x_1, x_2, \ldots, x_{m-1}, x_o$, forever.

In fact, there is a world of difference between the solutions of Equation (A.1.4) and those of Equation (A.1.18), as will be made clear in what follows. We will discover that the simple-looking dynamical system of Equation (A.1.18) does exhibit exceedingly complex and fascinating behavior, which we shall call *chaotic*. On the other hand, nothing exciting or complicated is to be expected of Equations (A.1.4), (A.1.7) or (A.1.13).

Not all nonlinear dynamical systems, therefore, display chaos. Can chaos occur in systems of ODEs? Yes, of course, provided we are prepared to lift ourselves off the plane and consider dynamics in three or more dimensions.[13,31,32]

A.1.2 The Geometry of Motion in Phase Space

As we saw in the previous section, a general dynamical system can be written either in the form

$$\frac{d\boldsymbol{x}}{dt} \equiv \dot{\boldsymbol{x}} = f(\boldsymbol{x}), \quad \boldsymbol{x}(t) \in \mathfrak{R}^m \tag{A.1.20}$$

if time is continuous, $t \in \mathfrak{R}$, or

$$\boldsymbol{x}_{n+1} = g(\boldsymbol{x}_n), \quad \boldsymbol{x}_n \in \mathfrak{R}^m \tag{A.1.21}$$

$n = 0, 1, 2, \ldots,$ if t is discrete, cf. Equations (A.1.16) and (A.1.17). In both cases, for existence and uniqueness of solutions, we require $f = (f_1, \ldots, f_m)$ and $g = (g_1, \ldots, g_m)$ to be C^r functions, with $r \geq 1$.

Clearly, a great number of physical, chemical, or biological models can be described by systems of equations, like (A.1.20) or (A.1.21). Let us consider two famous examples:

(i) The Lorenz equations[42,58]

$$\begin{aligned} \dot{x} &= \sigma(y - x) \\ \dot{y} &= -rx + y - xz \quad (x, y, z) \in \mathfrak{R}^3 \\ \dot{z} &= -bz - xy \end{aligned} \tag{A.1.22}$$

representing an oversimplified model of Rayleigh–Bénard convection[55] and

(ii) The Hénon map[33]

$$\begin{aligned} x_{n+1} &= y_n + 1 - \mathrm{a}x_n^2 \\ & \qquad\qquad (x_n, y_n) \in \mathfrak{R}^2 \\ y_{n+1} &= bx_n \end{aligned} \tag{A.1.23}$$

which, in the conservative case ($|b| = 1$), describes the (x, y) displacements of protons from their circular path in a high-energy accelerator.[3]

However, over and beyond their applicability to specific physical systems, these models are important because they exhibit practically all chaotic phenomena found in low-dimensional systems.

One of the first questions we wish to ask about such a dynamical system is this: What is the behavior of its solutions, as $t \to \infty$? Will they end up at some equilibrium point, do they eventually relax to some simple, periodic motion, or will they execute bounded, quasiperiodic oscillations? Are there any other possibilities?

It doesn't sound as if we are asking too much; yet we have posed some highly nontrivial questions. We realize, of course, that finding the general solution of Equations (A.1.20) or (A.1.21), even in such simple-looking examples as Equations (A.1.22) or (A.1.23), is presently beyond our expectations. So we turn to the next most interesting goal: What can we say about the *properties* of the dynamics for (arbitrarily) long times?

The approach we will adopt in order to answer such questions will be to a large extent geometrical. Let us think of the solutions of a dynamical

system as *orbits*, or *curves* in their m-dimensional phasespace, parametrized by the time variable t.
Each point

$$\boldsymbol{x}(t) = (x_1(t), x_2(t), \ldots, x_m(t)) \in \Re^m \quad \text{(A.1.24)}$$

is then taken to correspond to a *state* of the system in $\Re^m$. Given an initial point $\boldsymbol{x}(0)$, there is one and only one orbit passing by that point. Locally, of course, and for short times the dynamics is smooth and the properties of the orbits vary continuously in phase space. Our interest, however, is in the more global behavior of the motion, for long times.

Rather than following orbits in continuous time, let us consider an $(m-1)$-dimensional *surface of section* Σ, embedded in $\Re^m$, which is transverse to the flow, i.e., is intersected by most orbits at nonzero angles, see Figure A.3a. We expect that the properties of the dynamics in $\Re^m$ will

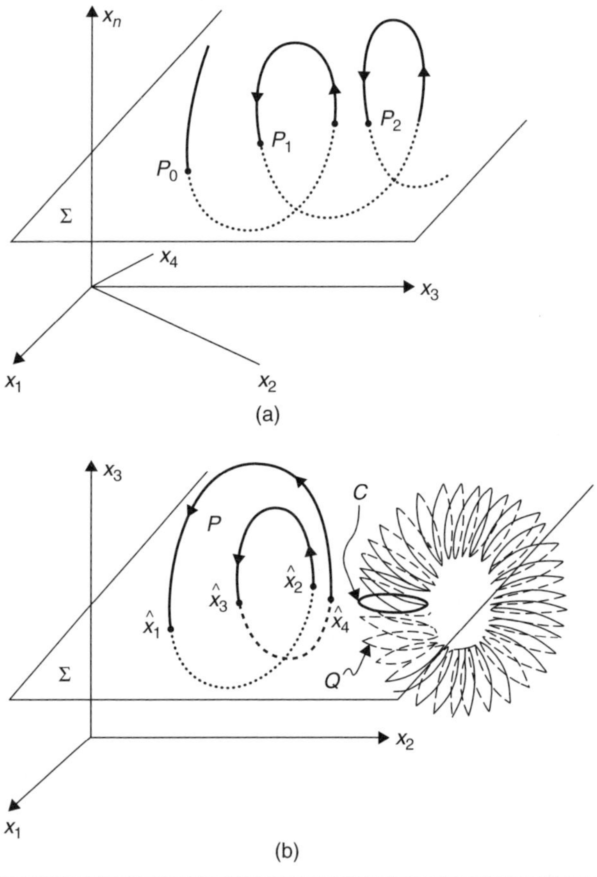

Figure A.3 (a) Phase space of an m-dimensional dynamical system with an $(m-1)$-dimensional surface of section Σ. (b) Phase space for $m=3$ showing a periodic (P) and a quasi-periodic orbit (Q).

be reflected by the behavior of the intersections of the orbits with Σ in the following way: Take $m=3$ and suppose that Equation (A.1.20) has *a periodic* solution that closes on itself, as in the case of the curve P in Figure A.3b. Clearly, this closed orbit will intersect Σ at a finite number of points $\hat{x}_1, \hat{x}_2, \ldots, \hat{x}_p$ creating a periodic orbit of period p on Σ. Now imagine a *quasiperiodic* orbit Q in $\mathfrak{R}^m$, which never closes. This solution is expected to intersect Σ at an *infinite* number of points, generating thus, as $t \to \infty$, a smooth, close curve C in Σ.

Now introduce some dissipation in the system, due to which phase space volumes *contract* in time, just as areas did in our example in Equation (A.1.11) of the previous section, see Equation (A.1.12). This means that the periodic and quasiperiodic orbits mentioned above will become *attractors*, as most orbits will tend toward them, as $t \to \infty$. Owing to their simple form (a finite number of points or a smooth, closed curve) they are called simple or regular attractors.

What else is there? How much "worse" can things get? We may well imagine a highly complicated curve, weaving its intricate path in $\mathfrak{R}^3$. However, in the end it will have to intersect the plane Σ along some curve and the orbit itself, locally, will necessarily look everywhere one-dimensional, right? Wrong.

One of the major consequences of Lorenz's pioneering work, in the early 1960s, was that three-dimensional simple-looking dynamical systems, like Equation (A.1.22), can have orbits, which are so intricately "woven," that there seems to be no end to their structure, even after many magnifications.[31,58] In fact, these orbits tend, as $t \to \infty$, to a very complicated subset of $\mathfrak{R}^3$ consisting of the unstable invariant manifold of saddle equilibrium points of the system, see Figure A.4.

In two-dimensional flows, like Equation (A.1.13), invariant manifolds of saddle points are simple curves (see W^s and W^u in Figure A.2). In

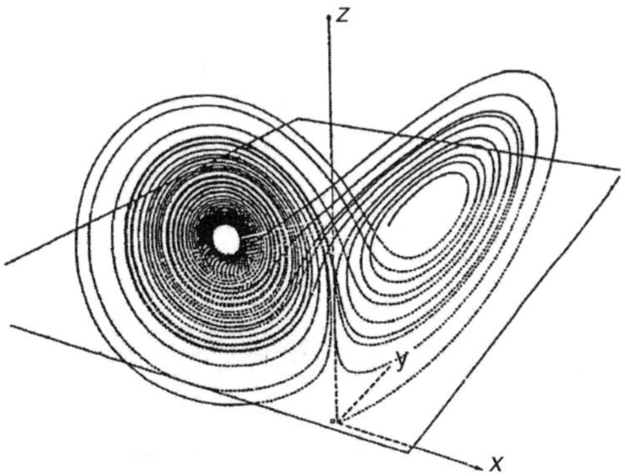

Figure A.4 The Lorenz strange attractor, Equation (A.1.22), with $\sigma=10$, $b=8/3$, $r=28$.

three-dimensions, however, they can become surfaces, which stretch out, bend, and fold over infinitely many times, within a bounded region of phase space. Since these invariant surfaces exhibit structure at all scales, they end up having a dimension higher than two (but less than three) and are called *strange attractors*.

If we wanted to study one such strange attractor in detail, we could take cross sections of the one due to Lorenz (see Figure A.4) and start blowing them up, integrating the equations on the computer for long times. This would be too time-consuming, however. It is a lot simpler to iterate Hénon's map of Equation (A.1.23) many times, e.g., in the case $a=1.4, b=0.3$ where it is known that such a strange attractor appears to exist, see Figure A.5.

Therefore, as we see from the enlargements in Figures A.5b to d, strange attractors are seen, upon successive magnifications, to possess a self-similar structure. We call this property *self-similarity under scaling.*

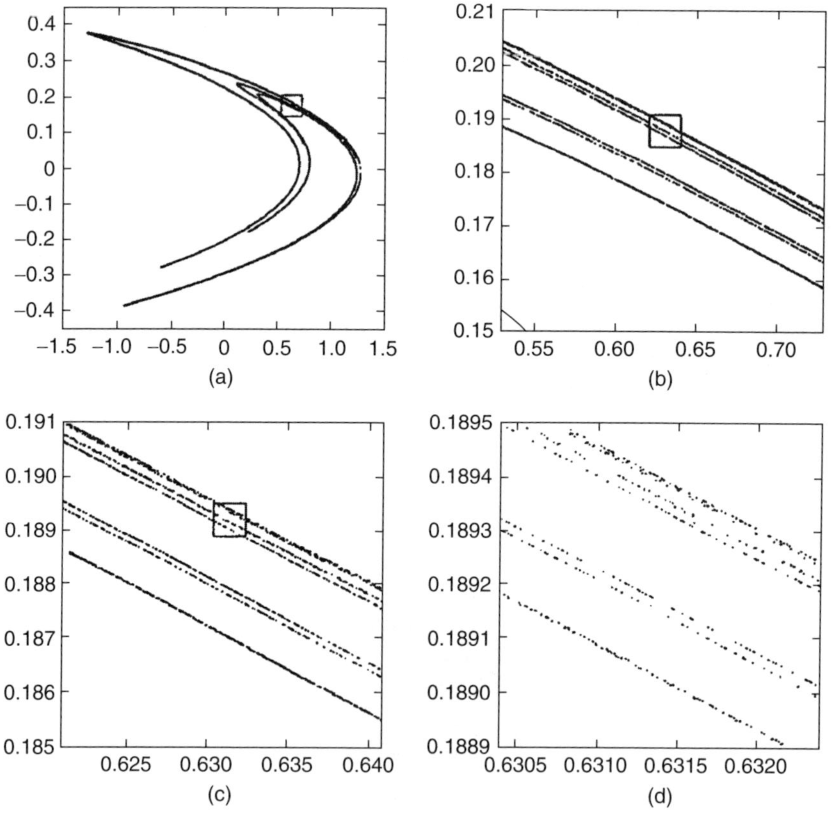

Figure A.5 (a) The Hénon strange attractor, Equation (A.1.23), with $b=0.3$ and $a=1.4$. (b), (c), and (d) show magnifications of the box in the previous figure, demonstrating the fractal structure of the attractor.

It is fundamentally a geometric property and is connected with the fact that strange attractors have *noninteger* dimension and form sets which are called *fractals*.[46] (See Section A.5 of this Appendix.)

Let us turn now to the second fundamental property of strange attractors, which is a purely dynamical one: It concerns the *exponential* separation of nearby orbits on the attractor and was the main new result pointed out by Lorenz in his famous paper.[42] This means that the dynamics of such systems in phase space are *unstable* or repelling, at least in one direction vertical to the flow. Thus, most orbits, which start very close to each other, diverge rapidly and find themselves at diametrically distant points on the attractor.

A grave consequence of this fact is that the slightest error in the choice of initial conditions will grow exponentially and lead to very different regions than those visited by the "true" orbit. This is what is known as "extremely sensitive dependence on initial conditions," "chaotic behavior" or, simply, "chaos."

You will often find this phenomenon referred to as "deterministic chaos," because it is an inherent property of the dynamics and has nothing to do with externally imposed randomness or noise. The computer errors mentioned above are unavoidable, since they merely reflect the globally unstable nature of the motion.

In short, chaotic orbits are not computable. The slightest uncertainty in the knowledge of their initial conditions is destined to grow exponentially in time. This is often described by the statement "small causes produce large effects" and is expressed by the famous picture of a butterfly flapping its wings in Beijing and influencing the weather in Paris.[26]

There is nothing uncertain, of course, about the equations of motion. They are completely deterministic. Their solutions are not. As a consequence of this, Laplace's famous philosophical premise of a deterministic universe, whose past and future are readily accessible from the knowledge of its natural laws, is put into serious question. Chaos teaches us that even the simplest deterministic systems can have *properties which are as random as a coin toss*, or a spin of the roulette wheel.

Finally, we saw in this section that complicated *geometric* properties of attractors in phase space and complicated *dynamical* properties of the motion are intimately related. Space and time assume in chaotic dynamics also a dual role and become faces of the same coin. It will be impossible, from now on, to separate one from the other.

A.1.3 Fundamental Concepts of Chaos

We have thus discovered, from our introduction to nonlinear dynamical systems, that the two most fundamental concepts of chaos are the following:

Self-similarity under scaling
Extremely sensitive dependence on initial conditions

The first one contains the encouraging message that if an object looks complicated on a large scale, its complexity may be due to the superposition, at all scales, of patterns which become self-similar in the limit of infinite magnification.

It is indeed an appealing idea. Imagine if we could really explain the onset of turbulence by the simultaneous appearance of an infinite succession of eddies of smaller and smaller size:

> Big whorls have little whorls
> which feed on their velocity
> and little whorls have lesser whorls
> and so on to viscosity.
>
> (L.F. Richardson: see Reference 26)

Consider the possibility that the branching of air channels in our lungs or of the arteries in our kidneys occurs in such a way that each new "generation" of branches is but a scaled-down version of the previous one. It would be very nice if life were that simple.

Nature, of course, is considerably more complicated. Still, the idea of scaling works rather well as a first approximation: There are many physical phenomena which are characterized by self-similarity under scaling, even though they may require an infinity of scales for their description.[49] In fact, several "routes to chaos" or "chaotic transitions" have been discovered where self-similarity plays a crucial role (see Sections A.3 and A.4).

Once we have determined that certain phenomenona exhibit self-similarity, the next question is: What mathematical techniques are available that would be appropriate for their study? It is at this point that we must appeal to the theory of critical phenomena of statistical mechanics where the method of *renormalization* has proved most useful in similar circumstances.[18,43,52]

What is renormalization? In the study of dynamical systems, it is first the discovery of a functional equation which describes how the dynamics becomes self-similar at the transition to chaos. This equation is *universal* in the sense that it is the same for a large class of systems which display the same kind of chaotic transition.

One then looks for nontrivial solutions of this (nonlinear) functional equation, which would correspond to what happens at the parameter values where the transition to chaos occurs. Finally, linearizing the equation about that solution, one studies the eigenvalues of the associated linear operator and examines whether they lie inside the unit circle.

Those that lie outside correspond to *unstable* directions in function space, which must be avoided if the chaotic transition is to occur. If there is only a small, finite number of such eigenvalues (with magnitude larger than one), then most likely the transition can be realized in the laboratory and will have physical significance. Moreover, one looks among these eigenvalues, as well as other scaling constants in the functional equation, for *universal numbers* characterizing this transition to chaos.

Let us turn now to the discussion of the second fundamental concept of chaos: the extremely sensitive dependence on initial conditions. We have already seen that it is the outcome of the interaction between two factors: the unstable dynamics of chaotic regions and our imperfect knowledge of the state of our system. Remove any one of these factors and there would be no chaos—no intrinsic, deterministic chaos, that is.

Imagine you shoot a white billiard ball toward a group of differently colored balls on a frictionless table. Imagine also that all collisions between the balls (and the walls of the table) are elastic, so that, once set in motion, the balls never stop. Now start changing slightly the direction at which you shoot the white ball and start taking movies of these experiments. You will soon discover, after the first few frames, that these movies are all different from each other!

What happened? Clearly, the slightest variation in your initial conditions quickly leads to different angles of collision, with some balls missing entirely some others and thus pursuing vastly deviating trajectories in every experiment. The end result, in every case, is a very different arrangement of colors and paths after only a very small number of collisions. This is chaos.

To give an idea of the sensitivity of this experiment, we need only remark that, if you were standing next to the table, the gravitational attraction between your body and the balls (an extremely weak force to be sure) begins to have a significant effect on the dynamics after about nine or ten collisions!

Now, suppose you don't care about the details of the motion and wish to consider certain "averaged" features of the dynamics. You will then find, if you wait long enough, that all parts of the table are visited "democratically" by the same number of balls and by all colors!

Thus we have come all the way from individual orbits and purely deterministic dynamics to a *statistical* description of a large number of possibilities. In chaotic systems, *uncertainty* of the initial state, however small, is bound to grow very fast and lead naturally from single particles to probability distributions and from determinism to randomness.

How can this property of chaos be measured? The most relevant quantity to compute is the *rate* of exponential separation of trajectories, measured by the *Lyapunov exponents* of the orbits in the different phase space directions. Let us consider, for simplicity, one-dimensional maps

$$x_{n+1} = f(x_n), \quad x_n \in \mathfrak{R} \tag{A.1.25}$$

$n = 0, 1, 2, \ldots$, and write the orbit resulting from an initial condition x_o, in the form

$$x_n = f^n(x_0) \equiv \underbrace{f(f(\ldots f(x_0)\ldots))}_{n} \tag{A.1.26}$$

Suppose now that we want to follow a neighboring orbit, starting at $x'_o = x_o + \epsilon$, for ϵ small,

$$x'_n = f^n(x'_o) = f^n(x_o + \epsilon)$$

and measure how the distance between these orbits changes after many iterations,

$$|f^n(x_o + \epsilon) - f^n(x_o)| \propto \epsilon e^{\lambda(x_o)n}, \quad n \gg 1 \tag{A.1.27}$$

The rate of separation between these two orbits, $\lambda(x_o)$, is called the Lyapunov exponent and is, in general, a function of x_o. In the limit of $\epsilon \to 0$ and $n \to \infty$ we thus obtain

$$\lambda(x_o) = \lim \frac{1}{n} \log \left| \frac{df^n(x_o)}{dx_o^n} \right| = \lim_{n\to\infty} \frac{1}{n} \sum_{i=o}^{n} \log |f'(x_i)| \tag{A.1.28}$$

Therefore, we conclude that if $\lambda(x_o) > 0$, for a "dense" set of initial conditions, the corresponding orbits will be expected to exhibit chaotic behavior, while if $\lambda(x_o) < 0$ one expects convergence of the orbits onto a regular attractor (the case $\lambda(x_o) = 0$ is generic in *integrable* conservative systems in which nearby orbits separate linearly from each other).

In m-dimensional dynamical systems there are, of course, m Lyapunov exponents. However, we are usually interested in the largest one among them, $\lambda_{\max} = \max\{\lambda_i / 1 \le i \le m\}$, which determines the rate at which the norm of the distance between nearby trajectories changes in time. If $\lambda_{\max} > 0$ and is independent of $x(0)$ in some region $D \in \Re^m$, then we observe chaotic behavior with statistical properties that are shared by most orbits in that region.[18,31]

Example 1

Consider the $1 - D$ map of the unit interval

$$x_{n+1} = f(x_n) = \begin{cases} ax_n, & 0 \le x_n \le \frac{1}{2} \\ a(1 - x_n), & \frac{1}{2} \le x_n \le 1 \end{cases} \tag{A.1.29}$$

with $a > 0$. Obviously, for any x_0

$$\lambda(x_0) = \lim_{n\to\infty} \frac{1}{n} \sum_{i=1}^{n} \log |f'(x_i)| = \log a$$

Thus for $a < 1$ the iterates of Equation (A.1.29) behave regularly, while for $a > 1$ the map displays chaotic behavior for all x_0.

It should be pointed out that, in general, Lyapunov exponents cannot be obtained analytically, as in the simple case above. Usually they are estimated numerically after fairly long computations. Evidently, the more

$\lambda_i > 0$ one finds and the larger their magnitude, the more chaotic the system is. In fact Lyapunov exponents are important *invariants* of the dynamics as they are not affected by smooth coordinate transformations. Still, they cannot reveal by themselves the proportion of phase space occupied by chaotic orbits. To estimate the size of the chaotic regions, one must resort to other more graphical means of exploring global features of the dynamics, like e.g., Poincaré surfaces of section.

A.2 Local Bifurcations in Discrete Dynamics

A.2.1 Parameter-Dependent Systems

The equations of motion of dynamical systems often include certain parameters with important physical significance: mass, spring constants, damping coefficients, nonlinearity strengths, moments of inertia, etc. Varying the values of these parameters, generally, has the effect of varying only *quantitatively* the properties of orbits, while the qualitative picture of the dynamics in phase space remains unchanged.

There exist, however, in many systems, certain *critical values* through which if a parameter is varied, sudden *qualitative changes* occur in the dynamics. Such changes are called *bifurcations*.

Consider, for example, our old friend—the anharmonic oscillator,

$$\ddot{x} = \mu x - x^3 = -\frac{dV}{dx} \tag{A.2.1}$$

with cubic nonlinear force, and let the coefficient of the linear term, μ, be a parameter, which can be varied. The potential function $V(x)$ from which this equation is derived is

$$V(x) = -\frac{1}{2}\mu x^2 + \frac{1}{4}x^4 \tag{A.2.2}$$

Clearly, if $\mu < 0$, $V(x)$ is everywhere positive and concave upward (see Figure A.6a) yielding bounded oscillations around $x = 0$, for all initial conditions in the $(x, \dot{x})$ phase space, as shown in Figure A.6b. The same is true for $\mu = 0$.

However, for $\mu > 0$ something very interesting happens: The system can no longer perform small oscillations about the origin, as that has become an unstable equilibrium (see Figure A.6). Instead, it now has the "choice" of oscillating about two new stable equilibria, $\bar{x}_{\pm} = \pm\sqrt{\mu}$, which have been "born" (bifurcated) out of the origin, at the bifurcation value $\mu = \mu_o = 0$ (Figure A.6d).

There are some familiar features in this picture which we have already encountered in Section A.1.1. These are the separatrices, $S_{\pm}$, dividing different kinds of motion in phase space, occurring about the saddle point at

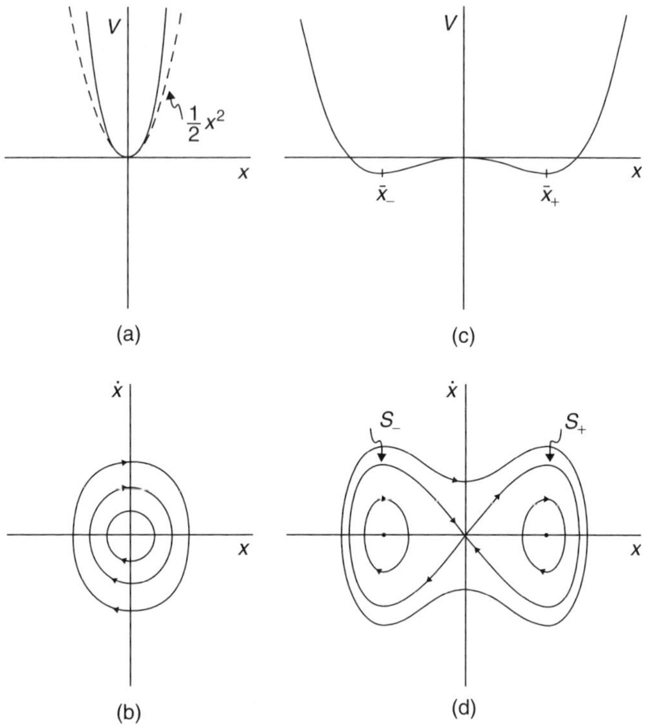

Figure A.6 Plot of the potencial $V(x)$, Equation (A.2.2) and the corresponding portrait in the $(x, \dot{x})$ phase space, respectively, for (a),(b): $\mu < 0$ and (c),(d): $\mu > 0$.

$x = \dot{x} = 0$ and the two centers at $\bar{x}_{\pm}$. Near these points, the dynamics are very well approximated by the linearized equations of motion [cf. Equation (A.1.9) and the discussion that follows]. The only case, in which linear analysis fails us here is at $\mu = 0$, where, after all, the linear term in Equation (A.2.1) vanishes!

Bifurcations are called *local* or *global* depending on whether, at the critical value $\mu = \mu_o$, the dynamics change at a "small scale" or "large scale" part of phase space. Clearly, what we have described above is an example of a local bifurcation.

As an example of a global bifurcation, let us examine what would happen if we included in Equation (A.2.1) a "dissipative" term proportional to the velocity $\dot{x}$:

$$\ddot{x} = \mu x - \mathrm{b}\dot{x} - x^3 \tag{A.2.3}$$

For $\mathrm{b} > 0$ the separatrices $S_{\pm}$ in Figure A.6d split in exactly the same way as in Figure A.2 for Equation (A.1.13). Indeed, for all $\mu > 0$, at $\mathrm{b} = 0$, an important change occurs that has a global effect on the dynamics of the

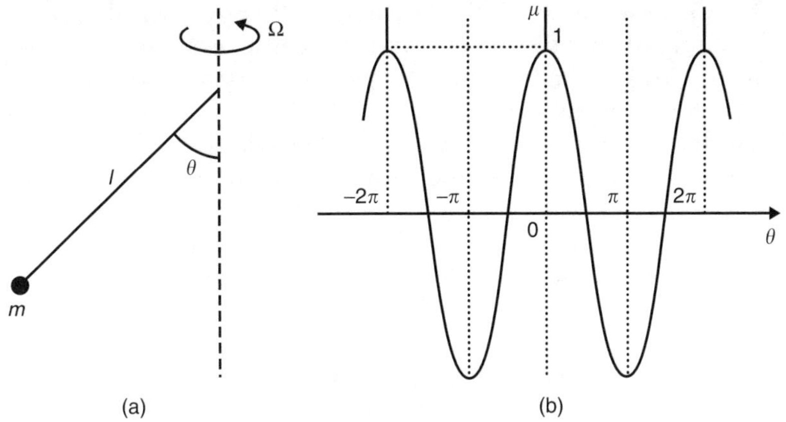

Figure A.7 The pendulum (a) and its bifurcation diagram (b) in the $(\bar{\theta}, \mu)$ plane.

problem: Orbits, evolving in large-scale regions of phase space, may now enter in (or exit from) the vicinity of the two equilibrium points at $\bar{x}_{\pm}$.

Thus, bifurcations are seen to introduce new phenomena and make the dynamics in phase space more complicated. Not much complexity, however, is to be expected from one-dimensional or two-dimensional flows. As we already mentioned in Section A.1, we need three-dimensional flows (or discrete dynamical systems of any dimension) to observe chaotic behavior.

As we will demonstrate in the sequel, chaos is closely related to bifurcations. In fact, by carefully studying different kinds of possible bifurcations we will be able to better investigate and understand a variety of chaotic phenomena that frequently arise in physical systems.

Example 2

Consider a pendulum of mass m, hanging by a weightless cord of length l from a support that can make the plane of oscillations of the pendulum rotate about a vertical axis with angular frequency Ω, as shown in Figure A.7a. Summing all the torques acting on the mass particle leads to the equation of motion

$$\ddot{\theta} + \left(\frac{g}{l} - \Omega^2 \cos\theta\right) \sin\theta = 0$$

Defining the parameter $\mu = g/(l\Omega^2)$ we can plot the positions of the equilibrium points $\bar{\theta}$ of the system in the $(\bar{\theta}, \mu)$ plane, Figure A.7b, obtaining the so-called *bifurcation diagram*.

Clearly, for Ω small (μ large) the pendulum will oscillate about the $\bar{\theta} = 0$ (vertical position), which is stable. However, at $\mu = \mu_o = 1$, a bifurcation occurs such that, for $0 \leq \mu < 1$, the pendulum can no longer oscillate about the vertical position which, though an equilibrium, becomes unstable.

Instead, two new stable equilibria emerge, $\bar{\theta} = \pm\theta_0$, such that $\mu = \cos\theta$ (see Figure A.7b), and the pendulum will oscillate around these angles. In the limiting case of infinite angular frequency, $\mu = 0$, the pendulum rotates in the horizontal plane $\theta_0 = \pi/2$.

Since chaos does not occur in a planar dynamical system like Equation (A.2.3), it is of great interest to study flows that are three (or, higher) dimensional. Moreover, as was explained in Section A.1.2, the chaotic properties of phase space orbits, in continuous time, are generally reflected by the behavior of their intersections with a surface of section, in discrete time.

Thus, it is natural to consider discrete dynamical systems in which we can find complex behavior even in one space dimension. They are the simplest systems exhibiting chaos and have the additional advantage that they are much easier to study computationally since they are described by difference (rather than differential) equations.

In particular, we will discuss here one-dimensional discrete systems (one-dimensional maps) of the form:

$$x_{n+1} = f(x_n), \quad n = 0, 1, 2, \ldots \tag{A.2.4}$$

where f is a piecewise continuous and differentiable function. For simplicity, we shall restrict the iterates of our mapping:

$$x_1 = f(x_o),\ x_2 = f(f(x_o)) = f^2(x_o), \ldots, x_n = f^n(x_o), \ldots \tag{A.2.5}$$

(for all n) within some interval $I \in \mathfrak{R}$, so as to avoid unbounded solutions and will even scale x so that $I = [0, 1]$.

Consider, for example, the quadratic mapping Equation (A.1.18) of Section A.1.1, which, after scaling $x_n \rightarrow ax_n$, becomes

$$x_{n+1} = f(x_n) = ax_n(1 - x_n), \quad n = 0, 1, 2, \ldots \tag{A.2.6}$$

(see also Example 1). Note that $f: [0, 1] \rightarrow [0, 1]$, for $0 \leq a \leq 4$, and let us study the dynamics of Equation (A.2.6) by plotting the function of Equation (A.2.6) in the (x_n, x_{n+1}) plane of Figure A.8, for different values of $a = 0.5, 2, 3.5$.

This can best be visualized geometrically by picking an initial x_o on the x-axis, following the line $y = f(x_o)$ to the point it crosses the diagonal, then up the line $x = f(x_o) = x_1$ to $y = f(x_1)$, back to the diagonal to find x_2, and so on. The final result is the identification of every orbit $\{x_n/n = 0, 1, 2, \ldots\}$ with a path made up of horizontal and vertical segments, joining the function $y = f(x)$ to the diagonal, as shown in Figure A.8. Where do these paths lead, as $n \rightarrow \infty$? Clearly, for $a = 1/2$ and $a = 2$, they lead to the points $\bar{x}_1 = 0$ and $\bar{x}_2 = 1/2$, respectively. These are fixed points of the map, since they satisfy

$$\bar{x}_i = f(\bar{x}_i) \Rightarrow \begin{cases} \bar{x}_1 = 0 \\ \bar{x}_2 = (a - 1)/a \end{cases} \tag{A.2.7}$$

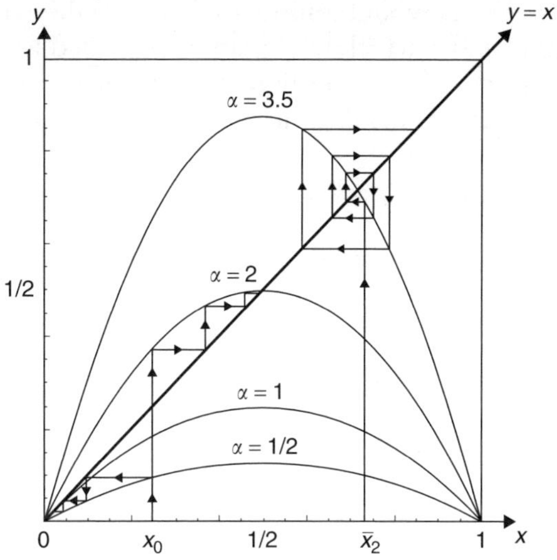

Figure A.8 The function $y = ax(1-x)$ for several values of the parameter *a*, in the unit square.

Small variations of the orbits about them [i.e., setting $x_n = \bar{x}_i + \xi_n$, $|\xi_n| \ll 1$ in Equation (A.2.6)] obey the linear equation

$$\xi_{n+1} = f'(\bar{x}_i)\xi_n, \quad n = 0, 1, 2, \ldots \tag{A.2.8}$$

If $|f'(\bar{x}_i)| < 1$, then $\bar{x}_i$ is a stable or attracting fixed point, while for $|f'(\bar{x}_i)| > 1, \bar{x}_i$ is unstable or repelling (see also the following and Equation (A.1.19) in Section A.1.1). Now we can explain the behavior of the orbits in Figure A.8. Since

$$f'(\bar{x}_1) = a, \quad f'(\bar{x}_2) = 2 - a \tag{A.2.9}$$

$\bar{x}_1 = 0$ is stable for $a = 1/2$ and all orbits will be attracted to it, as $n \to \infty$. However, for $a > 1$, an important change of behavior, or *bifurcation*, occurs near $x = 0$. All orbits are now repelled from $\bar{x}_1$ and, for $a = 2$, say, are attracted by the second fixed point, which in this case is located at $\bar{x}_2 = 1/2$, with $|f'(\bar{x}_2)| = 0 < 1$.

And what about the third case, $a = 3.5$? There, another bifurcation has occurred, near $x = \bar{x}_2$ this time. This second fixed point has also become unstable since, at this value of a, $|f'(\bar{x}_2)| = 1.5 > 1$, cf. Equation (A.2.9). Geometrically, this is manifested by the orbits forming an expanding "rectangular spiral" about $\bar{x}_2$, as shown in Figure A.8.

Are these two bifurcations (occurring at $a = 1$ and $a = 3$) of the same type? Evidently not, since f has no stable fixed point, for $a > 3$, to which orbits can be attracted. So, what happens to the dynamics in this case? Can we develop a theory that would distinguish between these bifurcations and provide conditions under which they may be expected to occur?

In order to answer these questions, we begin by writing our one-dimensional map in the general form

$$x_{n+1} = g(x_n, \mu), \quad n = 0, 1, 2, \ldots \tag{A.2.10}$$

where $\mu \in \mathfrak{R}$ is a parameter of the problem and study the dynamics near fixed points of Equation (A.2.10), $x_n = x$, satisfying

$$f(x, \mu) = g(x, \mu) - x = 0 \tag{A.2.11}$$

and near bifurcation points $\mu = \mu_o$, where

$$\left| \frac{\partial g}{\partial x_n}(x, \mu_o) \right| = 1 \tag{A.2.12}$$

We will find that there are four possible kinds of such bifurcations, depending on the leading terms of the Taylor expansion of g near its fixed points (called the "normal form" of the bifurcation). They are:

(a) The *saddle-node* bifurcation, with "normal form"

$$g(x_n, \mu) = \mu + x_n - x_n^2 \tag{A.2.13a}$$

(b) The *transcritical* bifurcation, with "normal form"

$$g(x_n, \mu) = x_n + \mu x_n - x_n^2 \tag{A.2.13b}$$

(c) The *pitchfork* bifurcation, with "normal form"

$$g(x_n, \mu) = x_n + \mu x_n - x_n^3 \tag{A.2.13c}$$

(d) The *period-doubling* bifurcation, with "normal form"

$$g(x_n, \mu) = -x_n - \mu x_n + x_n^3 \tag{A.2.13d}$$

In the remainder of this section, we will briefly describe these bifurcations and give explicit conditions such that g may be brought into one of these "normal forms" near its fixed points. Then, in Section A.3, we will discuss transitions to chaos, related to bifurcations (*a*) and (*d*) above and a third chaotic transition related to the *bifurcation of invariant curves* out of the origin occurring in two-dimensional maps $g: \mathfrak{R}^2 \rightarrow \mathfrak{R}^2$ called Hopf–Naimark–Sacker bifurcation (see Reference 61).

A.2.2 The Saddle-Node Bifurcation

Consider the mapping

$$x_{n+1} = g(x_n, \mu) = \mu + x_n - x_n^2 \tag{A.2.14}$$

whose equilibrium points are given by the equation

$$f(x, \mu) = g(x, \mu) - x = 0 \tag{A.2.15}$$

This means that

$$\mu = \mu(x) = x^2 \tag{A.2.16}$$

is a curve of fixed points of Equation (A.2.14). The slope of g at these fixed points,

$$\frac{\partial g}{\partial x}(x, \mu) = 1 - 2x \tag{A.2.17}$$

attains the bifurcation value of 1 at $x = 0$ where, due to Equation (A.2.16), we have $\mu = \mu_o = 0$. The type of bifurcation that occurs here, at $\mu_o = 0$, is shown in the bifurcation diagram of Figure A.9a, where we have plotted the curve of Equation (A.2.16) in the (μ, x) plane.

Note that the fixed points of Equation (A.2.16) are *stable* for $x > 0$ since the slope of g at these points is smaller than 1 [see Equation (A.2.17)]. Of course, we only consider x and μ small ($|\mu| \ll 1, |x| \ll 1$) throughout this analysis so that possible higher order terms in a Taylor expansion of g in μ, x_n, about $(0, 0)$ can be neglected in Equation (A.2.14).

Similarly, for $x < 0$, we find that the fixed points are unstable. Thus, we have, for $\mu > 0$, the coexistence of a saddle (unstable) and a nodal (stable) fixed point where none existed for $\mu < 0$. We say, therefore, that a *saddle-node* bifurcation has occurred in this system at $(\mu_o, \bar{x}) = (0, 0)$.

Given a general $g(x_n, \mu)$, with a fixed point at $(0, 0)$, i.e., $g(0, 0) = 0$, where

SN1: $$\frac{\partial g}{\partial x}(0, 0) = 1 \tag{A.2.18}$$

holds, how can we know whether a saddle-node bifurcation does occur at that point? From the above analysis, it is clear that a single curve must pass by that point, i.e., Equation (A.2.15) must have a unique solution $\mu = \mu(x)$, cf. Equation (A.2.16). According to the implicit function theorem (cf. e.g., Reference 61) this happens if and only if

SN2: $$\frac{\partial g}{\partial \mu}(0, 0) \neq 0 \tag{A.2.19}$$

Moreover, this curve must have a maximum (or minimum) at $(0, 0)$ in order to produce a bifurcation diagram, similar to the one depicted in Figure A.9a. This means that $\mu = \mu(x)$ must satisfy

$$\frac{d\mu}{dx}(0) = 0 \quad \text{and} \quad \frac{d\mu^2}{dx^2}(0) \neq 0 \tag{A.2.20}$$

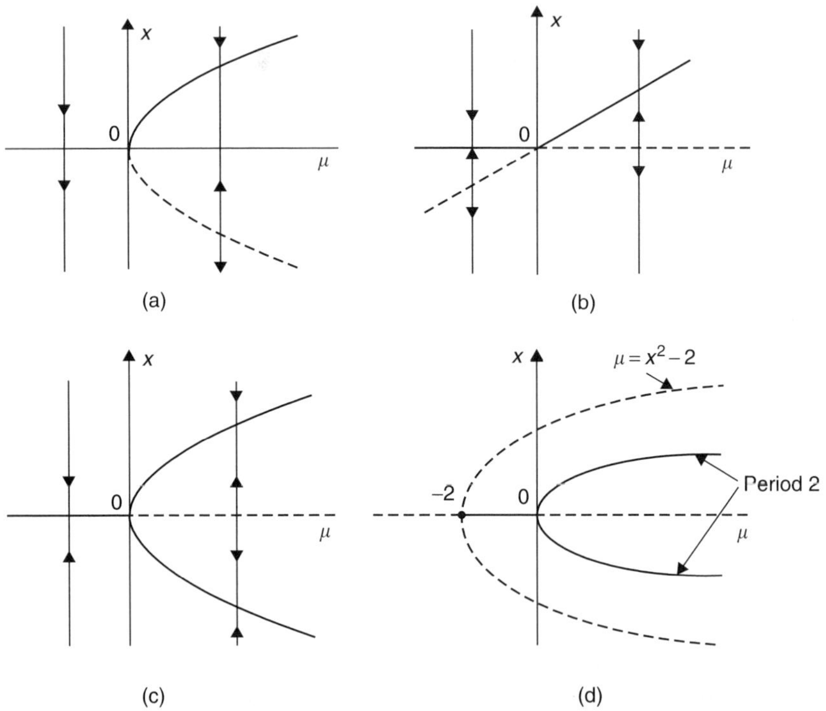

Figure A.9 Bifurcation diagrams for (a) the saddle-node, (b) transcritical, (c) pitchfork, and (d) period-doubling bifurcations of discrete dynamical systems. Solid curves denote stable orbits, while dashed denote unstable ones. Arrows on vertical lines indicate the direction of motion at the corresponding values of μ.

Differentiating Equation (A.2.15) with respect to x and setting $x=0$, i.e.,

$$\frac{\partial g}{\partial x}(0,0) + \frac{\partial g}{\partial \mu}(0,0)\frac{d\mu}{dx}(0) = 1$$

we find that the first condition in Equation (A.2.20) is automatically satisfied by virtue of Equations (A.2.18) and (A.2.19).

The second relation in Equation (A.2.20), however, does lead to a new condition, as can be seen after one more differentiation of Equation (A.2.15), at $x=0$, which gives

$$\frac{\partial^2 g}{\partial x^2}(0,0) + 2\frac{\partial^2 g}{\partial x \partial \mu}(0,0)\frac{d\mu}{dx}(0) + \frac{\partial g}{\partial \mu}(0,0)\frac{d^2\mu}{dx^2}(0) = 0$$

This, together with all the other results obtained so far, leads to the third and final condition for a saddle-node bifurcation:

SN3: $$\frac{\partial^2 g}{\partial x^2}(0,0) \neq 0 \qquad \text{(A.2.21)}$$

A.2.3 The Transcritical Bifurcation

Let us examine now the case Equation (A.2.13b) where μ enters multiplicatively in one of the linear terms:

$$x_{n+1} = g(x_n, \mu) = x_n + \mu x_n - x_n^2 \tag{A.2.22}$$

Here the equation of fixed points of Equation (A.2.15) has two solutions:

$$x = 0, \quad \mu(x) = x \tag{A.2.23}$$

in the (μ, x) plane [see Figure A.9b]. $(0, 0)$ is again a bifurcation point since the condition of Equation (A.2.18) holds here also:

T1: $$\frac{\partial g}{\partial x}(0, 0) = 1 \tag{A.2.24}$$

On the other hand, the fact that Equation (A.2.15) has more than one solution implies

T2: $$\frac{\partial g}{\partial \mu}(0, 0) = 0, \tag{A.2.25}$$

instead of Equation (A.2.19). Moreover, since one of these solutions is $x = 0$, we can write the equation of fixed points of the map in the form

$$g(x, \mu) - x = xh(x, \mu) = 0 \tag{A.2.26}$$

where now $\partial h(0, 0)/\partial \mu \neq 0$ (since $h(x, \mu) = 0$ has only one solution). This means that $\partial g(x, \mu)/\partial \mu \propto x$ as $(x, \mu) \to (0, 0)$, i.e.,

T3: $$\frac{\partial^2 g}{\partial x \partial \mu}(0, 0) \neq 0 \tag{A.2.27}$$

is valid. The final condition which must hold for a transcritical bifurcation, is that the solution $\mu = \mu(x)$ of $h(x, \mu) = 0$ does *not* have a critical point at $x = 0$, i.e., $d\mu(0)/dx \neq 0$. Differentiating Equation (A.2.26) with respect to x twice, we find that this leads to

T4: $$\frac{\partial^2 g}{\partial x^2}(0, 0) \neq 0 \tag{A.2.28}$$

The stability interchange between the curves of fixed points at $(0, 0)$ is easily explained in terms of the value of $|\partial g(x, \mu)/\partial x|$.

In closing, we should point out that we have included the transcritical bifurcation here only for the sake of completeness. As will become evident below, this type of bifurcation is not as important as the other 3, since it hasn't been connected as yet with one of the familiar and frequently occurring transitions to large-scale chaotic behavior.

A.2.4 The Pitchfork Bifurcation

Now we turn to the case where quadratic terms in x are absent near $(0, 0)$ and consider the map

$$x_{n+1} = g_n(x_n, \mu) = x_n + \mu x_n - x_n^3 \tag{A.2.29}$$

whose fixed point equation is also of the form of Equation (A.2.26). It has the solutions

$$x = 0, \quad \mu = \mu(x) = x^2 \tag{A.2.30}$$

drawn in the corresponding bifurcation diagram of Figure A.9c, from which the bifurcation derives its name. It follows, therefore, that the first three conditions for a pitchfork bifurcation P1, P2, and P3, are the same as T1, T2, and T3, cf. Equations (A.2.24), (A.2.25), and (A.2.27).

Unlike the transcritical bifurcation, however, the second curve of equilibrium points of Equation (A.2.30), in this case, satisfies

$$\frac{d\mu}{dx}(0) = 0 \quad \text{and} \quad \frac{d^2\mu}{dx^2}(0) \neq 0 \tag{A.2.31}$$

Differentiating twice the fixed point Equation (A.2.26), we find that Equation (A.2.25) together with the first equation in Equation (A.2.31) yield

P4:
$$\frac{\partial^2 g}{\partial x^2}(0, 0) = 0 \tag{A.2.32}$$

To exploit the second derivative criterion in Equation (A.2.31) we need to differentiate the fixed point equation once more. We thus arrive at the final condition for a pitchfork bifurcation

P5:
$$\frac{\partial^3 g}{\partial x^3}(0, 0) \neq 0 \tag{A.2.33}$$

The determination of the stability of the different branches of the curves in Figure A.9c is again straightforward.

Note that the bifurcations we encountered at the beginning of this chapter in two-dimensional flows [Equation (A.2.1)] at $x = 0$, $\mu = 0$ and the pendulum problem of Example 2, at $\theta = 0, \mu = 1$ are of pitchfork type.

A.2.5 The Period-Doubling Bifurcation

As we pointed out above, the first three types of bifurcations of one-dimensional maps also occur in one-dimensional flows. This is not the case with the fourth type of bifurcation, that of period-doubling, which requires at least a three-dimensional phase-space for its realization in continuous time dynamical systems.

Consider the mapping

$$x_{n+1} = g(x_n, \mu) = -x_n - \mu x_n + x_n^3 \tag{A.2.34}$$

whose fixed point equation can be written in the form

$$g(x, \mu) - x = x[x^2 - (2 + \mu)] \tag{A.2.35}$$

Note that, also in this case, there are two solutions to Equation (A.2.35),

$$x = 0, \quad \mu = \mu(x) = x^2 - 2 \tag{A.2.36}$$

Here, however, the second solution intersects the μ-axis of the (μ, x) bifurcation diagram at $\mu = \mu_o = -2$ (see Figure A.9d). Indeed, it is easy to show that, at this point the μ_o system of Equation (A.2.34) undergoes a pitchfork bifurcation, at which $x = 0$ becomes stable for $\mu > \mu_o$, while the fixed points $\mu = x^2 - 2$, appearing at $\mu = \mu_o$ are all unstable for $\mu > \mu_o$.

Now, observe that the fixed point $x = 0$ destabilizes again at $\mu = 0$, since the derivative

$$\frac{\partial g}{\partial x}(0, \mu) = -1 - \mu$$

becomes larger than one in magnitude for $\mu > 0$. However, no stable fixed point exists in that range to attract the orbits repelled by $x = 0$. This is exactly the situation we encountered in the quadratic map of Equation (A.2.6), with regard to its fixed point $\bar{x}_2$ at $a > 3$. What happens in this case?

To find out, instead of Equation (A.2.34) we need to consider the one-dimensional "doubled" map $x_{n+1} = g^2(x_n, \mu) = g(g(x_n, \mu))$, whose leading terms in a Taylor expansion about $(0, 0)$ are

$$x_{n+1} = g^2(x_n, \mu) = x_n + \mu(2 + \mu)x_n - 2x_n^3 + O(4) \tag{A.2.37}$$

[$O(4)$ = terms of order 4 or higher in x and μ]. Note that, whereas in terms of the original map [Equation (A.2.34)] the origin is characterized by $\partial g(0, 0)/\partial x = -1$, in the case of g^2 we have

$$\frac{\partial g^2}{\partial x}(0, 0) = 1 \tag{A.2.38}$$

Indeed, it can now be easily checked that the function $g^2(x, \mu)$ also satisfies

$$\frac{\partial g^2}{\partial \mu}(0, 0) = 0, \quad \frac{\partial^2 g^2}{\partial x \partial \mu} = 2 \neq 0$$
$$\frac{\partial^2 g^2}{\partial x^2}(0, 0) = 0, \quad \frac{\partial^3 g^2}{\partial x^3}(0, 0) = -12 \neq 0 \tag{A.2.39}$$

which imply that Equation (A.2.37) undergoes a pitchfork bifurcation at $(0, 0)$, according to the results of Section A.2.4 and Figure A.9d.

In this bifurcation, however, the curve of fixed points of Equation (A.2.37) which passes by $(0,0)$, i.e., $\mu=\mu(x)=-1+\sqrt{1+2x^2}$, where $x=g^2(x,\mu)$, is a curve of *period 2* points of $g(x,\mu)$ since $x'=g(x,\mu)$, $x=g(x',\mu)$, $x\neq x'$. Denoting periodic points by a "hat," we thus conclude that, for $\mu>0$, there exist pairs of points $(\hat{x}_1,\hat{x}_2)$ given by

$$\hat{x}_1=\left[\mu+\frac{\mu^2}{2}\right]^{1/2},\qquad \hat{x}_2=-\hat{x}_1 \tag{A.2.40}$$

with

$$\hat{x}_1=g(\hat{x}_2,\mu),\quad \hat{x}_2=g(\hat{x}_1,\mu) \tag{A.2.41}$$

which constitute a one-parameter family of periodic orbits of Equation (A.2.34) of period 2.

A.3 Three Routes to Chaos

Nature, no doubt, has many ways by which it changes from a quiet to a turbulent state: Slow-moving air currents can start to circulate faster and faster and form vortices in the atmosphere, which often develop into devastating hurricanes. Quietly flowing rivers sometimes gather speed through narrow passages and become rough and violent rapids which can defeat even the most experienced boatmen or swimmers! Surely, no scientific theory, method, or approach available today, including chaotic dynamics, can provide a satisfactory description of such marvellously complicated phenomena.

Nonlinear dynamics and the theory of chaos, however, have made significant progress in the past two decades toward improving our understanding of *some* transitions from a regular to a chaotic state, which can be observed in laboratory experiments. These "transitions to turbulence" or "routes to chaos" are extremely simple and occur under highly controlled conditions. Nevertheless, they are *real* and offer the first *detailed* view of how some complexities in nature arise.

Science usually proceeds inductively from the particular to the general, from a specific example to a large class of systems sharing the same property. It attempts to organize knowledge by determining conditions under which all systems belonging to the same category are expected to exhibit *universal* behavior. The next step, of course, is to develop the proper mathematical tools by which each class of systems can be analyzed in a unified way.

This is exactly what has occurred with the following "routes to chaos" identified by chaotic dynamics:

(a) The period-doubling sequence
(b) The intermittency phenomenon
(c) The breakup of quasiperiodic motion.

In this section, we will discuss the conditions under which a system can be expected to exhibit one or more of the above transitions to the chaotic state. It is instructive, however, to start by listing a number of important common features shared by all these transitions.

(i) They are each connected with a specific type of *local bifurcation.*
(ii) They are accurately described by different kinds of *one-dimensional maps.*
(iii) They can all be studied by *renormalization* techniques which reveal their universal scaling structure.

Furthermore, all three routes have been found to occur quite frequently in numerical simulations of a large variety of nonlinear systems and have been observed in many laboratory experiments.

Our purpose here is not to describe these routes to chaos in detail. The interested reader will find excellent accounts of them in the literature, e.g., References 4, 16, 51. Rather, we intend to stress their main properties and discuss their differences and similarities in a way that may be helpful to a student or a starting researcher in this area of nonlinear dynamics.

A.3.1 The Period-Doubling Route

We already mentioned, at the beginning of Section A.2.2, the way by which one of the fixed points of the quadratic mapping of Equation (A.2.6), $\bar{x}_2 = 1 - 1/a$, undergoes a destabilization at $a = 3$. Since, at that point, $f'(\bar{x}_2) = -1$, cf. Equation (A.2.9), we expect that a period-doubling bifurcation occurs at $a = 3$. We could, of course, check whether such a bifurcation takes place, by following the theory of Section A.2.5 and testing whether conditions of Equation (A.2.39) hold for the doubled map

$$x' = f^2(x) = a^2x - a^3x^2 - x^3(a^2 - 2a^3) - a^3x^4 \tag{A.3.1}$$

at the critical point $(x, a) = (2/3, 3)$ (see Example 5).

We prefer, however, to follow a different approach here: Let us plot the function of Equation (A.3.1) in the unit square $[0, 1] \times [0, 1]$ at a value of $a > 3$, say $a_1 = 1 + \sqrt{5} \cong 3.236$. As we see in Figure A.10a, this is a value at which the orbit of period 2 has maximum stability, i.e.,

$$a = a_1\colon \lambda_2 = (f^2)'(\hat{x}_i) = f'(\hat{x}_1)f'(\hat{x}_2) = 0, \quad i = 1, 2, \ldots \tag{A.3.2}$$

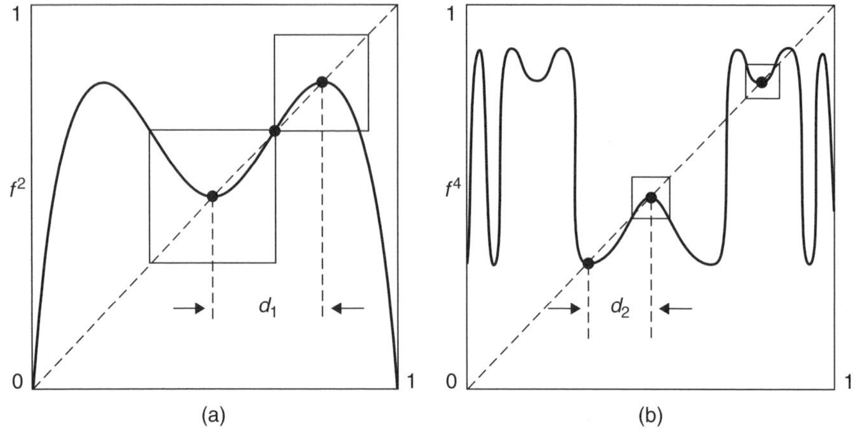

Figure A.10 (a) The second iterate of $f(x) = ax(1-x)$ with the period 2 points $\hat{x}_1$, $\hat{x}_2$. (b) The fourth iterate of $f(x)$, with the period 4 points at maximum stability.

since $f'(\hat{x}_i) = 0$, $\hat{x}_i$ being the points of period 2 (see also the discussion at the beginning of Section A.2.2).

What has happened is that a period 2-cycle has been born at $a = 3$ by a bifurcation of the type described in Section A.2.5 (and Example 4) with stability index $\lambda_2 = 1$, see Equation (A.3.2). As α increases, the "hills" become higher and the "valley" deeper, until $\lambda_2 = 0$ at $a_1 = 1 + \sqrt{5}$. Then as a continues to grow, we reach the value $a = 1 + \sqrt{6}$ where $\lambda_2 = -1$ and the 2-cycle also undergoes a period-doubling bifurcation.

To see this, we shall examine again the doubled-map of $f^2(x)$, i.e., $x' = f^4(x)$. Plotting this 16th-degree polynomial in the unit square we observe that every local maximum (minimum) of f^2 is replaced by one local minimum (maximum) and two maxima (minima) (see Figure A.10b). In fact, in Figure A.10b we have also raised a slightly to $a = a_2 \geq 1 + \sqrt{6}$, where the 4-cycle is superstable, i.e.,

$$a = a_2\text{: } \lambda_4 = (f^4)'(\hat{x}_i) = \prod_{i=1}^{4} f'(\hat{x}_i) = 0, \quad i = 1, 2, 3, 4 \tag{A.3.3}$$

Clearly, in this way, it appears that at some $a \geq a_2$, $\lambda_4 = -1$, a new period-doubling bifurcation will occur to a period 8-cycle and so on.

At the same time, as each period 2^{m+1} cycle is born, a new unstable orbit (of period 2^m) is added to the "repellers" of the system, which, of course, preserve their instability as the hills and valleys of $f^{2^m}(x)$ continue to grow higher (respectively deeper) with increasing a. On the other hand, even though there is always an "attractor," i.e., a stable period 2^{m+1}-cycle, as $m \to \infty$, this orbit becomes so long and complicated that it is indistinguishable from a "chaotic" trajectory over finitely long times.

But chaos can only be defined over infinitely long times! How can we find out what happens in the limit $m \to \infty$? This is the question we shall return to in the next section when we discuss Feigenbaum's application of renormalization methods to the chaotic transition via period-doubling bifurcations.

At this point, let us make a crucial observation concerning Figure A.10 and Figure A.12. Note that the unit square of Figure A.8 at $a=2$ is also present at the center of Figure A.10a in an "upside-down" and reduced scale version! Moreover, one bifurcation later, it is still there, at the center of Figure A.10b, turned again upside-down and further reduced in scale. A similar transformation has taken place with the square about the $\hat{x}_2$ point of Figure A.10a, at the next bifurcation, shown in Figure A.10b.

This is, in fact, the first indication of *self-similarity under scaling* characterizing period-doubling bifurcations, which we are going to describe extensively in Section A.4.1. For the time being, it is instructive to proceed numerically, increasing a by small steps and letting the map itself $x' = f(x) = ax(1-x)$ (after discarding the first few transient iterates) "draw" its own attractor on an (a, x) bifurcation diagram for $1 \leq a \leq 4$.

The result is a highly intricate and fascinating "tree" structure, a poor representation of which is shown here in Figure A.11. In fact, no pictorial representation of this "tree" can ever do justice to all the wonders, which take place at infinitely many points inside its thick "foliage"!

Now let us go back to Figure A.10 and observe that if we denote by d_m the distance from the point at the central extremum to its nearest point on a (superstable) period 2^m orbit, then d_m provides an estimate of the size of the square about the central extremum, shown in Figures A.12a and b. Furthermore, let us denote by a_m the value of a at which the 2^m-cycle is superstable.

It was Feigenbaum[21–23] who first observed that these quantities obey scaling laws

$$d_m \propto \alpha^m, \quad a_\infty - a_m \propto \delta^{-m}, \quad m \to \infty \tag{A.3.4}$$

whose scaling constants

$$\alpha = 2.502907875\ldots, \quad \delta = 4.6692016091\ldots \tag{A.3.5}$$

are *universal* for all maps $x' = f(x)$ of the unit interval $I = [0, 1]$ (i.e., $f: I \to I$), with a single quadratic maximum. For more mathematical details on this topic the reader is encouraged to look at Ref. 14–17, while for period-doubling in 2-dimensional, conservative maps, Ref. 8 and 30 are recommended.

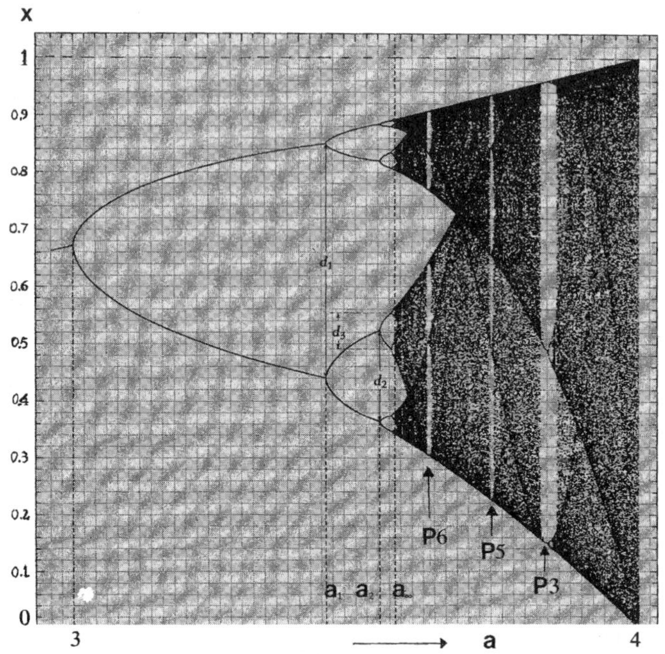

Figure A.11 The bifurcation diagram of Equation (A.2.6) showing the sequence of period-doubling bifurcations limiting on a_∞. Windows of periodic orbits of period 3, 5, and 6 are also visible (marked by P3, P5 and P6, respectively).

A.3.2 Onset of Chaos via Intermittency

One observes sometimes that a physical system exhibits, over a certain range of parameter values, r say, the following behavior: One (or more) of its variables, oscillating periodically in a regular and predictable way, suddenly start to display brief "chaotic bursts" of irregular oscillations, intermittent between long intervals of simple quasiperiodic motion.

Moreover, these chaotic bursts are seen to appear when r "crosses" a certain critical value r_c and become more and more frequent as r is varied further and further away from r_c. This situation is depicted in Figure A.12, as observed in a Rayleigh–Bénard experiment, and is called the *intermittency* route to chaos.[7]

In fact, intermittency can also be observed in the solutions of low-dimensional dynamical systems, described by $m(\geq 3)$ first-order ODEs, as e.g., the Lorenz Equation (A.1.21), discussed in Section A.1.2. The critical parameter value in that example is $r_c \cong 166.0$ (σ, b having the usual values, $\sigma = 10, b = 8/3$, for which a strange attractor is observed at $r \gtrsim 25$ (see Reference 58).

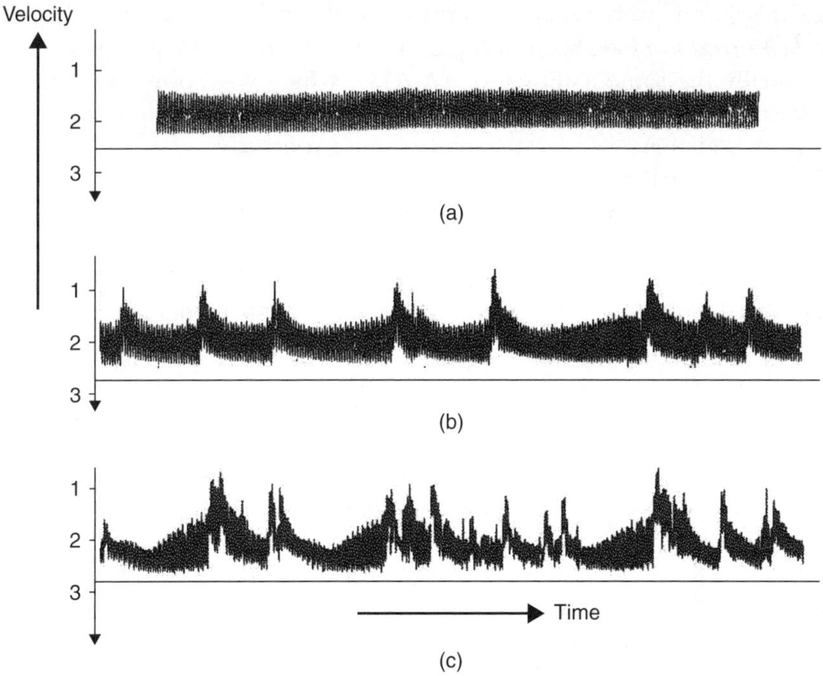

Figure A.12 Intermittency in a Rayleigh–Bénard experiment, showing: (a) the laminar phase, for $r \lesssim r_c$. (b) $r \gtrsim r_c$: chaotic bursts appearing intermittently between laminar intervals, and (c) chaotic bursts become a lot more frequent, when the parameter r is further increased from r_c. (After P. Bergé et al., *J. Phys. (Paris) Lett.*, 41, L344, 1980.)

If we now consider one of the Lorenz variables, say $y(t)$, at intersection points $y_n = y(t_n), n = 0, 1, 2, \ldots,$ with $z = \text{const.}$, and plot successive intersections of an orbit with the y_n, y_{n+1} plane, we observe the following phenomenon:[50] The points (y_n, y_{n+1}) appear to lie on a curve which for $r < r_c$ intersects the diagonal at two points, at $r = r_c$ is tangent to the diagonal, while for $r > r_c$ it has no points in common with the $y_n = y_{n+1}$ line. Pomeau and Manneville modeled this phenomenon by the one-dimensional map

$$y_{n+1} = f(y_n) = \epsilon + y_n + a y_n^2 \tag{A.3.6}$$

$\epsilon = r - r_c$ ($\epsilon \leq 0, \epsilon = 0$ and $\epsilon > 0$), as shown in Figure A.13. Note that for $\epsilon < 0$, $f(y_n)$ intersects the diagonal line at two points labeled N and S in Figure A.13a. They are clearly fixed points of the map, one of them being stable (the "node" N) and the other unstable (the "saddle" S).

These two points coalesce into a single one at $\epsilon = 0$ and cease to exist altogether for $\epsilon > 0$ (see Figures A.13b and c). The reader should recall that this is precisely what we termed, in the previous section, a *saddle-node bifurcation* (see Section A.2.2). In fact, the map in Equation (A.3.6) is essentially the same as Equation (A.2.14), which was shown to constitute a "normal form" for such bifurcations in one-dimensional systems.

How is all this related to the intermittency route to chaos? The $r < r_c$ case of periodic oscillations corresponds to $\epsilon < 0$, where all orbits are attracted to the periodic orbit crossing the y_n, y_{n+1} plane at N. Of course at $\epsilon > 0$, this orbit no longer exists. However, if ϵ is small enough, orbits can be trapped for a long time between $f(y_n)$ and the $y_n = y_{n+1}$ line before escaping from this "corridor" by one of the chaotic bursts mentioned above (see Figure A.13c).

Clearly, as long as the orbits stay within that "corridor," the system exhibits quasiperiodic oscillations similar to the periodic ones occurring for $\epsilon < 0$. Then, after exiting from the right side, orbits return (due to dissipative effects) back to this region, reenter the corridor at another point, and exhibit a new "laminar phase" whose duration depends, of course, on the point of reentry.

What do we do now that we have a map, Equation (A.3.6), which appears to give a good *qualitative* picture of the dynamics? Can we use it to get some quantitative information about the system? Let us note that since, near the critical state ($\epsilon \geq 0$), the y_{n+1} and y_n values are very close to each other and the time between them is very short compared with the duration of a laminar interval, we may replace Equation (A.3.6) by a first-order ODE

$$\frac{dy}{dl} = \epsilon + ay^2 \qquad \text{(A.3.7)}$$

where l measures time along one of these intervals.

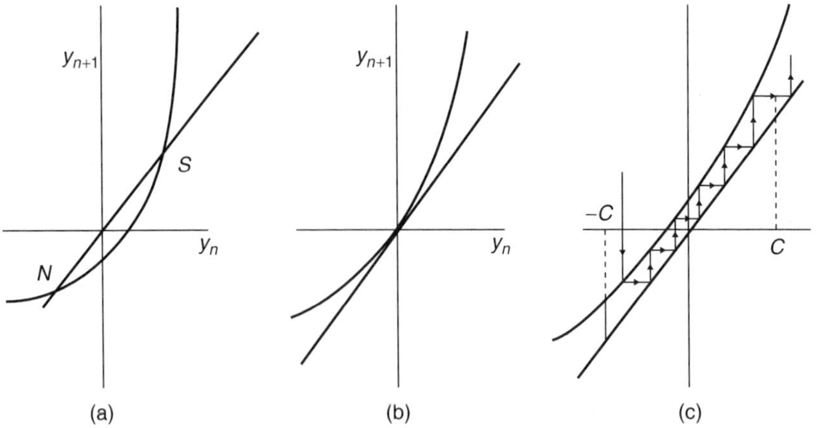

Figure A.13 Plot of the function $y_{n+1} = f(y_n)$, see Equation (A.3.6). (a) $\epsilon < 0$, (b) $\epsilon = 0$, and (c) $\epsilon \gtrsim 0$.

Now let us put some reasonable bounds on the extent of the laminar region (say from $y=-c$ to $y=c$, see Figure A.13c) and assume that the probability of an orbit entering at any point $y_{in} \in [-c, c]$ is an *even* function of y_{in}, i.e.,

$$P(y_{in}) = P(-y_{in}) \qquad \text{(A.3.8)}$$

We may thus calculate the average length (or time duration) $\langle l \rangle$ of these laminar intervals by first integrating Equation (A.3.6) from $y=y_{in}$ to $y=c$, to obtain

$$l(y_{in}) = \int_{y_{in}}^{c} \frac{dy}{\epsilon + ay^2} = \frac{1}{\sqrt{a\epsilon}} \left\{ \tan^{-1}\left[\frac{c}{\sqrt{\epsilon/a}}\right] - \tan^{-1}\left[\frac{y_{in}}{\sqrt{\epsilon/a}}\right]\right\} \qquad \text{(A.3.9)}$$

and then evaluate

$$\langle l \rangle = \int_{-c}^{c} dy_{in}\, P(y_{in})\, l(y_{in}) = \frac{1}{\sqrt{a\epsilon}} \tan^{-1}\left[\frac{c}{\sqrt{\epsilon/a}}\right] \qquad \text{(A.3.10)}$$

using Equation (A.3.9), $\int_{-c}^{c} P(y_{in}) dy_{in} = 1$, and the fact that the second integral vanishes since its integral is an odd function.

Now, for finite a, Equation (A.3.10) implies

$$\langle l \rangle \propto \epsilon^{-1/2}, \quad \epsilon \to 0 \qquad \text{(A.3.11)}$$

which is a physical quantity that can be measured in the laboratory. Of course, for a given system we don't know the proportionality constant in Equation (A.3.11), but we can verify whether the exponent is $-1/2$, in the following way: Measure the average length of the laminar intervals at two values of the parameter $\epsilon_1 = r_1 - r_c \geq 0$ and $\epsilon_2 = r_2 - r_c \geq 0$, $\langle l \rangle_1$ and $\langle l \rangle_2$, respectively, and check whether the relation

$$\langle l \rangle_1 / \langle l \rangle_2 = (\epsilon_2/\epsilon_1)^{1/2}$$

is satisfied within experimental error. If this is true, then the phenomenon you are studying is very likely intermittency of the above type and, therefore, a transition to chaos is imminent.

A.3.3 The Breakup of Quasiperiodic Motion

We finally turn to a third route to chaos, which was studied extensively in the 1980s analytically as well as numerically and experimentally: the "breakup" of quasiperiodic orbits (or "tori") of low-dimensional dynamical systems. In keeping with the other two chaotic transitions outlined in this section, this one is also connected with a specific type of bifurcation

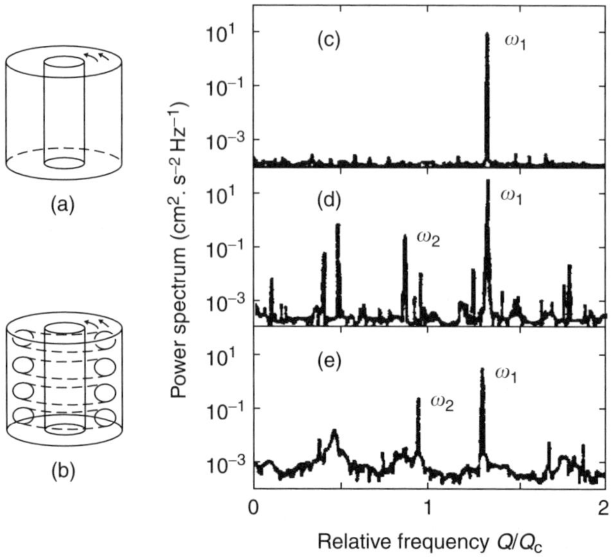

Figure A.14 Onset of chaos via the Ruelle–Takens scenario in a Couette experiment of a rotating fluid, (a) and (b). Note how the appearance of one frequence ω_1 in (c) is followed by a second one, ω_2, in (d) and then by a continuous power spectrum in (c), as the relative speed between the cylinders is increased. (After H. Swinncy and J.P. Gollub, *Physics Today* 31, 41, 1978.)

(the Hopf bifurcation) and has likewise been treated by renormalization techniques.

The main idea is the following: Suppose you are observing a physical system perform oscillatory motion. If it is a fluid, you may be measuring its temperature or velocity at some point and find that it is periodic in time with frequency ω_1. When Fourier analyzed, the power spectrum of your measurements should look a lot like the one shown in Figure A.14c.

Now, as you vary a parameter of the system, μ, you may notice that, for $\mu > \mu_o$, the motion turns from periodic to *quasiperiodic*, as a second frequency appears (see Figure A.14d), which in general, is not related to ω_1 by a simple rational dependence. In phase space, this would mean that the orbit now lies on a *torus*, which intersects a two-dimensional surface of section along a *smooth* closed curve [see Figure A.15a (whereas, before, when there was only ω_1, it intersected this surface at a single point)].

A.4 Renormalization and Universality

In Section A.1.3 we spoke about one of the cornerstones of chaotic dynamics: the fundamental concept of *self-similarity under scaling*. We briefly

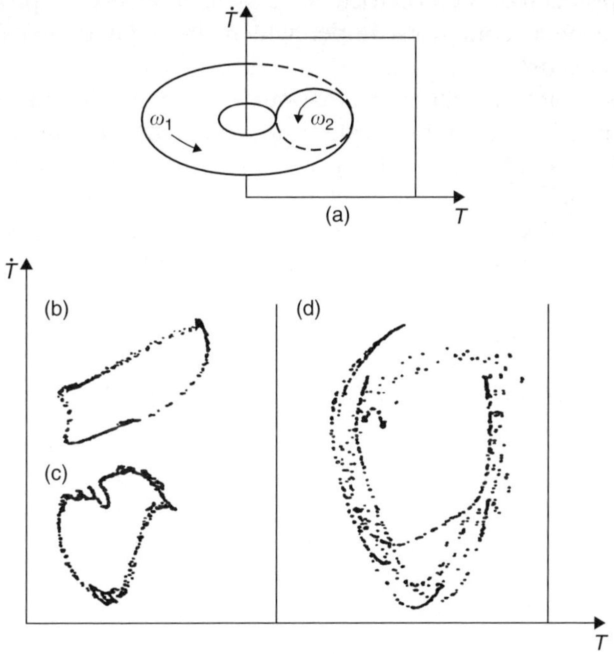

Figure A.15 Breakup of quasiperiodic motion on a $T, \dot{T}$ surface of section of a Rayleigh–Bénard experiment, where T is the temperature and $\dot{T} = dT/dt$. The invariant curve of the intersection points of the orbit with this surface in (a) develops folds in (b), (c), and eventually breaks up in (d). (After M. Dubois et al., *J. Phys. (Paris) Lett.*, 43, L295, 1982.)

discussed it there in connection with some very complicated boundaries between coexisting attractors of simple dynamical systems. Among them, one finds examples of objects, widely known as *fractals*, which will be discussed in more detail in Section A.5 of this Appendix.

Thus, in Section A.3.1, we came across the phenomenon of self-similarity again in connection with the transition to chaos via period-doubling bifurcations. In particular, we noted, with regard to Figure A.8 and Figure A.10, that there are certain geometric features in these pictures (e.g., the inset central square in Figure A.10a) which are apparently reproduced, at smaller scales, after every bifurcation. Furthermore, according to numerical results, this scaling appears to become *exact*, as the number of bifurcations $m \to \infty$.

We may, therefore, ask: Are there analogous scaling phenomena associated with the other two routes to chaos, i.e., intermittency and the breakup of quasiperiodic motion?

And in the event that such self-similar phenomena and scaling laws do exist, how can they be studied mathematically? Is there perhaps a common approach or method to analyze their "universality" properties and determine general conditions under which these phenomena would be expected to occur?

In this section we will show that such a method is indeed available and is inspired by a technique called *renormalization*, which has proved very useful in several areas of physics and in particular in the theory of phase transitions in statistical mechanics.[48] We will start by describing its application to the sequence of period-doubling bifurcations in Section A.4.1 and then proceed to outline its main steps in the other two routes to chaos, in Sections A.4.2 and A.4.3.

A.4.1 Renormalization of the Period-Doubling Route

As we discussed already in Section A.3.1, in connection with the period-doubling bifurcations of the quadratic map of Equation (A.2.6), the necessary conditions for the realization of this route to chaos are: the unimodality of the map (i.e., the presence of only one quadratic maximum) and the negativity of the Schwartzian derivative.[15,17,55]

Let us, therefore, consider the one-dimensional quadratic mapping

$$x' = r(1 - 2x^2) = rf(x) = g(r, x) \qquad \text{(A.4.1)}$$

which has a (single quadratic) maximum and is also (for simplicity) taken to be an even function of x.

Let us also assume that the first few period-doubling bifurcations have already taken place (see e.g., Figure A.10) after which we have plotted, at some appropriate magnification, the function

$$g^{2^{n-1}}(r_n, x) = \underbrace{g(g(\ldots g(r_n, x)\ldots))}_{2^{n-1}} \qquad \text{(A.4.2)}$$

near $x = 0$ in Figure A.16a. The parameter r is at the value r_n, at which the periodic orbit of period 2^n [some of whose points are denoted by A, B, C, D in Figures A.16a and b] is *superstable* [recall the definition of superstability in Equation (A.3.2)]. If for the same parameter value we also plot the doubled function

$$g^{2^n}(x) = g^{2^{n-1}} \cdot g^{2^{n-1}}(x) \qquad \text{(A.4.3)}$$

in Figure A.16b we will observe that, as in Figure A.10 every extremum is replaced, upon doubling, by one of the opposite kind together with two extrema of the same kind, one on either side. Note that the A, B, C, D points of the 2^n-periodic orbits are at the extrema of g^{2^n}, as required by superstability.

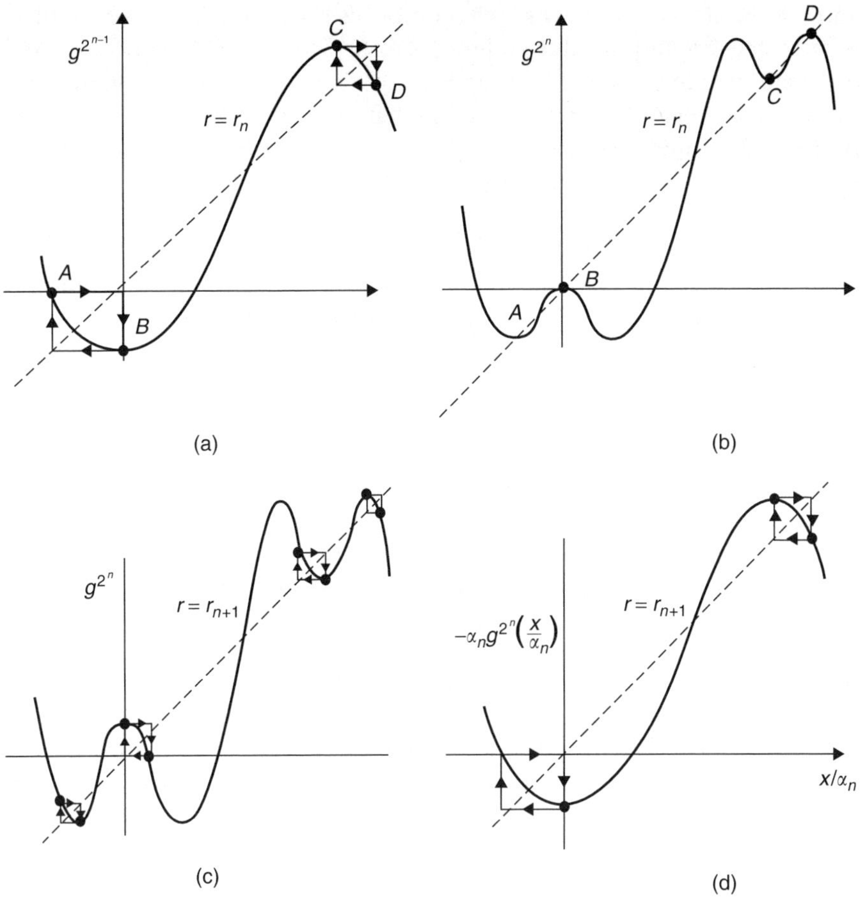

Figure A.16 The function g^{2^n} scaled appropriately at the $r = r_n$ and $r = r_{n+1}$— the parameter values of superstability of the periodic orbits of period 2^n, in (a), (b), and period-2^{n+1} in (c) and (d). Note the similarity between (a) and (d).

Now increase r_n to r_{n+1}: The 2^n-periodic points A, B, C, D, have turned unstable and the orbits are attracted to points of period-2^{n+1}, which are superstable (see Figure A.16c). In fact, this resemblance between Figure A.16a and Figure A.16c can be made more evident by scaling the x-axis by a constant $-1/\alpha_n$ (<0) and the y-axis by $-\alpha_n$, cf. Figures A.16a and d. This is self-similarity under scaling in all its glory!

It was Feigenbaum's remarkable discovery that the scaling constant, α_n, which reveals this similarity, approaches a specific value as n is increased,

$$\lim_{n \to \infty} \alpha_n = \alpha = 2.502907875\ldots \tag{A.4.4}$$

and—more importantly—that this number is *universal* in the sense that it is the same for all period-doublings of *any function* $g(r, x)$, near a single quadratic maximum (provided its Schwartzian derivative is negative[15,17]). In fact, what Feigenbaum showed was that in the limit $n \to \infty$ of the above process, the function $\alpha^n g^{2^n}(r_\infty, x/\alpha^n)$ tends to a universal function $g(x)$, satisfying the functional equation

$$g(x) = -\alpha g(g(x/\alpha)), \quad g(0) = 1, \quad g'(0) = 0 \tag{A.4.5}$$

independent of the paramenter $r(r,\ r_\infty,$ where the chaotic transition occurs). Finally, if we consider the rate of convergence of $r_n \to r_\infty$, as n increases, a second *universal* constant is found:

$$\lim_{n\to\infty} \frac{r_n - r_{n+1}}{r_{n+1} - r_{n+2}} = \delta = 4.6692016091\ldots$$

cf. Equations (A.3.4) and (A.3.5). Of course, the numerical values of r_∞, here (or a_∞, in the case of the logistic map $f(x) = ax(1-x)$, see Section A.3.1) at which the transition to chaos occurs, have nothing "universal" about them. They are particular to each individual system.

Earlier, in Section A.3.1, we mentioned the experimental results of Libchaber and Maurer, who were the first to demonstrate that an actual, physical system (a thin layer of fluid heated from below) would become chaotic following the period-doubling route[39,40] (see Fig. A.17). They were able to measure the magnitudes of the odd Fourier amplitude modes, $A_{2k+1}^{(n)}$, of a fluid velocity component of period 2^n, at bifurcation to 2^{n+1}-periodicity, and demonstrate that, when *averaged* over k, they scale by the approximate relation

$$\left|A_{2k+1}^{(n+1)}\right|_{\text{av}} = \mu^{-1}\left|A_{2k+1}^{(n)}\right|_{\text{av}}, \quad \mu = 0.152$$

Remarkably enough, this number μ can also be derived by Feigenbaum's universality theory (References 21 and 23), and is found to be related to the universal number α by the simple formula

$$\mu^{-1} \sim \frac{1}{4\alpha}\sqrt{2\left(1 + \frac{1}{\alpha^2}\right)} = 0.1525\ldots$$

This was only the beginning of an "avalanche" of examples of physical, chemical and biological processes that appeared in the scientific literature, exhibiting transition to chaos via period-doubling bifurcations, as predicted by the above theory. "Chaos" as a new scientific discipline had come of age and chaotic dynamics was rapidly becoming a subject in its own right. And in this development, renormalization and universality had undoubtedly played a principal role in successfully unraveling some of the self-similar aspects of nature.

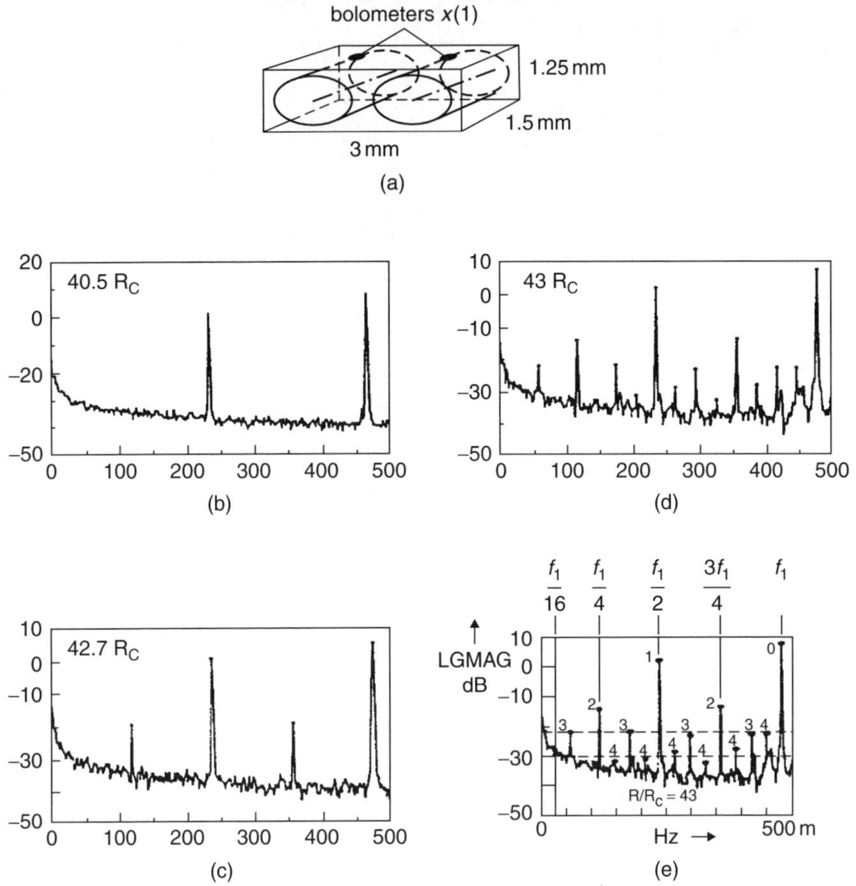

Figure A.17 Period-doubling bifurcations in a Rayleigh–Bénard experiment (a). Note how the periodic motion in (b) has bifurcated to one of half the frequency, in (c), which in turn has led in (d) to an orbit of one quarter of the original frequency. (After: Ref. 40 and 55.)

A.4.2 Renormalization of the Intermittency Route to Chaos

Following the success of the renormalization method in the period-doubling route to chaos, it was natural for researchers to look for a similar approach to explain the other two chaotic transitions via intermittency and torus breakup. These attempts were also successful and helped to further establish the importance of self-similarity and scaling in nonlinear phenomena.

Let us start with intermittency: The first question one wishes to ask, of course, is this: Where is the evidence of self-similarity here? How can one use scaling to "renormalize" the dynamics, as was done, for example,

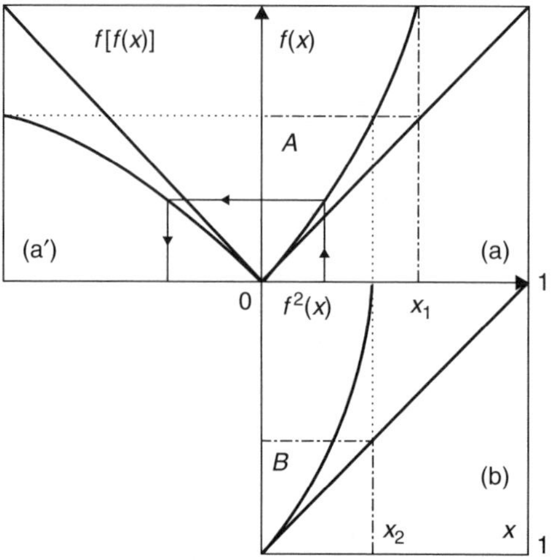

Figure A.18 Schematic plot of the function $f(x)$ (a) in Equation (A.4.6) and (b) the function $f^2(x)$. Note the similarity between the boxes A and B, in (a) and (b), respectively.

in the case of period-doubling bifurcations, in Figure A.16? After all, there is only one bifurcation here (of the saddle-node type), at which the transition to chaos occurs.

It was Hu and Rudnick[35] who first observed that a certain self-similarity under function-doubling (or, composition) occurs here, with regard to the intermittency map:

$$x' = f(x) = x + a|x|^z, \quad z > 1 \tag{A.4.6}$$

cf. Equation (A.3.6), at criticality, i.e., at $\epsilon = 0$. To see how this self-similarity arises, let us plot in Figure A.18a the function of Equation (A.4.6) for $a > 0$, in the unit square $[0, 1] \times [0, 1]$. Note that using the point x_1 at which $f(x_1) = 1$ we can define a square A (with dashed sides), as in Figure A.18a.

Now, consider the point x_2 at which $f^2(x_2) = 1$ and observe that it plays the same role for square A as x_1 does for the original unit square. Thus, it may be used to define a second square B (with dashed sides), within Figure A.18b, which is similar to square A of Figure A.18a, upon scaling by a positive number α_1, say.

Repeating this doubling process n times, we define a renormalization operator

$$\mathcal{T} f^{2^n}(x) = \alpha_n f^{2^n}(f^{2^n}(x/\alpha_n)) \sim g^*(x), \quad n >> 1 \tag{A.4.7}$$

whose action on Equation (A.4.6) is numerically seen to have a fixed point

$$\mathcal{T} g^*(x) = g^*(x) = \alpha g^*(g^*(x/\alpha)), \quad g^*(0) = 0, \quad g^{*'}(0) = 1 \tag{A.4.8}$$

with $\alpha_n \to \alpha > 0$, as $n \to \infty$, a constant whose value depends on z. This constant, α, may be thought of as "universal" in the sense that it is the same for all $f(x)$ (see the analogous functional equation (A.4.5) obtained by Feigenbaum in the period-doubling scenario, discussed earlier).

A.4.3 Renormalization of the Breakup of a Golden Torus

J. Greene was the first to recognize that the breakup of invariant tori of the area-preserving, standard map

$$r' = r - \frac{K}{2\pi} \sin 2\pi\theta$$
$$\theta' = \theta + r' \tag{A.4.9}$$

could be studied by following the stability properties of nearby periodic orbits, as a parameter of the problem K is varied.[29] As is well-known, according to the KAM theorem (see also Reference 41) "most" invariant tori of an area-preserving map (which are straight lines $r = \text{const.}$, at $K = 0$) are preserved for $K \neq 0$ as smooth, continuous curves in the (r, θ) plane, extending from $r = 0$ to $r = 1$ in the unit square phase space $[0, 1] \times [0, 1]$, see Figures A.19a and b.

These invariant curves are characterized by an irrational rotation number w which should not be "too close" to a rational in the sense that its continued fraction approximation, e.g., w_n, cf. Equation (A.3.35), converges "very slowly" to w. Thus, a natural choice for such an irrational is the *golden mean*, w^*, whose continued fraction approximants, $w_n = F_n/F_{n+1}$, converge to w^* at the slowest possible rate, i.e., with all $m_i = 1$.

Even these tori, however, do not persist for all $K > 0$. In fact, there is a critical $K_c(w)$ at which an invariant curve (of rotation number w) is expected to lose smoothness and "break up," thus allowing orbits to "leak" through its holes [for $K > K_c(w)$] and wander chaotically over large-scale regions of the (r, θ) phase plane, (see Figure A.19c). This is one reason why one often refers to torus "breakup" as a transition to *large-scale chaos* in 2 degrees of freedom Hamiltonian systems, frequently modelled by area-preserving maps.

Greene postulated (and then went on to show numerically) that the golden mean torus, with $w = w^* = (\sqrt{5} - 1)/2$ is the *last one* to be destroyed as K is increased. He studied the way by which periodic orbits with $w_n = F_n/F_{n+1}, n >> 1$, destabilize, altering the scaling properties of distances between nearby points in their vicinity. Thus, he was able to obtain a very accurate estimate of the threshold for the chaotic transition of the golden mean curve, $K_c(w^*) = 0.971635\ldots$.

Greene's results were later developed into a complete renormalization treatment of the breakup of the golden mean torus by MacKay.[44,45] A finite

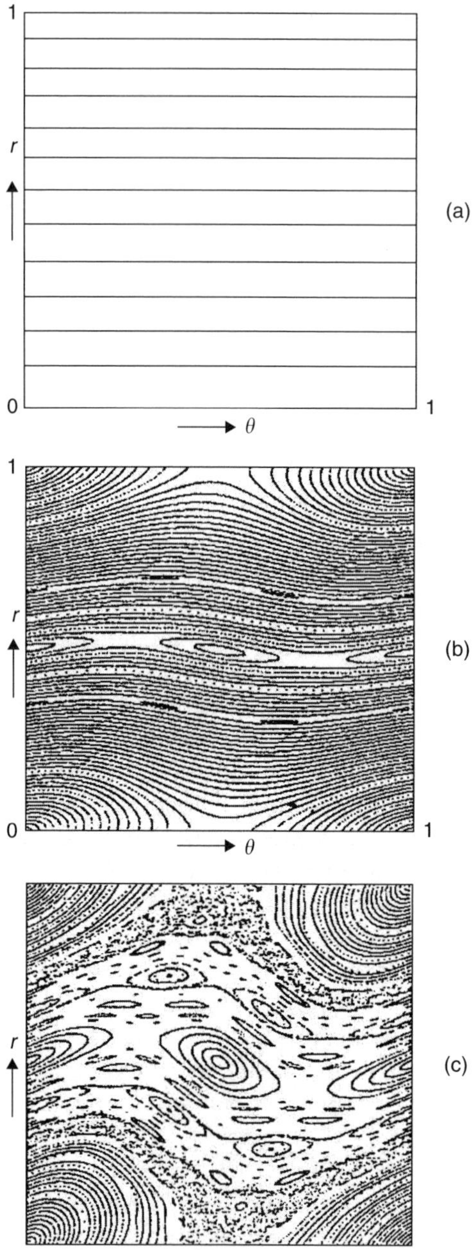

Figure A.19 Plot of the iterates of the area-preserving map of Equation (A.4.49) for many initial conditions $\theta_o = 0$ and $0 < r_o < 1$ at (a) $K = 0$, (b) $K = 0.32$, and (c) $K = 0.9 < K_c(w^*) = 0.971635$. Note that in (c), the golden mean quasiperiodic orbit and its symmetric one (above and below the three islands in the middle of the figure) still exist as smooth curves extending from $\theta = 0$ to $\theta = 1$, across the (r, θ) unit square.

number of unstable directions were identified and related to universal scaling constants of the problem. At the same time, a somewhat different renormalization approach to the breakup of KAM tori was developed by Escande and Doveil (see References 19 and 52). Thus, this type of chaotic transition, at least for the important class of the so-called noble winding numbers (whose continued fraction representation ends with $m_i = m$, $\forall i > k$) and for 2 degrees of freedom Hamiltonian systems, is by now well understood (see also References 41, 44).

However, the situation in higher-dimensional systems is noticeably less clear. In recent years, there have been several attempts to study torus breakup in Hamiltonian systems with more than 2 degrees of freedom. We mention here, e.g., the results of Bollt and Meiss[6] on the application of Greene's criterion to four-dimensional symplectic maps and Laskar's frequency analysis method.[38]

Until today, our understanding of the transition from quasiperiodic to chaotic motion in higher-dimensional conservative systems is far from complete. A lot remains to be done, especially with regard to the dynamics of orbits in the vicinity of "broken" tori, near the "boundary" between regular and chaotic regions. This is where the very important problem of slow (Arnol'd) diffusion arises regarding the motion of orbits through thin chaotic layers.[10,41]

In my view, Arnol'd diffusion will remain for a long time an interesting topic of research, by virtue of its direct relevance to a number of problems of physical interest like beam stability in particle accelerators,[3] plasma confinement,[44] and the stability of the solar system.[38]

A.5 The Theory of Fractals

From the contents of this Appendix, so far, the reader must have become familiar with the fact that most nonlinear phenomena found in nature often display an "erratic," "irregular," or "unpredictable" evolution in time which we call *chaotic behavior.* In particular, in our description of the various *transition* to chaos, we introduced concepts such as "self-similarity under scaling" and "universal" scaling constants (as well as functions) in much the same way as we do in problems of critical phenomena of statistical mechanics.

So, one would be justified in asking if it is possible to study, in a similar way, the *chaotic morphology* of nonlinear systems in *space*. In other words, it would be interesting to investigate the properties of objects (or sets) which are characterized by an extremely complicated "structure," down to the smallest scales at which they can be observed.

Consider, for example, a map of the coast of Norway. If we wanted to measure its length we would realize, after taking into consideration

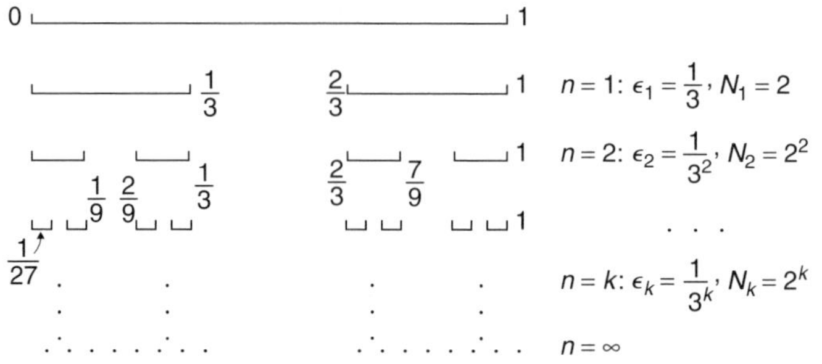

Figure A.20 The Cantor ternary is the set of points left in [0, 1], at the $n \to \infty$ limit of the above process of elimination of the middle third of each line segment at every step $n = 1, 2, 3, \ldots$

smaller and smaller "fjords," that its length tends to infinity! (Really, why is it that we keep discovering new seaside locations in Greece for our summer vacations?)

Take a look at a simple sea sponge, or the dendritic multibranched network of blood vessels of the human kidney. Is it by chance that nature chooses such complicated objects to perform its functions? Or is it perhaps that the cleaning of blood in the kidney or the absorption of nutrients by a sponge are better achieved through their mazy geometrical structure?

In this part of the Appendix, we will examine carefully this kind of objects (sets of points, line segments, surfaces, etc.) which display a complexity of the type described above and are called *fractal*, according to the terminology of Mandelbrot.[52] In Section A.5.1, we shall study the role of self-similarity and scaling in the construction of these sets and introduce the concept of fractal dimension, which will be compared with the mathematically more rigorous Hausdorff dimension.

Then, in Section A.5.2, we shall consider singular probability distributions, often encountered in physical, chemical, or biological processes called *multifractals*. We introduce the concept of generalized dimensions, often used for a more detailed analytical description of the statistical properties of these distributions. In particular, we concentrate our discussion on the correlation dimension D_2 and describe its application to physical systems for which only time series of measurements of one single variable is available.

A.5.1 Self-Similarity and Fractal Dimension

In Sections A.3 and A.4, we came across the concept of self-similarity under scaling in connection with "routes to chaos" occurring via different types of bifurcations. Here, we shall make extensive use of this concept starting with the construction of the Cantor ternary set, which we exhibit in Figure A.20.

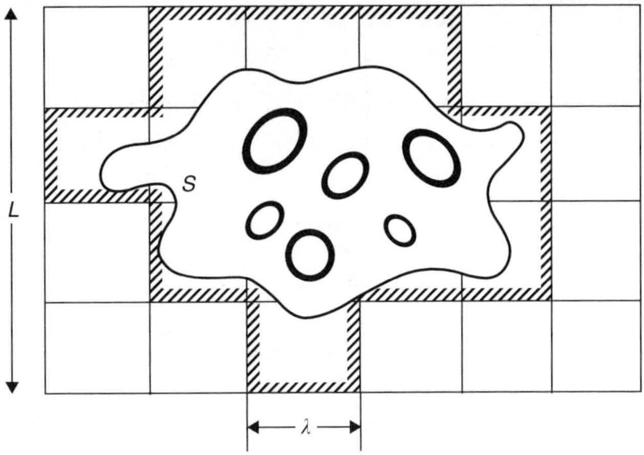

Figure A.21 A complicated figure *S*, embedded in two dimensions and its "covering" of a number of square "boxes" of side λ.

Observe that, at each step of this procedure, $n \to n+1$, there exist parts of the sets of 2^n and 2^{n+1} line segments, which are exactly similar to each other by a scale of $r = 1/3$.

Let us assume that an object S of physical interest is fully embedded in two dimensions, see, e.g., the shaded area in Figure A.21.

If L is a first estimate of the "size" of S, we divide the square of side L that contains S into a grid of smaller square boxes of side $\lambda < L$ and define $N(L, \lambda)$ as the minimum number of boxes of side λ that covers S completely. For example, in Figure A.21, these boxes are contained in the dashed frame of the object and $N(L, \lambda) = 13$.

Of course, when λ is large the covering described above cannot fully take into account all the "holes" in the interior of the object or all the "cavities" on its boundary of size less than λ. Thus, if we want to describe S more accurately, we must choose

$$\lambda \to 0 \quad \text{and} \quad N(\lambda, L) \to \infty$$

so that the δ-measure of the object,

$$M_\delta = N(L, \lambda)\lambda^\delta, \quad \delta > 0 \tag{A.5.1}$$

is *finite*.

What is this exponent δ? Clearly, if the object is two-dimensional after a (finite) number of magnifications, it ceases to display more detailed structure and yields $N \sim (L/\lambda)^2$. So, in this case $\delta = d = 2$, i.e., it is the dimension of the space in which the object has been initially embedded. But let us assume, in general, that the value of δ (that ensures $M_\delta < \infty$) is not known and is computed by the following procedure:

$$M_\delta \xrightarrow[\substack{\lambda \to 0 \\ N \to \infty}]{} \begin{cases} 0 & \text{for } \delta > D_0 \\ 0 < c < \infty & \text{for } \delta = D_0 \\ \infty & \text{for } \delta < D_0 \end{cases} \tag{A.5.2}$$

In this way, we determine the important exponent

$$D_0 = \inf_\delta \{\delta\colon M_\delta = 0\} = \sup_\delta \{\delta\colon M_\delta = \infty\} \tag{A.5.3}$$

which is called the *fractal dimension* of S. Let us denote now by d (=2 in the above example) the *embedding dimension*, i.e., the minimum integer dimension of a Euclidean space, which contains the object entirely. Finally, introducing dimensionless units, we define the variable

$$\epsilon = \frac{\lambda}{L}$$

whence, with $N(L, \lambda) = N(\epsilon)$, we obtain the important formula

$$N(\epsilon) \propto \epsilon^{-D_0} \quad \text{or} \quad N(\epsilon) = c\epsilon^{-D_0}, \quad \epsilon \to 0 \tag{A.5.4}$$

through which one usually defines the fractal dimension D_0, where $c > 0$ is an unknown constant that depends on the specific problem.

Example 1

The Ternary Cantor Set.

Consider the ternary Cantor set shown in Figure A.20. Since at the nth step of the procedure we may "cover" the set with a (minimum) number of "boxes" $N_n = 2^n$ of length $\epsilon_n = 3^{-n}$, we get from Equation (A.5.4):

$$2^n \propto 3^{nD_0} \quad \text{or} \quad 2^n = c3^{nD_0}, \quad n \to \infty$$

Taking logarithms on both sides of this relation and dividing by n, we find in the limit $n \to \infty$

$$D_0 = \frac{\log 2}{\log 3} = 0.63092975\ldots \tag{A.5.5}$$

What is the meaning of this number? Simply that the ternary Cantor set of Figure A.20, embedded in a space of dimension $d = 1$ (the interval [0, 1]) does not consist of isolated points! From this one cannot conclude, of course, that the total "length" (Lebesque measure) of the set is positive. In fact, in the case at hand, this "length" is exactly zero since

$$l = \lim_{n \to \infty} N_n \epsilon_n = \lim_{n \to \infty} \left(\frac{2}{3}\right)^n = 0$$

As we shall see in more detail in Section A.4, sets like the one described in Figure A.20 are called *Cantor sets*, in general, and are characterized by a number of amazing properties: They are bounded, closed, *perfect* (*each of their points* is a limit point of the set), and although they have, in general, zero measure, they consist of an uncountable number of points.

Example 2

Koch's Curve.

Remove the middle third of a line segment of length 1 and replace it by two segments of length 1/3 forming an equilateral triangle (as shown in Figure A.22). Repeat this procedure for each of these four segments by adding segments of length 1/9, 1/27, and so on.

In this way, at the *n*th step of the construction one observes that the resulting "curve" is entirely covered by $N^n = 4^n$ segments of length $\epsilon_n = 1/3^n$. Thus, from Equation (A.5.5), just like in Example 1, the fractal dimension of Koch's curve is easily computed (taking logarithms) to be

$$D_0 = \frac{\log 4}{\log 3} = 1.2618595\ldots \tag{A.5.6}$$

in the limit $n \to \infty$.

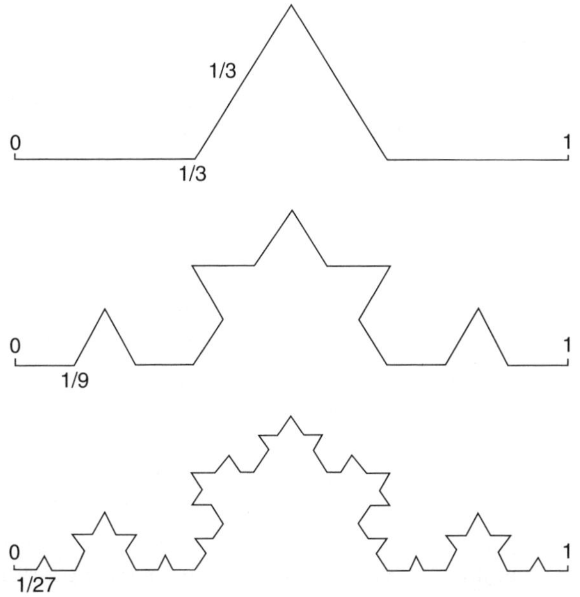

Figure A.22 Koch's curve.

It is obvious from this result that the above procedure does not describe a simple curve. Since $1 < D_0 < 2$, the resulting crooked "*line*" is "something more" than a one-dimensional curve and occupies "something less" than a portion of the (two-dimensional) plane. Observe that even though it is constrained in a finite area of the plane it turns out to have *infinite length*:

$$l = \lim_{n\to\infty} N_n \epsilon_n = \lim_{n\to\infty} \left(\frac{4}{3}\right)^n = \infty$$

Another important property of the above continuous "curve" is that it is composed of an infinite number of angles (where it is not possible to define the tangent) and thus, since no straight line segments are left in the end, it is nowhere differentiable! Remember that the Brownian motion, i.e., the motion of gas molecules observed under the microscope, has a similar form, since the time between succesive molecule collisions is extremely short.

The sets we have described above, characterized by a dimension $D_0 \leq d$ and self-similarity under scaling, are called *fractals*, a word first coined by Mandelbrot.[46] More precisely, if $D_0 < d$, they are called thin fractals, while if $D_0 = d$, they are called fat fractals. As we shall see below, however, the dimension D_0 sometimes offers an inadequate characterization of a fractal. Whenever possible it is better to use the more rigorous concept of Hausdorff dimension D_H.

Example 3

Consider Sierpinski's "gasket" constructed by the procedure shown in Figure A.23. We leave it as an exercise for the reader to compute the fractal dimension D_0 of the resulting set, as $n \to \infty$, and the total area occupied by the Sierpinski gasket in the plane ($D_0 = \log 3/\log 2$, area $= 0$).

As we saw in the examples of the Cantor ternary set, Koch's curve, and Sierpinski's gasket, one way to construct fractals is to begin with an initial object of "length" (or "diameter") $L = 1$, divide it into M pieces similar to the initial one but smaller by a scale r (<1), then divide each piece into M smaller parts by the same scale r, and so on (see Figure A.24).

Suppose now that at the nth step of this procedure we cover the resulting set with $N(\epsilon)$ boxes of "diameter" ϵ, while we denote by $N_1(\epsilon)$ the number of boxes that cover (at this step and with the same ϵ) *one* of the M pieces, in which the set has been divided. Clearly, the following holds:

$$N(\epsilon) = MN_1(\epsilon) \tag{A.5.7}$$

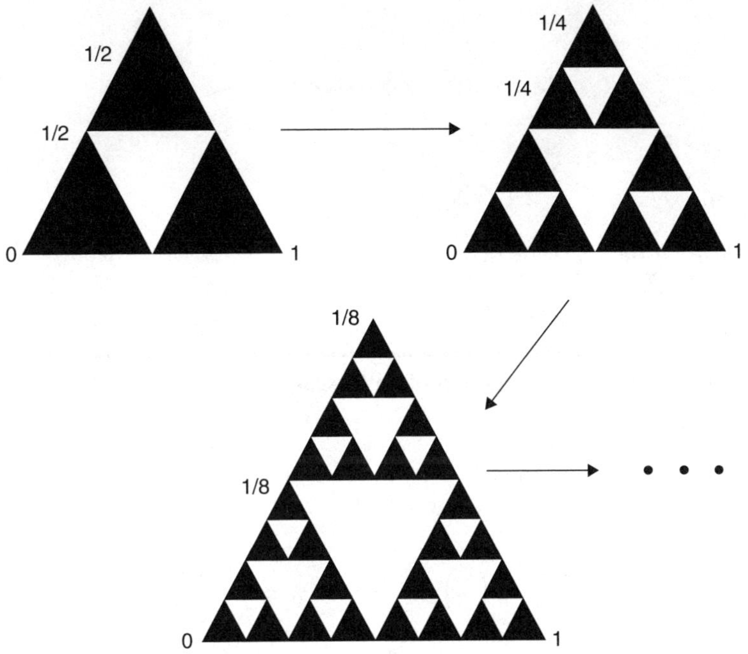

Figure A.23 Sierpinski's gasket.

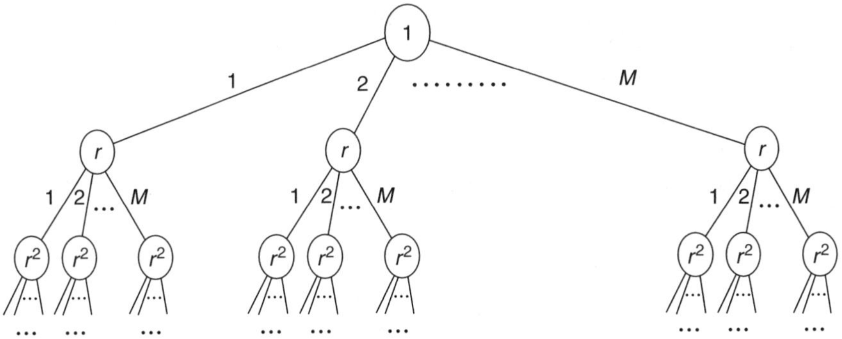

Figure A.24 Construction of a simple fractal of one scale r.

At this point, we make crucial use of the concept of self-similarity under scaling as follows: these $N_1(\epsilon)$ boxes of the nth iteration are none other than the $N(\epsilon')$ boxes of the *previous iteration*, but with diameter $\epsilon' = \epsilon/r$, namely

$$N_1(\epsilon) = N\left(\frac{\epsilon}{r}\right) \tag{A.5.8}$$

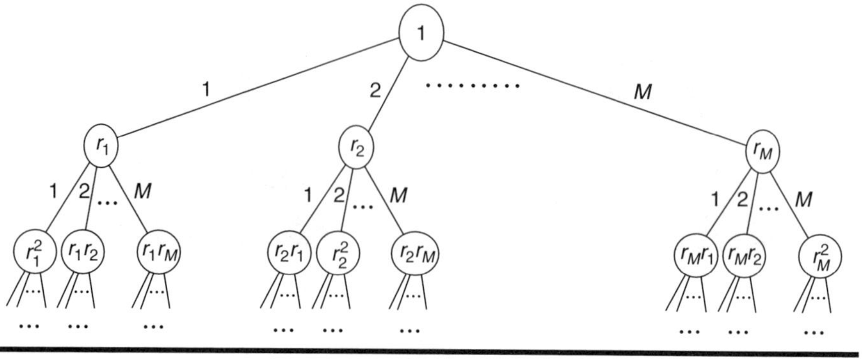

Figure A.25 Iterative construction of an *M*-scale fractal.

Substituting now Equations (A.5.8) in (A.5.7) and using the fractal dimension formula [Equation (A.5.4)] we find

$$\epsilon^{-D_0} = M\left(\frac{\epsilon}{r}\right)^{-D_0} \Longrightarrow D_0 = \frac{\log M}{\log(1/r)} \tag{A.5.9}$$

Formula of Equation (A.5.9) holds generally for every fractal constructed by a deterministic iterative procedure with *one scale*. We can apply it directly to fractals such as those discussed earlier, without having to repeat every time the same mathematical operations (e.g., for the Cantor ternary set, $M = 2$ and $r = 1/3$, while for Koch's curve, $M = 4$ and $r = 1/3$).

But what do we do in the case of multiscale fractals? Let us take, for example, a fractal of M scales, like the one constructed by the procedure shown in Figure A.25.

Let us denote here also by $N_i(\epsilon)$ the minimum number of boxes (of "size" ϵ) required to cover the ith part of the set, $i = 1, 2, 3, \ldots, M$. We thus have

$$M(\epsilon) = \sum_{i=1}^{M} N_i(\epsilon) = \sum_{i=1}^{M} N\left(\frac{\epsilon}{r_i}\right) \tag{A.5.10}$$

[see Equations (A.5.7) and (A.5.8).] Using Equation (A.5.4) again we find (for $\epsilon \to 0$)

$$\epsilon^{-D_0} = \sum_{i=1}^{M}\left(\frac{\epsilon}{r_i}\right)^{-D_0} \Longrightarrow \sum_{i=1}^{M} r_i^{D_0} = 1 \tag{A.5.11}$$

It is obvious that D_0 can no longer be computed analytically, in general, as in the case of one-scale fractals. It is necessary, therefore, to turn to numerical methods for the solution of our problem.

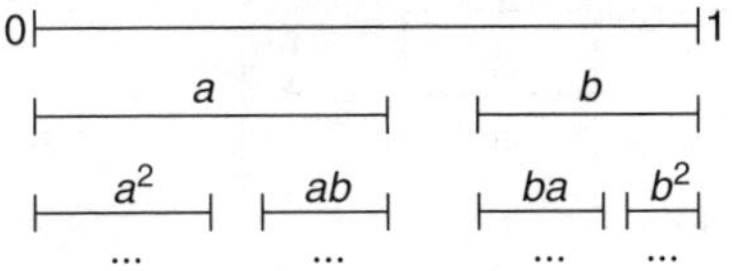

Figure A.26 Cantor ternary set of two scales, *a* > *b*.

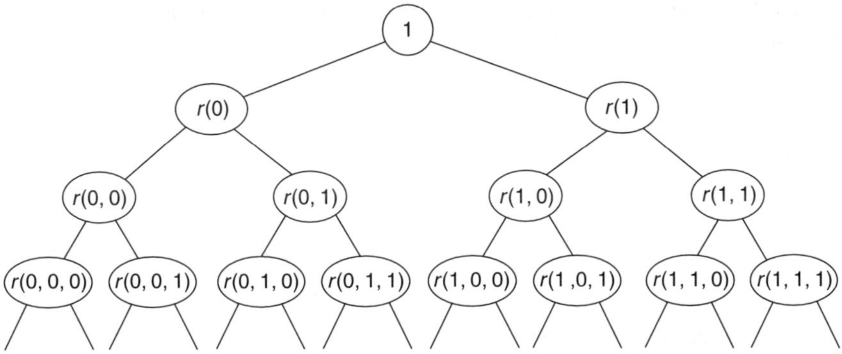

Figure A.27 Construction of a fractal in which different scales appear at each step.

For example, let us assume that we want to find the fractal dimension of a Cantor ternary set of 2 scales, $r_1 = a$, $r_2 = b$ (with $a + b \leq 1$) like the one of Figure A.26.

Using formula (A.5.11), we immediately obtain

$$a^{D_0} + b^{D_0} = 1 \tag{A.5.12}$$

It is not necessary, of course, in every construction of a fractal by successive divisions to have scales already known at the first step. In general, one may wish to introduce new scales at each step. Let us consider, for example, such a procedure shown in Figure A.27 with $M = 2$.

One way to describe the emergence of new scales is to define, at the nth step, the ratio of these scales,

$$\sigma\,(i_1, i_2, \ldots, i_n) = \frac{r\,(i_1, i_2, \ldots, i_n)}{r\left(i_1, i_2, \ldots, i_{n-1}\right)}, \quad i_j = 0, 1 \tag{A.5.13}$$

with respect to those of the previous step. Converting now the above binary sequences to decimal numbers, $0 \leq t < 1$ (in a one-to-one fashion), we

(i_1, i_2)	t	$\sigma_2(t)$
(0, 0)	0	a
(0, 1)	1/4	b
(1, 0)	1/2	a
(1, 1)	3/4	b

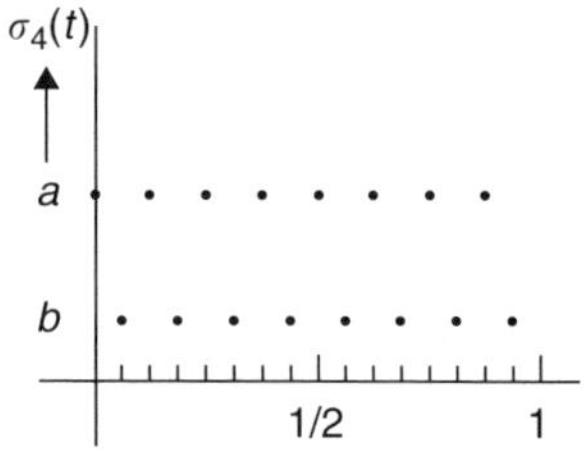

Figure A.28 The table of the values of $\sigma_2(t)$ and the scaling function $\sigma(t) \approx \sigma_4(t)$ of the ternary set of Figure A.26.

introduce the *scaling function*

$$\sigma_n(t) = \sigma\,(i_1, i_2, \ldots, i_n), \quad t = \sum_{j=1}^{n} \frac{i_j}{2^j} \tag{A.5.14}$$

which we wish to study, of course, in the limit of this procedure

$$\sigma(t) = \lim_{n \to \infty} \sigma_n(t) \tag{A.5.15}$$

This function, $\sigma(t)$, is very interesting; it provides us with a graphical representation of the distribution of scales of the fractal, as two new branches appear out of each branch of the previous step. If some scale dominates during the construction, this will become clear by the behavior of $\sigma(t)$ in $0 \le t < 1$.

Let us take, for example, the Cantor ternary set of 2 scales (see Figure A.26). At the first step we have $r(0) = a$, $r(1) = b$, and the scaling function of Equation (A.5.14) takes the value $\sigma_1(0) = a$, $\sigma_1(1/2) = b$. At the second step we get four values for $\sigma_2(t)$, as shown in the table of Figure A.28.

It is evident that, as n increases, more points $\sigma_n(t) = a$ or b are added to the figure and $\sigma_n(t)$ approaches a scaling function $\sigma(t)$ which is everywhere discontinuous and takes only two values a and b:

$$\sigma_n(t) = \begin{cases} a & \text{for } t = \frac{k}{2^{n-1}} \\ b & \text{for } t = \frac{2k+1}{2^n} \end{cases} \qquad k = 0, 1, 2, \ldots, 2^{n-1} - 1 \tag{A.5.16}$$

See Figure A.29.

Let us now turn to a more interesting example of a scaling function, related to our familiar transition to chaos through period doubling. In particular, let us consider the mapping

$$x_{n+1} = 1 - \lambda x_n{}^2, \quad x_n \in [0, 1] \tag{A.5.17}$$

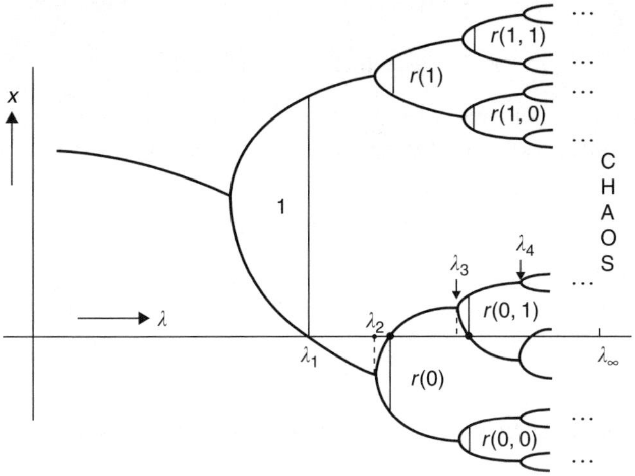

Figure A.29 Schematic diagram of period doubling bifurcations of the mapping Equation (A.5.17).

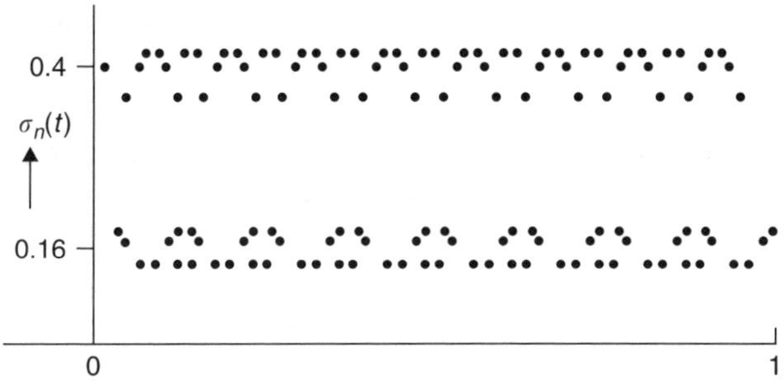

Figure A.30 Approximation of the scaling function $\sigma(t)$ of period doubling by $\sigma_7(t)$.

and draw the "tree" of period doubling orbits of Equation (A.5.17) schematically in Figure A.29. Note in the x-direction the line segments $[r(0), r(1)]$, $[r(0,0), r(0,1), r(1,0), r(1,1)]$, etc., at the points $\lambda = \lambda_2, \lambda_3, \ldots, \lambda_m$, where the orbit of period 2^m has maximum stability (the latter is not essential, as we could also have taken intervals at the bifurcation points).

Computing now numerically the scaling function $\sigma_n(t)$, cf. Equations (A.5.13) and (A.5.14), for the above intervals, we get a set of points that, as $n \to \infty$, seems to tend to an *everywhere discontinuous* function $\sigma(t)$, which we draw approximately [using $\sigma_n(t)$ for $n = 7$] in Figure A.30.

As the first approximation, we see from Figure A.30 that the chaotic attractor of Feigenbaum (with $\lambda = \lambda_\infty$ in Figure A.29) is characterized by two scales:

$$r_1 \cong 0.4 \quad \text{and} \quad r_2 \cong r_1^2 \cong 0.16 \tag{A.5.18}$$

Using the ideas developed above, one could attempt to find an approximate estimate of $\overline{D}_0$ of the fractal dimension of this set (for $\lambda = \lambda_\infty$) using

$$r_1^{\overline{D}_0} + r_2^{\overline{D}_0} = r_1^{\overline{D}_0} + r_1^{2\overline{D}_0} = 1$$

See Equations (A.5.11) and (A.5.12). This gives $\overline{D}_0 = 0.525 \ldots$ which differs by only 2.5% from the exact value of the dimension of the Feigenbaum attractor $D_0 = 0.5388 \ldots$.[27]

A.5.2 Multifractals and Generalized Dimensions

As we saw in the previous section, fractals are static "figures," embedded in a space $\mathfrak{R}^d$, whose parts can be distinguished from each other by their "densities" in $\mathfrak{R}^d$. In the study of physical systems very often we are interested in the *distribution* of a certain quantity (like mass, electric charge, and energy) over a point set in $\mathfrak{R}^d$. The set of these points (which may or may not be a fractal) is called the *support* of the distribution.

Let us consider, as an example, the iterative procedure by which the unit interval [0, 1] is divided in three equal parts, each of which is then divided in three smaller equal parts and so on. At each step, the probability assigned to the right and the left part is p_1 times the previous probability, while, for the middle part, the corresponding probability is $p_2 = 1 - 2p_1$, with $p_1 < p_2$. Drawing the first three steps in the construction of this probability distribution (also called the Farmer–Cantor distribution), we get the picture of Figure A.31.

Observe that as the number of steps n increases, the "boxes" of length $\epsilon_n = (1/3)^n$ in which the support of the distribution is divided are characterized by the probabilities

$$P_n^{(m)} = p_1^m p_2^{n-m}, \quad m = 0, 1, 2, \ldots, n \tag{A.5.19}$$

By inspection of Figure A.31 and using Equation (A.5.19), it can be seen that the number of "boxes" at the nth step of the construction of the Farmer–Cantor distribution with probability Equation (A.5.19) is

$$N_m^{(n)} = \binom{n}{m} 2^m = \frac{n!}{m!\,(n-m)!} 2^m \tag{A.5.20}$$

Taking the logarithm of $N_m^{(n)} P_m^{(n)}$, using Stirling's formula and differentiating with respect to m, we find the $m = m_1$ value at which

$$N_{m_1}^{(n)} P_{m_1}^{(n)} = \text{maximum}, \quad \text{for } n \to \infty \tag{A.5.21}$$

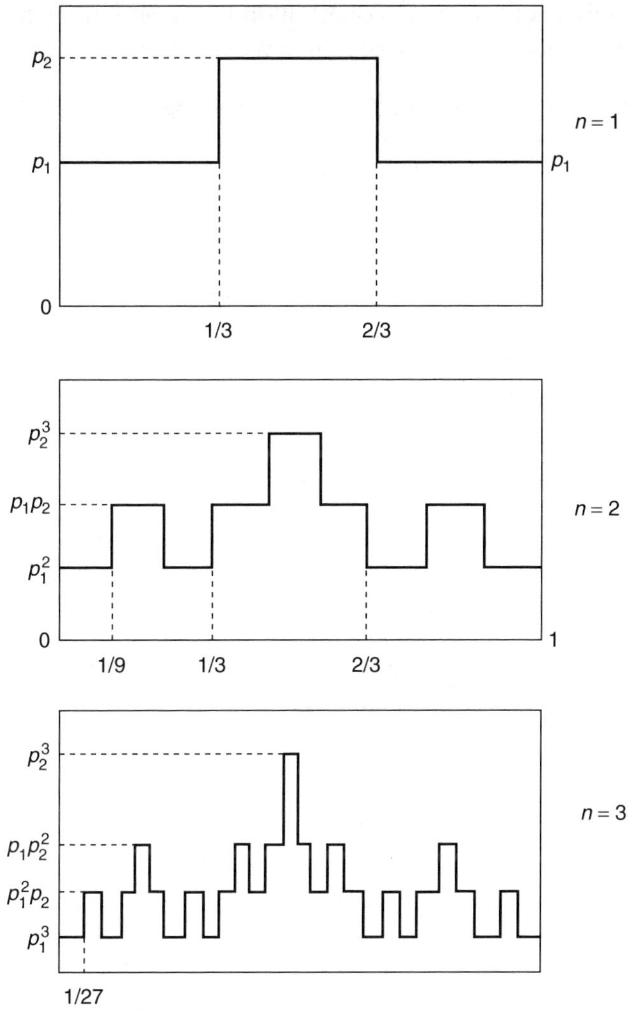

Figure A.31 The first three steps of the construction of the Farmer–Cantor distribution.

to be $m_1 = 2p_1 n$. This gives the largest contribution from these "boxes" as $\epsilon_n \to 0$.

Obviously, from the definition of the total probability of the distribution at each step we have

$$\sum_{m=0}^{n} N_m^{(n)} P_m^{(n)} = 1 \tag{A.5.22}$$

As $n \to \infty$, however, the main contribution to the sum [Equation (A.5.22)] comes from the above m_1 boxes, hence we have

$$N_{m_1}^{(n)} P_{m_1}^{(n)} \sim 1, \quad \text{as } n \to \infty \tag{A.5.23}$$

See Equation (A.5.21), with $m_1 = 2p_1 n$.

Assume now that the points of these m_1 "boxes" constitute a set of (fractal) dimension f_1, i.e.,

$$N_{m_1}^{(n)} \propto \epsilon_n^{-f_1}, \quad n \to \infty \tag{A.5.24}$$

See Equation (A.5.4). Substituting now from Equation (A.5.20) with $\epsilon_n = (1/3)^n$ in Equation (A.5.24) and taking logarithms as before, we find

$$f_1 = \frac{2p_1 \log p_1 + p_2 \log p_2}{\log (1/3)} \leq 1 \tag{A.5.25}$$

So the main contribution to the above probability distribution comes from a fractal set of dimension f_1 [see Equation (A. 5.25)]. If we now draw f_1 as a function of p_1, we find a curve that goes through $(0, 0)$ and has a maximum $f_1 = 1$ for $p_1 = 1/3$ (what happens as $p_1 \to 1/2$ and $p_2 \to 0$?).

Suppose now that we want to study the contribution of other subsets with probability different from the maximum. This can be achieved by defining the quantity

$$X_q^{(n)} \equiv \sum_{m=0}^{n} N_m^{(n)} \left[P_m^{(n)} \right]^q, \quad -\infty < q < \infty \tag{A.5.26}$$

where the *selection parameter* q plays the following role: for large $q > 0$ the main contribution to Equation (A.5.26) comes again from "boxes" of large probability. But for $q < 0$ (and $|q| \gg 1$) it is the "boxes" of *lower probability* that contribute to the sum of Equation (A.5.26).

Distributions like the one we have studied above, which are characterized by Equation (A.5.26) where the main contribution comes from subsets with different fractal dimensions, are called *multifractals*. As is clear from Figure A.31, such distributions display evident "structure within structure" over successive magnifications and self-similarity under scaling.

Let us consider now a dynamical procedure that creates iteratively a sequence of points $x_n \in \Re$, $n = 0, 1, 2, \ldots$, which form a bounded set S. Furthermore, let us cover this set with a number of "boxes" $M(\epsilon)$ of side ϵ, and let us denote by N_i the number of points x_n which fall within the ith box for $0 < n < N$. In this way, we may define a *probability distribution* for this set

$$p_i = \lim_{N \to \infty} \frac{N_i}{N}, \quad i = 1, 2, \ldots, M(\epsilon) \quad (\epsilon \to \infty) \tag{A.5.27}$$

as well as the Renyi *partition function*[65]

$$X_q(\epsilon) = \sum_{i=1}^{M(\epsilon)} p_i^q, \quad -\infty < q < \infty \tag{A.5.28}$$

using the probabilities in Equation (A.5.28), as we did for the Farmer–Cantor distribution [see Equation (A.5.26)]. q here also plays the role of a selection parameter which helps us distinguish different subsets of S with different contributions to the sum of Equation (A.5.28).

On the other hand, the generalized dimensions D_q of a distribution are given by the formula

$$D_q = \lim_{\epsilon \to 0} \frac{1}{q-1} \frac{\log \sum_{i=1}^{M(\epsilon)} p_i^q}{\log \epsilon} \tag{A.5.29}$$

See Equation (A.5.28). Observe that by setting $q=0$ in Equation (A.5.29) we obtain the fractal dimension of the set S, D_0, as defined in Equations (A.5.4) and (A.5.7) of the previous section. But what is the meaning of the other dimensions D_q of the set S?

The above strategy allows us to write the sum of Equation (A.5.28) in the following, more convenient form:

$$\frac{1}{N}\sum_{j=1}^{N} \tilde{p}_j^{\,q-1} \equiv C_q(\epsilon) = \frac{1}{N}\sum_{j=1}^{N}\left[\frac{1}{N}\sum_{i=1}^{N} \Theta\left(\epsilon - \left|\boldsymbol{x}_i - \boldsymbol{x}_j\right|\right)\right]^{q-1} \tag{A.5.30}$$

In (A.5.30), the $\tilde{p}_j$ is the probability that the *actual orbit* visits the jth box and each box is considered separately with probability 1/N. $\Theta(\xi)$ is the usual Heaviside step function

$$\Theta(\xi) = \begin{cases} 1 & \text{for } \xi \geq 0 \\ 0 & \text{for } \xi < 0 \end{cases} \tag{A.5.31}$$

In practice, it is much easier to check how many points $\boldsymbol{x}_i$ lie at a distance equal to or smaller than ϵ from each $\boldsymbol{x}_j$, substitute this number in Equation (A.5.30), and thus compute the $C_q(\epsilon)$ function. Therefore, the generalized dimension of Equation (A.5.29) can now be computed with no great difficulty from the formula

$$D_q = \lim_{\epsilon \to 0} \frac{1}{q-1} \frac{\log C_q(\epsilon)}{\log \epsilon} \tag{A.5.32}$$

where we have used Equation (A.5.30) for the $C_q(\epsilon)$.

Of particular usefulness, for many applications, is the correlation dimension D_2 that results from Equations (A.5.30) and (A.5.32) for $q=2$:

$$D_2 = \lim_{\epsilon \to 0} \frac{\log C_2(\epsilon)}{\log \epsilon} \tag{A.5.33}$$

where

$$C_2(\epsilon) = \frac{1}{N^2} \sum_{i=1}^{N} \sum_{j=1}^{N} \Theta\left(\epsilon - \left|\boldsymbol{x}_i - \boldsymbol{x}_j\right|\right) \tag{A.5.34}$$

is the so-called correlation integral. This dimension D_2, which is known to be a lower bound of the fractal dimension D_0 ($D_2 \leq D_0$, see Equation (A.5.29)), was first used for the study of one-variable (scalar) time series

$$\{x(t), x(t+\Delta t), \ldots, x(t+(N-1)\Delta t)\} \tag{A.5.35}$$

by Grassberger and Procaccia.[31] Such time series result, for example, from the numerical integration of differential equations with many dependent variables, one of which is $x(t)$ (where Δt is the integration step). Of greater interest, however, are time series [Equation (A.5.35)] in which $x(t)$ is an experimentally measured quantity like, for example, temperature, atmospherical pressure, electrical potential of the heart, amount of rainfall, or intensity of an earthquake, while Δt is the time interval between two successive observations.

According to a theorem by Takens,[60] if, instead of the (generally unknown) vector of the dependent variables of a natural phenomenon

$$\boldsymbol{Y}(t) \equiv \{x_1(t), x_2(t), \ldots, x_d(t)\}$$

one uses the vector of the time-shifted values of *one variable*

$$\boldsymbol{X}(t) \equiv \{x(t), x(t+\tau), \ldots, x(t+(d'-1)\tau)\} \tag{A.5.36}$$

for a suitable choise of the delay $\tau = k\Delta t$, $k \in N$, the resulting orbit, $\boldsymbol{X}(t)$, is *topologically equivalent* to $\boldsymbol{Y}(t)$ in the sense that the former results from the latter after a number of continuous variable transformations, for $d' \geq 2d+1$ (in practice, however, as many applications show, it is sufficient to use $d' = d$ in Equation (A.5.36)).

This means that if we know the correct embedding dimension d of a certain physical phenomenon and properly select the delay time τ, the geometrical (and dynamical) properties of the $\boldsymbol{Y}(t_i)$ and $\boldsymbol{X}(t_i)$ sets, $i = 1, 2, \ldots, N$ (for N sufficiently large), are identical. So, Grassberger and Procaccia suggested the following algorithm for the study of the properties of a dynamical process $\boldsymbol{Y}(t)$, for which we know only the time series of a scalar quantity $x(t)$.

For $i = 1, 2, \ldots, N$ form the vectors

$$\boldsymbol{X}_i^{(m)} = \{x(t_i), x(t_i + \tau), \ldots, x(t_i + (m-1)\tau\} \tag{A.5.37}$$

with $m = 1, 2, 3, \ldots$ and $t_i = t_0 + i\Delta t$, choosing a time period $\tau = k\Delta t$ sufficiently big so that the components of $\boldsymbol{X}_i^{(m)}$ are not strongly correlated (in time) with each other. This can be done, for example, for values of τ near the first minimum of the mutual information function.[25] Consequently, we substitute the $\boldsymbol{X}_i^{(m)}$ from Equation (A.5.37) to Equation (A.5.34), calculate the correlation integral

$$C_2(\epsilon, m) = \frac{1}{N^2} \sum_{i=1}^{N} \sum_{j=1}^{N} \Theta\left(\epsilon - \left|\boldsymbol{X}_i^{(m)} - \boldsymbol{X}_j^{(m)}\right|\right) \tag{A.5.38}$$

and via Equation (A.5.33) find the dimension

$$D_2(m) = \lim_{\epsilon \to 0} \frac{\log C_2(\epsilon, m)}{\log \epsilon} \tag{A.5.39}$$

By drawing now $D_2(m)$ vs. m, one finds that $D_2(m)$ converges to the actual correlation dimension of the orbit,

$$D_2(m) \to D_2 \quad \text{as} \quad m \to d$$

and furthermore $D_2(m) = D_2$ for $m > d$.

In this way, we are able to compute both the D_2 dimension of a deterministic dynamical process as well as the (minimum) embedding dimension d for the set of points $\boldsymbol{X}(t_i)$ we are interested in. Of course, this can be done with no particular difficulty if the dimension d is finite and relatively small ($d \leq 10$) so that the number of data necessary for its determination is not prohibitively large. As examples, we cite here Refs. 5, 9 where the above approach was applied to time series derived from Electrocardiogram recordings and weather temperature data respectively. In both cases, evidence of a low-dimensional attractor was found and was used: (a) to characterize different individuals and (b) to make short term temperature predictions.

It is obvious that this method is particularly useful for the study of *chaotic time series* for which a Fourier spectrum analysis or other traditional data analysis methods are unable to determine the number of variables governing the dynamics. By applying the above procedure to chaotic orbits of simple dynamical systems, such as the Hénon mapping and the Lorenz equations, Grassberger and Procaccia found $D_2(\text{Hénon}) \cong 1.21$ and $D_2(\text{Lorenz}) \cong 2.05$. These dimensions computed by time series of only one variable (x_n for Hénon and $x(t)$ for Lorenz) are identical with those

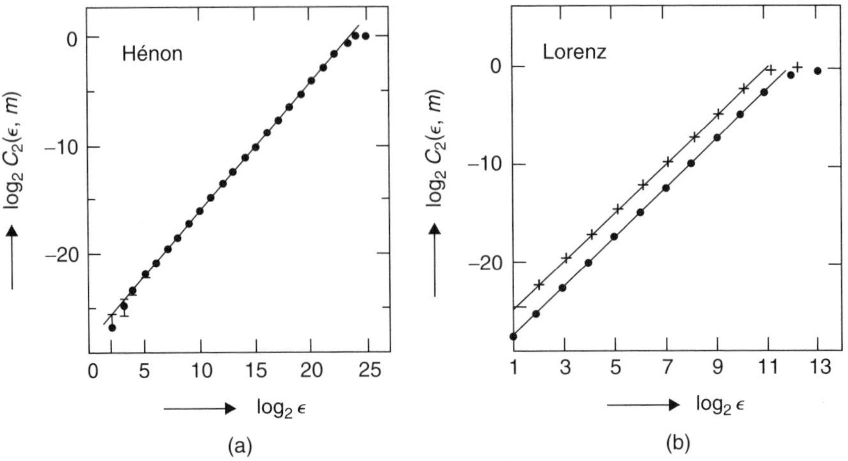

Figure A.32 The function $\log_2 C_2(\epsilon, m)$ over $\log_2 \epsilon$ for m large enough so that the correlation dimension D_2 can be distinguished: (a) for the Hénon mapping and (b) for the Lorenz model, at parameter values where the motion occurs on a strange attractor, see Figures A.4 and A.5.

computed using in Equations (A.5.38) and (A.5.39) the complete vectors $\boldsymbol{Y}_n = (x_n, y_n)$ and $\boldsymbol{Y}(t) = \{x(t), y(t), z(t)\}$ of the solutions of these two systems (see Figure (A.32)).

However, it is necessary at this point to warn the reader who is eager to apply these methods to his favorite physical, biological, or economic system that in most realistic cases (such as the weather, heartbeat, earthquakes, or the stock market), one finds only an *indication of low dimensional chaotic dynamics*. Usually there is significant *noise* present in the data, which makes the time evolution nondeterministic and unpredictable.

There exist by now excellent books describing this nonlinear analysis of chaotic time series and its limitations is (see References 37 and 58). Comprehensive and readable textbooks on the topics described in this Appendix have also been published[49], while, very recently, an encyclopedia has been printed[56], containing explanatory lemmas on most of the new concepts of chaos and fractals.

A.6 A New Science or a New Understanding of Nature?

Is complexity viewed as a discipline encompassing chaos and fractals a true science? In sofar as science is a collection of principles, theories, methods and results which can be tested experimentally, I believe it is. Moreover, it is a valid science, since several of its predictions have already been verified in the laboratory on a variety of real, physical systems.

Is it a new science? This is more difficult to answer, but may not be a relevant question, after all. Indeed, rather than wondering how to call it, shouldn't we be asking ourselves instead what is new about complexity?

In my view, the extreme sensitivity of the dynamics in phase space on a dense set of initial conditions, or simply the manifestation of chaos in *low-dimensional* systems, is an important new development. The realization that there can be complete loss of information and unpredictability in simple dynamical systems strikes a mortal blow to determinism and makes (at least inside chaotic regions) the practice of following individual orbits virtually obsolete.

One consequence of this is that the details and specific form of the equations of motion do not really matter. In fact, depending on the form of the lowest order nonlinear terms near an equilibrium point, we can divide low-dimensional dynamical systems into different classes. Within each class, similar bifurcation phenomena, and transitions to chaos are observed, sharing the same universality properties.

Furthermore, it has been widely observed in many physical systems that the dynamics in the full set of phase space dimensions is only transient. Eventually all orbits tend to lie on a low-dimensional attractor whose *geometric* structure depends on the parameters of the system and is deeply connected with the dynamical properties of the motion.

Another important discovery is that complexity does not always mean disorder. There are patterns and rules in nature, which by themselves are simple, but when applied in different combinations and at various scales can produce complicated behavior and exceedingly intricate spatiotemporal patterns.

In some cases, we can decode these puzzling messages and identify the universal laws by which they are constructed. As long as we can find the proper magnifying glass (often provided by an appropriate renormalization operator) all complexity will be removed and order will be restored.

Of course, this simple yet elegant view of nature does not always work, but the mere fact that in many cases it explains complex phenomena observed in the laboratory is already a victory. It is one step toward understanding those most intriguing manifestations of nature, which show no evident symmetry, regularity, or discernible structure.

We are still far from understanding the turbulence of sea waves or the precise way in which clouds are formed and trees grow. We still cannot predict when a heart attack will occur or an earthquake will strike. But we have understood one thing: That there is no single mathematical formalism or unique set of equations that are the most "appropriate" for the description of nonlinear phenomena.

Especially when dealing with chaotic behavior, it is important to determine first its fundamental properties and the geometry of its dynamics in phase space. Then one looks for the proper mathematical framework. And if smooth, differentiable functions or Euclidean geometry turn out to be inadequate, it may very well be that we have to turn to highly singular functions and fractal geometry.

"Old" mathematics served us well in the analysis of nearly linear systems with regular, periodic behavior. Most likely, however, the understanding of irregular, aperiodic, or chaotic phenomena will require a "new" kind of mathematics. Already significant progress has been made in that direction, aided by recent advances in sophisticated numerical computations.

Complexity also teaches us to abandon certain illusions we might have had, for example, about accurately forecasting the weather. We know now that long-term prediction in chaotic dynamics is impossible and we understand why. However, even though detailed knowledge of the evolution of individual orbits is often inaccessible, the probabilistic description of the dynamics may be significantly improved by the study of an underlying low-dimensional attractor, should one turn out to exist.

Nature and life are filled with irregular phenomena, sudden changes of behavior, and unexpected events, which we call "surprises," because they appear to occur at random. We may not yet understand them, but at least we have ceased to consider them as "magical" or "mysterious" and look at them through a crystal ball.

We have come to grips with complexity and begun to subject it to serious scientific scrutiny and investigation. We no longer feel apprehensive or indifferent toward complex phenomena. In fact, we have even started to discover their aesthetic qualities and potential for artistic expression.[48] And perhaps, in this way, we have learned something new about ourselves.

Acknowledgements

Besides the fundamental results of many researchers on whose work the science of complexity is based, I have taken the liberty to include in this Appendix some modest contributions of my research group at the Center of Research and Applications of Nonlinear Systems of the University of Patras. I wish to take the opportunity to thank all my students and colleagues who have helped significantly to this effort. I am particularly indebted to Mr. Christos Apokis for his skillfull assistance in the technical preparation of this manuscript.

Last, but not least, I want to thank Professor Peter Stavroulakis for inviting me to submit this Appendix to the volume at hand and hope that in this way I have helped to make the theory of chaos and fractals more accessible to the wider engineering community.

References

1. Arrowsmith, D. and Place, C.N., *Introduction to Dynamical Systems*, Cambridge University Press, Cambridge, 1990.
2. Barnsley, M.F., *Fractals Everywhere*, Academic Press, Boston, 1988.
3. Bazzani, A., Mazzanti, P., Turchetti, G., and Servizi, G., Normal forms for Hamiltonian maps and nonlinear effects in particle accelerators, Il *Nuovo Cim.*, 102B(1), 51, 1988.
4. Bergé, P., Pommeau, Y., and Vidal, C., *Order Within Chaos*, John Wiley & Sons, New York, 1986.
5. Bezerianos, A., Bountis, T., Papaioannou, G., and Polydoropoulos, P., Nonlinear time series analysis of electrocardiograms, *Chaos*, 5(1), 95, 1995.
6. Bountis, T., Fundamental concepts in classical chaos: Part I, *Open Syst. and Inform. Dyn.*, 3, 23–95, 1995.
7. Bountis, T., Fundamental concepts in classical chaos: Part II, *Open Syst. and Inform. Dyn.*, 4, 281–322, 1997.
8. Bountis, T., Period – doubling bifurcations and universality in conservative systems, Physica 3D, 577, 1981.
9. Bountis, T., Karakatsanis, L., Papaioannou, G., and Pavlos, G., Determinism and noise in surface temperature data in Greece, *Ann. Geophys.*, 11, 947, 1993.
10. Bountis, T. and Kollmann, M., Diffusion rates in a 4-D mapping model of accelerator dynamics, *Physica*, 71D(1,2), 122, 1994.
11. Bountis, T., Casati, G., and Procaccia, I., Eds, *Complexity: A unifying direction in science*, Intern. J. Bifurc. Chaos, to appear, June 2006.
12. Casati, G. and Chirikov, B., *Quantum Chaos*, Oxford University Press, Oxford, 1995.
13. Chirikov, B.V., A universal instability of many dimensional oscillator systems, *Phys. Rep.*, 52(5), 264, 1979.
14. Collet, P. and Eckmann, J.-P., *Iterated Maps of the Internal as Dynamical Systems*, Birkhäuser, Boston, 1980.
15. Collet, P., Eckmann, J.-P., and Lanford, O., Universal properties of maps on an interval, *Commun. Math. Phys.*, 76, 211, 1980.
16. Cvitanovic, P., Ed., *Universality in Chaos,* 2nd ed., Adam Hilger, Bristol, 1989.
17. Devaney, R., *An Introduction to Chaotic Dynamical Systems*, Benjamin-Cummings, Menlo Park, California, 1986.
18. Eckmann, J.-P. and Ruelle, D., Ergodic theory of chaos and strange attractors, *Rev. Mod. Phys.*, 57, 617, 1985.
19. Escande, D.F., Stochasticity in classical Hamiltonian systems: universal aspects, *Phys. Rep.*, 121(3,4), 166, 1985.
20. Falconer, K.J., *Geometry of Fractal Sets*, Cambridge University Press, Cambridge, 1985.
21. Feigenbaum, M.J., Quantitative universality for a class of nonlinear transformations, *J. Stat. Phys.*, 19, 25, 1978.
22. Feigenbaum, M.J., The universal metric properties of nonlinear transformations, *J. Stat. Phys.*, 21, 669, 1979.

23. Feigenbaum, M.J., The transition to aperiodic behavior in turbulent systems, *Commun. Math. Phys.*, 77, 65, 1980.
24. Feigenbaum, M.J., Kadanoff, L.P., and Shenker, S., Quasiperiodicity in dissipative systems: A renormalization group analysis, *Physica*, 5D, 370, 1982.
25. Fraser, A.M. and Swinney, H.L., Independent coordinates for strange attractors from mutual information, *Phys. Rev.*, A33, 1134, 1986.
26. Gleick, J., *Chaos: Making of a New Science*, Viking, New York, 1987.
27. Grassberger, P., On the Hausdorff dimension of fractal attractors, *J. Stat. Phys.*, 19, 25, 1981.
28. Grassberger, P. and Procaccia, I., On the characterization of strange attractors, *Phys. Rev. Lett.*, 50, 346, 1985.
29. Greene, J., A method for determining a stochastic transition, *J. Math. Phys.*, 20, 1183, 1979.
30. Greene, J., MacKay, R.S., Vivaldi, F., and Feigenbaum, M.J., Universal behavior of area – preserving maps, *Physica*, 3D, 468, 1981.
31. Guckenheimer, J. and Holmes, P.J., *Nonlinear Oscillations, Dynamical Systems and Bifurcations of Vector Fields*, Springer, New York, 1983.
32. Helleman, R.H.G., in *Fundamental Problems in Statistical Mechanics*, vol. 5, Cohen, E.G.D., Ed., North Holland, Amsterdam, 1980.
33. Hénon, M., A two-dimensional mapping with a strange attractor, *Commun. Math. Phys.*, 50, 69, 1976.
34. Hentchel, H.G.E. and Procaccia, I., The infinite number of generalized dimensions of fractals and strange attractors, *Physica*, 8D, 435, 1983.
35. Hu, B. and Rudnick, J., Exact solutions of the Feigenbaum renormalization equations for intermittency, *Phys. Rev. Lett.*, 48, 1645, 1982.
36. Kadanoff, L.P., *From Order to Chaos*, World Scientific, Singapore, 1993.
37. Kantz, H. and Schreiber, T., *Nonlinear Time Series Analysis*, Cambridge, University Press, 1999.
38. Laskar, J., Froeschlé, C., and Celetti, A., *Physica*, 56D, 253, 1992; see also: Laskar, J., *Icarus*, 88, 266, 1990; Laskar J. et al., *Nature*, 361, 608 and 615, 1993.
39. Libchaber, A. and Maurer, J., Une experience de Rayleigh – Bénard de géometrie réduite: Multiplication, accrochage e démultiplication de fréquences, *J. Phys. (Paris) Coll.*, 41, C3-51, 1980.
40. Libchaber, A. and Maurer, J., in *Nonlinear Phenomena at Phase Transitions and Instabilities*, Riste, T., Ed., Plenum Press, London, 1982.
41. Lichtenberg, A. and Lieberman, M., *Regular and Stochastic Motion*, Springer, New York, 1983.
42. Lorenz, E.N., Deterministic nonperiodic flow, *J. Atm. Sci.*, 20, 130, 1963.
43. Ma, S.K., *Modern Theory of Critical Phenomena*, Benjamin, New York, 1976.
44. MacKay, R.S. and Meiss, J.D., Eds., *Hamiltonian Dynamical Systems*, Adam Hilger, Bristol, 1987.
45. MacKay, R.S., *Renormalization Approach to Area-Preserving, Twist Maps*, World Scientific, Singapore, 1994.
46. Mandelbrot, B.B., *The Fractal Geometry of Nature*, Freeman, New York, 1982.

47. Nicolis, G. and Prigogine, I., *Exploring Complexity*, Freeman, New York, 1989.
48. Peitgen, H.-O. and Richter, P.H., *The Beauty of Fractals*, Springer-Verlag, Berlin, 1986.
49. Peitgen, H.-O., Jürgens, H., and Saupe, D., *Chaos and Fractals: New Frontiers of Science*, Springer-Verlag, New York, 1992.
50. Pomeau, Y. and Manneville, P., Intermittent transition to turbulence in dissipative dynamical systems, *Commun. Math. Phys.*, 74, 189, 1980.
51. Rasband, S.N., *Chaotic Dynamics of Nonlinear Systems*, John Wiley, New York, 1990.
52. Reichl, L.E., *The Transition to Chaos in Conservative Classical Systems: Quantum Manifestations*, Springer, New York, 1992.
53. Ruelle, D. and Takens, F., On the nature of turbulence, *Commun. Math. Phys.*, 20, 167, 1971.
54. Schroeder, M., *Fractals, Chaos and Power Laws*, Freeman, New York, 1991.
55. Schuster, H.-G., *Deterministic Chaos: An Introduction*, 2nd ed., VCH, Weinheim, 1988.
56. Scott, Alwyn, Ed., *Encyclopedia of Nonlinear Science*, Routledge, Taylor and Francis, New York, 2004.
57. Shenker, S., Scaling behavior in a map of a circle onto itself: Empirical results, *Physica*, 5D(2,3), 405, 1982; see also: Shenker, S. and Kadanoff, L.P., Critical behavior of a KAM trajectory: Empirical results, *J. Stat. Phys.*, 27, 631, 1982.
58. Sparrow, C., *The Lorenz Equations: Bifurcations, Chaos and Strange Attractors*, Springer, New York, 1982.
59. Sprott, J., *Chaos and Time Series Analysis*, Oxford University Press, 2002.
60. Takens, F., Detecting strange attractors in fluid turbulence, *Lecture Notes in Math.*, 898, Springer, Berlin, p. 366, 1981.
61. Wiggins, S., *Introduction to Applied Dynamical Systems and Chaos*, Springer, New York, 1990.

Appendix B

Mathematical Models of Chaos

Christos H. Skiadas

CONTENTS

B.1 Introduction

Chaotic theory is included between the youngest of the sciences, and has become one of the most advanced fields in contemporary research. This new theory originated from the early 20th century mathematics and physics, but mainly began to emerge in the 1960s. The development of the computer technology, helped mathematicians and scientists in many fields to discovered that the determinism was not so simple. Even simple deterministic systems when applied to real life cases give rise to highly complex forms, something that the classical Newtonian determinism cannot explain.

The twentieth century science will be known for only few theories: between them chaos, relativity, quantum mechanics, and cosmology play an important role. Chaotic forms are evident everywhere around the world and the universe, from the shape of galaxies and chaotic nebulae to the currents of the ocean and the flow of blood through fractal blood vessels to the branches of trees and the effects of turbulence.

Today chaos changed to a full science of its own. It is multidisciplinary as many scientific fields are involved and has received widespread publicity. Chaotic science and related developments promises to continue to yield absorbing scientific information. The future of the modern science will be highly influenced by the growth and widespread of chaotic literature.

Chaos came as a new aspect of determinism. Galileo and mainly Isaac Newton[1] developed the "Newtonian determinism" and many scientific and quasi-scientific disciplines are grounded in the "Newtonian mechanistic" paradigm, which developed from the 16th to the 19th centuries. The famous law of gravity and the laws of planetary motion along with subsequently discovered laws of physics allowed scientists to face at the universe as a purely deterministic system. A French mathematician Laplace assumed that it could be possible to forecasts the future events in a pure deterministic way. Another Frenchman Auguste Comte, a social philosopher in 19th century was a leader in the creation of Positivism, a science-based social philosophy of determinism. According to this theory the prediction of future natural and social phenomena and the management of the world around us is possible by based in the future advancement of science and the abilities of the experts.

The Newtonian determinism was challenged in 20th century by the Einstein's relativism and Heisenberg's quantum mechanics. Then matter and space as absolute entities where called into question. The modern science started to challenge every old theory. Then the chaotic theory came following the development of the computers and the related techniques of modeling and simulation.

Chaotic theory is defined as the study of complex nonlinear dynamical systems or complex systems that are expressed by recursion and higher mathematical algorithms and are dynamic non-constant and non-periodic. Chaotic theory includes the quantitative and qualitative study of unstable non-periodic behavior in deterministic nonlinear dynamical systems.

Chaos can also be viewed as including stochastic properties. By applying the same definition in both stochastic and deterministic chaos, the mathematicians formed a bridge between the two sciences that were regarded as mutually exclusive until then. The science of chaos is related to the study of deterministic systems that are so sensitive to measurement that their output appears random.

Why is chaos useful in Telecommunications? Although chaos is unpredictable, it is in most cases deterministic. If two nearly identical chaotic systems of the appropriate type are impelled, or driven, by the same signal, they will produce the same output. A variety of interesting technologies for communications are based on this phenomenon.

The chaotic systems frequently have many complicated space paths or trajectories that "attract" the solutions of the system and are called "attractors". Attractors may be multidimensional because systems can have many different state-space variables that vary in three dimensions. The attractors are usually called "strange attractors" because of their form. There are very useful because, even in very complicated chaotic systems, they give us information about the space where the chaotic system may vary.

The development of chaos theory is mainly due to computers. In studying chaos millions of recursive calculations are needed. Edward Lorenz[2] a pioneer in using computers to forecast weather trends, he accidentally found in 1960 that in certain kinds of nonlinear equations, or systems based on simultaneous nonlinear equations, the results display sensitive dependence on initial conditions. This is a distinguishing feature of chaotic systems.

At the end of the nineteenth century Henri Poincaré[3] was aware of the fact that the orbits of three bodies moving under a central force due to gravity are quite complicated and change drastically with small changes of initial conditions. Now we know that to solve this problem we need a set of more than three coupled differential equations and the paths of the solutions have a chaotic character. Poincaré[4] tried solutions for the restricted three-body problem, when the one mass is relatively very small related to the other two. He understood that even in this simpler case the character of the motion still was very complicated. He tried to find a theorem to explain more generally the phenomenon and to establish a theory related to the chaotic paths in a system of differential equations.

The three-dimensional paths in a system of non-linear differential equations could be chaotic. Instead, the two-dimensional paths are not chaotic and this is the result of the famous Poincaré–Bendixon theorem. In other words chaos cannot occur in a two-dimensional autonomous system

$$\dot{x} = f(x, y)$$
$$\dot{y} = g(x, y)$$

where $\dot{x}$ and $\dot{y}$ are the first time derivatives. The above system can take the form of a two-dimensional equation for x and y

$$\frac{dy}{dx} = \frac{g(x, y)}{f(x, y)}$$

and we can find phase paths in the plane. Assuming that we have determined the equilibrium points at which both $f(x, y) = 0$ and $g(x, y) = 0$, the Poincaré–Bendixon theorem says that a phase path must either (a) terminate at an equilibrium point, or (b) return to the original point, giving

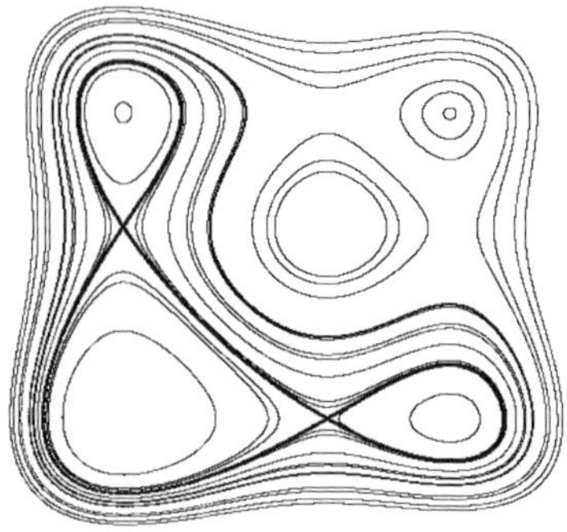

Figure B.1 The (x, y) diagram.

a close path, or (c) approach a limit cycle. Then, there are no-chaotic solutions. A simple such system is

$$\dot{x} = (y - a)(y - b)(y - c)$$

$$\dot{y} = (x - a)(x - b)(x - c)$$

With parameter selection $a = 0.1$, $b = 0.6$, and $c = 0.9$, Figure B.1 illustrates this case. The more complicated paths pass from the equilibrium unstable points.

The Russian mathematician Lyapunov realized that a single number could be used to represent the change caused by a perturbation. He divided the size of the perturbation at one instant in time by its size a moment before. He then performed the same computation at various intervals and averaged the results.

This quantity, now known as the Lyapunov multiplier, describes how much, on average, a perturbation will change. If the Lyapunov multiplier is less than one, the perturbations die out, and the system is stable. If the multiplier is greater than one, however, the disturbances grow, and the system is unstable. All chaotic systems have a Lyapunov multiplier greater than one and so are always unstable. This is the root of the unpredictability of chaos.

Edward Ott, Celso Grebogi and James A. Yorke of the University of Maryland[5] developed a scheme (OGY) in which a chaotic system can be encouraged to follow one particular unstable orbit through state space. The scheme can therefore take advantage of the vast array of possible behaviors that make up the chaotic system.

To start, one obtains information about the chaotic system by analyzing a slice of the chaotic attractor. After the information about this so-called Poincaré section has been gathered, one allows the system to run and waits until it comes near a desired periodic orbit in the section. Next the system is encouraged to remain on that orbit by perturbing the appropriate parameter. The strength of this method is that it does not require a detailed model of the chaotic system but only some information about the Poincaré section. It is for this reason that the method has been so successful in controlling a wide variety of chaotic systems.

Ott, Grebogi and Yorke[4] showed that the presence of chaos could be an advantage in controlling dynamic behavior. Because chaotic systems are extremely sensitive to initial conditions, they also react very rapidly to implemented controls.

Instead the simplest three-dimensional non-linear system of differential equations the Rössler[6] model gives easily chaotic paths for appropriate selection of the parameters. This system has equations

$$\dot{x} = -y - z$$

$$\dot{y} = x + ey$$

$$\dot{z} = f + xz - mz$$

The system has only one non-linear term in the last equation. However, when the parameters have values $e = f = 0.2$ and $m = 5.7$ chaotic paths appear with the form presented in Figure B.2.

The Rössler signal is nearly periodic and resembles a sine wave whose amplitude and wavelength have been randomized somewhat from one cycle to the next. Simply taking a periodic signal and adding chaos to it can make such signals, which have come to be known as pseudo-periodic, in general. If several identical systems whose attractors are all period two or greater are driven by the appropriate pseudo-periodic signal, they will all get in step.

Radical changes appear when the differential equation systems are replaced by systems of difference equations or simple difference equations

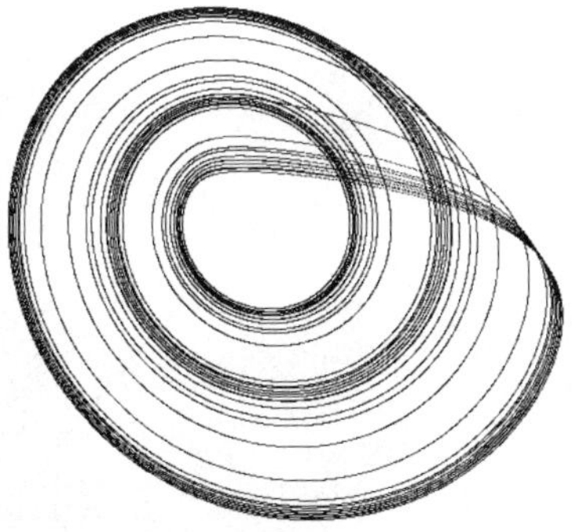

Figure B.2 The Rössler diagram.

with non-linear terms. If we replace the above Rössler system by a difference equation analogue of the form

$$x_{t+1} = x_t - (y_t + z_t)/6$$
$$y_{t+1} = y_t + (x_{t+1} + ey)/6$$
$$z_{t+1} = z_t + (0.35 + x_{t+1}z_t - 3.5z_t)/4$$

then, for six values of the parameter e (0.29, 0.30, 0.31, 0.32, 0.33, and 0.34) interesting chaotic forms like sea-shells result presented in Figure B.3.

As the mathematical development of methods and tools of analysis for centuries concentrated on differential mainly and not to difference equations it is not surprising that today, the most popular chaotic difference equation model, the logistic model,

$$x_{t+1} = bx_t(1 - x_t)$$

was only "invented" in the last three decades. This very simple model shows period doubling and then chaotic behavior when the parameter b is higher than a characteristic value.

On the other hand, the differential equation analogue of this model was very popular

$$\dot{x} = bx(1 - x)$$

This differential equation, as it was expected, gives simple finite solutions without complications and chaotic behavior. The model, called Logistic, was applied in demography from 1838 by Verhulst.

The discussion about models and modeling includes deterministic, stochastic and chaotic models and also, chaotic analysis, model building and simulation.

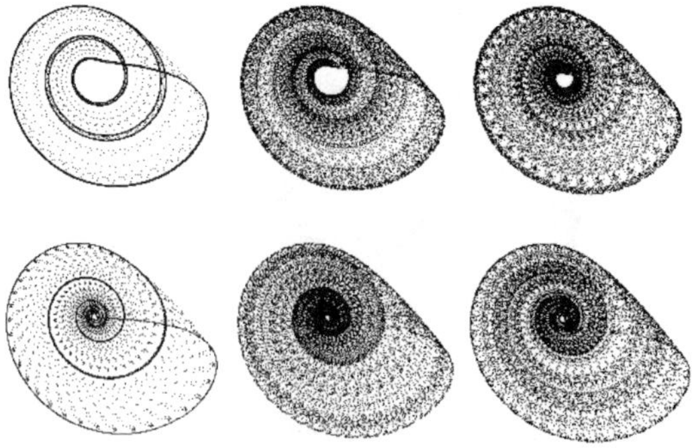

Figure B.3 A discrete analogue of the Rössler model.

B.2 The Logistic Map

The logistic map arises from the logistic difference equation

$$x_{t+1} = bx_t(1 - x_t)$$

This equation has extensively analyzed during the last decades and became the basis of numerous papers and studies in various disciplines. The analysis of this relatively simple iterative function showed that a reach variety of chaotic principles and rules where inherent. The study of the period doubling bifurcations specifically, that appear when the chaotic parameter b changes from one characteristic value to another gave rise to the development of a theory that lead to a universal law in order to explain the phenomenon, at least in the families of quadratic maps.

In the following, we present the basic properties of the logistic map based on the classical analysis proposed by Feigenbaum[7] and others. However, new approaches are used, by applying ideas from a geometric point of view of the subject.

A simple geometric presentation of the properties of the logistic map is illustrated in Figure B.4.

Given the value of the parameter b, the graph of all the values that the logistic map could take is a parabola passing from the origin (0, 0) and the point (0, 1). In Figure B.4 a heavy line marks this curve. The maximum of the parabola is obtained at (1/2, b/4). This is easily obtained by calculating the first derivative of the logistic function that is

$$x'_{t+1} = b(1 - 2x_t)$$

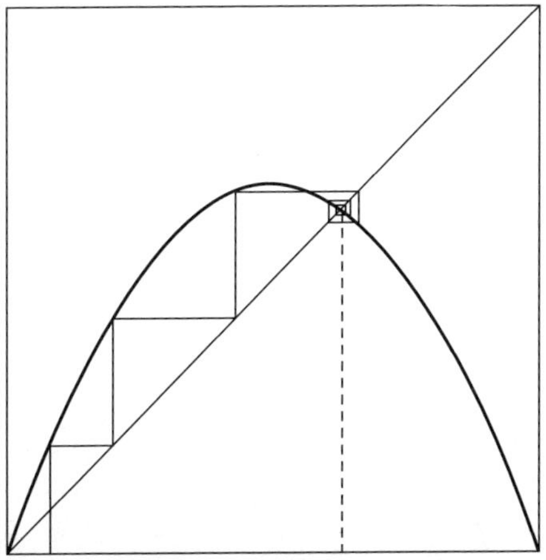

Figure B.4 The Logistic Map.

and equating it to zero. Hence, $x = b/4$. By taking into account that $0 \le x \le 1$, it is clear that the parameter is $0 < b \le 4$. Interesting properties of the logistic map can be illustrated geometrically inside a square with sides equal to 1.

The first attracting point is located at $x_{t+1} = x_t = x = 1 - 1/b$. This is obtained from the equality $x = bx(1 - x)$, which provides two points at (0, 0) and $(1 - 1/b,\ 1 - 1/b)$. Only the second point is an attracting point. This point is located on the main diagonal of the square where $x_{t+1} = x_t$ and exists when $b > 1$. Starting from an initial value $x_0 = 0.08$ the values of x_t, x_{t+1} are computed and presented in the above graph; The parameter $b = 2.7$.

If we introduce the inverse function of the logistic map, as this is illustrated geometrically in the following graphs, we have a convenient way to estimate the attracting point for a given value of the chaotic parameter b. At some higher values of b the attracting point no-longer attracts and two new attracting points appear. In the same graph the second order diagram (x_t, x_{t+1}) appears. The parameter is $b = 2.7$ for Figure B.4. When $b = 3$, all the curves pass from the point (2/3, 2/3); see Figure B.5. The slope, the value of the first derivative, at this point is the same, equal to -1, for both (x_t, x_{t+1}) and (x_{t+1}, x_t) graphs and equal to 1 for the (x_t, x_{t+2}) graph. $b = 3$ is the upper limit of the chaotic parameter for which the first attracting point exists.

The period-4 bifurcation appears in Figure B.6. For better presentation and, in order to locate the characteristic points we introduce the inverse (x_{t+3}, x_t) along with the (x_t, x_{t+4}) diagrams. The parameter b ranges within $1 + \sqrt{6} < b < 3.5440903$. There are four characteristic values for x.

Models expressing **Logistic** behavior are the **sinus** and a model proposed by May.[8] Models with different behavior than that of the quadratic maps are the Gaussian type models.

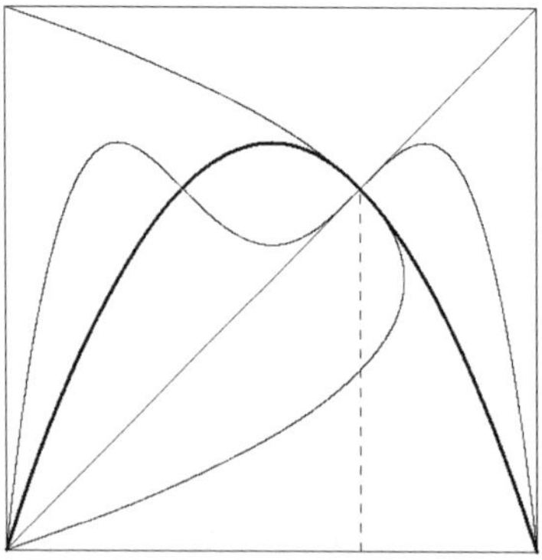

Figure B.5 Logistic map, $b = 3$.

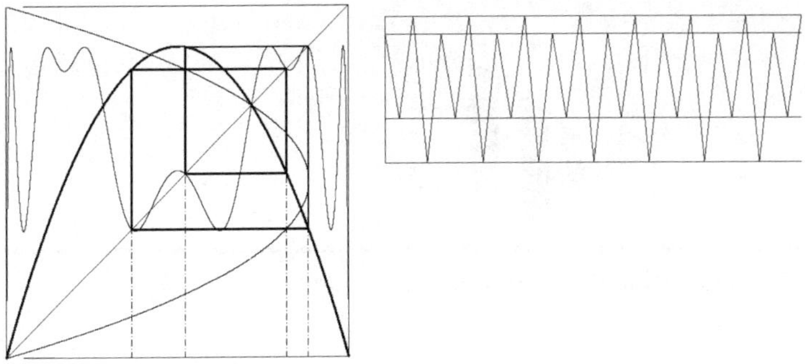

Figure B.6 Order-4 bifurcation of the Logistic Map. Real time and phase space diagrams.

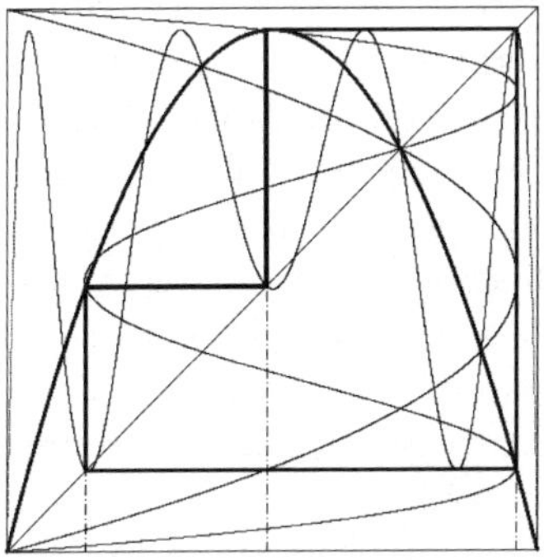

Figure B.7 Order-3 bifurcation.

The logistic map arises from the logistic difference equation

$$x_{t+1} = bx_t(1 - x_t)$$

The analysis of this relatively simple iterative function showed that a reach variety of chaotic principles and rules where inherent. Especially, the study of the period doubling bifurcations that appear when the chaotic parameter b changes from one characteristic value to another gave rise to the development of a theory that lead to a universal law in order to explain the phenomenon, at least in the families of quadratic maps.

As is illustrated in the next bifurcation diagrams of the logistic and the Gaussian model there are clear differences between the two models (see Figure B.7 and Figure B.8). However, there appear some similarities in the chaotic windows.

The Australian physicist Robert May[8,9] was modeling population fluctuations, using the nonlinear "logistic equation"

$$\mathbf{y} = \mathbf{rx}(1 - \mathbf{x})$$

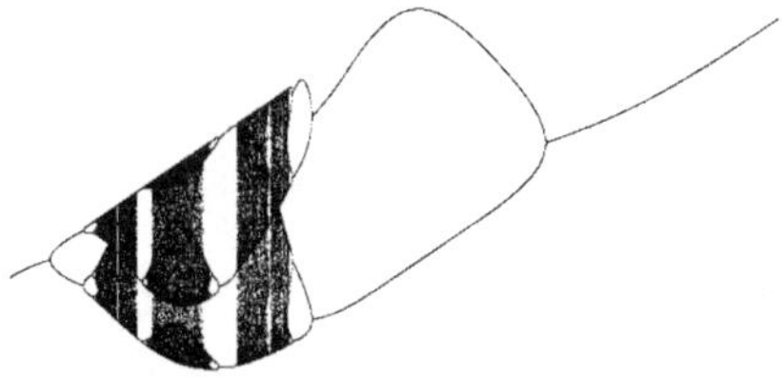

Figure B.8 Bifurcation diagram of the Gaussian model; $a = 7.5$.

where **y** equals the preceding value of **x** (ranging from 0 to 1) in an iterated series.

The virtue of using this form of an equation (in which one quantity is multiplied by its reciprocal) in population studies is that the results are "environmentally" self-limiting. This is in contrast to the unrealistic exponential equations used by British economist Thomas Malthus, who fretted that the Earth's human population would grow without limit.

This logistic equation underwent period doubling, meaning that the results shifted from a constant stream to two cycles, four cycles, etc., as the **r** parameter was increased, and to May's surprise, it yielded chaotic results when the parameter **r** (ranging from 1 to 4) crossed a certain threshold — about 3.58. This was clear evidence of deterministic stochastic (random) behavior, seeming oxymoron!

Some models of the Gaussian form show special chaotic behavior mainly due to the quadratic exponent of the equations of these models. A simple Gaussian equation is the following

$$x_{t+1} = b + e^{-ax_t^2}$$

The model, for $a = 4$ and $-1 < b < 1$, has a first and then a second period doubling bifurcation and then, it returns to the original state in a reverse order. A further increase of the parameter a of the quadratic term, $a = 7.5$ gives a foul sequence of period doubling bifurcations and order-3 saddle-node bifurcations as is illustrated in Figure B.8.

A rich bifurcation diagram appears when the Gaussian model is combined with the logistic to form the G–L model

$$x_{t+1} = b - x_t(1 - x_t) + e^{-ax_t^2}$$

When the parameter $a = 7.5$ the bifurcation diagram in the range $-1.1 < b < 0.7$ has two main bifurcation regions with period doubling bifurcations and order-3 and order-5 saddle-node bifurcations.

B.3 Delay Models

The delay models can be expressed by a large variety of equations. The delay models expressed by the difference equation (m is the delay)

$$x_{t+m} = f(x_{t+m-1}, x_{t+m-2}, \ldots, x_t)$$

and the delay models expressed by the differential equation

$$\dot{x} = f(x_t, x_{t-T})$$

where T is the parameter expressing the delay are the most used general forms of delay models.

With some variations more complicated models arise.

Delay models can be found in several real life cases when the future development of a phenomenon is influenced by the past. The complexity of the delay system is higher when the delay is large and the equations used are highly non-linear. The models listed below express such chaotic behavior.

An appropriate differential equation analogue to the delay difference equation of the logistic type is given by the following iterative formula. This method is based on the linear spline approximation. By using this formula we can find adequate solutions for several delay equations' forms. The delay parameter is m, s is the intermediate step and the iteration step is d. The delay equation is included only in the first equation. It is the last term to the right of the first equation. By changing this delay equation or replacing it by another many interesting cases arise.

$$x_1 = x_1 + bx_1(1 - x_m)d$$

$$x_2 = x_2 - (m/s)(x_2 - x_1)d$$

$$\ldots$$

$$x_m = x_m - (m/s)(x_m - x_{m-1})d$$

The application which follows is based on the following parameter set: $b = 1.22$, $d = 0.01$, $s = 3$ and the delay is $m = 14$. Figure B.9 illustrates the (x_{m+1}, x_1) and (x_1, t) diagrams. The differential equation solved by the above procedure has the form

$$\dot{x}_m = bx(1 - x_m)$$

The simulation producing Figure B.10 is based on another delay equation form expressed by

$$\dot{x}_m = bx_m(1 - x^2)$$

The parameters selected are $m = 15$, $s = 6$, $d = 0.001$, and $b = 2.6$. The two graphs of Figure B.10 represent the (x_{m+1}, x_1) and (x_1, t). The first

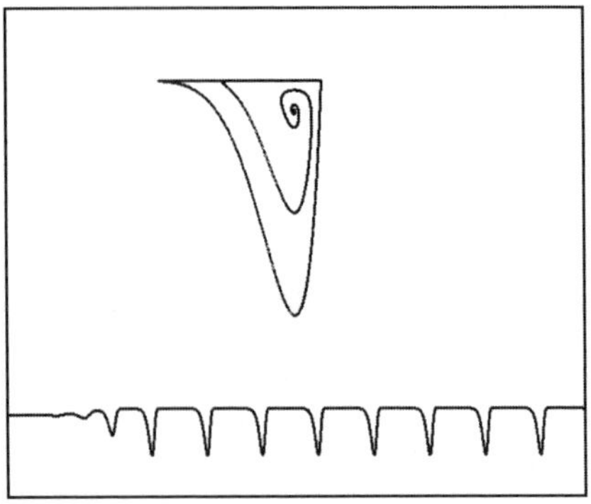

Figure B.9 The delay Logistic.

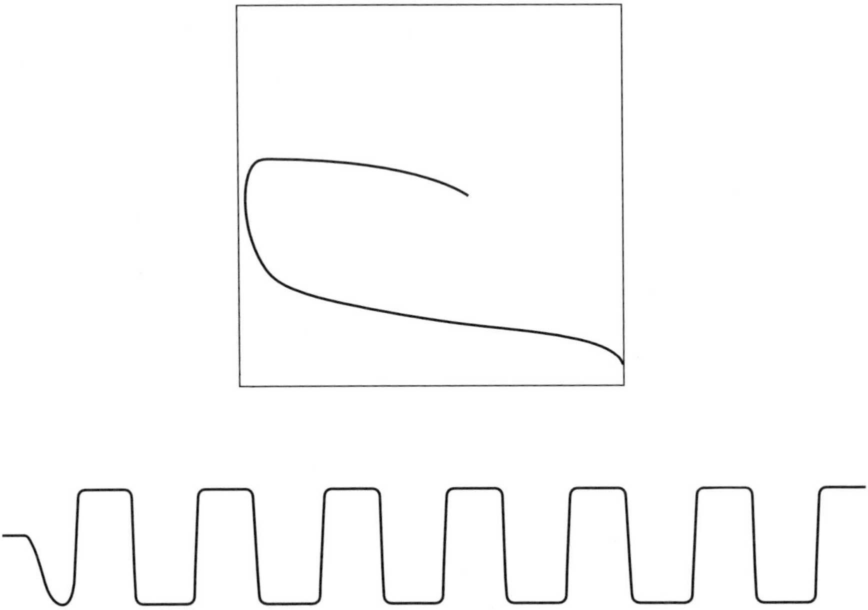

Figure B.10 Delay model: a square pattern.

graph is an almost perfect square. By appropriate enlargement one finds that the corners are not very sharp, as this is due to the selection of the parameters and the integration step. However it is a very good formula expressing a square in a continuous equation form. The oscillation turns also to be one of the step-form.

In both cases presented the value of the parameter b selected is close to the upper limit. However, one can have several intermediate cases with particular interest.

Figure B.11 **The Hénon attractor; $b = 0.3$, $a = 1.4$.**

B.3.1 The Hénon Model

Another form of the delay logistic model (see Figure B.11) is expressed by the following difference equation

$$x_{t+1} = bx_{t-m} + ax_t(1 - x_t)$$

When the delay parameter $m = 1$ the last equation reduces to an equation equivalent to that of the Hénon model, that is

$$x_{t+1} = bx_{t-1} + ax_t(1 - x_t)$$

The analysis on this map, for the case of $b = -1$, shows that it has a stable fixed point at

$$x = 1 - 2/a$$

which turns into the period-2 orbit with

$$x = \frac{a + 2 \pm \sqrt{(a - 6)(a + 2)}}{2a}$$

In some cases a system of two difference equations is replaced by a simple difference delay equation with the same properties. This is the case of the Hénon model and other related models, extensions and variations, such as the Holmes model.

Hénon[10,11] proposed a two-dimensional difference equation model with very interesting behavior which is now considered as one of the standard models in the chaotic literature

$$x_{t+1} = 1 - ax_t^2 + y_t$$
$$y_{t+1} = bx_t$$

The equivalent delay difference equation of the Hénon system is

$$x_{t+1} = bx_{t-1} + 1 - ax_t^2$$

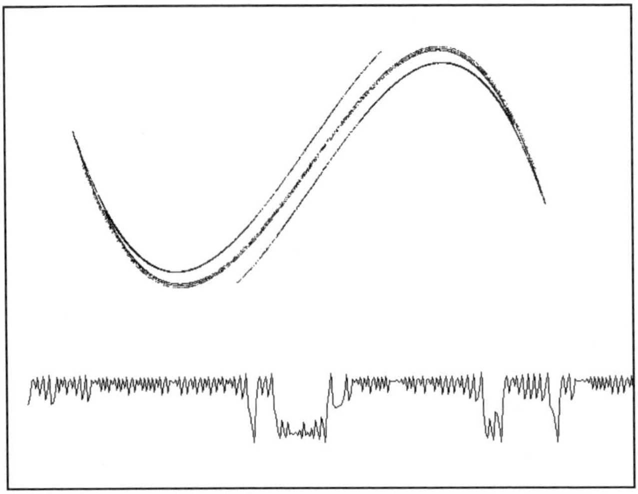

Figure B.12 The Holmes model.

The Holmes model is illustrated in Figure B.12.

Interesting properties has also another model proposed by Holmes with equation

$$x_{t+1} = bx_{t-1} + ax_t - x_t^3$$

When the parameters of the Holmes model take the values $a = 2.76$, $b = -0.2$ the map is presented in Figure B.12.

A cosine model with more complicated properties to that of the Hénon model is the model expressed by the system of two difference equations:

$$x_{t+1} = by_t + a\cos(x_t)$$
$$y_{t+1} = x_t$$

A delay *cosine* equation takes the form

$$x_{t+1} = bx_{t-1} + a\cos(x_t)$$

The last expression retains some of the properties of the Hénon map but it contains also other interesting properties as are that presented in Figure B.13 where the parameters take the values $a = -0.4$ and $b = -1$. The map is an area-preserving map as the Jacobian is $J = b = -1$. The influence of the cosine function in the delay equation is present. The various forms are repeated for many times by following 45° and 135° degrees of symmetry.

Figure B.13 Chaotic patterns of the delay Cosine function.

B.3.2 Hamiltonian Systems

The majority of two-dimensional Hamiltonian systems are conservative systems, that is, systems that have a non-trivial first integral of motion. The first integral must satisfy the following equation:

$$\frac{d}{dt}f(x,y) = \dot{x}f_x(x,y) + \dot{y}f_y(x,y)$$

where $f_x = \partial f/\partial x$, $f_y = \partial f/\partial y$.

The first integral is useful because of the connection between its level curves and the trajectories of the system. When f is a first integral, it is constant on every trajectory of the original system $(\dot{x},\ \dot{y})$. Thus every trajectory is part of some level curve of f. Hence each level curve is a union of trajectories. The first integral is so named because it is usually obtained from a single integration of the differential equation

$$\frac{dy}{dx} = \frac{Y(x,y)}{X(x,y)}$$

If the solutions of this equation satisfy $f(x,y) = c$, then f is a first integral of $(\dot{x},\ \dot{y})$.

The simplest Hamiltonian system is given by the following set of two differential equations

$$\dot{x} = y$$
$$\dot{y} = -x$$

This system easily gives a differential equation of the previous form, that is

$$\frac{dy}{dx} = -\frac{x}{y}$$

which has solutions satisfying

$$x^2 + y^2 = h$$

where h is a positive constant. However, with $f(x, y) = x^2 + y^2$ the first equation above is satisfied for all (x, y) in the plane. Thus the last equation for $f(x, y)$ is a first integral on the whole plane and so the above system is a conservative system. The Hamiltonian is

$$H(x, y) = \frac{1}{2}(x^2 + y^2)$$

The level curves are co-centric circles and the radius is defined by the set of initial conditions (x_0, y_0).

The conservative systems play an important role in mechanical problems. The equations of motion of such systems can be constructed from their Hamiltonian. For example, a particle moving in one dimension, with position coordinate x, momentum y and Hamiltonian $H(x, y)$ has equations of motion

$$\dot{x} = \frac{\partial H(x, y)}{\partial y}, \quad \dot{y} = -\frac{\partial H(x, y)}{\partial x}$$

The Hamiltonian $H(x, y)$ is a first integral for the last system of differential equations because

$$x\frac{\partial H}{\partial x} + y\frac{\partial H}{\partial y} = \frac{d}{dt}H = 0$$

The analysis provides enough information to draw the phase portrait of the system in Cartesian coordinates. The resulting patterns are symmetric or no-symmetric depending on the selection of the equations of the system.

We proceed to the following analysis of a system of two differential equations providing an egg-shaped form. The system is expressed by the following equations

$$\dot{x} = y(y - b)$$
$$\dot{y} = x$$

This system has two equilibrium points when $(x = 0,\ y = 0)$ and $(x = 0,\ y = b)$. The first point is a stable point as it is associated with imaginary eigenvalues $\lambda_{1,2} = \pm i\sqrt{b}$. Instead, the second point is unstable as it can

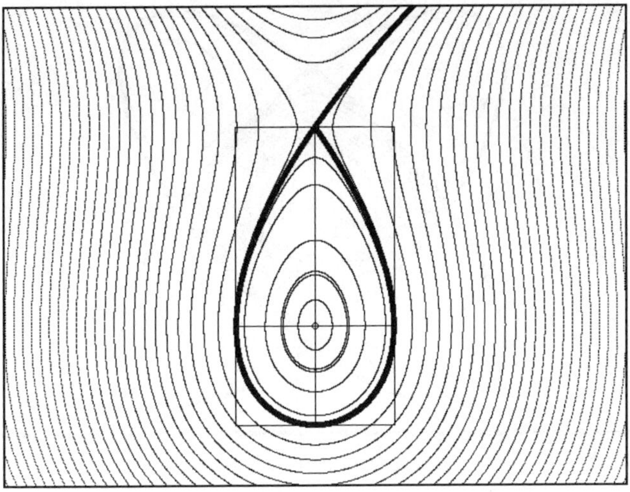

Figure B.14 An egg-shaped Hamiltonian form.

be verified by observing the corresponding real eigenvalues $\lambda = \pm\sqrt{b}$. The tangent at this point is

$$\frac{dy}{dx} = \frac{1}{\sqrt{b}}$$

The system can be solved explicitly as immediately results

$$\frac{dy}{dx} = \frac{x}{y(y-b)}$$

Consequently the solution is

$$\frac{y^3}{3} - b\frac{y^2}{2} = \frac{x^2}{2} + h$$

where h is the integration constant.

As the Hamiltonian in this case is of the form

$$H(x,y) = -\frac{x^2}{2} + \frac{y^3}{3} - b\frac{y^2}{2}$$

This constant has the following value in the maximum limit of the egg-shaped form, $h = -b^3/6$. This form, which is marked by a heavy line in Figure B.14, is a characteristic level curve. The sharp corner of the egg-shaped form is located at (0, b), whereas the point (0, $-b/2$) is the point where the maximum speed is achieved. This speed is $\dot{x} = 3b^2/4$. The

Figure B.15 A 4-type symmetric Hamiltonian form.

equation for the periphery of the egg-shaped form is

$$\frac{y^3}{3} - b\frac{y^2}{2} = \frac{x^2}{2} - \frac{b^3}{6}$$

When $y = 0$ the maximum ($x = \sqrt{b^3/3}$) and minimum ($x = -\sqrt{b^3/3}$) values for x are obtained.

More complicated forms appear by adding new terms in the right hand side of the Hamiltonian equations of the two-dimensional system. The 4-type symmetric form illustrated in Figure B.15 is based on the following set of differential equations

$$\dot{x} = -y(y^2 - b^2)$$

$$\dot{y} = x(x^2 - b^2)$$

There are five equilibrium points, one located at the center of coordinates (0, 0) and the other four located at the symmetric points (b, b), $(b, -b)$, $(-b, b)$, and $(-b, -b)$. There are also four unstable points located at $(0, b)$, $(0, -b)$, $(b, 0)$, and $(-b, 0)$.

The integration of the resulting differential equation for x and y provides the following equation

$$\frac{x^4 + y^4}{4} - b^2\frac{x^2 + y^2}{2} = h$$

The integration parameter at the critical limit takes the value $h = -b^4/4$. In Figure B.15 a heavy line marks the critical level curves. Many interesting models can be developed using rotation and translation which can find applications in many disciples especially in biological systems in fluid flow and in crystal formation to name a few as we shall see in the following section.

B.3.3 Rotations/Translations Based Models

A circle in parametric form is represented by

$$x_n = r_n \cos(\theta_n)$$
$$y_n = r_n \sin(\theta_n)$$

where $r = \sqrt{x^2 + y^2}$ and θ_n is the rotation angle.

If the system from step n at time t is rotated to step $n+1$ at time $t+1$ with an angle $\Delta\theta$ the above equations yield:

$$x_{n+1} = r_n \cos(\theta_n + \Delta\theta)$$
$$y_{n+1} = r_n \sin(\theta_n + \Delta\theta)$$

or

$$x_{n+1} = r_n[\cos(\theta_n)\sin(\Delta\theta) - \sin(\theta_n)\sin(\Delta\theta)]$$
$$y_{n+1} = r_n[\cos(\theta_n)\sin(\Delta\theta) + \sin(\theta_n)\cos(\Delta\theta)]$$

By using the parametric equations of the circle yields the rotation map in difference equation:

$$x_{n+1} = x_n \cos(\Delta\theta) - y_n \sin(\Delta\theta)$$
$$y_{n+1} = x_n \sin(\Delta\theta) + y_n \cos(\Delta\theta)$$

When the rotation is followed by a translation equal to a parallel to the X-axis the last equations lead to the following rotation-translation difference equations

$$x_{n+1} = a + x_n \cos(\Delta\theta) - y_n \sin(\Delta\theta)$$
$$y_{n+1} = x_n \sin(\Delta\theta) + y_n \cos(\Delta\theta)$$

$$x_{n+1} = a + b[x_n \cos(\Delta\theta) - y_n \sin(\Delta\theta)]$$
$$y_{n+1} = b[x_n \sin(\Delta\theta) + y_n \cos(\Delta\theta)]$$

A convenient formulation of the last iterative map expressing rotation and translation is by using the following difference equation

$$f_{n+1} = a + bf_n e^{i\Delta\theta}$$

where

$$f_n = x_n + iy_n$$

The determinant of this map is $\det J = b^2 = \lambda_1\lambda_2$. This is an area contracting map if the Jacobian J is less than 1. The eigenvalues of the Jacobian are λ_1 and λ_2, and the translation parameter is a.

In the case selected above rotation is followed by a translation along the X axis. Applying a translation only to one axis instead to both X and Y has the advantage of dropping one parameter while this do not affect the results in the majority of the cases studied.

A very interesting property of the map is the existence of a disk F of radius $ab/(1-b)$ centered at $(a, 0)$ in the (x, y) plane. The points inside the disk remain inside under the map.

The fixed points obey the following equations

$$\left(x - \frac{a}{1-b^2}\right)^2 + y^2 = \frac{a^2b^2}{(1-b^2)^2}$$

and

$$\cos(\Delta\theta) = \frac{1}{2b}\left(1 + b^2 - \frac{a^2}{r^2}\right)$$

The first equation defines a circle K with radius $R = ab/(1-b^2)$, centered at $[a/(1-b^2),\ 0]$ in the (x, y) plane. The second equation can be solved provided that the angle $\Delta\theta$ is a function of r.

The first part of the analysis is based on the assumption that no-area contraction takes place so that the parameter $b = 1$. In this case the angle of the disk F tends to infinity and thus, all the space of the plane (x, y) is considered. An immediate finding from the analysis is that the symmetry axis of the map is that expressed by the line $x = a/2$. This is different from the symmetry axis of the differential equation analogue where the symmetry axis is the line $x = 0$. In the difference equation case the egg-shaped form is moved to the right from the original positions of the system of coordinates. The simplest way to find the axis of symmetry is to search for the points where $x_{n+1} = x_n$ and $y_{n+1} = y_n$. This leads to the following

$$x = \frac{a}{2}$$

and

$$y = \frac{a}{2}\cot\left(\frac{\Delta\theta}{2}\right)$$

It is clear that there is no-symmetry axis in the X direction as y does not show stable magnitude, as it is a function of the angle $\Delta\theta$. However an approximation of y is achieved when the angle $\Delta\theta$ is small. Then

$$y \approx \frac{a}{\Delta\theta} = \frac{MG}{a^2}$$

This point ($x = a/2$, $y = MG/a^2$) is an equilibrium point. However, this is not a stable equilibrium point. The system easily may escape to infinity when y is higher from this last value.

Figure B.16 and Figure B.17 illustrate a rotation-translation case with parameters $a = 2$, $MG = 100$. In Figure B.17 a number of paths inside the egg-shaped formation are drown. All the paths are analogous to those obtained by the differential equation analogue studied above but they are moved to the right at a distance $x = a/2$. A chaotic attractor like a bulge appears around the center of coordinates. An enlargement of this attractor is illustrated in Figure B.17. This is a symmetric chaotic formation with axis of symmetry the line $x = a/2$.

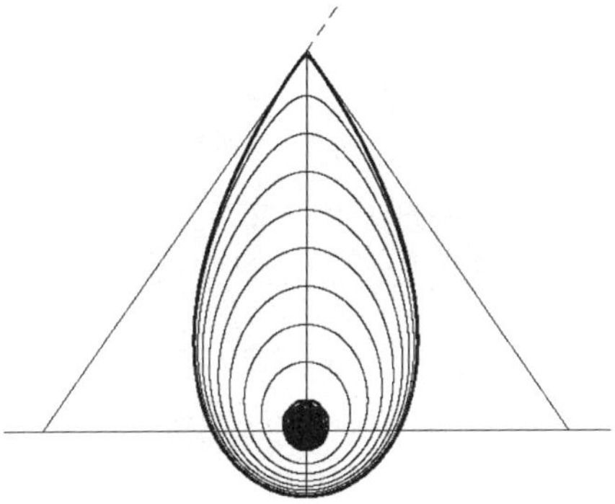

Figure B.16 Rotation-translation.

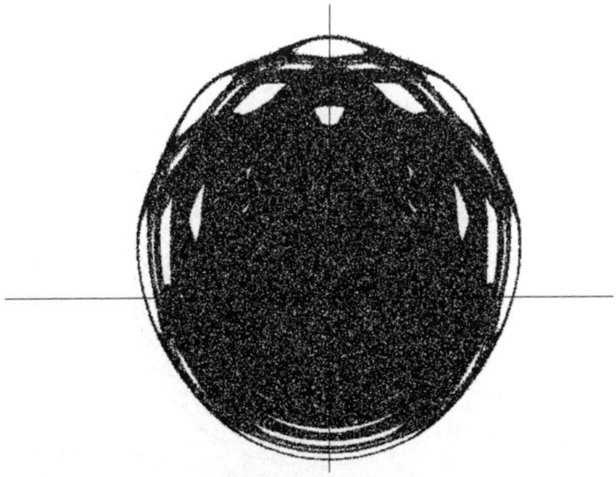

Figure B.17 The central Chaotic bulge.

Illustration of the chaotic bulge of the rotation-translation model. The parameters selected are, $a=1$ for Figure B.18 and $a=0.8$ for Figure B.19 and $MG=100$ for both cases. The rotation angle is $\Delta\theta=\sqrt{MG/r^3}$.

In Figure B.18 third (small cross) and fourth order equilibrium points appear indicated by a cross. In this case the following relations hold: $x_{n+3}=x_n$ and $y_{n+3}=y_n$, and $x_{n+4}=x_n$ and $y_{n+4}=y_n$. Accordingly in Figure B.19 two equilibrium cases appear, one of second order (indicated by small cross) where $x_{n+2}=x_n$ and $y_{n+2}=y_n$ and one of the third order where $x_{n+3}=x_n$ and $y_{n+3}=y_n$.

In both cases an algorithm is introduced for the estimation of the equilibrium points. The convergence is quite good provided that the starting values are in the vicinity of the equilibrium points.

Rotation models express a large variety of systems in various disciplines. The simple rotation in the plane, followed by a translation, does not lead to chaotic forms. The rotation equation in the usual form is based on linear

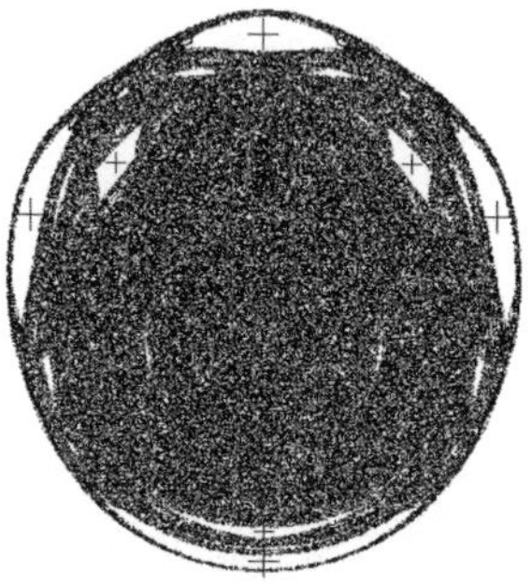

Figure B.18 The Chaotic bulge ($a=1$).

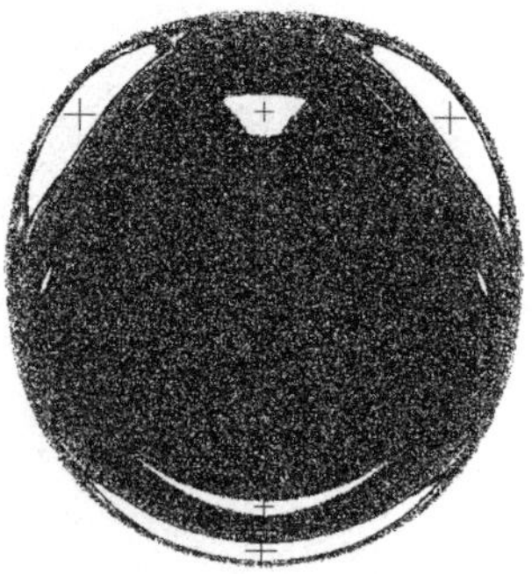

Figure B.19 The chaotic bulge ($a=0.8$).

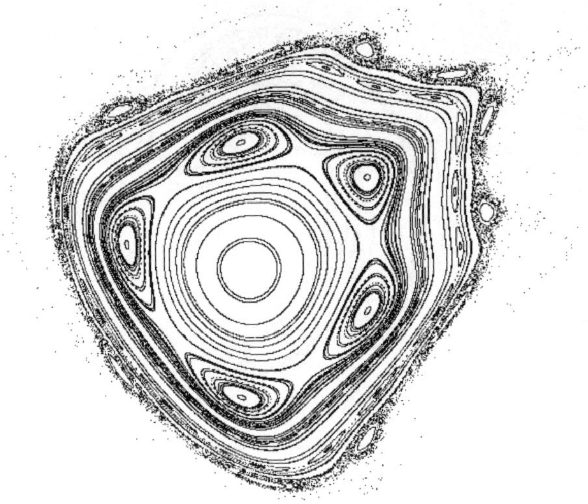

Figure B.20 A rotation chaotic map.

functions as

$$x_{t+1} = x_t \cos(\theta_t) - y_t \sin(\theta_t)$$
$$y_{t+1} = x_t \sin(\theta_t) + y_t \cos(\theta_t)$$

However, the introduction of non-linear terms in the above system gives chaotic solutions. First by adding a quadratic term in both equations, in order to retain the Jacobian of the system equal to unity and to preserve space, the system becomes non-linear.

$$x_{t+1} = x_t \cos(\theta) - y_t \sin(\theta) + x_t^2 \sin(\theta)$$
$$y_{t+1} = x_t \sin(\theta) + y_t \cos(\theta) - x_t^2 \cos(\theta)$$

The rotation angle is stable (in our example $\theta = 1.3$). The system gives an interesting two-dimensional map with chaotic regions and islands of stability inside these regions (see Figure B.20).

Another idea to add non-linearity is by assuming that the rotation angle θ is a non-linear function as is the case of the famous **Ikeda** attractor. The model is based on a system of two rotation equations with a translation parameter a added in the first equation, a space parameter b and a non-linear rotation angle of the form

$$\theta_t = c - \frac{d}{1 + x_t^2 + y_t^2}$$

Figure B.21 The Ikeda attractor.

The rotation-translation system is of the form

$$x_{t+1} = a + b[x_t \cos(\theta_t) - y_t \sin(\theta_t)]$$
$$y_{t+1} = b[x_t \sin(\theta_t) + y_t \cos(\theta_t)]$$

The original system proposed by Ikeda to explain the propagation of light into a ring cavity gives an interesting attractor for the following values of the parameters $a=1$, $b=0.83$, $c=0.4$, and $d=6$. The attractor is illustrated in Figure B.21.

This map has the Jacobian $J=b^2$. When $b<1$ several chaotic and non-chaotic patterns are illustrated in the various simulations. The Ikeda map is one of the standard maps in the chaotic literature.

Ikeda model is one of the most popular models that include rotation and translation and have non-linear terms that, for appropriate values of the parameters, give rise to chaotic patterns and attractors.

The properties of the chaotic maps are so rich and many times unexpected. The shape of the Ikeda attractor in the previous standard form depends on the selection of the values of the parameters. By changing these parameters various very interesting new forms appear. The aim is to go further from the classical stability analysis. We try to find mechanisms to design simple or complicated and chaotic forms and to pass from a qualitative to a quantitative point of view. The equations used are 3-dimensional and the original inspiration comes from the Rössler model.

The main tools for shape formation are **rotation** and **translation** as is the Ikeda case above, but also **reflection** with translation based on the set of equations

$$x_{t+1} = a + b[x_t \cos(2\theta_t) + y_t \sin(2\theta_t)]$$
$$y_{t+1} = b[x_t \sin(2\theta_t) - y_t \cos(2\theta_t)]$$

If the same function as in the Ikeda case is used for the angle θ_t

$$\theta_t = c - \frac{d}{1 + x_t^2 + y_t^2}$$

the reflection case gives an attractor (Figure B.22) which is divided in two new separate forms (Figure B.23). The parameters for the first case are $a=2.5$, $b=0.9$, $c=1.6$, and $d=9$. In the second case the parameter $a=6$ while the other parameters remain unchanged. The two attractors have centers located at $(x=a,\ y=0)$ and at $(x=a+ab\cos(\theta),\ y=ab\sin(\theta))$, where $\theta=c-d/(1+a^2)$. These cases simulate the split of a vortex in two separated forms.

The rotation equation applied in the Ikeda case but now with $b=1$, is a transformation preserving space. By assuming that the rotation angle is a

Figure B.22 Reflection with 1 image.

Figure B.23 Reflection with 2 images.

function of the distance $r_t = \sqrt{x_t^2 + y_t^2}$ interesting forms arise. Such a case based on the following function for θ

$$\theta_t = c + \frac{d}{r_t^2}$$

gives the form illustrated in Figure B.24 in a (x, y) diagram. There is a symmetry axis at $x = a/2$ and one equilibrium point in the central part of Figure B.24 located at $x = a/2, y_{t+1} = y_t$. The five equilibrium points in the periphery are of the fifth order $(x_{t+5} = x_t,\ y_{t+5} = y_t)$. One of these points is also located on the $x = a/2$ axis, whereas the other four are located in symmetric places regarding this axis. The parameters are $a = 3,\ c = 1,\ d = 4$.

Also the Ikeda model with $b = 1$ provides interesting forms with the same axis of symmetry $x = a/2$. Figure B.25 illustrates this case with $a = 1,\ c = 5$, and $d = 5$. In Figure B.25 there are two pairs of order-2 equilibrium points, five fifth-order and eight eighth-order equilibrium points.

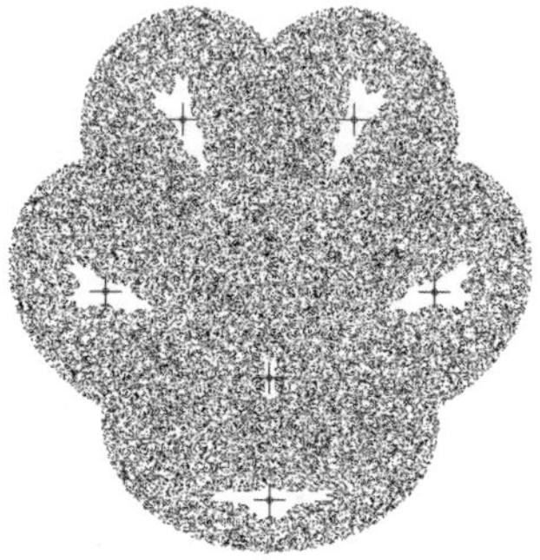

Figure B.24 Symmetric forms.

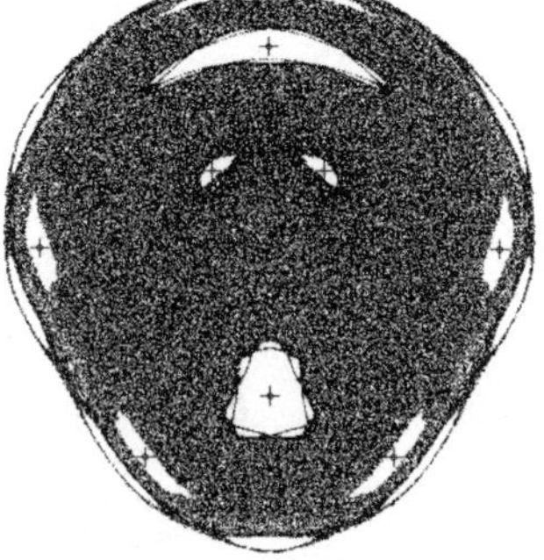

Figure B.25 Symmetric forms.

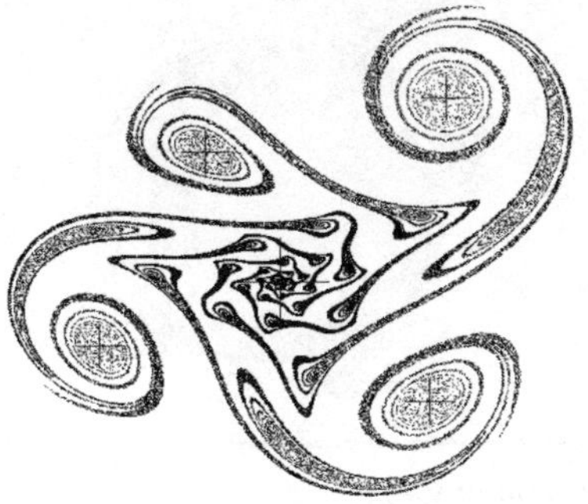

Figure B.26 Flower.

Figure B.27 Flower forms.

Another category of patterns includes forms which we call "flowers." By using the rotation formula with rotation angle

$$\theta_t = c + \frac{d}{r_t^{\delta}}$$

$b = 0.9$ and $a = c = d = 2\pi/k$, with $k = 17/12$ and $\delta = 2$ the following graph appears (Figure B.26). When $\delta = 3$ and $a = 2\pi/(3k)$, $c = d = 2\pi/k$ Figure B.27 results.

B.3.4 Other Patterns and Chaotic Forms

By using the rotation formula with $b = 1$ and an equation for the rotation angle, which takes into account a parallel movement of the coordinates equal to $s = 4$ according to the relation

$$\theta_t = c - \frac{d}{1 + (x_t - s)^2 + (y_t - s)^2}$$

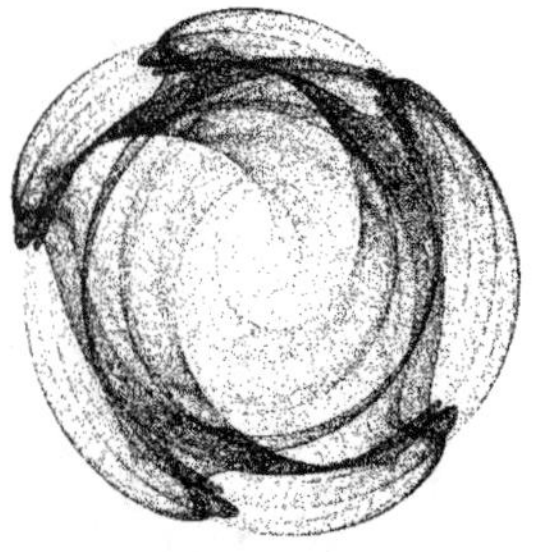

Figure B.28 An artistic drawing.

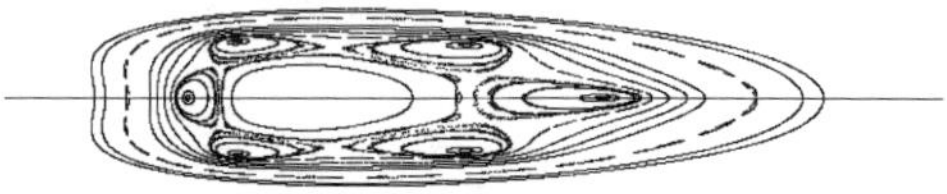

Figure B.29 A pattern of a ship.

a formation like the outer parts of a super-nova explosion appears (Figure B.28). This is also an artistic drawing presenting 5 dolphins in a circle. The parameters are $c = 2.8$, $d = 13.22$.

A graph simulating the upper-view of a ship is based on a set of rotation equations of the form

$$x_{t+1} = -a - (x_t - a)\cos\theta + y_t \sin\theta / r_t$$

$$y_{t+1} = -x_t r_t \sin\theta - y_t \cos\theta$$

$$r_t = \sqrt{0.5(x_t^2 + \sqrt{x_t^4 + 4y_t^2})}$$

The symmetry axis is located at $x = a(\cos\theta - 1)/2$. The parameters selected are $\theta = 2$, $a = 2.8$ and the related illustration appears in Figure B.29.

The Ushiki model gives a form of two separated attractors (Figure B.30) for parameter values $a = 3.67$, $b = 3.1$, $c = 0.06$, and $d = 0.4$. The equations of the model are

$$x_{t+1} = (a - x_t - by_t)x_t$$

$$y_{t+1} = (c - y_t - dx_t)y_t$$

The Tinkerbell model's iterations formulate six chaotic attractors (Figure B.31). The model is based on the following set of equations

$$x_{t+1} = x_t \cos\theta_t - y_t \sin\theta_t + 1 - 0.8x_t z_t$$

$$y_{t+1} = x_t \sin\theta_t + y_t \cos\theta_t$$

$$\theta_t = 5.5 - 1/\sqrt{x_t^2 + y_t^2 + z_t^2}$$

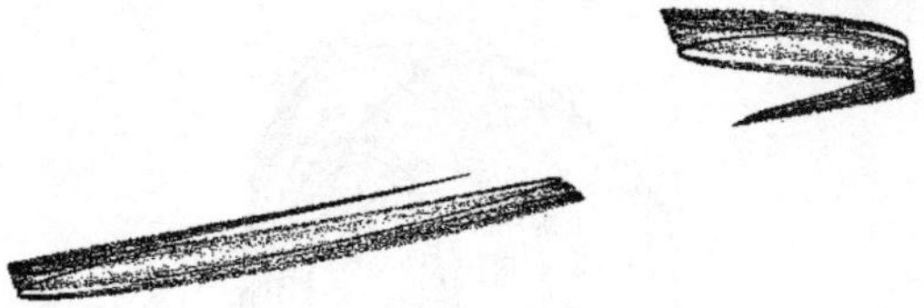

Figure B.30 The Ushiki attractor.

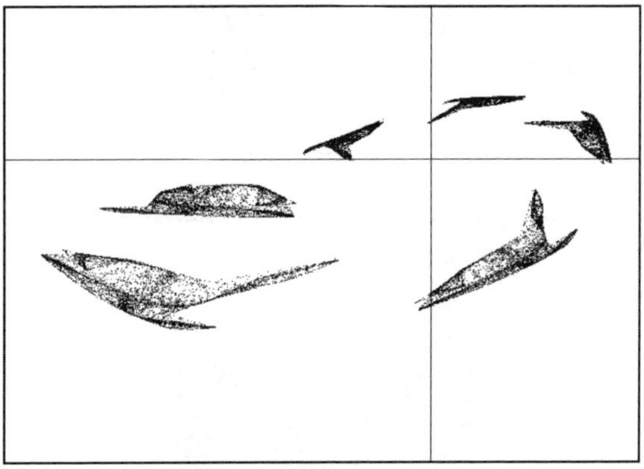

Figure B.31 The Tinkerbell attractor.

B.4 Three-Dimensional Models

The simplest three-dimensional chaotic models are based on a set of three linear differential equations on which we add a non-linear term. Such system is the following Arneodo model expressed by the differential equations

$$\dot{x} = y$$
$$\dot{y} = z$$
$$\dot{z} = -(z + sy - mx + x^2)$$

where s and m are parameters.

This model was proposed by Arneodo et al. in their study of the simplest systems, which exhibit multiple periodicity and chaos. The Jacobian of the model is $J = -m + 2x$. There are two stationary or equilibrium points. One is the origin (0, 0, 0) and the other is $(m, 0, 0)$.

For the following application the parameters selected are $s = 3.8$ and $m = 7.5$. The initial values where set equal to $x_0 = 0.5$, $y_0 = 1$, $z_0 = 1$ and the integration step $d = 0.001$. The (x, y) diagram appears in Figure B.32.

The three-dimensional and higher dimensional models include also the Lorenz, the Rössler and the Autocatalytic models and extensions

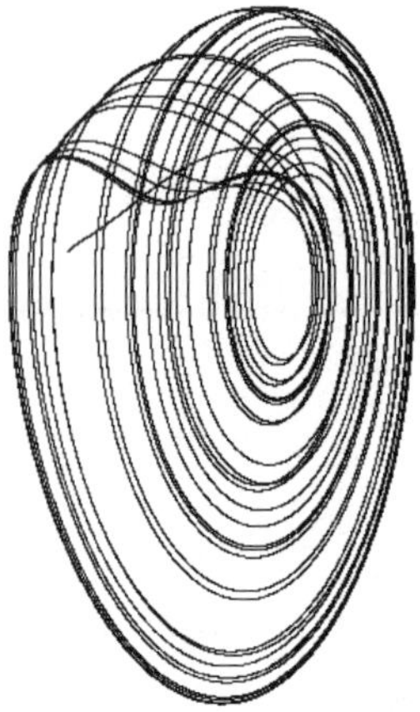

Figure B.32 The Arneodo model.

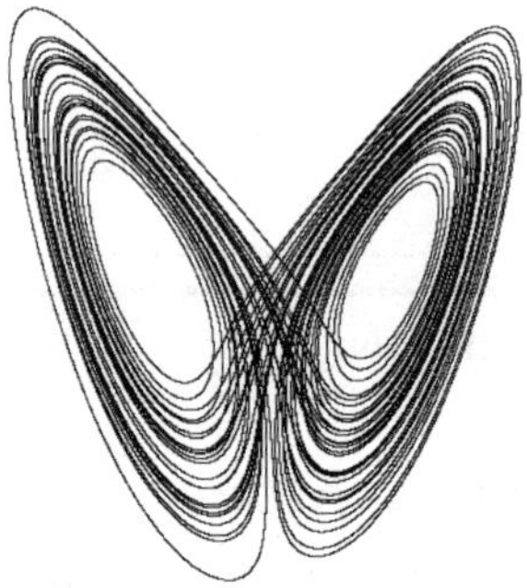

Figure B.33 The Butterfly attractor.

and variations of the above models. A view of the Lorenz attractor is illustrated in Figure B.33.

Ed. Lorenz proposed an idealized model of a fluid like air or water, the warm fluid below rises and the cool fluid above sinks, setting up a clockwise or counterclockwise current. The Prandtl number σ, the Rayleigh and the Reynolds number r, and b are parameters of the system. The width of the flow rolls is proportional to the parameter b. The variable x is proportional to the circulatory fluid flow velocity. If $x > 0$ the fluid rotates clockwise. The variable y is proportional to the temperature difference between ascending and descending fluid elements, and z is proportional to the distortion of the vertical temperature profile from its equilibrium. The equations of this three-dimentional model are

$$\dot{x} = -\sigma x + \sigma y$$
$$\dot{y} = -xz + rx - y$$
$$\dot{z} = xy - bz$$

The equilibrium point (0, 0, 0) exists for all r, and for $0 < r < 1$, it is a stable point, while, for $r \geq 1$ is unstable. The origin corresponds to a fluid at rest. Two new equilibrium points exist for $r \geq 1$, $c_+ = (\sqrt{b(r-1)}, \sqrt{b(r-1)}, (r-1))$ and $c_- = (-\sqrt{b(r-1)}, -\sqrt{b(r-1)}, (r-1))$, representing convective circulation (clockwise and counterclockwise flow). At the equilibrium points $c_\pm$ the Lorenz model has pure imaginary eigenvalues $\lambda = \pm i\sqrt{2\sigma(\sigma+1)/(\sigma-b-1)}$ at $r_h = \sigma(\sigma+b+3)/(\sigma-b-1)$. Using the standard settings $\sigma = 10$, $b = 8/3$, the Hopf bifurcation point is $r_h = 24\frac{14}{19} \approx 24.73684$. For $r_h < r$ the three equilibrium points of the Lorenz model are unstable. Lorenz found numerically that the system behaves "chaotically" whenever the Rayleigh number r exceeds the critical value r_h, that is, all solutions appear to be sensitive to initial conditions, and almost all of them are apparently neither periodic solutions nor convergent to periodic solutions or equilibrium points.

In Figure B.33, the Lorenz attractor presented in a two-dimensional (x, z) view. This is the most common illustration of the famous attractor. Perhaps the famous "Butterfly effect" associated to the presence of chaos owes its name to this figure.

References

1. Newton, I., *Philosophiae Naturalis Principia Mathematica*, London: S. Pepys, 1686.
2. Lorenz, E., Deterministic nonperiodic flow, *J. Atmos. Sci.*, 20, 130–141, 1963.
3. Poincaré, H., *Les Méthodes Nouvelles de la Mécanique Céleste*, Paris: Gauthiers-Villar, 1893.
4. Poincaré, H., *Science and Method*, Paris: Flammarion, 1912.
5. Ott, E., Grebogi, C., and Yorke, J.A., Controlling chaos, *Phys. Rev. Lett.*, 64, 1196–1199, 1990.
6. Rössler, O.E., An equation of continuous chaos, *Phys. Lett. A*, 57, 397–398, 1976.
7. Feigenbaum, M., Quantitative universality for a class of nonlinear transformations, *J. Stat. Phys.*, 19, 25–52, 1978.
8. May, R.M., *Stability and Complexity in Model Ecosystems*, Princeton: Princeton University Press, 1973.
9. May, R.M., Simple mathematical models with very complicated dynamics, *Nature*, 261, 459–467, 1976.
10. Hénon, M., Numerical study of quadratic area-preserving maps, *Quart. Appl. Math.*, 27, 291–311, 1969.
11. Hénon, M., A two-dimensional mapping with a strange attractor, *Comm. Math. Phys.*, 50, 69–77, 1976.

Index